BIOCHEMISTRY RESEARCH TRENDS

BIOPROCESS SCIENCES AND TECHNOLOGY

BIOCHEMISTRY RESEARCH TRENDS

Additional books in this series can be found on Nova's website under the Series tab.

Additional E-books in this series can be found on Nova's website under the E-books tab.

BIOCHEMISTRY RESEARCH TRENDS

BIOPROCESS SCIENCES AND TECHNOLOGY

MIN-TZE LIONG
EDITOR

Nova Science Publishers, Inc.
New York

For permission to use material from this book please contact us:
Telephone 631-231-7269; Fax 631-231-8175
Web Site: http://www.novapublishers.com

Additional color graphics may be available in the e-book version of this book.

LIBRARY OF CONGRESS CATALOGING-IN-PUBLICATION DATA

Bioprocess sciences and technology / [edited by] Min-Tze Liong.
p. ; cm. -- (Biochemistry research trends)
Includes bibliographical references and index.
ISBN 978-1-61122-950-9 (hardcover)
1. Biotechnology. 2. Agricultural biotechnology. I. Liong, Min-Tze. II. Series: Biochemistry research trends series.
[DNLM: 1. Biodegradation, Environmental. 2. Biological Products. 3. Biotechnology. WA 671]
TP248.2.B37496 2011
660.6--dc22

2010044703

Published by Nova Science Publishers, Inc. † New York

CONTENTS

PREFACE

Chapter I – Microbial insecticides offer an efficient and environmentally friendly alternative to chemical insecticides. Natural insecticides have induced a global paradigm shift over the last two decades, where the current contribution of microbial products is a mere 1% of a multibillion (US $) pesticide market. Entomopathogenic bacterium *Bacillus thuringiensis* (Bt) accounts for more than 90% of this biopesticide market. With an increasing number of strains and subspecies, Bt could be used against a number of insects, while the high specificity of Bt makes it an ideal biopesticide. The identification of the structure of Bt Cry toxins and the understanding of their mode of action has opened up the possibility of utilizing protein engineering to help improve Bt pesticides.

Chapter II – The biomass energy production system is an ideal recycling method for environmental conservation through prevention of global warming. It is important to pay attention to bio energy production from non food materials. Cellulose and hemicellulose resources are the most abundant and under-utilized materials that can be used as raw materials for alcohol fermentation. Butanol is the most promising bio fuel having attractive characters, compared to other substitutes. The technology of butanol production had several bottlenecks, with the most critical factor attributed to inherent toxicity of butanol for solventogenesis microorganisms. The aspects of microbial cultivation in anaerobic bioreactors is essential to speed up cellulolytic conversion as well as solventogenesis.

Chapter III – The usage of synthetic plastics has resulted in mounting wastes accumulation in our environment. Many developed countries have started to consider replacing synthetic plastics with environment-friendly and biodegradable polymeric materials such as polyhydroxyalkanoate (PHA). PHA has received much attention lately as this material shares many similarities to synthetic plastics in terms of product performance and processability. PHA is generally produced by bacterial fermentation using various carbon feedstocks. Oil-based substrates such as palm oil products have generated increasing interest due to its suitability for high yield PHA production. The development and commercialization of palm oil agro-industry in Malaysia has made palm oil products readily available to be utilized as carbon feedstock for PHA production.

Chapter IV – Photosynthetic microorganisms (PSMs), known as cyanobacteria and algae, have emerged as promising platforms for the production of renewable biofuels. PSMs have evolved unique metabolic capabilities that allow the production of a greater diversity of biofuels. Their rapid generation times and extensive natural diversity make water-oxidizing PSMs particularly well suited for the physiological and genetic manipulations required to optimize the conversion of sunlight, water, and CO_2 into forms of chemical energy that can be used as transportation fuels. The application of genetic engineering to enhance biofuels production in PSMs is advancing rapidly, in order to improve algae and cyanobacteria as a feedstock for the production of biohydrogen, starch-derived alcohols, diesel fuel surrogates, and/or alkanes.

Chapter V – This chapter reviews the implementation of biomass from different algae species for the development of therapeutic and preventive pharmaceutical medications. The widely employed blue-green microalga *Spirulina platensis* is used as an example of scientifically-based approach to develop new pharmaceutical medications. By means of instrumental neutron activation analysis, the possibility of targeted introduction of trace elements into *Spirulina* bio-complexes while maintaining its natural beneficial properties has been shown. The possibility of precise dosage of elements in Se-, Cr- and iodine-containing substances based on *Spirulina platensis* biomass is demonstrated. The advantages of biotechnological methods developed to ensure the quality of new pharmaceutical medications are discussed.

Chapter VI – Algae and cyanobacteria, as photosynthetic organisms, are cultivated in mineral nutrition media bubbled with carbon dioxide. In order to enhance the yield of algal biomass attention is paid to the processes hydrodynamics, heat transfer and mass transfer in the algal device. Centrifugation or filtration is the main method of microalgae harvesting. As a whole, the algal biomass is a valuable nutritional additive for humans and animals. The biochemical composition of algae comprises all essential amino acids and minerals in the right natural ratio. Their qualities as food and forage are complete with the abundant presence of vitamins, valuable oils and phytosterols. Algal biomass, algal oils, hydrocarbons and hydrogen from microalgae are often pointed as proper renewable source of fuel.

Chapter VII – Microalgae are important constituents of many ecosystems. Comprehensive analysis and nutritional studies have demonstrated that the algal proteins are of high quality and comparable to conventional vegetable proteins. Nutritional supplements produced from microalgae have been the primary focus of microalgal biotechnology for many years. Biomass production with the substantial exclusion of accompanying processes has been the subject of new development, the production of single-cell protein or microbial biomass. As the cells are capable of synthesizing all amino acids, they can provide the essential ones to humans and animals. This chapter will evaluate the properties, production systems and applications of SCP from algae for human consumption.

Chapter VIII – The current bioethanol production at industrial scale relies on ethanol from sugars and starch, known as "first generation bioethanol". Bioethanol produced from lignocellulosic biomass is called "second generation bioethanol". It offers several advantages included lower cost feedstocks and higher green house gases (GHG) reduction. One of the main configurations in second generation ethanol production is simultaneous saccharification and fermentation (SSF) where enzymatic hydrolysis and sugar fermentation are run in the same vessel at the same time, which would be called consolidated bioprocessing (CBP). This chapter includes revision about fermenting microorganism for second generation bioethanol production, emphasizing the importance of obtaining efficient xylose fermenting microorganisms. Moreover, recent advances in obtaining successful microorganisms for CBP are also reported.

Chapter IX – This chapter discusses on the science and technology in the production of SCP from yeast. The contents begin with a brief introduction on yeast and fundamental knowledge related to physiology. The resources for production of yeast-based SCP that include those conventional and non-conventional as well as renewable and non-renewable sources will be covered. Issues related to the toxicity and safety of yeast-based SCP associated with nucleic acid, digestibility, toxic and possible carcinogenic compounds are also highlighted. The conventional processes involved in the production of yeast based SCP, optimal parameters for SCP production, the growth kinetics of yeast as well as process design and control are also discussed.

Chapter X – This chapter provides an overview of wood degrading basidiomycetes with pharmacological effects. Since these mushrooms are rare in nature, artificial cultivation of fruit bodies has been established, as well as a biotechnological cultivation of fungal biomass in bioreactors, both on solid substrates and in liquid media. Bioreactor design and process parameters are presented and discussed. An industrially important medicinal mushroom *Ganoderma lucidum* - Ling zhi or Reishi is described as an outstanding example, with emphasis on the cultivation possibilities, structure and properties of pharmacologically active compounds, pharmacological effects, and medicinal applications.

Chapter XI – Chemical insecticides to control *Aedes aegypti* adversely affect the environment and also promote emergence of resistant mosquitoes. Plant compounds with damage effect on immature stage forms or adult insects can be an alternative to synthetic insecticides. This chapter shows essential oils, organic solvent extracts and lectins from plants as enviro-friendly agents against *A. aegypti* and other arthropods. Larvicidal mechanisms proposed for essential oils, organic solvent extracts and lectins are also reported. This chapter contributes to the study of alternative strategies to control arthropods. Additionally, it reports the medicinal and industrial uses of plant compounds with biotechnological applications.

Chapter XII – Plants are gaining widespread acceptance for the large-scale production of recombinant proteins as they provide a number of advantages over conventional systems. Strategies to improve the recombinant protein yield in plants include the development of novel promoters, the improvement of protein stability and accumulation, and the improvement of downstream processing technologies. There are already several plant-produced proteins on the market including one at large scale. Several plant-derived recombinant pharmaceutical proteins are reaching the final stages of clinical evaluation, and more are in the development pipeline. In this review, recent progress in molecular farming will be highlighted, including the potential for commercial drug development and production.

Chapter XIII – Cysticercosis is a parasitic disease caused by *Taenia solium* which affects rurally bred pigs and humans in developing countries. Since pigs are indispensable intermediate host, it is conceivable to curb transmission by reducing pig cysticercosis through vaccination. The authors have developed an anti-cysticercosis vaccine (S3Pvac) comprising three peptides originally identified in *Taenia crassiceps*. In search of reducing vaccine production costs, a version of S3Pvac recombinantly expressed in filamentous phages has been developed. However, the parenteral administration of the vaccine was found to be a major logistic and economic difficulty. An orally administered vaccine would elude these difficulties, especially for the prevention of mucosally transmitted parasites as *Taenia solium.* Transgenic plants are ideal systems to produce oral vaccines, thus this chapter also discusses on the expression of the S3Pvac peptides in transgenic embryogenic papaya cells.

Chapter XIV – Progress in plant biology and genetic engineering identified plants as an excellent resource for efficient production of recombinant proteins. The public health's view on plants as naturally safe and pathogen-free is also being embraced. The current growing interest in plant-derived vaccines and other pharmaceuticals is hopefully gaining attention to technology from the industries. Pharmaceutical plant biotechnology is now getting a thorough tune up. Highlights of this research are summarized here with references to the most recent reviews and pivotal literature sources which enable the reader to get particulars on the prevailing strategies and trends in this field.

Chapter XV – Plants have co-evolved with their pests and pathogens. Consequently, they display resistance against harmful organisms manifested through various mechanisms including production of toxic chemicals that may also be used as useful biocontrol agents. With the realization of the harmful impacts of chemicals used in agriculture and medicine, chemicals extracted from plants have been viewed as the best alternatives. Although botanicals have often been suggested as an alternative for chemical control, it is unlikely that botanicals alone would be able to produce the required level of control. Identification of active compounds may lead to suitbale chemical synthesis and the use of biotechnological approaches to enhance productivity and to minimize cost of production.

Chapter XVI – Chitin is a linear homopolymer of N-acetylglucosamines conjugated by β-1,4-glycosidic bonds. Insect exoskeleton consists of chitin highly complexed with protein, lipids and other components. Structural rigidness often limit the growth of organisms, thus an elaborate process of molting is evolved, which depends on the existence and balance of some certain hormone, and subsequently catalyzed by chitin-remodeling enzymes including chitin synthases and chitinolytic enzymes. Coordination of chitin synthesis and its degradation requires strict control of the participating enzymes during development, which are thus ideal targets for novel insecticides. The preparation of these enzymes is essential for biochemical studies. Moreover, since chitin is the second most abundant but underutilized biomass, chitin degradation by chitinolytic enzymes is of biotechnological importance.

Chapter XVII – The past two decades have witnessed a phenomenal rise in the prominence of mammalian cells as hosts for the production of recombinant protein therapeutics such as monoclonal antibodies, cytokines, enzymes and growth hormones. The quality and efficacy of these proteins are far superior when produced in mammalian cells as opposed to microbial or yeast cells. This review will focus on: (i) strategies for development of high-yielding protein expression cell lines; (ii) current trends in bioprocess development; (iii) challenges and opportunities facing biomanufacturing in the era of globalization; and (iv) the promise of 'omics' and cell engineering strategies for future drug development.

Chapter XVIII – The continuing presence of heavy metals in industrial wastes is of great concern. Even though current methods of removing heavy metals offer cheap and efficient mode of waste management, the generation of toxic sludge, obnoxious gases and chemical spillage makes it imperative to have an alternative treatment technique. The exploitation of microbial ability to survive in harsh environments such as wastewaters polluted with heavy metals is an alternative solution for heavy metals removal. These microbes would normally nullify the toxicity of heavy metals via the combination of oxidation, reduction, accumulation or adsorption. However, before this technology is attempted at the industrial scales, various process arrangements need to be investigated at the laboratory, pilot and semi – commercial scales.

Chapter XIX – Heavy metals in the environment invariably enter the food chain and cause health hazard mainly through bioaccumulation. Although, the most desirable action is to avoid heavy metals entering into food chain, it is not practical as the agricultural practices and industrial activities release many undesirable heavy metals into the environment. The physical or chemical remedial measures are expensive in large scale programs. Bioremediation where microorganisms are playing a major role is the most acceptable, cheaper and efficient system that is available at present. Heavy metal binding peptides of microbial origin, chemisorption or biosorption, complexation, adsorption on microbial surfaces or chelator assisted extraction of heavy metals are the currently used technology.

Part 1: Bacteria as a Source of Raw Material

In: Bioprocess Sciences and Technology
Editor: Min-Tze Liong
ISBN 978-1-61122-950-9

Chapter 1

BACILLUS THURINGIENSIS BIOPESTICIDES: KEY STRAINS AND TOXINS, MASS CULTURE, FORMULATION STRATEGIES AND IMPROVEMENTS

***T. C. Bamunuarachchige*[1,*], *D. M. J. C. Dissanayaka*[2], *N. A. A. D. Wickramarathna*[3] *and G. A. Udayantha*[4]**

[1]Department of Biological Science, Faculty of Applied Sciences, Rajarata University, Sri Lanka
[2]Microbiology Lab, Teaching Hospital, Anuradhapura, Sri Lanka
[3]Rubber Research Institute, Department of Agriculture, Sri Lanka
[4]Ministry of Health, Sri Lanka

ABSTRACT

In a world with a forever increasing conscience on environmental pollution, microbial insecticides offer an efficient, environmentally friendly alternative to the chemical insecticides. The use of chemical insecticides over the decades has lead to many a complex problem, such as development of resistance, killing of beneficial insects, accidental and intentional poisoning etc. Given these drawbacks, natural insecticides such as botanicals and microbial insecticides have induced a global paradigm shift over the last two decades. All microbial groups such as bacteria, fungi, nematodes and the acellular viruses are utilized as insecticides. Although the future of microbial insecticides seems promising, the current contribution of microbial products is a mere 1% of a multibillion (US $) pesticide market. Entomopathogenic bacterium *Bacillus thuringiensis* (Bt) accounts for more than 90% of this biopesticide market. With an ever increasing number of strains and subspecies, Bt could be used against a number of insect orders such as lepidoptera, diptera, coleoptera, malophaga etc. The high specificity of Bt makes it an ideal biopesticide, as it is generally non-toxic to fish, birds, amphibians, reptiles and mammals. However, Bt shows some drawbacks such as narrow host range, build up of resistance and susceptibility to environmental factors such as exposure to UV. Both formulation strategies and genetic engineering could be utilized to successfully overcome these drawbacks. The identification of the structure of Bt Cry toxins and the understanding of their mode of action has opened up the possibility of utilizing protein engineering to help improve Bt pesticides and take them beyond the current obstacles.

1. Introduction

Insects offer a number of benefits to mankind which includes pollination, control of pests and the production of commercially valuable commodities, such as honey and wax. However, some members of the class insecta have become a menace, mainly through the reduction of crop yields by infestations and acting as vectors for many a disease such as malaria, dengue, sleeping sickness etc. Although people practice a number of methods to control these menacing pests, the application of synthetic chemical pesticides has been the most effective and popular method over the last five decades. However, the problems associated with the overuse of chemicals, such as resistance build up, high production costs, pest resurgence, death of natural enemies and beneficial insects, and negative impacts on the environment, have created a conscience shift towrds more environmentally friendly biological control methods. There are many promising biocontrol agents such as bacteria, fungi, viruses, nematodes and parasitoids. However, the use of the entomopathogenic bacterium *Bacillus thuringiensis* (Bt) accounts for more than 90% of the biopesticide market.

Bacillus thuringiensis, is a ubiquitous, endospore forming, gram positive, rod shaped soil bacterium which is harmless to humans, mammals, amphibians, reptiles and birds. Hence, it offers an ideal environmentally friendly solution to insect pest control. Although Bt provides an ideal solution for control of agricultural pests, with toxicity against many an insect order such as lepidoptera, diptera, coleoptera and malophaga, the ability of some subspecies, especially *B. thuringiensis* sbsp. *israelensis,* to control mosquitoes is a major contribution towards improving human living conditions throughout the world.

2. Diversity of *Bacillus Thuringiensis*

Until 1977, only 13 *Bt* strains had been described. All 13 subspecies were toxic only to some species of lepidopteran larvae. In 1977 the first subspecies toxic to dipteran species was found, and the first discovery of strains toxic to species of coleopteran larvae followed in 1983. Subsequently strains have been found with toxicity to several other insect orders, such as hymenoptera, homoptera, orthoptera, and mallophaga. The spectrum is ever growing with the identification of strains toxic to nematodes, mites, and protozoa [1]. With the regular addition of new strains and subspecies, currently Bt contains more than thousand members. These strains are categorized into about 67 serotypes according to the flagella antigen type.

3. Toxins of Bt

Bt produces a number of toxins. These include the α exotoxin, β exotoxins, Vip proteins and Vip related proteins, phospholipase C, parasporins, hemolysins, diarrheal type enterotoxins and δ endotoxins. These toxins contribute to the bacterial virulence together with the proteases and chitinases.

3.1. α Exotoxin

One of the thermo labile extra cellular factors observed by the early research groups of Bt. α Exotoxin causes lysis of haemocytes of insects and displays toxicity against mice. α Exotoxin is sensitive to trypsin, and pH. pH values above 10 and below 3.5 are known to destroy them.

3.2. β Exotoxins

There are two types of β-exotoxins, type I β exotoxin and the type II β exotoxin. Type I β exotoxin is a thermo stable molecule produced during the vegetative growth of Bt. It is toxic to insects, invertebrates and vertebrates and hence has detrimental effects on nontarget organisms [2]. The type I β exotoxin is primarily active against dipteran species, but it is also active against coleopteran, lepidopteran, and some nematode species. Type I β exotoxin is an adenine nucleotide analogue [3] that has been found to interfere with RNA polymerase [4]. Thus, it is believed that this molecule inhibits the synthesis of RNA by competing with ATP for binding sites, thereby affecting insect molting and pupation and causing teratological effects at sublethal doses [5]. Type I β exotoxin also displays some toxicity to mammalian cells [2] and has been banned from public use based on advice of the World Health Organization [6]. In 1989 a novel β exotoxin was identified. This exotoxin was named as type II β exotoxin and is believed to be a Uracil analogue of the type I exotoxin [7].

3.3. Vips and Vip Related Proteins

Vegetative insecticidal proteins or Vips were first identified in 1990 and the number has increased steadily since then. These are secreted during the vegetative growth like the β exotoxins. Vips do not exhibit any similarity with the Cry or Cyt toxins (see 4.8). Currently, all Vip-related sequences that have been described, fall into three different families, Vip1,Vip2, and Vip3 [8]. Among the Vips, Vip1Aa1 and Vip2Aa1 are very active against corn rootworms, particularly *Diabrotica virgifera virgifera* and *Diabrotica longicornis* [9, 10]. Vip3 proteins have a different host range, they are primarily toxic to lepidopteran pests [11, 12]. Vips target the insect midgut epithelium. Apoptosis was initially suggested as a mode of action, but it was recently shown that like Cry toxins, activated Vip3A toxins are pore-forming proteins, capable of making stable ion channels in the membrane. Vip3A proteins need to be activated by proteases prior to recognition at the surface of the midgut epithelium by specific membrane proteins [13].

3.4. Phospholipase C (Lesithinase)

Some strains of Bt and *Bacillus cereus* produce thermo labile, extra cellular enzymes known as phopholipase c or lesithinase. Phospholipase C is known to penetrate cell membranes and attack the cell substrate, causing cell sloughing from the basal lamina. This is brought about by the cleavage of inositol phospholipids found in the plasma membranes. The binding of phospholipase C to phosphatidylcholine (PC) surfaces and the enzymatic activity seems to be related to the tyrosine residues which provide a dimerization interface close to the membrane binding region of the molecule [14].

3.5. Parasporins

Parasporins are novel Bt crystal proteins with cytocidal activity, first identified by Mizuki and others in 2000 [15]. These proteins are non hemolytic and found in non insecticidal strains of Bt. Due to their remarkable ability to preferentially kill human liver and colon cancer cells, parasporins have gained much recognition as promising molecules in both biological and medical fields. Although research on the mode of action of parasporins at the molecular level is still limited, work on parsporin-2 has provided certain clues. Accordingly, N terminal and C terminal processing produces the active parsporin-2 with a size of 30kda and this active toxin is a cell-discriminating, membrane-targeting and pore-inducing toxin that subsequently causes irreversible intracellular decay in cancer cells [16].

3.6. Hemolysins

The group of proteins known to lyse the vertebrate erythrocytes, produced by some Bt species and *Bacillus cereus* are termed as hemolysins. These include the Hly-II characteristic of Bt [17] and some Cyt proteins such as the Cyt K (*B. cereus*), Cyt 1Aa of *B. thuringiensis* sbsp *israelesis* and Cyt1Ab1 of *B. thuringiensis* subsp. *Medellin.*

3.7. Diarrheal Type Enterotoxins

Some Bt strains are known to produce *B. cereus* type diarrheal enterotoxins. However, most of the food borne enterotoxin diarreal outbreaks are attributed to *B. cereus*. Recent studies indicate that due to the difficulty of morphological and 16s DNA based differentiation of *B. cereus* from Bt, many of these cases may be falsely attributed to *B. cereus* [20]. The two well characterized enterotoxins of *B. cereus* are haemolysin BL (HBL) and the non-haemolytic enterotoxin (NHE). Both are three-component toxins requiring expression of all three components for virulence [21, 22]. These two enterotoxins typically give relatively mild and short-lived diarrhoeal syndrome. Another protein known as Cyt K is known to be involved in food poisoning very rarely [23]. Cyt K has been identified in *Bacillus thuringiensis* sbsp. *kurstaki* and *B. thuringiensis* sbsp. *aizawai* while hbl genes have been identified in *B. thuringiensis* sbsp. *kurstaki, B. thuringiensis* sbsp. *aizawai, B. thuringiensis* sbsp. *israelensis* and *B. thuringiensis* sbsp. *tenebrionis* [24]. Due to these evidences at the molecular level, *Bacillus thuringiensis* needs to be reevaluated in terms of diarrheal enterotoxins.

3.8. δ Endotoxin

The δ endotoxin is the most important toxin produced by Bt in terms of its success as a biopesticide. This is largely due to its specificity and non-toxicity to humans and other animals. Bt δ endotoxin is contained within a large crystal structure known as the parasporal crystal, synthesized during bacterial sporulation. Parasporal crystal contains mainly protein (about 95%) and a small percentage of carbohydrate. This crystal is generally an aggregate of proteins which are known as crystal or Cry proteins. However, in certain subspecies such as *B. thuringiensis* sbsp. *israelensis*, the Cyt proteins also contribute to the crystal. Cyt proteins are also known to act synergistically with the Cry proteins to enhance the latter's activity. In fact the synergy between Cyt toxins and Cry toxins are far greater than any synergy among the Cry toxins themselves [18, 19]. To date more than 300 *cry* genes have been identified and the number is still increasing. Different subspecies of Bt are specific to different orders of class insecta based on their Cry protein composition of the parasporal crystal (Table 1).

Table 1. Cry toxin composition, serotype and target insect orders of different sub species of *B. thuringiensis*

B. *Ihuringiensis* strain or subspecies	Cry composition	Target insect order	Serotype
B.t. kurstaki HD-1	Cry1Aa1, Cry1Ab2, Cry2 Aa1 and Cry2Ab1	Lepidoptera and Diptera	3
B.t; israelensis	Cry 4A, Cry4B, Cry 10A, Cry 11A and Cyt 1A	Diptera	14
B.t. tenebrionis	Cry 3Aa1,	Coleoptera	8
B.t jegathesan	Cry11Ba1, Cry19Aa1	Diptera	28
B.t. aizawai	Cry1Aa, Cry1Ab, Cry1Ac, Cry1 C, Cry1Da and Cry2A	Lepidoptera and Diptera	7

3.8.1. Mode of Action of the Bt δ Endotoxin

The mode of action of Bt has been elucidated mainly based on the lepidopteran toxic Cry proteins. Accordingly, when the insects ingest Bt spores, the parasporal crystal is released in the insect midgut and the protoxin becomes solubilized due to the alkaline pH of the insect midgut. The solubilized protein is generally a protoxin and it needs to be activated by the insect proteases. Once activated, the toxin binds to the specific receptors of insect midgut epithelial cell membrane. This binding leads to the creation of an ion channel leading to the excessive loss of cellular ATP and electrolytes (Figure 1). Subsequently the cellular metabolism of the insect ceases, the insect stops feeding, becomes dehydrated leading to eventual death. Although the Cry toxin mediated pore formation is generally accepted as the mechanism that leads to the death of the insect, it has been recently suggested that the toxicity could be related to G-protein mediated apoptosis following receptor binding [25].

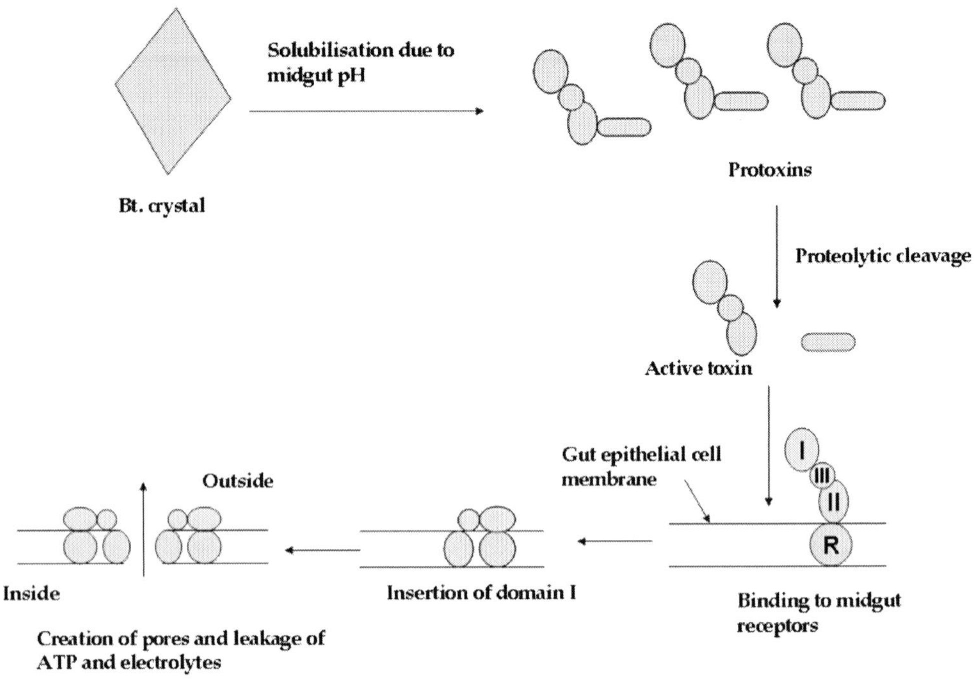

Figure 1. Mode of action of Bt δ-endotoxin: The ingested Bt crystal get solubilized in the insect midgut due to high basic pH. The solubilized protoxins are activated by the proteolytic cleavage carried out by the insect proteases. The domain II of the active toxin binds to the receptors on the epithelial cell membrane. Domain I of the toxin is inserted to the membrane and two such insertions creates a pore leading to loss of electrolytes and ATP. Adapted from Knowels (1994) [32].

3.8.2. The Tertiary Structure of Cry Proteins

According to the general structure of Cry proteins (Figure 2) it consists of three domains. The N-terminal domain (domain I) is a bundle of seven α-helices in which the central helix-a5 is hydrophobic and is encircled by six other amphipathic helices. This helical domain is responsible for membrane insertion and pore-formation. Domain II consists of three antiparallel β-sheets with exposed loop regions, and domain III is a β-sandwich [26, 27, 28, 29, 30, 31]. Domain II is responsible for binding with the receptor, and according to one of the suggested models, domain III acts as the hinge that allows insertion of domain I into the membrane [32].

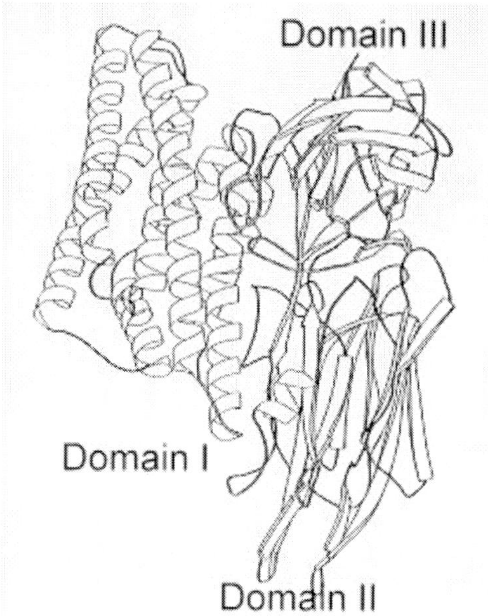

Figure 2. The three domain structure of δ-endotoxin Cry 3Aa. Based on Li et al. (1991) [26].

3.8.3. Classification of Cry Toxins

Initially the Cry toxins were grouped into four major classes, Cry I, Cry II, Cry III and Cry IV based on the activity of the toxins on insect orders. The Cry I toxins were specific to lepdoptera, Cry II were specific to lepidoptera and diptera, Cry III were specific to coleoptera and Cry IV were specific to diptera. These proteins were further divided into subclasses (A,B, C etc) and subgroups (a,b, c etc.) according to the DNA sequence of the gene (eg. Cry IA (a) of *Bacillus thuringiensis* subsp. *kurstaki* HD-1). However, due to the subsequent identification of a large number of Cry proteins, this system became obsolete. In order to accommodate the ever increasing Cry proteins, a new classification system was proposed, on the basis of percentage homology of the amino acid sequences [33]. This classification system uses a four part naming system, with three basic demarcations set at 95, 78 and 45% homology. The name given to a new toxin depends on the boundary at which it enters the system. If it enters below 45%, it is assigned a new primary number (Arabic number eg. Cry9). If it enters between 45% and 78%, it is given a secondary ranking (uppercase letter eg. Cry1A). If the entry is between 78% and 95%, it is given a tertiary ranking (lower case eg. Cry1Ab) and if it enters above the 95% homology boundary it is given a quaternary ranking.

4. Applications of *Bacillus Thuringiensis* Toxins

Traditionally Bt strains have been used as biopesticides to control agricultural pests. The majority of these were used against the lepidopteran pests. With the identification of strains and subspecies toxic to other insect orders, the usage expanded in agriculture. The identification of dipteran toxic strains opened up a new avenue in the control of vector borne diseases. These strains especially *B. thuringiensis* sbsp. *israelensis* is widely used to control mosquito vectors responsible for the spread of diseases such as dengue, malaria and sleeping sickness and filaria. The dawn of the recombinant technology, especially the advent of transgenics brought in another exciting prospect of producing the toxin in other organisms. In order to reduce the insect pest damage, transgenic plants harboring Bt *cry* genes were produced. These include potato, tomato, rice, canola, corn, soybean, cotton any many more. After successful field trials many of these were approved for commercial release.

The newest promising aspectof the Bt toxins is the possible applications in the medical field due to the identification of several Bt toxins known as the parasporins which are capable of to selectively destroying hepatic and colon cancer cells [16].

4.1. Production of *Bacillus Thuringiensis* Based Biopesticides

In addition to the general drawbacks (see 6), large scale production of Bt based biopesticides faces many challenges such as, minimizing the costs, optimizing the conditions to increase the product yield, designing the appropriate bioreactor, maintaining the efficiency of the toxins, competition with the synthetic pesticides, avoiding contaminations etc. These challenges need to be addressed properly at each stage of the production to avoid product failure.

4.1.1. Culture Media

Luria Bertani (LB) medium is the standard laboratory medium used for *Bacillus thuringiensis*. LB costs approximately US $ 15.1 per 10 L and thus is commercially costly. The nutrient requirement of Bt cultures during vegetative growth, transition and sporulation should be properly understood in order to ensure a high yield without compromising the production cost. In batch culture, during the vegetative growth phase, nitrogen and carbon sources are catabolized via the Embden–Meyerhof–Parnas pathway with major intermediates like acetate, and other metabolic intermediates of glucose. The latter are metabolized during the transition and sporulating phase, via the TCA cycle [34]. Zouari et al. (1998) demonstrated that in a gruel hydrolysate-based medium, a combination of amino acid sources and ammonium sulphate significantly enhanced bioinsecticide production [35]. Ozkan et al. (2003) found that the addition of ammonium ions (NH_4^+) during exponential growth could have an important influence over δ-endotoxin production and sporulation [36]. However, a strong negative impact on toxin yield was observed by the addition of $(NH_4)_2SO_4$ at the beginning of the sporulation phase [37]. In agricultural countries, complex agricultural wastes and byproducts (molasses, soy bean seed meals, fodder yeast, corn steep liquor, cheese whey, coconut water) can be used as alternative media. Poopathy et al. (2003) introduced a potato based culture medium for Bt cultures with a relatively lower cost of US $ 0.11 per 10L [38].

Another interesting approach towards a low cost nutrient source to culture Bt is to utilize large quantities of sludge that results from municipal wastewater treatment [39]. Not only does this provide a cheap nutrient source but also a solution to a pressing environmental problem. Both primary and secondary sludge are known to produce different spore and endotoxin yields. Carbon and nitrogen content of the wastewater sludge also affects the yields of endotoxin and spores. The optimum C:N ratio in secondary and mixed (a mixture of primary and secondary) sludge for Bt was found to be 7.7 and 9.9, respectively [40]. However, high flocculation nature of sludge creates a severe O_2 transfer problem to Bt which gets embedded in the sludge particles. This problem could be overcome with the addition of surfactants such as Tween 80 and ALTOX 848. However, some surfactants like Span 20 could adversely affect the entomotoxicity (Tx) of Bt [41]. Surfactants also help reduce the foaming that may result due to high protein content of sludge. On the other hand, use of hydrolysed sludge could reduce the viscosity and improve the O_2 transfer, leading to higher Tx values. However, it is advisable to use hydrolysed sludge directly for fermentation to avoid increase of viscosity due to sterilization [39].

Coconut water also has a great potential as a culture medium for Bt pesticides. The major chemical constituents of coconut water are sugars (glucose, fructose, sucrose) and minerals while minor ones are vitamins and amino acids. Young coconuts contain primarily reducing sugars while mature nuts contain approximately 90% of the total sugar as sucrose. One of the main setbacks of utilizing coconut water is yeast contaminants, as they out compete Bt in a high sugar medium. A study reported that a medium consisting ground silk worm (*Bombyx mori*) pupae could be a relatively cheap alternative (4.32 US $) compared to milk whey and soybean milk, soybean protein and nutrient broth medium (standard) [42].

4.1.2. Fermentation Types

Batch fermentation is the predominant fermentation process for Bt biopesticides as the toxins and the spores are produced at the stationary phase. However with the advent of defective spore formers, producing Cry toxins constitutively, there have been instances of continuous culture processes as well. Fermentation process can be further divided into submerged or liquid fermentation and semisolid or solid–state fermentation (SSF). Submerged fermentation deals with the utilization of liquid culture media which is stirred or aerated frequently. Solid-state fermentation is carried out with microbes grown on solid material with limited amounts of water. SSF has several advantages over submerged fermentation. These include relatively simple medium (which is composed of solid material and a small amount of water), low wastage of water, easy aeration, occupation of relatively smaller space and low cost drying and storage of product. However, transfer resistance, heat gradients and steep gaseous concentration may adversely affect SSF performance. The moisture content in a SSF should be adjusted to favor the bacterial strain. Capalbo et al (2001) developed a simple solid-state method for the production of *Bacillus thuringiensis* subsp. *tolworthi* to control fall armyworm in maize [43].

4.1.3. Type of Bioreactors

Submerged fermentation processes are carried out with stirred tank bioreactors (STB) or tower type bioreactors. When compared to mechanically agitated STBs, the tower type shows poorer mass transfer and mixing abilities [44]. The tower types could be further categorized into Air lift bioreactors (ALBs), bubble columns and airlift bioreactors with net-draft tubes

(Figure 3). Out of these three, ALBs with net-draft tubes have shown, shorter mixing time, better liquid circulation and higher O_2 transfer rates [45]. Fu et al., (2003) showed that double net draft tube reactor produced higher biomass as well as Bt toxin, mainly due to its higher oxygen transfer coefficient. In the bubble column run, lower dissolved O_2 values were observed, which contributed to the lower biomass production [46]. With superior performances, net-draft tube bioreactors show much potential as industrial scale fermenters for Bt production. Hongzhang et al (2002) designed a novel industrial scale ($70m^3$) bioreactor that could improve the O_2 transfer and heat removal of solid-state fermentation processes [47]. This bioreactor provided internal air circulation in addition to the external air circulation. Solid material swells and shrinks due to air pulsations which provide sufficient O_2 and room for propagation, allowing efficient heat transfer.

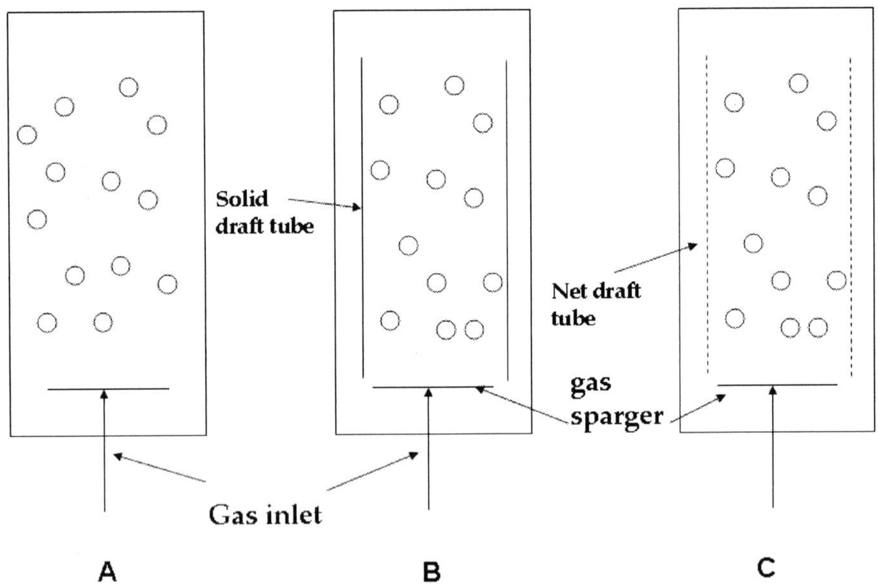

Figure 3 Tower type bioreactors: (A) bubble column bioreactor that doesn't contain a draft tube, (B) airlift bioreactor with a solid draft tube and (C) airlift bioreactor with net draft tube. Adapted from Fu et al. (2003) [46].

4.1.4. Factors Affecting the Production Efficiency

Glucose

Generally, much better bacterial growth, spore formation and toxin production are acheived during cultivation in lower glucose concentrations. Amin et al. (2008) reported that appreciable amount (18.6 g/L) of residual glucose was left in cultivation broth even when the bacterium was cultivated in a medium achieving the highest toxin concentration [48]. This is an alarming finding because of the amount of glucose wasted without being converted to toxins. They also reported that as much as 22% of the material gets converted to by products and suggest that the feed rate should be adjusted appropriately to ensure better utilization of glucose for the production of Bt endotoxins.

Nitrogen Source

Delta endotoxin synthesis and entomotoxic potential is affected by the Nitrogen source available in the medium. Rossa et al. (1990) investigated the influence of different combinations of organic and inorganic Nitrogen sources and C:N ratios on growth and δ endotoxin production by *B. thuringiensis* sbsp. *israelensis* [49]. They reported that the δ-endotoxin concentration increased when the medium was supplemented with $(NH_4)_2SO_4$. However, the δ-endotoxin: biomass dry weight ratio was unaffected by different C:N ratios. They also showed that the partial replacement of yeast extract as organic Nitrogen source by $(NH_4)_2SO_4$ significantly increases the δ endotoxin production. Kraemer-Schafhalter and Moser (1996) concluded that NH_4^+ ion, used as an inorganic Nitrogen source was necessary for synthesis of intracellular protein, cell wall components and the delta-endotoxin [50].

pH

It has been reported that vegetative growth of *Bacillus* sp. causes a build up of acetic and pyruvic acids, as a consequence of glucose metabolism. Acetic acid and pyruvic acid, thus produced, are later used during the onset of sporulation. These products cause the medium pH to decrease and become acidic. When glucose or the easily biodegradable Carbon source is depleted, the produced acids or any other complex protein source such as waste material in the medium are then consumed as secondary Carbon sources, causing an increase in the pH value of the broth [51]. Such pH fluctuations could adversely affect the final yield of the toxin. Therefore, the regulation of pH with the help of pH controlling agents during Bt fermentation is an essential requirement of Bt pesticide production. Replacement of usual pH control agents such as NaOH and H_2SO_4 with C and N providers such as NH_4OH and CH_3COOH has found to increase the entomotoxicity of the Bt products [52].

5. Drawbacks of *Bacillus Thuringiensis* as a Biopesticide

Products containing Bt has potential to irritate the eyes and skin. Although Bt has many advantages as a biopesticide, it shows several drawbacks as discussed below.

5.1. Specificity and Resistance Build up

Most Bt strains contain a limited number of Cry toxins. Hence they are highly specific and have a narrow host range. Therefore, a farmer who is applying Bt as a biopesticide gets protection only against targeted pests. A mixture of Bt strains or the combination with other control methods are needed. Furthermore, having a limited number of Cry proteins in the biopesticide makes it vulnerable for build up of insect resistance [53]. The proposed mechanisms of insect resistance against Bt involve; loss of binding sites to midgut receptors, mutations in proteolytic enzymes that affect processing of the protoxin and quick wound recovery [54]. Hence, to minimize the insect resistance evolvement, it would be ideal to utilize multi toxin strains such as *B. thuringiensis* sbsp. *israelensis*, *B. thuringiensis* sbsp.

kurstaki, *B. thuringiensis* sbsp. *azaiwai*, *B. thuringiensis* sbsp. *tenbrionis*, *B. thuringiensis* sbsp. *sotto*, and *B. thuringiensis* sbsp. *entomocidus*. In fact, *B. thuringiensis* sbsp. *israelensis* which contains a number of toxin proteins such as Cry 4A, Cry4B, Cry 10A, Cry 11A and Cyt 1A, is the most resilient Bt strain against insect resistance.

5.2. Survival of Bt in the Environment

Bt crystals are highly sensitive to UV which rapidly inactivates the protoxins. The crystal proteins of Bt are inactivated within few hours to days in the phyllosphere and soil [55]. However, the endospores can remain viable in the environment for more than 2 years [56]. The experimentally determined half-lives of endospores in soil is in the range of 100–200 days [57]. However, *Bacillus thuringiensis* germination and growth in soil is limited by nutrient availability [58], studies conducted in non-supplemented soil have indicated a decrease in the number of cells, and the maintenance of the population mainly in the spore form [59]. Hence it seems that Bt cannot multiply under most soil conditions. In general Bt crystal protein is only active up to 48 hours in water and Bt strains are susceptible to intense sunlight and pollution. So their persistence in aquatic environments is highly limited. However, certain field trials and laboratory experiments have shown that Bt remains significantly viable in sterile sewage and lake water even after 10 days with viability losses of about 40% [60]. This study has not considered the impact of the aquatic microbial component which could affect the survival of Bt through competition.

5.3. Toxicity to Non-Target Animals

Bacillus thuringiensis was originally registered in 1961 as a General Use Pesticide (GUP). As discussed under the Bt toxins, the production of β exotoxins and *B. cereus* type enterotoxins by some strains of Bt make them toxic to non-target animals. Therefore, any new strain has to be properly screened against these toxins before their inclusion in pest control programs, especially when there have been instances of confirmed enterotoxin gene presence in commercial Bt insecticides [24]. β exotoxins are generally screened and quantified using high performance liquid chromatography (HPLC). However, the much cheaper fly assays (feeding Bt to house flies and checking for responses) can also be successfully used to cut down the costs. The identification and classification of the enterotoxin genes has paved the way for quick screening of these with the help of techniques such as Multiplex PCR [20].

Bt crystals are also considered as potential irritants of the eye and skin and the Bt products are grouped as a slightly toxic pesticide under Toxicity Class III. There are two case records of Bt.-caused infections in humans: in one instance, a young Agricultural worker splashed a commercial Bt product on to his right eye. He developed a corneal ulcer which was healed later with the application of subconjunctival injections of gentamicin and cefazolin sodium [114]. In the other case, a farm worker in Oregon, who was handling *B. thuringiensis* sbsp. *kurstaki*, was splashed in the eye with the formulation. He suffered an eye infection and irritation which was cleared up after applying a steroid cream [115]

5.4. Shelf-life of Bt Pesticides

The commercial products Bt contain about 2.5 x 10^{11} viable spores per gram and their shelf-life depends very much on the type of formulation. Generally, oil-based concentrates are more stable than those containing water, e.g. 2 vs. 1.5 years at 25° C [61]. Usually manufacturers find it difficult to maintain formulated products containing water for longer than 18 months without refrigeration [62]. Even though it is generally accepted that the Bt products lose some of their effectiveness when stored for more than 6 months, Boisvert and Boisvert (2001) have shown that two water based commercial products were able to maintain good shelf-life during a period of three years [63]. The temperature at which the products are stored also seems to make an impact on the shelf-life. Although Ignoffo *et al.* (1982) mentioned that *B. thuringiensis* sbsp. *israelensis* formulations, both liquid and primary powder, can lose between 85 and 99% activity when tested against *Aedes aegypti* after 28 days of storage at 50°C [64], other authors have shown that the storage temperature did not have an effect on the insecticidal activity of *B. thuringiensis* sbsp. *israelensis* crystals. Dempah (1979) claimed that temperature could not be considered as a degradation factor because a 24-h exposure at 80° C of a primary powder did not cause any loss of activity [65]. However, it is widely accepted that the number of freeze – thaw cycles causes a decrease in shelf life [63, 66].

6. Formulation Strategies

In pest control, Bt needs to be formulated using the dormant or spore form, which contains crystalline toxin. Formulation provides a platform to include a number of additives. Hence, it can be used as a strategy to overcome the main drawbacks of Bt. In fact some even suggest the comparatively high efficiency of certain commercial products may have a contribution from the inert material in the formulation [67]. Generally, formulation serves a number of functions such as;

- Stabilization during production, distribution and storage
- Aid handling and application of the product for easy delivery to the target
- Protection from harmful factors thereby increase persistence
- Enhancement of activity

Typical Bt formulations include wettable powders, spray concentrates, liquid concentrates, dusts, baits, and time release rings. Most Bt products contain insecticidal crystal proteins (ICP) and viable spores, but in some Bt products, the spores are inactivated. The potency of commercial products is a critical factor and it is given as international units per milligram (iu/mg). Potency is a measure of the number of effective killing units of the product (crystal proteins, spores, etc.) in bioassays of cabbage loopers compared to the killing power of a Bt reference strain with an arbitrary potency value. Therefore, high potency number is an indication of the concentration of the product. However, lack of international standards has created a situation where producers give potency against different species of test insects, making it difficult to compare products.

6.1. Challenges for Formulations and Additives as Solutions

The challenges faced by Bt pesticide formulations vary with the climatic conditions and the habitat to be applied. Bt is mainly applied to agricultural fields and aquatic bodies. However, there can also be forest applications during large pest outbreaks. In an agricultural field, the major worrying factors are sunlight and rain. Exposure to sunlight decreases residual insecticidal activity of Bt due to decreased viability of spores and degradation of crystal proteins. In vegetation Bt spores have a very short half-life of about 3.8 hours. The lethal wavelengths which kill Bt spores are up to 380nm with a peak at 300nm. The UV-A/B range (280–380 nm) of sunlight reaching earth's surface is considered responsible for substantial photo-degradation of crystals and consequent loss of toxicity [68, 69]. However, there have been suggestions that spores are killed best at 400nm (visible light) due to much greater amount of energy compared to UV [70]. There is also a strain variation in the sensitivity to UV, especially to UV-B, and strains such as *B. thuringiensis* sbsp. *kurstaki* are more sensitive to UV-B than *B. thuringiensis* sbsp. *israelensis*, this difference may be attributed to the efficiency of the DNA repair systems and the composition and relative abundance of small acid-soluble proteins (SASP) [71]. *B. thuringiensis* sbsp. *israelensis* contains different kinds of SASPs and their relative abundance on the spore coat is also low [72].

However, shading is impractical for crops grown over a large area. A number of ingredients could be added as sunscreens which protect the Bt formulations by selective absorption, blockage, reflection, negation of active oxygen radicals and conversion into visible light. These sunscreens include the vitamins (eg. folic acid, p-aminobenzoic acid), amino acids, activated charcoal, optical brightners (eg. Tinopal LPW), pigments such as melanin, chlorophylls and dyes such as Congo red and Erio Acid Red XB100. Addition of synthetic screens may incur negative impacts on soil microbes. Hence, the use of natural compounds such as melanin [73, 74] which traps UV through free radicals and the use of UV-B photo protective cyanobacterial pigments [75, 116] seem a more safe option.

Rain helps physical removal of spores and insecticidal proteins thereby reducing their residual activity. This can be reduced with the addition of stickers such as skimmed milk, corn starch, cereal flour, gelatin, pectin, glutan etc. Among these, pectin and gluten are more effective in preventing wash offs and are also capable of providing protection from UV [76, 77]. Stickers act as water availability buffers as they absorb water during daytime and release slowly during the night. Starch, sugarcane powder, glutan and corn starch also provides additional support by acting as feeding stimulants and improving the palatability of Bt spores [78, 79]. This helps overcome the problem of cessation of feeding by the insect larvae before ingesting a lethal dose of the toxin. However, addition of pectin may reduce the palatability [80].

Most commercial formulations also contain a substantial amount of inert materials. Some of these inert materials are intended to facilitate the dispersion of the crystals, maintain the proper particle size, prevent clumping during storage and adsorption onto particulate material. Overall, they should play a role in maintaining the toxic crystals in a state where availability can be sustained for maximum efficacy of a given formulation. [62, 81].

Bt formulations are applied to aquatic systems, mainly to control mosquito larvae and black fly larvae under health programs. The major challenges in an aquatic environment are the sunlight and pollutants, the effect of UV portion of sunlight, however, is made negligible by the low levels of penetration. The feeding zone of the target insect larvae provides another challenge to Bt formulations. Bt spores bind to organic matter and are quickly deposited on the bottom while the Crystal proteins are heavy entities which tend to sink rapidly once the spores are ruptured. This rapid sinking needs to be avoided because the steady deposit of debris could cover spores and crystals over time and make them ineffective even against the bottom feeders. Products that remain as suspensions below the surface are mostly effective against *Aedes* larva, which is a column feeder as it tends to feed actively up and down the whole depth of a shallow body of water. The sinking products are effective against the *Culex* larva which scavenges along substrate surfaces especially at the bottom. The same formulations may not be that effective against *Anopheles* larvae, which primarily feed at the surface-air interface of water. They may not ingest a lethal quantity of toxic particles in the relatively short period of time taken by particles that sink from the surface layer. Hence, multiple formulation types are required to handle mosquito larvae of different feeding zones. Buoyant types are required to handle the anophilines while the suspended types are required to control *Culex*, and sinking types to handle *Aedes*. The buoyant type formulations include the oil based types and slow releasing and fast releasing low density floating type formulations. Heavy winds could move the light floating types windward in unobstructed waters. Hence, the particle density should be adjusted in such areas to counter this problem. The monomolecular surface films or monolayers are ideal buoyant formulations as they float but do not get carried away by winds. They are ideal additives for Bt formulations because they can kill stages, such as pupa and adult which cannot be killed by the toxins. Monolayers make O_2 unavailable for larvae and interfere with the breathing of larvae and pupae. They also prevent egg laying on the surface by adults and could also destroy them by drowning.

Control of black fly larvae provides an additional challenge because unlike mosquito larvae they are found in fast moving water bodies. These are controlled by pouring bacterial suspensions into the water at consecutive points from which they are carried downstream [62].

6.2. Bt Formulation Types

6.2.1. Dry Formulations

Dry formulations include dusts, granules and capsules. These are used less frequently than liquid counterparts due to their lower efficacy and handling difficulties. Bt dusts are free flowing solid pesticide formulations in which the technical powder (Bt spore or crystals) is incorporated into, or coated on, a solid carrier, such as refined clay or talcum powder. In order to prevent the particles from aggregating, a free flow agent is added (eg Wessalon S).

Granules have a larger particle size than the dusts and are much easier to handle and apply. Technical powder can be made bound to the surface of the carrier material ideally with the aid of stickers or they can be made by mixing into a liquid carrier that can gel after drying, allowing the trapping of the technical powder in a matrix. Such a matrix would have at least some amount of the active technical powder on the surface. Capsules could also be produced upon coating of a core element (eg. alginate capsules), while water dispersible granules (WDG) are produced from readily ruptured granules in water. Pellets or tablets are solid forms with a diameter more than 6 mm. Tablets are easy to apply in small water bodies and containers, making them immensely valuable as a formulation in urban mosquito control programs. Tablets are known to be more efficient in controlling *A. aegypti* than other formulation types [82]. A persistence study on C4P1 tablet formulation has shown that it is capable of controlling *A. aegypti* larvae in shaded areas from 45 to 54 days with an acceptable duration of 13 to 35 days in sunny areas as well [83].

6.2.2. Liquid Formulations

Liquid formulations of Bt include emulsions, oil-based concentrates and water based concentrates. Water based concentrates are inferior to oil-based concentrates with relevance to storage. Hence, these are used very rarely except in developing countries. Oil-based concentrates need to contain finely ground colloids in order to prevent sedimentation. Oil-based products have so many advantages such as low viscosity, and low evaporation. Additionally the low moisture content provides easier stabilization of activity.

6.2.3. Wettable Powders and Water Dispersible Granules

Early commercial products of Bt were wettable powders due to the easiness of preparation. However, it soon became evident that these formulations have a difficulty in mixing with water and that they have a comparatively large particle size which makes them inefficient. One way to overcome this is the addition of filler ingredients. The filler ingredients (carriers) prevent agglomeration during storage which helps to reduce loss of wettability and clogging of spray nozzles. Unlike carriers of dry formulations fillers of wettable powders should allow rapid dispersion in water as a uniform suspension. Skimmed milk has been identified as excellent filler for wettable powders. In WDG, particles are held together by binders which some times can act as stickers during application. Dispersants such as surfactants and sugars, added to these formulations allow rapid release of particles making WDG superior to wettable powders.

7. Genetic Engineering of *Bacillus Thuringiensis*

Genetic engineering can be utilized to overcome many drawbacks and barriers of Bt bio pesticides. These include improving the host range, overcoming resistance build up, increasing the product yield, improving efficiency and persistence, overcoming of niche barriers and cutting down the cost of production.

7.1. Improving the Host Range

The host range of a Bt pesticide depends on the type of Cry proteins produced. Many strains contain one or few Cry proteins which lead to a narrow host range. In order to broaden the host range, so that one strain could be made effective against many pests, Cry proteins active against different insect orders should be bought together in one strain. Alternatively hybrid or dual acting toxins can be produced by the fusion of two genes with different specificities. Another method is to introduce foreign toxins, especially *B. sphaericus* toxins to *Bacillus thuringiensis* through transformation techniques.

7.2. Producing Strains Carrying Several Cry Toxins with Different Species Specificities

Although natural transfer of *cry* genes between Bt strains is facilitated by conjugation, some transfers are restricted in nature due to factors found in the plasmids, for example Cry1C protein is only found in *B. thuringiensis* sbsp. *izawai* and *B. thuringiensis* sbsp. *entomocidus* [84] and pBtoxis of *B. thuringiensis* sbsp. *israelensis* has been transferred only to two strains out of 15 tested, that too with low transfer rate of about 10^{-8} transconjugants/recipient [85]. Therefore, in order to obtain novel strains that could achieve broad spectrum activity, *cry* genes with different species specificities can be cloned into vectors and transferred to wild type Bt. Ideally the activity of both the foreign and the native genes should not be affected by the process. *Bacillus thuringiensis* subsp. *kurstaki* HD1 contains at least five crystal protein genes, including *cryIA*(*a*), *cryIA*(*b*), *cryIA*(*c*), *cryIIA*, and *cryIIB* [86, 87]. These *cryIA* and *cryII* genes are found in many *B. thuringiensis* subsp. *kurstaki* strains, and the encoded proteins are active against lepidopteran insect larvae such as *Heliothis virescens* and *Manduca sexta.* However, *B. thuringiensis* subsp. *kurstaki* never contains *cry1C* which confers pathogenesity to *Spodoptera* species [88]. Transfer of *cry1C* to *B. thuringiensis* subsp. *kurstaki* from *B. thuringiensis* sbsp. *aiswai* has been successfully achieved by Kalman and others [89]. According to their results, this recombination hasn't created any sporulation related abnormalities and both the parent strains and the recombinants have shown the same sporulation efficiencies. Where as the 50% lethal concentration (LC 50) values against *Spodoptera exigua* have been improved remarkably (lowering by 6 fold) by the recombination.

7.2.1. Hybrid Toxins

Although hybrid toxins are expected to broaden the host range by combining different specificities, they are more valuable in terms of increasing potency. A hybrid toxin produced from 2.2kb of *cry1C* and 1.3kb of *cry1Ab* produced a mature toxin similar to Cry1C after proteolytic cleavage. This toxin was found to be more active than Cry1C. This phenomenon was attributed to the increased stability to proteolytic cleavage provided by the Cry1Ab portion [90].

7.2.2. Introduction of B. Sphaericus Toxins to Bacillus Thuringiensis

Since both *B. thuringiensis* sbsp. *israelensis* and *Bacillus sphaericus* produce mosquitocidal proteins, toxins from *B. sphaericus* provide a gene reservoir that can be utilized to produce recombinant Bt strains with improved host range. The two species have different host ranges, Bt is more toxic to *Ades* and *Culex* larvae while *B.sphaericus* is more toxic to *Anophilus* and *Culex*. *B. sphaericus* can also survive better than Bt in polluted water. Hence, a recombinant strain of these two species would be immensely beneficial. Park et al (2000) produced a recombinant strain that produces Cyt1A and Cry 2B of *B. thuringiensis* sbsp. *israelensis* together with the *B. sphaericus* binary (bin) toxin [91]. This recombinant strain was produced with the help of two plasmid vectors carrying the same origin of replication. Hence, the stability and the copy number of the plasmids may change with time. This recombinant strain was capable of improved toxicity to C. *quinquefasciatus larvae* with a LC 50 of 1.7ηg/ml. The toxicities of the natural *B. thuringiensis* sbsp. *israelensis* and *B. sphaericus* for the same larvae were 7.9 and 12.6 ηg/ml respectively.

7.3. Increasing the Yield of Cry Toxins

Most of the Cry proteins are produced at the stationary phase during the sporulation. Hence, only a portion of the growth cycle is dedicated for crystal production. It would be advantageous to have the toxins produced during the vegetative stage also, in order to increase the yield and cut down the production time. Furthermore, this approach would allow the use of continuous fermentation process instead of the conventional batch fermentation. This has the potential to cut down the production costs significantly, because continuous culture requires less expensive, small scale bioreactors.

Cry toxins which are produced during sporulation are governed by one or two sporulation depended promoters. Transcription from these promoters is initiated by σ^E and σ^k like factors [92]. For instance, the *cry1 Aa* gene is transcribed during the sporulation phase from two sporulation-specific promoters [93, 94, 95]. In *B. thuringiensis* subsp. *israelensis,* the *cry4A, cry4B, cry11A* and *cyt1A* genes are transcribed during mid-sporulation from promoters recognized by the σ^{35} RNApolymerase. Transcription of the *cyt 1 A* and *cry1 1 A* genes continues into late-sporulation from promoters recognized by the σ^{28} -RNA polymerase [96, 97, 98, 99, 100]. *cry 3A* is an exception to this type of regulation; it is governed by a vegetative promoter, which is believed to be initiated by a σ^A like factor. Understanding of these factors has provided means to increase yields through various strategies. One strategy would be to clone the *cry* gene without its native promoter into a vector and let the gene be driven by a constitutive promoter. This would allow continuous synthesis of the toxin which would eventually increase the yield. This approach could be further refined by the introduction of the gene construct into the choromosome of a sporulation defective bacterium. This would guarantee that the gene will not be lost with time during continuous fermentation. Another approach is the placement of the *cry* genes under the regulation of *cry3A* promoter and transfer to a nonsporulating Bt mutant.

Typical Cry 1 protein is known to produce a very large crystal but several other important Cry proteins contribute to smaller crystals, these include the Cry 3 class [101] and the Cry 2 class [86]. Several factors are believed to be responsible for this situation. These include the existence of only a single promoter in some instances [92], poor mRNA stability [100, 102] and degradation of nascent polypeptides. In addition to the σ^A like factor involved in initiation, another element, the Shine Dalgarno Sequence found just downstream of the 5' end of the *cry3A* transcript (STAB-SD), is also responsible for stability of the transcript.

Expression of *cry3A* under dual sporulation dependant *cytA* promoters with the STAB-SD increased the yield per unit of culture medium by 12.7 fold compared to the wild strain, and by 1.4 fold compared to the commercial product with higher copy number [103]. Clones with and without the STAB-SD construct clearly showed that the stability provided by this element to the ribosome transcript complex, is the primary reason for yield increase. The Cry

3A crystals produced in this study were similar or even larger than the Cry1 bypyramidal crystals. The added advantage is that the Cry 3 crystal is purer due to the lack of the large C-terminal domain of Cry1 types.

Another method of increasing the Cry toxin yield comes from the idea that energy and material used for the production of the large C-terminal domain of the toxin could be diverted to the production of the active toxin, because it does not contribute to the active toxin. Such gene constructs without the sequence coding for the C-terminal domain are called truncated genes. However, as discussed earlier such an attempt could yield low expression and the truncated proteins may not form inclusions due to low levels of truncated gene expression, poor stability of the truncated mRNA and proteins and inefficient crystallization. Several elements could be utilized to overcome these problems. These elements include the 5' STAB-SD mRNA stabilizing sequence of *cry3A* [104], 3' stem-loop structures that could stabilize *cry* transcripts [105], and helper proteins that enhance translation and/or crystallization. These helper proteins are the 20- kDa chaperone-like protein encoded by *orf3* of the *cry11A* operon [100, 106] and the 29-kDa protein encoded by *orf2* of the *cry2A* operon [107], which serves both as a molecular chaperone and a scaffolding protein facilitating Cry2A crystal formation. Park et al. (2000) showed that a strain with the 5' STAB-SD sequence, the 3' terminator coding for the stem-loop structure and the 20kDa and 29 kDa helper proteins, produced the highest yield [91]. This experiment also proved that the 3' stem loop coding sequence should be in the natural 5'-3' direction for the production of the Cry1 and that the two helper proteins are involved in the enhancement of the net synthesis of crystals.

7.4. Preventing Build up of Resistance

When Bt pesticides are repeatedly applied or when they persist in the environment, such as under lowlight conditions found in storage rooms, the insects are likely to develop resistance, especially if the applied product contains fewer Cry proteins or similar Cry proteins. Insects develop resistance mainly through the receptor modifications. If the pesticide has many different Cry proteins, such as the spectrum found in *B. thuringiensis* sbsp. *israelensis*, resistance development could be curtailed. This could be achieved by the transfer of unrelated *cry* genes to commercially applied strains. It has been found that the remarkable performance of *B. thuringiensis* sbsp. *israelensis* against resistance build up is partly due to the number of different *cry* genes present and partly due to the presence of *cyt1A*, a gene coding for a hydrophobic toxin with a completely different mode of action to that of the Cry proteins. Cry proteins usually bind to glycoprotein receptors of the insect midgut epithelial cells whereas the Cyt1A binds with the lipid component of the membrane, especially the unsaturated fatty acids [108]. So the transfer of the *cyt1A* to other mosquitocidal strains would make them more resilient against development of resistance.

Another option is to produce toxins that could bind to more than one type of receptors making it difficult to develop resistance through receptor modification, because of the need to change many receptors instead of one. This could be achieved through hybrid toxins.

Modifications to insect midgut proteases is another mechanism by which insects develop their resistance against Bt. The rationale behind this method is that, when these enzymes are modified they will not cleave the protoxins and aid in the production of active toxins. If stable truncated proteins without their C- terminal domains can be produced, the modifications to proteases would be of no use to the insect because the toxin no longer needs proteolytic cleavage for activation.

7.5. Improving Persistence and Breaking the Niche Barriers

The major problems of controlling surface or just below surface dwelling insects by Bt sprayed to aquatic environments, are the rapid sinking of crystals and sedimentation of spores and the inactivation of the crystals and spores due to sunlight. Although formulation could help alleviate this problem to an extent, it does not offer a reliable solution. If the *cry* gene could be transferred to an organism which is already found in this region and also acts as a food source of the insect to be controlled, it would offer a more sustainable solution. With this idea in mind scientists have produced a number of genetically modified organisms harboring the Bt toxin genes. Some cyanobacteria provide an ideal model as they have been identified as larval food sources [109]. However, field trials of some *cry* transferred cyanobacteria were not very promising due to low levels of *cry* expression. *Caulobacter crescentus,* a bacterium which can persist and grow in low nutrient environments and found predominantly closer to the water surface, is also an exciting prospect for *cry* transfer, especially due to their motility, which allows the spread of Cry toxins. Although one such attempt has expressed the toxin gene at a lower level compared to the wild Bt strains [110] there is potential for higher expression through the incorporation of better promoters and other regulatory elements.

Root infesting insects provide another niche barrier to the Bt sprays which are mainly applied to the leaves and shoots. In order to solve this problem, attempts have been made to transfer *cry* genes into bacterial strains which colonizes the rhizosphere. This idea has been tested with *Pseudomans fluorescens* that colonizes plant roots [111]. However, such strains should not carry antibiotic markers which could be transferred to other bacteria.

Although Bt is found in the phylloplane, it does not normally colonize foliage, as itis vulnerable to high UV exposure at the upper surface of leaves and could be washed away with rain. Methylotroph, *Methylobacterium extorquens* is a foliage colonizer which uses methanol produced by the leaves to synthesize a cell wall component. This phytosymbiont nature could be utilized to produce a more persisting foliar pesticide. Choi et al (2008) produced a recombinant *Methylobacterium extorquens* harboring *cry1Aa* gene [112]. They were able to observe Cry1Aa crystals with typical bypyramidal shape in the recombinant strains under the electron microscope. Furthermore, these strains showed rapid feeding inhibition and 100% mortality when bioassayed with *Bombyx mori.*

8. Protein Engineering

In the last two decades a number of modifications have been created in δ-endotoxins using a repertoire of protein engineering techniques such as, single or multiple amino acid changes in variable and conserved regions through site directed mutagenesis, exchange of restriction fragments between closely related *cry* genes or with other bacterial toxin genes and swapping of domains between toxin genes through PCR mediated cloning. The aspects considered include the ability to create pores, the stability of the proteins in the presence of proteases, and the binding with specific receptors. Understanding of the characteristics and functions of they Cry toxin domains has provided an opportunity to create novel toxins with enhanced toxicity and specificity. The defects found in some of the natural Cry toxins such as poor binding with the receptors, presence of cleavage sites for proteases and poor solubility can be rectified using protein engineering [113].

9. Future Aspects

With natural *cry* genes being identified frequently, there is an ample pool of genes that can be utilized for improvements. The integration of bioinformatics, especially protein modeling, has provided an exciting avenue to explore the horizons. As our knowledge about the Bt toxins is improved on a daily basis, the ways and means of handling the challenges such as resistance development against Bt and environmental persistence problems, needs to be revolutionized. Because of the constant use and persistence, a situation may arise, where resistance development becomes the norm rather than the exception.

Hence, the future may require a blend of novel approaches such as protein engineering to produce more efficient toxins and to widen the toxin spectrum, and conventional techniques such as use of alternative strains and other cultural practices to avoid continuous exposure to pesticides, to prevent resistance build up.

Identification of novel Bt toxins such as the parasporins, with prospective applications in selective control of cancer cells has opened up a new avenue in the field of medicine for Bt, which needs to be properly harnessed in the future. Furthermore, with the current tools that can identify even a single base change in a DNA sequence, there is an added responsibility to investigate the recent reports that suggest presence of enterotoxin genes and elements, in commercially applied Bt strains, to avoid confusions over mix-ups in food poisoning. Finally some recent reports suggest the importance of synergistic effect of insect gut microflora for pathogenesis of Bt. Future research is required in this area to clearly ascertain such an involvement, the proof of which might lead to better control of insect pests.

References

[1] Feitelson, J. S., Payne, J. & Kim, L. 1992. *Bacillus thuringiensis*: insects and beyond. *Bio/Technology*, *10*, 271–275.

[2] Mackedonski, V. V., Nikolaev, N., Sebesta, K. & Hadjiolov, A. A. 1972. Inhibition of ribonucleic acid biosynthesis in mice liver by the exotoxin of *Bacillus thuringiensis*. *Biochimica et Biophysica Acta*, *272*, 56–66.

[3] Farkas, J., Sebesta, K., Horska, K., Samek, Z., Dollijs, J. & Storm, F. 1969. The structure of exotoxin of *Bacillus thuringiensis* var. *gelechiae*. *Collection of Czechoslovak Chemical Communication*, *34*, 1118–1120.

[4] Sebesta, K. & Horska, K. 1970. Mechanism of inhibition of DNA-dependent RNA polymerase by exotoxin of *Bacillus thuringiensis*. *Biochimica et Biophysica Acta, 209*. 357–376.

[5] Burgerjon, A., Biache, G. & Cals, P. 1969. Teratology of the Colorado potato beetle, *Leptinotarsa decemlineata*, as provoked by larval administration of the thermostable toxin of *Bacillus thuringiensis*. *Journal of Invertebrate Pathology*, *14*, 274–278.

[6] WHO Guidelines 1999

[7] Levinson, B. L., Kasyan, K. J., Chiu, S. S., Currier, T. C. & Gonzalez, J. M. 1989. Identification of β-exotoxin production, Plasmids encoding β-exotoxin, and a new Exotoxin in *Bacillus thuringiensis* by using High-Performance Liquid Chromatography. *Journal of Bacteriology*, *172(6),* 3172-3179.

[8] Crickmore, N., Zeigler, D. R., Schnepf, E., Van Rie, J., Lereclus, D., Baum, J, Bravo, A. & Dean, D. H. 2010. "*Bacillus thuringiensis* toxin nomenclature". http:// www.lifesci.sussex.ac.uk/Home/Neil_Crickmore/Bt/. Accessed 02/02/2010

[9] Han, S., Craig, J. A., Putnam, C. D., Carozzi, N. B. & Tainer, J. A. 1999. Evolution and mechanism from structures of an ADP-ribosylating toxin and NAD complex. *Nature Structural Biology*, *6*, 932–936.

[10] Warren, G. W. 1997. Vegetative insecticidal proteins: novel proteins for control of corn pests. In: Advances in Insect Control: The Role of Transgenic Plants (edited by N. Carozzi and M. Koziel) Pp. 109–121. Taylor and Francis, London, United Kingdom.

[11] Bhalla, R., Dalal, M., Panguluri, S. K., Jagadish, B., Mandaokar, A. D., Singh, A. K. & Kumar, P. A. 2005. Isolation, characterization and expression of a novel vegetative insecticidal protein gene of *Bacillus thuringiensis*. *FEMS Microbiology Letters*, *243*, 467–472.

[12] Estruch, J. J., Warren, G. W., Mullins, M. A., Nye, G. J., Craig, J. A. & Koziel, M. G. 1996. Vip3A, a novel *Bacillus thuringiensis* vegetative insecticidal protein with a wide spectrum of activities against lepidopteran insects. *Proceedings of the National Academy of Sciences USA*, *93*, 5389–5394.

[13] Lee, M. K., Walters, F. S., Hart, H., Palekar, N. & Chen, J. S. 2003. The mode of action of the *Bacillus thuringiensis* vegetative insecticidal protein Vip3A differs from that of Cry1Ab δ-endotoxin. *Applied and Environmental Microbiology*. *69*, 4648–4657.

[14] Shi, X., Shao, C. Zhang, X., Zambonelli, C., Redfield, A. G., Head, J. F., Seaton, B. A. & Roberts, M. F. 2009. Modulation of *Bacillus thuringiensis* Phosphatidylinositol specific phospholipase C activity by mutations in the putative dimerization interface. *The journal of biological chemistry*, *284(23),* 15607-15618.

[15] Mizuki, E., Park, Y. S., Saitoh, H., Yamashita, S., Akao, T., Higuchi, K. & Ohba, M. 2000 *Clinical and Diagnostic Laboratory Immunology*, *7,* 625-634.

[16] Kitada, S., Abe, Y., Shimada, H., Kusaka, Y., Matsuo, Y., Katayama, H., Okumura, S., Akao, T., Mizuki, E., Kuge, O., Sasaguri, Y., Ohba, M. & Ito, A. 2006. Cytocidal actions of Parasporin-2, an anti-tumor crystal toxin from *Bacillus thuringiensis*. *Journal of Biological Chemistry*, *281*, 26350-26360.

[17] Budarina, Z. I., Sinev, M. A., Mayorov, S. G., Tomashevski, A. Y., Shmelev, I. V. & Kuzmin, N. P. 1994. Hemolysin II is more characteristic of *Bacillus thuringiensis* than *Bacillus cereus*. *Archives of Microbiology*, *161*, 252–257.

[18] Crickmore, N., Bone, E. J., Williams, J. A. & Ellar, D. J. 1995. Contribution of the individual components of the δ-endotoxin crystal to the mosquitocidal activity of *Bacillus thuringiensis* subsp. *israelensis*. *FEMS Microbiology Letters*, *131*, 249–254.

[19] Khasdan, V., Ben-Dov, E., Manasherob, R., Boussiba, S. & Zaritsky, A. 2001. Toxicity and synergism in transgenic *Escherichia coli* expressing four genes from *Bacillus thuringiensis* subsp. *israelensis*. *Environmental Microbiology*, *3*, 798–806.

[20] Abou Zeid, A. A. 2009. Molecular characterization and enterotoxin genes typing of local strains of *Bacillus Cereus*. *Australian Journal of Basic and Applied Sciences*, *3(1)*, 160-166.

[21] Lund, T. & Granum, P.E. 1997. Comparison of biological effect of the two different enterotoxin complexes isolated from three different strains of *Bacillus cereus*. *Microbiology*, *143*, 3329–3336.

[22] Lindback, T., Fagerlund, A., Rodland, M. S. & Granum, P. E. 2004. Characterization of the *Bacillus cereus* Nhe enterotoxin. *Microbiology*, *150*, 3959–3967.

[23] Lund, T., De Buyser, M. L. & Granum, P. E. 2000. A new cytotoxin from *Bacillus cereus* that may cause necrotic enteritis. *Molecular Microbiology*, *38*, 254–261.

[24] Scarano, C., Virdis, S., Cossu, F., Frongia, R., De Santis, E. P. L. & Cosseddu, A. M. 2009. The pattern of toxin genes and expression of diarrheal enterotoxins in *Bacillus thuringiensis* strains isolated from commercial bioinsecticides. *Veterinary Ressearch Communication*, 33(Suppl 1), S257–S260.

[25] Zhang, X., Candas, M., Griko, N. B., Taussig, R., & Bulla Jr., L. A., 2006. A mechanism of cell death involving an adenylyl cyclase/PKA signaling pathway is induced by the Cry1Ab toxin of *Bacillus thuringiensis*. *Proceedings of the National Academy of Sciences, 103*, 9897–9902.

[26] Li, J., Carroll, J. & Ellar, D. J. 1991. Crystal structure of insecticidal δ endotoxin from *Bacillus thuringiensis* at 2.5 Å resolutions. *Nature, 353*, 6347, 815-821.

[27] 27. Grochulski, P., Masson, L., Borisova, S., Pusztai-Carey, M., Schwartz, J. L., Brousseau, R. & Cygler, M., 1995. *Bacillus thuringiensis* CryIAa insecticidal toxin: crystal structure and channel formation. *Journal of Molecular Biology*, *254*, 447–464.

[28] Morse, R. J., Yamamoto, T. & Stroud, R. M., 2001. Structure of Cry2Aa suggests an unexpected receptor binding epitope. *Structure*, *9*, 409–417.

[29] Galitsky, N., Cody, V., Wojtczak, A., Ghosh, D., Luft, J. R., Pangborn, W. & English, L. 2001. Structure of the insecticidal bacterial δ-endotoxin CryBb1 of *Bacillus thuringiensis*. *Acta Crystallographica*, *D 57*, 1101–1109.

[30] Boonserm, P., Davis, P., Ellar, D. J. & Li, J. 2005. Crystal structure of the mosquito-larvicidal toxin Cry4Ba and its biological implications. *Journal of Molecular Biology*, *348*, 363–382.

[31] Boonserm, P., Mo, Angsuthanasombat, Ch. & Lescar, J. 2006. Structure of the functional form of the mosquito larvicidal Cry4Aa toxin from *Bacillus thuringiensis* at a 2.8-A° resolution. *Journal of Bacteriology*, *188*, 3391–3401.

[32] Knowles, B. 1994. Mechanism of action of *Bacillus thuringiensis* insecticidal endotoxins. *Advances in Insect Physiology,* 24, 275-308.

[33] Crickmore, N., Zeigler, D. R., Feitelson, J., Schnepf, E., Van rie, J., Llereclus, D., Baum, J. & Dean, D. H. 1998. Revision of the nomenclature for the *Bacillus thuringiensis* pesticidal crystal proteins. *Microbiology and molecular biology reviews*, *62(3)*, 807–813.

[34] Benoit, T. G., Wilson, G. R. & Baygh, C. L. 1990. Fermentation during growth and sporulation of *Bacillus thuringiensis* HD-1. *Letters in Applied Microbiology*, *10*, 15–18.

[35] Zouari, N., Dhouib, A., Ellouz, R. & Jaoua, S. 1998. Nutritional requirements of a strain of *Bacillus thuringiensis* subsp *kurstaki* and use of gruel hydrolysate, for the formulation of a new medium for deltaendotoxin production. *Applied Biochemistry and Biotechnology, 69*, 41–52.

[36] Ozkan, M., Dilek, F. B., Yetis, U. & Ozcengiz, G. 2003. Nutritional and cultural parameters influencing antidipteran delta-endotoxin production. *Research in Microbiology*, *154*, 49–53.

[37] Yang, X. M. & Wang, S. S. 2000. Phase-specific optimization of multiple endotoxin-protein production with genetically engineered *Bacillus thuringiensis*. *Biotechnology and Applied Biochemistry*, *31*, 71–76.

[38] Poopathi, S. & Kumar, A. K. 2003. Novel fermentation medium for the production of *Bacillus thuringiensis* serovar. *israelensis*, in mosquito control. *Journal of Economic Entomology*, *96*, 1039 – 1044.

[39] Brar, S. K., Verma, M., Tyagi, R. D., Vale´ro, J. R. & Surampalli, R. Y. 2005. Sludge based *Bacillus thuringiensis* biopesticides: Viscosity impacts. *Water Research*, *39*, 3001–3011.

[40] Vidyarthi, A. S., Tyagi, R. D., Vale´ro, J. R. & Surampalli, R. Y. 2002. Studies on the production of *B. thuringiensis* based biopesticides using wastewater sludge as a raw material. *Water Research*, *36*, 4850–4860.

[41] 41. Vidyarthi, A. S., Tyagi, R. D. & Valéro, J. R. 2001. Effect of surface active agents on the production of biopesticides using wastewater sludge as a raw material. *Water Science and Technology*, *44(10),* 253–259.

[42] Alves, L. F. A., Alves, S. B., Pereira, R. M. & Capalbo, M. F. 1997. Production of *Bacillus thuringiensis* Berliner var. *kurstaki* grown in alternative media. *Biocontrol Science and Technology*, *7*, 377 - 383.

[43] Capalbo, D. M. F., Valicente, F. H., Moraes, I. de O. & Pelizer, L. H. (2001). Solid-state fermentation of *Bacillus thuringiensis tolworthi* to control fall armyworm in maize. *Electronic Journal of Biotechnology*, *4(2),* 2-4.

[44] Lee, J. M. 1992. *Cell Kinetics and Fermenter Design. Biochemical Engineering.* Pp 170-174. Englewood Cliffs, NJ, USA: Prentice-Hall.

[45] Wu, W. T. & Wu, J. Y. 1990. Airlift reactor with net draught tube. *Journal of Fermentation and Bioengineering*, *70*, 359–361.

[46] Fu, C. C., Wu, W. T. & Lu, S. Y. 2003. Performance of airlift bioreactors with net draft tube. *Enzyme and Microbial Technology*, *33*, 332–342.

[47] Hongzhang, C., Fujian, X., Zhonghou, T. & Zuohu, L. 2002. A novel industrial level reactor with two dynamic changes of air for solid state fermentation. *Journal of Bioscience and Bioengineering*, *93(2),* 211-214.

[48] Amin, G., Alotaibi, S., Youssef, N. A. & Saleh, W. D. 2008. Bioinsecticide production by the bacterium *Bacillus thuringiensis* (pattern of cell growth, toxin production and by-product synthesis). *Journal of Basic and Applied Sciences*, *4(1),* 27-31.

[49] Rossa, C. A., Yantorno, O. M., Arcas, J. A. & Ertola, R. J. 1990. Organic and inorganic nitrogen source ratio effects on *Bacillus thuringiensis* var *israelensis* delta-endotoxin production. World Journal of Microbiology and Biotechnology, *6*, 27–31.

[50] Kraemer-Schafhalter, A. & Moser, A. 1996. Kinetic study of *Bacillus thuringiensis* in lab-scale batch process. *Bioprocess Engineering, 14*, 139–144.

[51] Harms, R. L., Martinez, D. R. & Griego, V. M. 1986. Isolation and characterization of coproporphyrin produced by four subspecies of *Bacillus thuringiensis*. *Applied Environmental Microbiology, 51*, 481–486.

[52] Vu, K. D., Tyagi, R. D., Vale′ro, J. R. & Surampalli, R. Y. 2009. Impact of different pH control agents on biopesticidal activity of *Bacillus thuringiensis* during the fermentation of starch industry wastewater. *Bioprocess and Biosystem Engineering, 32*, 511–519.

[53] Tabashnik, B. E., Finson, N., Groeters, F. R., Moar, W. J., Johnson, M. W., Luo, K. & Adang, M. J. 1994. Reversal of resistance to *Bacillus thringiensis* in *Plutella xylostella*. *Proceedings of National Academy of Science, USA, 91*, 4120-4124.

[54] Tabashnik, B. E. 1994. Evolution of resistance to *Bacillus thuringiensis*. *Annual Reviews of Entomology, 39*, 47-49.

[55] Glare, T. R. & O'Callaghan, M. 2000. *Bacillus thuringiensis*: *Biology, Ecology and Safety*. Chichester: John Wiley and Sons, Ltd.

[56] Smith, R. A. & Barry, J. W. 1998. Environmental persistence of *Bacillus thuringiensis* spores following aerial application. *Journal of Invertebrate Pathology, 71*, 263–267.

[57] Hansen, B. M., Damgaard, P. H., Eilenberg, J. & Pedersen, J. C. 1996. *Bacillus thuringiensis*. Ecology and environmental effects of its use for microbial pest control. Ministry of Environment and Energy, Denmark, Danish Environmental Protection Agency, Environmental Project 316.

[58] West, A. W., Burges, H. D., Dixon, T. J. & Wyborn, C. H. 1985. Survival of *Bacillus thuringiensis* and *Bacillus cereus* spore inocula in soil: effects of pH, moisture, nutrient availability and indigenous microorganisms. *Soil Biology and Biochemistry, 17*, 657–665.

[59] Vilas-Boas L. A., Vilas-Boas G. F., Saridakis H. O., Lemos M. V., Lereclus D. & Arantes O. M. 2000. Survival and conjugation of *Bacillus thuringiensis* in a soil microcosm. *FEMS Microbiology Ecology, 31*, 255–259.

[60] Furlaneto, L., Saridakis, H. O. & Arantes, O. M. N. 2000. Survival and conjugal transfer between *Bacillus thuringiensis* strains in aquatic environment. *Brazilian Journal of Microbiology, 31*, 233-238.

[61] Devisetty, B. N. 1988. Microbial formulations: opportunities and challenges. In: *Pesticide Formulations and Application Systems* (edited by Hovde, D. A. & Beestman, G. B.) Pp. 46- 64, Vol. 8. ASTM STP 98, American Society for Testing and Materials, Philadelphia.

[62] Burges, H. D. & Jones, K. A. 1998. Formulation of bacteria, viruses and protozoa to control insects. In: *Formulation of Microbial Biopesticides: Beneficial Microorganisms, Nematodes and Seed Treatments* (edited by Burges, H. D). Pp. 33-127, Dordrecht, The Netherlands: Kluwer Academic.

[63] Boisvert, M. & Boisvert, J. 2001. Storage stability of two liquid formulations of *Bacillus thuringiensis* subsp. *israelensis* and effect of freezing over time, *11*, 475- 482.

[64] Ignoffo, C. M., Garcia, C., Kroha, M. & Couch, T. L. 1982. High-temperature sensitivity of formulations of *Bacillus thuringiensis* var. *israelensis*. *Environmental Entomology, 11*, 409- 411.

[65] Dempah, J. 1979. Essais de *Bacillus thuringiensis israelensis* sur les moustiques. Rapport. Diplome E Â tudes AvanceÂes. Entomologie MeÂ dicale., FaculteÂ des Sciences, Paris XI, et Laboratoire ORSTOM, Bondy, France.

[66] Tousignant, M. E., Boisvert, J. L. & Chalifour, A. 1992. Reduction of mortality rates of *Bacillus thuringiensis* var. *israelensis* aqueous suspensions due to freezing and thawing. *Journal of the American Mosquito Control Association, 8*, 149- 155.

[67] Ahmedani, M. S., Khaliq, A & Haque, M. I. 2007. Scope of commercial formulations of *Bacillus thuringiensis* berliner as an alternative to methyl bromide against *trogoderma granarium* everts larvae. *Pakistan. Journal of Botany*, *39*(3), 871-880.

[68] Pozsgay, M., Fast, P., Kaplan, H. & Carey, P. R. 1987. The effect of sunlight on the protein crystal from *Bacillus thuringiensis* var. *kurstaki* HD1 and NRD12: a Raman spectroscopy study. *Journal of Invertebrate Pathology*, *50*, 246–253.

[69] Pusztai, M., Fast, P., Gringorten, H., Kaplan, H., Lessard, T. & Carey, P. R. 1991. The mechanism of sunlight-mediated inactivation of *Bacillus thuringiensis* crystals. *Biochemical Journal*, *273*, 43–47.

[70] Griego, V. M. & Spence, K. D. 1978. Inactivation of *Bacillus thuringiensis* spores by ultraviolet and visible light. *Applied Environmental Microbiology*, *35*, 906–910.

[71] Myasnik, M., Manasherob, R., Ben-Dov, E., Zaritsky, A., Margalith, Y. & Barak, Z. 2001. Comparative sensitivity to UV-B radiation of two *Bacillus thuringiensis* subspecies and other *Bacillus* sp. *Current Microbiology*, *43*, 140–143.

[72] Cucchi, A. & Rivas, C. S. 1994. Sensitivity of spores and growing cells of *Bacillus thuringiensis* var. *israelensis* and *Bacillus sphaericus* to osmotic variations. *Current Microbiology*, *28*, 123–127.

[73] Liu, Y. T., Sui, M. J., Ji, D. D., Wu, H. I., Chou, C. C. & Chen, C. C. 1993. Protection from ultraviolet irradiation by melanin of mosquitocidal activity of *Bacillus thuringiensis* var. *israelensis*. *Journal of Invertebrate Pathology*, *62*, 131–136.

[74] Saxena, D., Ben-Dov, E., Manasherob, R., Boussiba, S. & Zaritsky, A. 2002. A UV tolerant mutant of *Bacillus thuringiensis* subsp. *kurstaki* producing melanin. *Current Microbiology*, *44*, 25–30.

[75] Goetz, T., Windhoevel, U., Boeger, P. & Sandmann, G. 1999. Protection of photosynthesis against ultraviolet-B radiation by carotenoids in transformants of the cyanobacterium *Synechococcus* PCC7942. *Plant Physiology Rockville*, *120*, 599–604.

[76] Ramos, L. H. M., McGuire, M. R., Wong, L. J. G. & Franc, R. C. 2000. Evaluation of pectin, gelatin and starch granular formulations of *Bacillus thuringiensis. Southwestern Entomologist, 25(1),* 59-67.

[77] Behle, R. W., Mcguire, M. R. & Shasha, B. S. 1997. Effects of sunlight and simulated rain on residual activity of *Bacillus thuringiensis* formulations. *Journal of Economy Entomology*, *90*(6), 1560-1566.

[78] McGuire, M. R., Shasha, B. S., Lewis, L. C. & Nelsen, T. C. 1994. Residual activity of granular starch-encapsulated *Bacillus thuringiensis*. *Journal of Economy Entomology*, 87, 631-637.

[79] Rosas-Garcıa, N. M. 2006. Laboratory and field tests of spray-dried and granular formulations of a *Bacillus thuringiensis* strain with insecticidal activity against the sugarcane borer. *Pest Management Science*, 62, 855-861.

[80] Rosas-Garcia, N. M., Villegas-Mendoza, J.M. & Torres-Ortega, J. A. 2009. Design of a *Bacillus thuringiensis*-based formulation that increases feeding preference on

Spodoptera exigua (Lepidoptera: Noctuidae) Larvae. *Journal of Economy Entomology*, 102(1), 58-63.

[81] Boisvert, M. 2005. Utilization of *Bacillus thuringiensis* var. *israelensis* (*Bti*)-based formulations for the biological control of mosquitoes in Canada (edited by Côté, J.-C., Otvos, I. S, Schwartz, J.-L. & Vincent, C.) *6th Pacific Rim Conference on the Biotechnology of Bacillus thuringiensis and its Environmental Impact,* Victoria, BC.

[82] Becker, N. 2000. Bacterial control of vector-mosquitoes and black flies. In: *Entomopathogenic Bacteria: from Laboratory to Field Application* (edited by J. F. Charles, A. Delécluse & C. Nielsen-LeRoux), Pp. 383-396. Dordrecht: Kluwer Academic Publishers.

[83] de Melo-Santos, M. A. V., Sanches, E. G., de Jesus, F. J. & Regis, L. 2001. Evaluation of a new tablet formulation based on *Bacillus thuringiensis* serovar. *israelensis* for larvicidal control of *Aedes aegypti. Mem Inst Oswaldo Cruz, Rio de Janeiro, 96*, 1-2.

[84] Visser, B. 1989. A screening for the presence of four different crystal protein gene types in 25 *Bacillus thuringiensis* strains. *FEMS Microbiology Letters, 58,* 121–124.

[85] Hu, X., Hansen, B. M., Yuan, Z., Johansen, J. E., Eilenberg, J., Hendriksen, N. B., Smidt, L. & Jensen, G. B. 2005. Transfer and expression of the mosquitocidal plasmid pBtoxis in *Bacillus cereus* group strains. *FEMS Microbiology Letters, 245*, 239–247.

[86] Hofte, H. & Whiteley, H. R. 1989. Insecticidal crystal proteins of *Bacillus thuringiensis. Microbiology Review, 53*, 242–255.

[87] Kronstad, J. W. & Whiteley, H. R. 1986. Three classes of homologous *Bacillus thuringiensis* crystal-protein genes. *Gene, 43*, 29 – 40.

[88] Visser, B., van der Salm, T., van den Brink, W. & Folkers, G. 1988. Genes from *Bacillus thuringiensis* entomocidus 60.5 coding for insect-specific crystal proteins. *Molecular and General Genetics, 212*, 219–224.

[89] Kalman, S., Kiehne, K. L., Cooper, N., Reynoso, M. S. & Yamamoto, T. 1995. Enhanced production of insecticidal proteins in *Bacillus thuringiensis* strains carrying an additional crystal protein gene in their chromosomes. *Applied and Environmental Microbiology, 61(8),* 3063–3068.

[90] Caramori, T., Albertini, A. M. & Galzzi, A. 1991. In vivo generation of hybrids between two *Bacillus thuringiensis* insect-toxin-encoding genes. *Gene, 98*, 37-44.

[91] Park, H. W., Bideshi, D. K. & Federici, B. A. (2000). Molecular genetic manipulation of truncated Cry1C protein synthesis in *Bacillus thuringiensis* to improve stability and yield. *Applied and Environmental Microbiology. 66(10),* 4449–4455.

[92] Baum, J. A. & Malvar, T. 1995. Regulation of insecticidal crystal protein production in *Bacillus thuringiensis. Molecular Microbiology, 18*, 1–12.

[93] Wong, H. C., Schnepf, H. E. & Whiteley, H. R. 1983. Transcriptional and translational start sites for the *Bacillus thuringiensis* crystal protein gene. *Journal of Biological Chemistry, 258*, 1960–1967.

[94] Brown, K. L. & Whiteley, H. R. 1988. Isolation of a *Bacillus thuringiensis* RNA polymerase capable of transcribing crystal protein genes. *Proceedings of the National Academy of Sciences USA, 85*, 4166-4170.

[95] Brown, K. L. & Whiteley, H. R. (1990). Isolation of the second *Bacillus thuringiensis* RNA polymerase that transcribes from a crystal protein gene promoter. *Journal of Bacteriology. 172*, 6682-6688.

[96] Waalwijk, C., Dullemans, A. M., vanWorkurn, M. E. S. & Visser, B. 1985. Molecular cloning and the nucleotide sequence of the M, 28 000 crystal protein gene of *Bacillus thuringiensis* subsp. *israelensis. Nucleic Acids Research*, *13*, 8207-8217.

[97] Ward, E. S. & Ellar, D. J. 1986. *Bacillus thuringiensis* var. *israelensis* δ-endotoxin. Nucleotide sequence and characterization of the transcripts in *Bacillus thuringiensis* and *Escherichia coli. Journal of Molecular Biology*, *191,* 1-11.

[98] Yoshisue, H., Ihara, K., Nishimoto, T., Sakai, H. & Komano, T. 1995. Expression of the genes for insecticidal crystal proteins in *Bacillus thuringiensis*: *crylVA,* not *crylVB,* is transcribed by RNA polymerase containing σ^H and that containing σ^E. *FEMS Microbiology Letters*, *127*, 65-72.

[99] Yoshisue, H., Nishimoto, T., Sakai, H. & Komano, T. 1994. Identification of a promoter for the crystal protein-encoding gene *crylVB* from *Bacillus thuringiensis* subsp. *israelensis. Gene*, *137*, 247-251.

[100] Dervyn, E., Poncet, S., Klier, A. & Rapoport, G. 1995. Transcriptional regulation of *Bacillus thuringiensis* subsp. *Isruelensis cryIVD* gene operon. *Journal of Bacteriology*, *177*, 2283-2291.

[101] Donovan, W. P., Dankocsik, C. C. & Gilbert, M. P. 1988. Molecular characterization of a gene encoding a 72-kilodalton mosquito-toxic crystal protein from *Bacillus thuringiensis* subsp. *israelensis*. *Journal of Bacteriology*, *170*, 4732– 4738.

[102] de Souza, M. T. M., Lecadet, M. & Lereclus, D. 1993. Full expression of the *cryIIIA* toxin gene of *Bacillus thuringiensis* requires a distant upstream DNA sequence affecting transcription. *Journal of Bacteriology*, *175*, 2952–2960.

[103] Park, H. W, Baoxue, G. E., Bauer, L. S. & Federici, B. A. 1998. Optimization of Cry3A yields in *Bacillus thuringiensis* by use of sporulation-dependent promoters in combination with the STAB-SD mRNA sequence. *Applied and Environmental Microbiology*, *64(10)*, 3932–3938.

[104] Agaisse, H. & Lereclus, D. 1996. STAB-SD: a Shine-Dalgarno sequence in the 5' untranslated region is a determinant of mRNA stability. *Molecular Microbiology*, *20*, 633–643.

[105] Wong, H. C. & Chang, S. 1986. Identification of a positive retroregulator that stabilizes mRNAs in bacteria. *Proceedings of the National Academy of Sciences USA*, *83*, 3233–3237.

[106] Wu, D. & Federici, B. A. 1995. Improved production of the insecticidal CryIVD protein in *Bacillus thuringiensis* using *cry1A(c)* promoter to express the gene for an associated 20-kDa protein. *Applied Microbiology and Biotechnology*, *42*, 697–702.

[107] Winder, W. R. & Whiteley, H. R. 1989. Two highly related insecticidal crystal proteins of *Bacillus thuringiensis* subsp. *kurstaki* possess different host range specificities. *Journal of Bacteriology*, *171*, 965–974.

[108] Glick, B. R. & Pasternack, J. J. 2003. Microbial insecticides. In: *Molecular Biotechnology, Principles and Applications of Recombinant DNA*. Pp. 455-480. Washington, DC: ASM press.

[109] Thiery, I., Nicholas, L., Rippka, R. & de Marsac, T. 1991. Selection of cyanobacteria isolated from mosquito breeding sites as a potential food source for mosquito larvae. *Applied and Environmental Microbiology*, *57*, 1354-1359.

[110] Thanabalu, T., Hindley, J., Brenner, S., Oei, C. & Berry, C. 1992. Expression of the mosquitocidal toxins of *Bacillus sphaericus* and *Bacillus thuringiensis* subsp.

Israelensis by recombinant *Caulobacter crecentus*, a vehicle for biological control of aquatic insect larvae. *Applied and Environmental Microbiology, 58*, 905-910.

[111] Obukowicz, M. G., Perlack, F. J., Kretzmer, K. K., Mayer, E. J. & Watrud, L. S. 1986. Integration of the delta-endotoxin gene of *Bacillus thuringiensis* into the chromosome of root-colonizing strains of pseudomands using Tn5. *Gene, 45*, 327-331.

[112] Choi, Y. J., Gringorten, J. L., Be′langer, L., Morel, L., Bourque, D., Masson, L., Groleau, D. & Mı′guezl, C. B. 2008. Production of an insecticidal crystal protein from *Bacillus thuringiensis* by the methylotroph *Methylobacterium extorquens. Applied and Environmental Microbiology, 74(16)*, 5178–5182.

[113] Saraswathy, N. & Kumar, P. A. 2004. Protein engineering of δ-endotoxins of *Bacillus thuringiensis. Electronic Journal of Biotechnology, 7(2)*, 178-188.

[114] Samples, J R., and Buettner, H. 1983. Corneal ulcer caused by a biologic insecticide (*Bacillus thuringiensis*). Am. J. Ophthalmol. 95: 258-260.

[115] Green, M., Heumann, M., Sokolow, R., Foster, L.R., Bryant, R. and Skeels, M. 1990. Public health implications of the microbial pesticide *Bacillus thuringiensis*: an epidemiological study, Oregon, 1985-86. Am. J. Pubic Health. 80: 848-852.

[116] Ehling-Schulz, M., Bilger, W. and Scherer, S., 1997. UV-B-induced synthesis of photoprotective pigments and extracellular polysaccharides in the terrestrial cyanobacterium *Nostoc commune*. J Bacteriol, 179, 1940–1945.

In: Bioprocess Sciences and Technology
Editor: Min-Tze Liong
ISBN 978-1-61122-950-9

Chapter 2

BUTANOL
A TOTAL SUBSTITUTE FOR GASOLINE FROM AGRICULTURAL RESIDUE

J. M. R. S. Bandara** and H. V. P. Wijewardhana
Department of Agriculture Biology,
University of Peradeniya, Peradeniya, Sri Lanka

ABSTRACT

Fossil fuel is a magical substance which brought numerous privileges to industrialized societies. Alternatively, industrial, domestic and agricultural waste generation has become a detrimental fact. The materials emitted from fuel and waste affect the environment as well as human health. Global warming, pollution due to heavy metals and other toxic materials spread of viral and bacterial infectious diseases are some of the key effects of excessive use of fossil fuel, its derivatives and improper handling of waste.

It is important to pursue material and energ[1]y recycling, while minimizing waste, at global level. The biomass energy production system is an ideal recycling method for environmental conservation through prevention of global warming. The mechanism involves fixing of CO_2 *via* photosynthesis through metabolic machineries in plants and microorganisms. The use of bio energy system prevents further release of non recycled CO_2 and carbon sequencing through fixing in the biomass. It is important to pay attention to bio energy production from non food materials such as agricultural and municipal waste. The biomass- starch and sugars are obtained from food crops; hence inappropriate using as raw materials for energy, mainly in the aspects of food security, long term supply and cost. Cellulose and hemicellulose resources are the most abundant and under utilized materials, specially rice straw, bagasse, and forest wastes.

Cellulose and hemicelluloses are main structural materials of plant cell wall and can be used as raw materials for alcohol fermentation. Simply, cellulose and hemicelluloses

*Corresponding author: *J.M.R.S. Bandara* Email: bandara.sarath@gmail.com are converted to simple sugars. The process can be done either by enzymatic conversion or acid hydrolysis or microbial scarification. Microbial scarification is the most attractive method compared to others. Simple sugars can be converted to desired alcohols by fermentation. Butanol is the most promising bio fuel having attractive characters, compared to other substitutes such as ethanol, bio oil and bio diesel. The energy content, higher octane rating, low Reid Vapor Pressure (RVP), safer flash point and the ability to mix freely with gasoline are the most remarkable features. The technology of butanol production had several bottlenecks.

Most critical factor was inherent toxicity of butanol for solventogenesis microorganisms. With 1-2% concentration of butanol in the medium, the organism will be destroyed. Most researchers attempted to overcome this problem through the use of genetically modified organisms, but it was not an ultimate solution. Recent developments in the concept of bio refinery addressed most of the key issues of the process and made commercial viability of butanol production from lignocelluloses. The cultivable and uncultivable microorganisms in consortia used in the anaerobic bioreactor can remarkably speed up cellulolytic conversion as well as solventogenesis. Alternatively, harness of other energy products in the bio refinery resulted positive energy and material balance.

1. Introduction

Energy is a fundamental right for all human beings in the 21st century. It is known as the key factor for cost of utility items and political power. When the price of energy changes, the price of consumer items also fluctuate accordingly, irrespective of the item concerned. If a country is self-sufficient in energy, it is in a strong position in the global political regime. Most powerful countries have a political understanding with the countries rich in energy resources. Political instabilities have been created by energy trading alone. The leaders of political fronts talk about energy independency for their country's survival and maintenance of political power regionally as well as globally.

All energies originate from solar power and stored in different forms such as potential energy and biological energy. It is now well known that the fossil fuel also originated from biomass. Fossil fuel is the main attractive energy source during the 20th century. From the 3rd and 4th decades of the 20th century, fossil fuels have made incredible impact on the activities of the modern world. Fossil fuels are not only used as transport energy but also for power generation. In the twenty first century it is very hard to find a territory without electricity. The fossil fuels are not only limited to energy supply; by products are used for a wide range of activities including lubricant, fabric, fertilizer, packaging and various products including food colors and flavors. Annual consumption fossil fuel in the world is 1258 thousand million barrels [1] and it is predicted to increase by 54% between 2001 and 2025 [2]. The refined derivatives from fossil fuels are many. But Gasoline and Diesel are the most demanding due to their usage in internal combustion engines which generate power for motion as well as electricity.

The term renewable energy is used to describe a source of energy or power that has the capacity to replenish itself. Renewable energy or renewable power is by definition, renewable, such that this source of power can replenish itself over and over again. Renewable energy can sometimes be called infinite energy, because it relies on energy that is in infinite supply. Renewable energy is also considered as clean energy, because it does not produce toxins or pollutants that are harmful to the environment in the same manner that non-renewable energy does. Thus, renewable energy is also known as green or clean energy. Much of the world today is converting to renewable energy sources to meet the domestic and industrial power requirement in more environmentally friendly manner. The most commonly used types of renewable energy are hydro power, wind power, solar power, biogas, geothermal and biomass.

Fossil fuel however enjoys the benefit of cheaply available source and low cost of production. The current level of energy consumption for the production of marketable unit volume of fossil fuel is lower than biofuels at present. On the other hand the energy density of fossil fuel is very much higher compare to other sources of energy. However, the most undesirable feature of fossil fuel is environmental impacts mainly emissions of green house gases which contribute to global warming. Additionally Nox and Sox also contribute to atmospheric pollution.

Alternatively, industrial, domestic and agricultural waste generation has become a grave environment hazard. Environmental pollution has become a global problem during the last three to four decades. Global warming, pollution due to heavy metals and other toxic materials spread of viral and bacterial infectious diseases are some of the key effects of excessive use of fossil fuels, its derivatives and improper handling of waste. The health cost, cost of labor and cost on corrective measures of environmental effects are massive and affect the global economy as well. It is important to pursue material and energy recycling, while minimizing waste, at global level.

The biomass energy production system is an ideal recycling method for environmental conservation through prevention of global warming. The mechanism involves fixing of CO_2 via photosynthesis through metabolic machineries in plants and microorganisms. The use of bio energy system prevents further release of non recycled CO_2 and carbon sequencing through fixing in the biomass. It is important to pay attention to bio energy production from non food materials such as agricultural and municipal waste. Among the more popular eco-friendly alternatives to petroleum are ethanol, butanol, biodiesel and biogas.

2. BUTANOL

Butanol is a gasoline blend stock, and a more acceptable alternative to ethanol, a biofuel having a considerable contribution in meeting mandates for clean air standards and renewable fuel obligations. Butanol has become an attractive fuel substance which solves most of the technical problems comparative to ethanol. Butanol is an aliphatic 4-carbon alcohol that can be produced by fermentation using wide range of feed stocks as used in ethanol production.

The butanol production by microbial fermentation is well known and documented. The famous microorganism *Clostridium acetobutylicum* produce butanol with acetone and ethanol. Butanol was produced in their metabolic pathway by using simple sugars. The acetone production was first reported by Pasteur in 1861 AD [3]. During the World War I the acetone production was accelerated by the British army with the help of a well known chemist Chamin Weizmann during the 1914, mainly using the smokeless powder called cordite. Acetone was used as the colloidal solvent for nitro-cellulose, which used was used to manufacture cordite [3]. During this period maize was used as a substrate for production of acetone and butanol. By the end of the last world war, butanol was available in plenty as a byproduct and it is used as a solvent for auto paint industry at present. The butanol and its ester, butyl acetate was ideal solvents for nitrocellulose lacquer. In 1920 butanol fermentation was only the second in the importance to ethanol fermentation. In 1920-1923 the development of butanol industry got effected with bacteriophage infection. After 1936, butanol manufacturing plants were opened all over the world including Japan, India and South Africa. But during the 1950's and 60's the butanol production declined drastically. The reason for the decline in AB fermentation after the war was twofold. First, the petro-chemical industry grew at an unprecedented rate, and by the late 1950's competition between fermentation and chemical process had become very acute. Secondly, particularly in the United States, molasses was begun to be used in substantial amounts as cattle feed [3]. The last butanol fermentation factory was closed down in South Africa during 1982 [3].

Butanol could overcome many of the problems associated with the use of ethanol as a transportation fuel due to following features.

- Higher molecular weight than ethanol
- Lower vapor pressure
- Lower water solubility
- Higher energy density

The low vapor pressure and lower water solubility reveals that it can be more easily added to gasoline than ethanol. It is less corrosive than ethanol and therefore can be more easily transported in pipelines and used in vehicle engines in higher blend concentrations than ethanol without need for extensive modifications. Current standards in the U.S. and Europe allow a greater concentration of butanol in gasoline than allowed for ethanol. Ramey and Yang (2004) reported that a 10,000 miles trip was done across the United States without need for any engine modification in a '92 Buick, with complete replacement of gasoline with butanol [4].

Table 1. Energy densities, maximum yields and compatibility with existing fuel for various types of biomass and liquid fuels

	Molar energy density (kJ/mol)	Mass energy density (MJ/kg)	Volumetric energy density (MJ/kg)	Bulk density (kg/L)
Petro fuels				
Gasoline		42.3	34.7	0.72-0.78
Diesel		45.3	38.3	0.80-0.89
Biofuels feedstock				
woods	-	19.38	6-7.1	0.310-0.370
Glucose	2803	15.6	24.3	1.562
Syngas ($CO+H_2$)	569	18.95		-
Ethanol	1367	29.9	23.6	0.792
n-Butanol	2676	36.1	29.2	0.810

Source: 5.

Butanol has potential disadvantages too. Butanol is more toxic than ethanol and is reported to have a bad odor especially when hydrolyzed. While there are long term experiences in the use of ethanol in transportation as a primary fuel, butanol has no such history. In spite of butanol's long industrial usefulness as a solvent and for other purposes, it is yet to be accepted by the transportation industry as a mainstream alternative fuel.

3. Fermentation Strains

Several species are capable of producing significant amounts of neutral solvents. The strains used commonly for production of acetone and butanol are *Clostridium acetobutylicum.* A number of different species of butanol producing Clostridia are currently recognized based on differences in the type and ratio of the solvents produced. *C. beijerinckii (C. butylicum)* produces solvents in approximately the same ratio as *C. acetobutylicum*, but isopropanol is produced in place of acetone, while *C. aurantibutyricum* produces both acetone and isopropanol in addition to butanol. *C. tetanomorphum* is a species which produces almost equimolar amounts of butanol and ethanol but no other solvents.

The isolation of solvent-producing Clostridial strains has proved to be relatively easy since they are both spore formers and obligate anaerobes, with relatively simple growth requirements. Various methods used for isolation have been well documented and practiced. These bacteria have been found most commonly associated with living plant material and with decaying plant material or soil.

The choice of strains for industrial fermentation depends on the nature of the raw material used, the ratio of end products required, the need for additional nutrients and phage resistance. Improving the environmental tolerance of strains is very important in commercial production of solvents. It can facilitate profitable fermentative techniques, low investment for fixed capital and variable capital involvement in the project.

In general genetic modification of strain improvements are of two types:

a. Rational (Direct)

Straight forward manipulation of genes for required expression, especially when the genetic sequences and its impacts are known. As an example, Ramey and Yang (2004) reported the impact of down regulating acetate formation enzymes in the metabolism of *C.trybutyricum* and developed a new strain [4]. This process enhanced the capability of production of butyrate while cutting down acetate.

b. Random (Combinational)

This approach required special attention since physiological mechanisms of tolerance remain uncharacterized. It is time consuming and generally known as classical strain improvement. Some time it uses physical stress such as chemicals, heat, pH or ultra violet light for development of mutagens. The draw back of these systems is the non-transferability and intractability of resulting genotypic changes probably due to transient mutations that occurred.

Apart from these strain development techniques, Wijewardhana et al. (2010) introduced the use of a microbial consortium available under natural environmental conditions. With the use of consortium of bacteria, there was a notable improvement of productivity in the solventogenesis process. Consortium consists of culturable and unculturable strains, and requires use of metagenomic techniques to differentiate the activities of individual strains. The rapid and continuous production of solvents is mainly due to the co-metabolism within the consortium. It will be supported by the scavenging process of primary and secondary toxic materials in the reactor environment. In the consortium technique there is a possibility of developing new stains by classical strain improvement. On the other hand solventogenesis micro organisms have the opportunity to utilize hexose sugars as well as cellobiose, cellodextrains, and pentose sugars. It is reported that Clostridium bacteria are better adapted to cellobiose than to glucose [7].

Based on substrate utilization by a single strain of bacterium or a consortium of bacteria, two mechanisms could be identified. First process initially converts lignocellulosic materials into simple sugars and related intermediate cellulosic materials such as cellobiose and subsequently another set of micro organisms is needed for solventogenesis. Second, is the application of known single or mixed microbial strains for direct conversion of lignocellulosic materials into solvents and this is named as co-culture system. Yu et al. (1985) investigated a sequential co-culture approach for the conversion of solka floc or a mixture of solka floc and aspenwood xylan to solvents [8]. *C. thermocellum* was chosen for co-culture with *C. acetobutylicum* since it produces cellulose and xylanase enzymes. [3].

Improved microbial cultures and co-cultures are available for commercial production. Following companies are developing modified microorganisms for production of butanol, isobutanol, or related fuels.

- Arbor Fuels
- BUTALCO GmBH
- Bioenergy International
- Butamax (DuPont/BP)
- Cobalt Biofuels
- Gevo, Inc.
- Green Biologics, Ltd.
- METabolic Explorer
- TetraVitae Bioscience

(Source: 9)

4. Feedstock for Butanol Production

The creation of liquid transportation fuels from renewable biomass has been a long term goal [10]. The biomass is available in various forms as starch, sugars and cellulose. Most of the starch and sugars are from food crops; hence it is inappropriate to use as raw materials for energy which would lead to unstable situation in food supply. Additionally it affects the economic viability of biofuel production itself as starch substances have higher market value compared to other alternative substances.

Non cellulosic biomasses as well as cellulosic biomasses are available as the feed stocks for renewable Butanol production. Principle non cellulosic biomass is starch from various grains, sugars and root crops. Corn is the vital substance which used for bio fuel production in the United States and in Brazil sugar cane is used for ethanol production. Cassava is one of the root crops which can be utilized as feedstock for biofuel production. In early studies it was understood that various forms of sugars could be fermented by *Clostridia* spp. specially glucose, fructose, mannose, sucrose, lactose, starch and dextrin could be completely converted into solvents. Galactose, xylose, arabinose, raffinose, melezitose, inulin and mannitol were partially utilized and later studied proved that xylose and arabinose can be completely converted in to solvents. Cheese whey, apple pomace and algal biomass also can be converted in to solvents and they can be considered as non cellulosic biomass. Cheese whey has attracted special interest due to problem of disposal in cheese processing plants around the world. Lactose in cheese whey is a possible source of butanol production (Figure 1).

Agricultural residue is a well known biomass. Cellulosic feedstock is abundant in places where other liquid fuels, chiefly petroleum and it's by products, are not readily available [5]. Cellulose and hemicellulose resources are the most abundant and under utilized materials, specially rice straw, corn stove, wheat straw, bagasse, municipal waste and forest wastes. It is estimated that 7.5 x 1010 tons are annually synthesized through photosynthetic process [11]. Cellulose is the main structural material of plant cell wall and embedded in a hetero-matrix composed of xylan, other hemi-cellulose and lignin [12]. The potential of the use of cellulose as a raw material for alcohol fermentation is very high. The crystalline structure of cellulose is difficult to hydrolyze by nature and it is more difficult in the presence of hemicelluloses. The second generation biofuel production utilizes simultaneous saccharification and fermentation (SSF) or simultaneous saccharification and co-fermentation (SSCF) [12]. In both methods a series of complex processes occur, viz: cellulase production, hydrolysis of cellulose and hemicelluloses, fermentation of hexoses and pentoses. Consolidated bio processing (CBP) is the third generation biofuel production process which enables cellulase production; substrate hydrolysis and fermentation are in the single process system and microorganisms are used as converters. Celluloses and hemicelluloses are converted to simple sugars, an alternative for enzymatic conversion or acid hydrolysis. Microbial saccharification and fermentation is the most attractive method compared to others.

As a example Rice is one of the major staple foods of the world population and most of the available resources are committed to rice production. Currently cultivated improved rice varieties produce a similar amount of straw for each kilogram of rice and which is available as the major agricultural residue. In traditional varieties the ratio is different and it is more than one kilogram per kg of rice grain produced. The annual rice production of the world is 650 million metric tons. The average cellulose and hemicellulose content of the rice straw varies from 64% -69% dry weights. Hence 422 million metric tons of cellulosic materials are annually available as rice straw in the world. Further studies are required to separate and identify the microbial consortia which are capable of harnessing liquid fuel from biomass with the highest yield and rapid conversion rate. Table 3 indicates the major polymer content of rice straw. Similarly corn and wheat straw have cellulosic materials convertible into liquid biofuel. The major polymeric constituents in g/100g of rice straw (dry weight basis) are cellulose – 43.3, Lignin – 5.4, hemicelluloses – 25.1 and protein – 5.6 [13].

The breakdown of cellulose cellulolytic bacteria is the most important step of butanol production from agriculture residue. Complete degradation of cellulosic material is not achieved by a single species of microorganism. It is done by a series of microorganisms (known as well as unknown). Similarly non-cellulolytic bacteria also take part in the operation even in a cellulose rich environment. Most of the times during inter species hydrogen transfers; the hydrogen produced by cellulose degradation bacteria is oxidized to methane by methanogenic bacteria. Some of the bacteria act as electron sinks being energetically favorable for cellulolytic conversion. Cellulase as a single enzyme is not much effective to digest the complex cellulose structure. Microorganisms are required develop several strategies as single polypeptides with multiple cellulosic domain, or extracellular multienzyme complex.

Majority of cellulolytic micro organisms are found in the phylum Firmicutes (*firmus*, strong, and *cutis*, skin, referring to the cell wall) are a phylum of bacteria, most of which have Gram-positive cell wall structure) and belong to the class Clostrdia and the genus *Clostridium.* Lamed et al., (1983) discovered a multi-component cellulolytic exocellular complex of enzymes: cellulosome that have the potential for cellulose binding and degradation [14]. The constituent unit of cellulose is cellobiose residues, but not glucose. Cellulosome has capacity to hydrolyze biopolymer to its constructive blocks, disaccharide cellobiose and cellodextrin. On the other hand polycellulosomes are promoting the bacterium to adhere to the cellulose fiber and it has the capacity to degrade the non structural cellulose as well as structurally complexed crystalline cellulose. The degradation of cellulose is accomplished through the action of enzyme that include endo-1,4-β- glucanases and exo-1-4-β- glucanases [12].

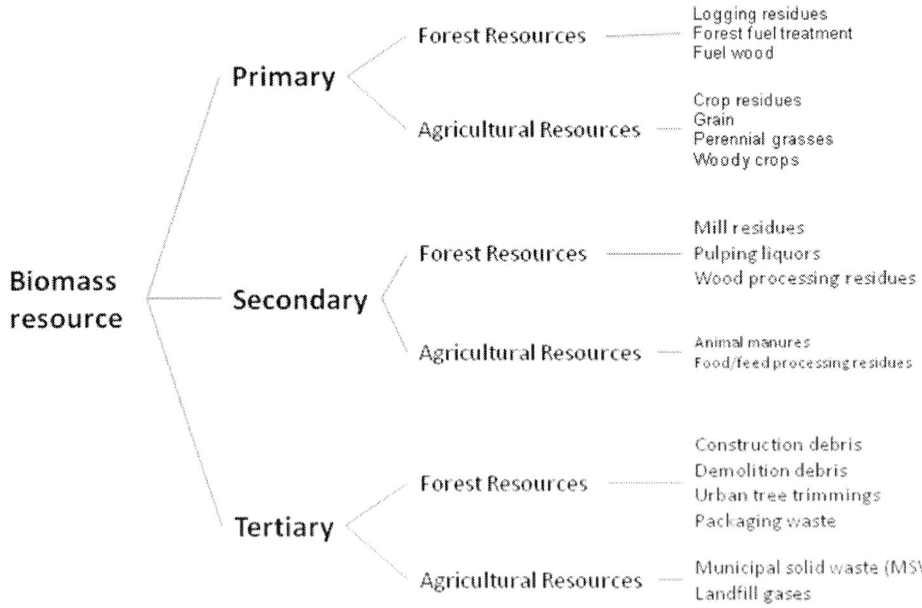

Figure 1. Biomass resource base.

5. Limitations and Challenges in Industrial Processing

The technical knowhow that existed in the twentieth century would have facilitated the batch process and continuous process as the main processing systems of butanol production. But none of those plants are in operation today. Online process technology was not available for process optimization of the butanol plants prior to 1982, the year in which the last functioning butanol production plant in South Africa was closed down. Therefore with the knowledge gap in the production optimization that existed then made it difficult to understand the key biological and chemical parameters of conventional butanol manufacturing process. Even in the conventional processing capital cost investment is very high. Gapes (2000) estimated that 26% and 64% of investment from the total capital cost is required for equipments in batch process and continuous process respectively [15].

The barriers in butanol production that hampered the industrial butanol production are

1 Toxicity of butanol to fermenting agent
2 Maintenance of material balance
3 Maintenance of energy balance.

These barriers are interrelated therefore it is absolutely imperative that one addresses all the problems simultaneously for a comprehensive ultimate solution. Most new inventions in this regard are limited to either laboratory model or pilot scale operation. It is too early to discuss the upcoming technological innovations.

One of the major drawbacks to acetone, butanol and ethanol (ABE) fermentation is the high cost of recovering the relatively low concentrations of accumulated solvents by distillation [3]. With the available technology and microbial strains, maximum yield of butanol obtained is less than 20 grams per liter in the fermentation due to inherent butanol toxicity. Typical solvent fermentation produces acetone, butanol, ethanol and water. Therefore, separation of butanol from rest of the constituents, acetone, ethanol and water of the fermentation mixture is achieved by fractional distillation. This separation technology requires higher amount of energy and it is not economically competitive with existing fossil fuels. The main reason for higher energy consumption in the distillation process is the high evaporation temperature of butanol (117° C) which is higher than that of the remaining constituents.

Apart from the improvement of bacterial strains through genetic manipulation, development of mutants and co-culture consortium etc., there are technologies developed to improve butanol separation. The process covered under US Patent no 5,753,474, the Dual Immobilized Reactors with Continuous Recovery (DIRCRTM) is a new process proposed. Patent for butanol production by fermentation based on the use of two separately immobilized cultures: *Clostridium tyrobutyricum* and *Clostridium acetobutylicum* [5] is a two stage processes. In the first stage biomass is converted into butyric acid, carbon dioxide and hydrogen. In the second stage butyric acid is converted into butanol. In 2008, Yang, disclosed the use of catalytic hydrogenation of butyric acid in to butanol.

Table 2. Stoichiometric Equations

Product	Stoichiometric Equations				
Acetone	$C6H12O6+H2O$	→	$CH3COCH3$	$+ CO2$	$+H2$
Butanol	$C6H12O6+H2O$	→	$C4H9OH$	$+ CO2$	$+H2O$
Ethanol	$C6H12O6+H2O$	→	$C2H5OH$	$+ CO2$	

Source: 15.

Wijewardhana et al. (2010), developed a mechanism for butanol production from agricultural residue. It is a process that requires low capital investment with a low cost of production, based on rice straw as the main substrate for fermentation.. They used a unique solvent separation system which supported the uninterrupted fermentation process. The culturable and unculturable microorganisms in the consortium used in the anaerobic bioreactor remarkably speeded up the cellulolytic conversion as well as solventgenesis in the butanol production process. The process system is under patent process.

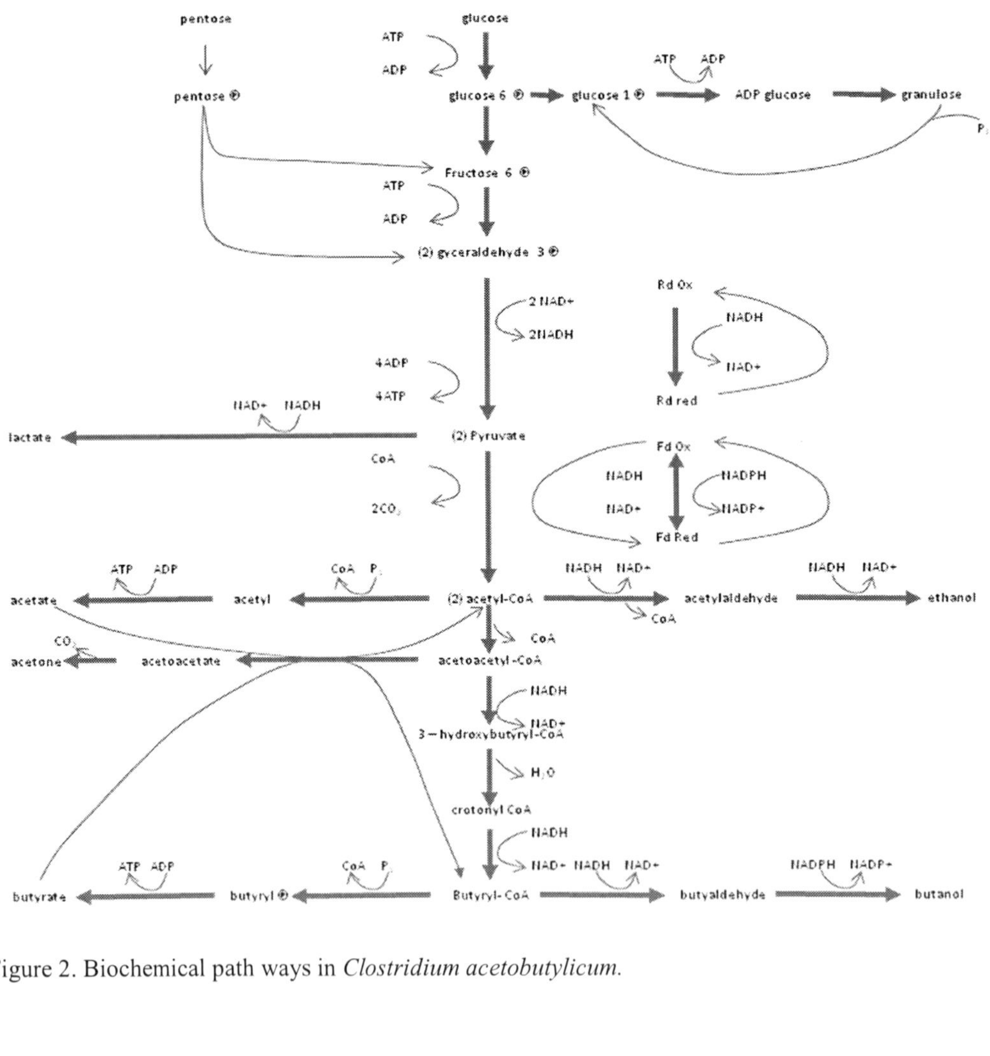

Figure 2. Biochemical path ways in *Clostridium acetobutylicum.*

Table 3 briefly indicates stoichiometric equations of solvents produced from glucose. Conventional *Clostridium acetobutylicum* fermentation process produces butanol, acetone, ethanol and hydrogen at a ratio of 6:3:1:0.4 respectively. The stoichiometry of solvents produced from cellulosic biomass is not properly understood yet due to the complexity of conversion and heterogeneity of cellulosic biomass. But Giallo et al. in 1983 noted that cellulolytic mesophilic *Clostridium* strains were better adapted to cellobiose than to glucose. Wijewardhana et al. (2010) observed a similar behavior in their developed microbial consortium. Further research is required to understand the material balance of solvent production from cellulosic biomass.

6. Biological Pathways and Determinants of Butanol Production

Acidogenesis of simple sugars by *Clostridium* species, produce hydrogen, carbon dioxide acetate and butyrate. With the decrease in pH of the medium during acidogenesis phase, microorganisms enter solventogenesis phase. During the solvent production phase, reassimilation of acids that occurred with continued consumption of carbohydrate, normally decrease the pH of culture medium. The relationship between the breakpoint in the pH of fermentation and the onset of solvent production, which occurs at the beginning of the second phase of the fermentation, had been identified. The biochemical pathways utilized for conversion of carbohydrates to hydrogen, carbon dioxide, fatty acid and solvents by *C. acetobutylicum* have been firmly established.

During glycolysis *C. acetobutylicum* and *C. beijerinckii* generate less ATP and more NAD(P)H than is required for biosynthesis and growth. The production of acids results in the generation of additional ATP. Only a portion of the reducing equivalents produced during glycolysis is consumed during the production of acids. Generally these Clostridia have the ability to produce hydrogen with most heterofermentative anaerobic bacteria, providing the cell an effective route for the disposal of both excess protons and electrons. Hydrogenase is the terminal enzyme in these cells, which enables use of protons as terminal electron acceptors.

It is theoretically possible for a cell to dispose all of the reducing equivalents produced during glycolysis in the form of hydrogen, thereby allowing the cell to generate the maximum amount of ATP. However, practically, this thermodynamically unfavorable reaction has only been observed when interspecies hydrogen transfer results in the direct transfer of hydrogen from hydrogen-producing species to a hydrogen-consuming species of bacteria. The concentration of hydrogen is apparently kept sufficiently low under these conditions to allow the redox potential of the $H^*/H2$ couple to be raised. It is evident that, only a portion of the excess reducing equivalents can be disposed of by hydrogen production among free-living solvent-producing *Clostridia.*

The ratio of acetate/butyrate produced appears to be directly regulated by the activity of this enzyme complex. The relative amounts of acetate and butyrate were independent of pH when the pH of the culture was varied between 4.8 and 6.8, but at a pH of > 6.8 a more efficient fermentation of glucose occurred, resulting in an increase in the production of acetate. The reaction catalyzed by NADH ferredoxin oxidoreductase also becomes thermodynamically more favorable when the partial pressure of hydrogen is lowered [17].

The shift from acid-producing metabolism to solvent-producing metabolism that occurred about midway in the AB fermentation was accompanied by a change in the ratios of hydrogen and CO2 produced. This indicated an increase in CO2 production and a decrease in hydrogen production during the solventogenic phase. A major proportion of the electron flow is directed to hydrogen production during the acidogenic phase, while the carbon flow is mainly directed to acid production, resulting in the maximum generation of energy.

The highest rate of hydrogen production associated with a high rate of glucose consumption occurred during the initial growth stage. The first decrease in the rate of hydrogen production coincided with a reduction in the metabolic activity and the growth rate of the culture. However, the specific hydrogenase activity associated with the whole cell did not alter during this stage, and the decrease in the rate of hydrogen production appeared to be due to the reduction in metabolic rate rather slim inhibition of hydrogenase activity. A second decrease in the rate of hydrogen production associated with a decrease in hydrogenase activity, coincided with the shift to solvent production and resulted in a progressive decrease in hydrogen production and a gradual reduction of the concentration of hydrogen in the headspace.

In understanding the way in which the production of solvents is initiated and maintained, it is important to identify the factors involved in activating the metabolic transition and the physiological state associated with the transition from the acidogenic to the solventogenic phase. The influence of pH has been recognized as a key factor in determining the outcome of AB fermentation. The *C. acetobutylicum* type culture strain ATCC 824 has been reported to produce good levels of solvents between pH 5.5 and 4.3 [18]. During industrial fermentation the pH of the fermentation medium was set at about pH 6.0, and at the initial phase of the fermentation the pH decreased to about 5.2 at the breakpoint, after which it increased, reaching about 5.8 at the end of the fermentation. The levels of solvents produced in the fermentation maintained at near-neutral pH were reported to be almost as high as those obtained in uncontrolled cultures in which the pH was allowed to drop to below 5.0. It has been demonstrated that, although a decrease in the pH is important in permitting the shift to solvent production to occur, pH itself is not the trigger [19, 20]. A number of other significant changes occur at the end of the initial phase of the fermentation in addition to the decrease in pH. The low pH is associated with an increase in the concentration of acid products, an increase in cell number, and a decrease in hydrogen production and the specific growth rate.

The generation of ATP by fermentation necessitates the use of organic compounds as terminal electron acceptors, and the reduced energy-rich organic acids produced as end products of metabolism are by their nature toxic to the cell. In their undissociated form, organic acids such as acetate and butyrate are able to partition in the cell membrane and behave as uncouplers which allow protons to enter the cell from the medium [3]. When the concentration of the undissociated acids becomes sufficiently high, they result in a collapse of the pH gradient across the membrane and cause a rapid decrease in the NTP/NDP ratio, which results in the total inhibition of all metabolic functions in the cell [21, 22]. The accumulation of acid end products and the associated decrease in pH at lower concentrations, results in a progressive decrease in the specific growth rate until cell growth is halted completely, although substrate utilization and cell metabolism continue [21, 22]. The shift to solvent production appears to be able to act as a detoxification mechanism, which allows the cell to avoid the inhibitory effects that would occur when acid products reach toxic levels of acid end products. Monot et al. (1984) reported that when the concentration of undissociated butyric acid reached a level of 0.5 to 0.g g/liter, growth was inhibited and the induction of solvents occurred when the concentration of undissociated butyric acid reached a level of 1.5 to 1.9 liter [18]. The key role of undissociated butyric acid could be correlated with the influence of pH and this compound seemed to be the essential factor in the regulation of solvent production. If the critical level could not be reached because of too high pH or too low initial substrate concentration, solvents were not produced [23]. Although a number of bacteria are known to maintain their internal pH at a more or less constant level above that of the external medium [24]. Instead, these bacteria maintain a limited, but more or less constant, pH gradient across the membrane as the pH of the external medium decreases. For the continued functioning or synthesis of some enzymes, maintenance of the internal pH above a threshold value may be essential since the activities of a number of enzymes are known to be pH sensitive. In addition, the maintenance of an internal threshold pH may be required for other metabolic functions, such as the uptake of carbohydrates and ammonia [25, 26]. The dissociation equilibrium inside the cell is affected by the maintenance of a higher internal pH. The internal concentration of acids would be dependent on the internal pH, the pH gradient across the membrane, and the rate of production of acid end products, all of which would affect the rate of diffusion of acids out of the cell [27]. Hence, the concentration of acids in the medium can only give an approximate estimation of the concentration of acids within the cell.

The effect of nitrogen limitation on the production of solvents is less clear-cut. In a study on the effect of ammonia limitation in batch culture, Long et al. (1984) [20] observed that when the concentration of ammonia in the medium was decreased to a level, which resulted in a decrease of biomass, the amount of glucose, utilized decreased and residual glucose remained at the end of fermentation.

The shift in metabolic activity, which occurs when cells switch from the acidogenic to the solventogenic phase has been shown to be accompanied, by a corresponding shift in the activity of the enzymes involved in the acid and solvent producing pathways. It has been identified that the activity of the four enzymes are involved in the pathway from acetyl-CoA to butyryl-CoA. After growth had ceased, the first three enzymes in this pathway (thiolase, 3-hydroxybutyryl dehydrogenase, and crotonase) coordinately expressed and exhibited maximum activity in solvent producing cells, but only low activity of the fourth enzyme (butyryl-CoA dehydrogenase) can be detected. Prior to butanol production in batch cultures, activity of butyraldehyde dehydrogenase and butanol dehydrogenase increased and then decreased with the decrease in biomass production. The uptake of acetate and butyrate during the solvent producing phase has been shown to be directly coupled to the production of acetone via acetoacetyl-CoA: acetate/butyrate:CoA transferase [28]. This enzyme has also been shown to exhibit a high level of activity in solvent-producing cells in both batch and continuous culture but was hardly detectable in acid-producing cells. The final step leading to the production of acetone occurs via the action of acetoacetate decarboxylase, and again the specific activity of this enzyme was found to be extremely low in acid producing cells but was increased by about 40-fold in solvent-producing cells [3, 28, 29, 30, 31].

The mechanisms that may be involved in the induction of enzyme biosynthesis and those which may be involved in the regulation of enzyme activity are important to be distinguished. The various factors involved in the induction of acetoacetate decarboxylase were investigated by Ballongue et al. [30]. They reported that linear fatty acids from C1 to C4 (formate, acetate, propionate, and butyrate) were able to function as inducers of enzyme biosynthesis, whereas linear acids from C5 to C7 were not.

At pH 4.8 in the presence of acid concentrations, induction was maximal comparable to those which occurred during the fermentation and conformed to the pKa of the acids. Although formic and propionic acids were effective, acetic, acetoacetic, and butyric acids were the best inducers. The anion itself is important in the induction mechanisms since no induction was obtained with dimethylsulfoxide. However, it is doubtful that whether the acids exert their effect directly or whether the effect occurs via some other intermediate. Acetoacetate decarboxylase was inhibited in growing cells and only became active once growth ceased. [30] This suggests that the induction of this enzyme is linked in some way to the inhibition of cell division or growth.

A decrease in the intracellular concentration of acetate and butyrate and their metabolic intermediates as well as a reduction in the levels of NADH and NADPH in the cell occurs with the switch to solvent production. Inhibitory effect of the low pH of the medium is reduced by the uptake and reutilization of acetate and butyrate from the medium.

The mechanism of acid utilization by means of the CoA transferase reaction would be energetically favorable as the energy in the thioester bond is conserved and transferred to the acids without the requirement for ATP utilization. It is not apparent what determines the ratio of acetate to butyrate consumed during reassimilation. In batch AB fermentation butyrate is normally consumed more rapidly than acetate [32].

All of the reducing equivalents from reduced ferredoxin would have to be diverted from hydrogen production to the production of reduced pyridine nucleotides to meet the requirements for butanol formation, if acids were not consumed during solvent production. Under these conditions, 50% of the reduced pyridine nucleotides generated from reduced ferredoxin would be required as NADH and 50% would be required as NADPH. Therefore it could be predicted that, when solvent production is coupled to acid reassimilation, the production of hydrogen would be reduced by about half and the bulk of the reduced pyridine nucleotide generated from reduced ferredoxin would need to be in the form of NADPH to satisfy the electron balance. Hence, the ratio of NADH/NADPH may be important in determining the ratio of acetate/butyrate consumed.

When acetate is consumed, a decrease in both butanol/acetone and the butanol/ hydrogen ratios occur with the reduction in the amount of butanol produced. Therefore, conditions that favor butyrate uptake should result in an enhancement of the butanol yield.

Due to the small number of ATP molecules generated during fermentation, the growth of anaerobes is considered to be limited by the rate of the energy-producing reaction [33]. The ATP molecules generated can be utilized in two different ways: first, in biosynthesis leading to cell growth; and second, in membrane energization. The amount of ATP available for biosynthesis would reduce due to a decrease in a rate-limiting step in ATP production, such as a decrease in flux through the glycolytic pathway or a decrease in sugar uptake. Second, a reduction in the amount of ATP available for biosynthesis could be due to the diversion of more ATP to membrane energization, due to an increase in either membrane permeability or ΔpH. To counter the effect of increased membrane permeability and to increase the ΔpH, the cell would be obliged to expend more energy for membrane energization toward the end of the acid-producing phase with the decrease in pH. A major consequence of the shift from acid to solvent production is a reduction in the net amount of ATP generated from approximately

3.3 mol of ATP per mol of glucose to 2 mol per mol of glucose [33]. As a direct outcome of the reduction in the amount of ATP available to the cell, the cell is no longer able to maintain normal vegetative growth.

In most batch culture studies the linkage between the switch to solvent production and the end of the exponential growth is evident. The relationship between the terminations of exponential growth and the induction of solvent formation is even more evident if cell number rather than optical density of the culture is used as a measure of growth. When the total cell count is increased with growth, it may be seen that cell division ceased around the time that the switch to solvent production occurred, although the optical density of the culture may continue to increase for a number of hours after this. Continued increase in the length or volume of the cells after cell division has ceased, the accumulation of storage products within the cell, or a change in the optical properties of the cell could be the reasons for continued increase in the optical density. Solvent-producing cells which have been re suspended in buffer and non growing immobilized cells have been used for the production of solvents.

The relationship between cell growth and the onset and maintenance of solvent production is less clear in continuous culture systems. However, irrespective of the limiting factors utilized, the kinetics of product formation follows a similar pattern. At high dilution, rates acid production predominates and the specific rate of acid production increases with increasing dilution rate [34,35, 36,). Extrapolations indicate zero production at low growth rates. The specific rate of glucose consumption also increases with increasing dilution rate, and extrapolation to zero indicates a maintenance requirement [37]. Both acids and solvents may be formed at intermediate dilution rate. Solvent production is favored and the highest yields and concentrations of solvents are obtained at low dilution rates. The specific rate of solvent production shows a different pattern from that of acid production and an increase with increasing dilution rate to a maximum point, after which increasing dilution rates result in a rapid decrease in the production rate [38]. Extrapolation shows solvent formation at zero growth rates. The most successful conditions for solvent production in continuous culture have utilized a low dilution rate combined with a low pH and phosphate or sulfate limitation in the presence of excess glucose [34]

Studies reveal that both acid production and solvent production can occur concurrently in both chemostat and turbidostat cultures under apparently steady-state conditions. The mean doubling times of the cells range from 6 to 30 h at low dilution rates, which have been utilized to obtain good yields of solvents. Growth and cell division in solvent-producing cells may be reduced or inhibited, and the loss of these cells through washout could be compensated for by the conversion of acid-producing cells into solvent-producing cells as in batch culture.

Ability to maintain a constant environment while the effect of a single independent variable is evaluated is the advantage of using continuous culture instead of batch culture. This advantage is dependent on both the ability to maintain a steady state during continuous culture and the physiological stability of the organism but permits rigorous evaluation of the physiology of the organism. According to the nature of butanol fermentation, it may not be possible to obtain a true steady state in a single-stage continuous system, which detracts from the advantages which can be obtained from continuous-culture studies.

7. Butanol Toxicity

Cell metabolism usually continues during the solvent-producing phase, until the concentration of the solvent reaches inhibitory levels of around 20 g /liter, after which further cell metabolism ceases. A major limitation in its use for the industrial production of solvents is the relatively low concentration of solvents produced during fermentation. Butanol is the most toxic and the only one produced in inhibitory concentrations, of the solvents produced during fermentation. Solvent production ceases when the concentration of butanol reaches about 13 g/liter in the industrial fermentation process [39,40,). It has been shown that the addition of 7 to 13 g of butanol per liter to cultures growing on hexose sugars resulted in a 50% inhibition of growth. Growth was inhibited totally at a concentration of 12 to 16 g/liter, and there appeared to be a threshold concentration of 4 to 4.8 g of butanol per liter below which no observable decrease in growth occurred. In cells grown on xylose, the inhibitory effect of butanol has been reported to be enhanced, and growth was inhibited totally when 8 g of butanol per liter was added to the medium [41]. Alternatively, the concentrations of acetone and ethanol do not appear to reach inhibitory levels during the fermentation. The addition of acetone and ethanol reduced growth by approximately 50% at a concentration of around 40 g/liter, and total growth inhibition occurred at a concentration of about 70 g of acetone and 50 to 60 g of ethanol per liter [43].

The mechanism of butanol toxicity is related to the hydrophobic nature of this compound. Along with other long-chain aliphatic alcohols, the primary effect of these molecules appears to be on the disruption of the phospholipids component of the cell membrane. Membrane fluidity is decreased by short chain aliphatic alcohols such as ethanol, whereas butanol and other longer-chain aliphatic alcohols have the opposite effect. Both the solubility of the alcohol in the membrane, and its effect on membrane fluidity increases with increasing chain length. The addition of low concentrations of butanol (<5 g/liter) appears to have no effect on the fluidity of extracted membranes of *C. acetobutylicum*, but the addition of higher (but sub inhibitory) concentrations of butanol (10 g/liter), caused a 20 to 30 %increase in the fluidity of lipid dispersal as determined by electron spin resonance, spin label analysis [44].

It could be expected that the increase in membrane fluidity in the presence of butanol results in the destabilization of the membrane and disruption of membrane-linked functions. The effects of butanol are complex and results in the inhibition or disruption of several interrelated membrane processes of which the sequence and relationship of these events are not known.

Butanol toxicity has also been linked to the autolytic degradation of solvent-producing cells in *C. acetobutylicum* P262 [45], and it was suggested that inhibitory concentrations of butanol were involved in the triggering of the release of cell-free autolysin during the solventogenic phase [46]. A pleiotropic autolysis-deficient mutant (*lyt-l*) isolated by Allcock et al. produced less autolysin than the parent strain and was more resistant to both its own and the parent strain autolysin [47]. Degradation of the solvent-producing *Clostridial* stage cells of the parental strain was enhanced by the addition of butanol (7 to 16 g/liter), but had little effect on the stability of the *lyt-l* mutant cells. In addition, enhanced butanol tolerance was exhibited by the vegetative cells of this mutant and this strain also produced slightly higher concentrations of both total solvents and butanol (14.2 g/liter compared with 13.3 g of butanol per liter).

It is assumed that an enhancement in butanol tolerance of the cells will result in the production of higher concentrations of solvents, as butanol toxicity appears to be the limiting factor in the amount of solvent which can be produced during AB fermentation. There is some evidence to suggest that tolerance to butanol may be enhanced by manipulation of the growth conditions during the fermentation. The addition of saturated fatty acids to the medium has been shown to result in an increase in the ratio of saturated fatty acids in the membrane [48]. Saturated fatty acid enrichment increased butanol tolerance by up to twofold, and cell growth and ATPase activity were also enhanced [42, 49].

Increase of membrane fluidity with increasing temperature suggests that a decrease in temperature during the solvent-producing phase might enhance butanol tolerance. This approach appears to have formed the basis of a patented process to enhance butanol production by decreasing the temperature from 300°C during the acidogenic phase of the fermentation to 240°C during the solventogenic phase of the fermentation [50].

A limited amount of success has been obtained to date with the attempts to obtain mutants which can tolerate and produce higher concentrations of butanol. This is perhaps not surprising in view of the complex multiple effects relating to butanol toxicity and may mean that multiple mutants will have to be isolated to achieve substantial increases in butanol production. But the selection of mutants which can grow in the presence of inhibitory concentrations of butanol has in general not resulted in the isolation of strains which produce substantially higher concentrations of butanol in the non growing solvent-producing phase. The increased butanol resistance was not specific for butanol, but extended to other alcohols. An important characteristic of the mutant was that it was stable and the butanol tolerance and increased yields were maintained over a period of several years in the absence of selective pressure. It is interesting to note, that none of the butanol tolerant mutants described so far have been able to produce solvent levels higher than the maximum solvent concentrations obtained in the industrial fermentation process.

The complex multiple effects relating to butanol toxicity coupled with our limited understanding of the physiological and biological factors involved in the production and tolerance of higher concentrations of butanol would appear to present severe limitations to the application of recombinant DNA technology for the production of increased concentrations of solvents.

As liquid transport fuel from cellulosic biomass it is very important to maintain the energy input and output. According to first Law of Thermodynamics (energy can neither be created nor destroyed) which represent the mass and energy balances. In terms of fermentation products cannot contain more mass or more energy in sum than the fermentation substrate. In fact since fermentation processes are exothermic the product will certainly contain less energy than the substrates. The mass and energy yields in turn determine the absolute, theoretical limit. Table 4 indicates the properties of input and out put of butanol fermentation from the cellulosic biomass process.

To over come the existing barriers of energy, mass and economical barriers the concept of biorefinery is introduced. In biorefinery concept, it is not only focus for liquid transport fuel production and also focus to the other possible usable by products. Alternatively, harness of other energy products of bio refinery resulted positive energy and material balance.

Table 3. Compound properties

Component / Reaction	s.g	Mr (g/mol)	Comb (J/g)	Comb (kJ/mol)
Glucose		180.16	15,544	2,800
Xylose		150.13	15,564	2,337
Ethanol	0.78	46.07	29,685	1,368
Acetone	0.79	58.08	30,797	1,789
1-Butanol	0.81	74.12	36,066	2,673
Hydrogen		2.02	141,375	286

Source: 15.

Biorefinery addressed most of the key issues of the process and made commercial viability of butanol production from cellulosic biomass. Biorefinery concept could be employed for rice straw that is currently used only as a feedstock. Butanol, acetone, ethanol and hydrogen could be harvested as the main products. Butanol and ethanol can be used as liquid transport fuels and hydrogen can be used as the energy source for internal requirements such as fractionate distillation and maintenance of the optimum working temperature of bioreactors. Carbon dioxide and hydrogen released as gases could be separated by membrane gas separators and Carbon dioxide could be used as an input for algal culture for the production biodiesel. Resultant distilled water can be used for industrial requirements. After extracting the cellulose and hemicelluloses from rice straw, remaining portion could be utilized as either manure for crop or as animal feed due to its high Nitrogen content.. In addition, available lignin could be used for other industrial purposes such as manufacture of partition boards etc. This alternative income from the byproducts could be used as a supportive economic gain for the main butanol production from rice straw. Energy industry is not limited to the supply of energy and it responsible for supplying the other important materials even though it causes environmental pollution. It is also a necessity for the renewable energy industry to meet the demand for other industrial inputs and forms of energy currently produced by the fossil fuel industry without being limited for the supply of transport liquid fuel.

References

[1] BP. 2009. *Statistical Review of World Energy*. URL: http://bp.com/statisticalreview accessed on 19th December 2009.

[2] EIA. 1998. Annual Energy Outlook 1999, with projections to 2020, DOE/EIA-0383. *Energy Information Administration*, Department of Energy, Washington D.C., USA

[3] Jones, D. T. & Woods, D. R. 1986. Acetone-butanol fermentation revisited. *Microbiological Reviews,* 50, 484-524.

[4] Ramey, D. & Yang, S. T. 2004. *Production of butyric acid and butanol from biomass – Final report,* US Department of Energy, Morgantown, WV USA.

[5] Fischer, C. R., Klein-Marcuschamer, D. & Stephanopoulos, G. 2008. Selection and optimization of microbial hosts for biofuels production, *Metabolic Engineering,* 10, 295-304.

[6] Wijewardhana H.V.P., J.M.R.S. Bandara K.F.S.T. Silva and Y.M.A.Y. Bandara. *Combting energy poverty of Sri Lanka,* Technologies for Development 8-10 February 2010 Lausanne, Swisterland

[7] Giallo, J., Gaudin, C., Belaich, J. P., Petitdemange, E. & Caillet-Mangin, F. 1983. Metabolism of Glucose and Cellobiose by Cellulolytic Mesophilic *Clostridium* sp. Strain H10. *Applied and Environmental Microbiology,* 45, 843-849.

[8] Yu, E. K. C., Chan, M. K. H. & Saddler, J. N. 1985. Butanol production from cellulosic substrates by sequential co-culture of *Clostridium thermocellum* and *C. acetobutylicum. Biotechnology Letter*, 7, 509-514.

[9] D. Glass Associates, Inc. 2010. Companies Developing Modified Microorganisms for Production of Butanol and Related Fuels. URL: http://dglassassociates.wordpress.com/2010/02/16/companies-developing-modified-microorganisms-for-production-of-butanol-and-related-fuels/ accessed on 16th February 2010

[10] Cysewski, G. R. & Wilke, C. R. 1976. Utilization of cellulosic materials through enzymatic hydrolysis. 1. Fermentation of hydrolysate to ethanol and single-cell protein. *Biotechnology and Bioengineering*, 18, 1297-1313.

[11] Monserrate, E., Leschine, S.B. & Canale-Parola, E. 2001. *Clostridium hungatei* sp. nov., a mesophillic, N2-fixing cellulolytic bacterium isolated from soil. *International Journal of Systematics, Evolution and Microbiology,* 51, 123-132.

[12] Carera, C. R., Sparling, R, Cicek, N. & Levin, D. B. 2008. Third generation biofuels via direct cellulose fermentation. *International Journal of Molecular Science,* 9, 1342-1360.

[13] Champagne, E.T. ed. (2004). Rice: Chemistry and Technology. 3rd ed. American Association of Cereal Chemists. St.Paul, Minnesota.

[14] Lamed, R., Setter, E. & Bayer, E. A. 1983. Characterization of a cellulose-binding, cellulose-containing complex in *Clostridium thermocellum. Journal of Biotechnology*, 156, 828-836.

[15] Gapes, R. H. 2000. The economics of acetone-butanol fermentation: theoretical and market considerations. *Journal of Molecular Microbiology and Biotechnology*, 2, 27-32.

[16] Yang, S.T., 2008, *Methods of production butanol US patent* 20080248540A1

[17] Gottschalk, G. 1979. *Bacterial metabolism (Springer series in microbiology).* New York: Springer Verlag.

[18] Monot F., Engassser J. M. & Petitdemange. H. 1984. Influence of pH and undissociated butyric acid on the production of acetone and butanol in batch cultures of *Clostridium acetobutylicum. Applied Microbiology and Biotechnology,* 19, 422-426.

[19] Gottschal, J. C. & Morris. J. G. 1981. Nonproduction of acetone and butanol by *Clostridium acetobutylicum* during glucose- and ammonium-limitation in continuous culture. *Biotechnology Letters,* 3, 525-530.

[20] Long, S., Jones, D. T. & Woods, D. R. 1983. Sporulation of *Clostridium acetobutylicum* P262 in a defined medium. *Applied and Environmental Microbiology,* 45, 1389-1393.

[21] Herrero, A. A. 1983. End-product inhibition in anaerobic fermentations. *Trends in Biotechnology,* 1, 49-53.

[22] Herrero, A. A., Gomez, R. F., Snedccor, B., Tolman, C. J. & Roberts, M. F. 1985. Growth inhibition of *Clostridium thermocellum* by carboxylic acids: a mechanism based on uncoupling by weak acids. *Applied Microbiology and Biotechnology,* 22, 53-62.

[23] Bahl, H., Andersch, W., Braun, K. & Gottschalk, G. 1982. Effect of pH and butyrate concentration on the production of acetone and butanol by *Clostridium acetobutylicum* grown continous culture. *European Journal of Applied Microbiology and Biotechnology,* 14, 17-20.

[24] Padan, E., Zllberstein, D. & Schuldiner, S. 1981. The proton electro-chemical gradient in *Escherichia coli. European Journal of Biochemistry,* 63, 533-541.

[25] Bowles, L. K., & Ellefson, W. L. 1985. Effects of butanol on *Clostridium acetobutyricum. Applied and Environmental Microbiology,* 50, 1165-1170.

[26] Roos, J. W., McLaughlin, J. & Pepoutsakis. E. T. 1984. The effect of pH on nitrogen supply, cell lysis, and solvent production in fermentations of *Clostridium acetobutylicum. Biotechnology and Bioengineering,* 27, 681-694.

[27] Fond, O., Matta Ammouri, G., Pefitdemange, H. & Engasser, J. M. 1985. The role of acids on the production of acetone and butanol by *Clostridium acetobutylicum, Applied Microbiology and Biotechnology,* 22, 195-200.

[28] Hartmanis, M. G. N., Kleson, T. & Gatenbeck, S. 1984. Uptake and activation of acetate and butyrate in *Clostridium acetobutylicum. Applied Microbiology and Biotechnology,* 20, 66-71.

[29] Andersch, W., Bahl, H. & Gottschalk, G. 1983. Level of enzymes involved in acetate, butyrate, acetone and butanol formation by *Clostridium acetobutylicum. European Journal of Applied Microbiology and Biotechnology,* 18, 327-332.

[30] Ballongue, J., Amine, J., Masion, E., Petitdemanage, H. & Gay, R. 1985. Introduction of acetoacetate decarboxylase in *Clostridium acetobutylicum. FEMS Microbiology Letter,* 29, 273-277.

[31] George, H. A. & Chen, J. S. 1983. Acidic conditions are not obligatory for onset of butanol formation by *Clostridium beijerinckii* (synonym, *C. butylicum*). *Applied Environment Microbiology,* 46, 321-327.

[32] Ross, D. 196l. The acetone-butanol fermentation. *Progress in lndustrial Microbiology,* 3, 73-85.

[33] Thauer, R. K., Jungermann, K. & Dekker, K. 1977. Energy conservation in chemotrophic anaerobic bacteria. *Bacteriological Reviews,* 4l, 100-180.

[34] Bahl, H. & Gottschalk, G. 1985. Parameters affecting solvent production by *Clostridium acetobutylicum* in continuous culture. *Biotechnology and Bioengineering,* 514, 217-223.

[35] Fick, M., Pierrot, P. & Engasser, J. M. 1985. Optimal conditions for long-term stability of acectone-butanol production cultures of *Clostridium acetobutylicum. Biotechnology Letters,* 7, 503-508

[36] Haggstrom, L. 1981. Immobilized cells of *Clostridium acetobutylicum* for butanol production. In: *Advances in Biotechnology,* Vol. 2 (edited by M. Moo-Young & C. W. Robinson) Pp. 79-83. Toronto, Canada: Pergamon Press.

[37] Haggstrom, L. 1985. Acetone-butanol fermentation and its variants. *Biotechnology Advances,* 3, 13-28.

[38] Hastings, J. H. J. 1978. Acetone-butyl alcohol fermentation. In: *Economic microbiology, vol. 2: primary products of metabolism* (edited by A. H. Rose). Pp. 31-45. New York, USA: Academic Press, Inc.

[39] Ryden, R. 1958. Development of anaerobic fermentation processes: acetone-butanol. In: *Biochemical Engineering* (edited by R. Steel) Pp. 125-148. London, UK: Heywood.

[40] Walton, M. T. & Martin, J. L. 1979. Production of butanol-acetone by fermentation, In: *Microbial Technology,* 1(2) (edited by H. J. Peppler & D. Perlman) Pp. 187-209. New York, USA: Academic Press. lnc.

[41] Ounine, K., Petitdemange, H., Raval, G. & Gay, R. 1985. Regulation and butanol inhibition of D-xylose and D-glucose uptake in *Clostridium acetobutylicum. Applied Environment Microbiology,* 49, 874-878.

[42] Costa, J. M. & Moreira, A. R. 1983. Growth inhibition for kinetics of the acetone-butanol fermentation. In: *Foundations of biochemical engineering: kinetics and thermodynamics in biological systems-ACS Symposium Series no. 207* (edited by H. W. Blanch, E. T. Papoutsakis & G. N. Stephanopoulos) Pp. 501-512, Washington, D.C., USA: American Chemical Society.

[43] Lung, J. C. Y., & D. l. C. Wang. 1981. Production of acetone and butanol by *Clostridium acetobutylicum* in continuous culture using free cells and immobilized cells. Proc. 2nd World Congr. Chem. Eng. 1:348-352

[44] Vollherbst-Schneck, K., Sands, J. A. & Montenecourt. B. S. 1984. Effect of butanol on lipid composition and fluidity of' *Clostridium acetobutylicum* ATCC 824. *Applied and Environmental Microbiology,* 47, 193-194.

[45] Van der Westhuizen, A., Jones, D. T. & Woods. D. R. 1982. Autolytic activity and butanol tolerance of *Clostridium acetobutylicum. Applied and Environmental Microbiology,* 44, 1277-1282.

[46] Barber, J. M., Robb, F. T., Webster, J. R. & Woods. D. R. 1979. Bacteriocin production by *Clostridium acetobutylicum* in an industrial fermentation process. *Applied and Environmental Microbiology,* 37, 433-437.

[47] Allcock, E.R., Reid, S. J., Jones, D. T. & Woods, D. R. 1981. Autolytic activity and an autolysis-deficient mutant of *Clostridium acetobutylicum. Applied and Environmental Microbiology,* 42, 929-935.

[48] Rogers, P. 1984. Genetics and biochemistry of *Clostridium* relevant to development of fermentation processes. *Applied Microbiology*, 31, 1-60.

[49] Brosseau, J. D., Yan, J. & Lo., K. V. 1986. The relationship between hydrogen gas and butanol production *Clostridium saccharoperbutylacetonicum. Biotechnology and Bioengineering,* 28, 305-310.

[50] Bramucci, Michael G. (Boothwyn, PA, US), Flint, Dennis (Newark, DE, US), Miller, Edward S. (Knoxville, TN, US), Nagarajan, Vasantha (Wilmington, DE, US), Sedkova, Natalia (Cherry Hill, NJ, US), Singh, Manjari (West Chester, PA, US), Van Dyk, Tina K. (Wilmington, DE, US) 2008 METHOD FOR THE PRODUCTION OF 2-BUTANOL United States 20080274525 http://www.freepatentsonline.com/y2008/0274525.html

In: Bioprocess Sciences and Technology
Editor: Min-Tze Liong

ISBN 978-1-61122-950-9

Chapter 3

PROSPECT AND SUSTAINABLE PRODUCTION OF POLYHYDROXYALKANOATE FROM PALM OIL

K. Bhubalan, J. Y. Chee and K. Sudesh*

Ecobiomaterial Research Laboratory, School of Biological Sciences,
Universiti Sains Malaysia, 11800 Penang, Malaysia

ABSTRACT

Increased and diversified socioeconomic activities of humankind in recent years have led to rapid consumption of natural resources. Synthetic plastics which are derived from finite resource have become an integral part of our lives. The ability to manipulate their structure and chemical composition has resulted in boundless applications of plastic materials. The usage of synthetic plastics is proportionally increasing with growing human population. This in turn has resulted in mounting wastes accumulation in our environment, which is often associated with global warming and destruction of ecosystems. Therefore, current trend in technology is geared towards the development of sustainable and environment-friendly processes and products. This includes the pre-production and post-consumption of products which exert negligible negative effects on the environment. Many developed countries have started to consider replacing synthetic plastics with environment-friendly and biodegradable polymeric materials such as polyhydroxyalkanoate (PHA). PHA has been receiving much attention because this material shares many si[1]milarities to synthetic plastics in terms of product performance and processability. Unlike synthetic plastics, the life cycle of PHA-based products is sustainable since it is primarily produced from renewable resources and is readily assimilated by microbial consortia upon disposal into the environment. PHA is generally produced by bacterial fermentation using various carbon feedstocks. Various fermentation strategies have been developed to increase the commercial viability of PHA as an alternative to some common plastics. Oil-based substrates such as palm oil products have generated interest lately due to its suitability for high yield PHA production. Selection of ideal carbon sources for the production of PHA is very much dependent on

*Corresponding author: *K. Sudesh* Email: ksudesh@usm.my

material costs and availability. The development and commercialization of palm oil agro-industry in Malaysia has made palm oil products readily available to be utilized as carbon feedstock for PHA production. In par with this, research on PHA production from bacterial fermentation using palm oil products is currently underway. This chapter will discuss current development of various PHA materials from palm oil products.

1. Introduction

Synthetic polymeric materials are probably one of the most useful and indispensable creation of mankind to date. The importance of these classes of polymers is evident from their domination in the commodity market, the established technologies related to plastics manufacturing and the ever growing public demand for synthetic plastics related products [1]. Modern plastics such as polyethylene, polypropylene, polystyrene, polyethylene terephthalate and polyvinyl chloride are strong, inexpensive, easily processable and durable [2]. Designed to have long life span with constant product quality and performance, they are inert to various chemicals as well as microbial degradation. However, the durability attributes of these plastics causes major concerns when they enter the waste stream.

Accumulation of recalcitrant materials in the environment is evident all over the world including in Malaysia as natural ecosystems such as rivers, lakes and mangroves are constantly polluted with ever mounting plastic wastes. It was estimated that the global consumption of petroleum-based plastics in 2006 was 205 million tones and was projected to reach 230 million tonnes in 2009, which accounts for an increase of around 25 million tonnes in just three years [3]. Large landfills are needed to cater for the ever increasing deposition of plastic waste materials. Besides, increasing amount of buried petrochemical-based plastics is hazardous to the environment as they cause soil infertility [4]. On the other hand, recycling seems to be a better alternative but categorization of a wide variety of plastic is a time-consuming and labour intensive process.

In the not so distant future, depletion of non-renewable fossil resources is inevitable because of ever increasing consumption by the world's fast-growing population. Due to the concerns such as preservation of natural resources and environmental sustainability, development of bio-based polymers which may or may not be biodegradable is necessary to reduce our reliance on petrochemical-based plastics. Bio-based materials such as polyamides, polysaccharides, polyphosphates, polyphenols, polyisoprenoids and polyhydroxyalkanoate are potential candidates for future substitution of synthetic materials [5]. Among them, PHA has received constant attention as a promising substitute because PHA polymers possess thermoplastic properties, are completely biodegradable as well as biocompatible and can be produced from renewable resources [6,7].

PHA is comprised of various types of hydroxycarboxylic acids and is synthesized by numerous microorganisms as intracellular carbon and energy storage materials [8]. PHA possess unique thermoplastic properties which resemble petroleum-based plastics [9]. Most PHA can be tailored into stiff packing materials or highly elastic elastomers. The chemical and physical properties of resulting PHA polymers can be influenced by the functionalized groups in the side chain of monomers such as, halogen, carboxyl, hydroxyl, epoxyl and phenoxy [10,11]. PHA is completely degraded into carbon dioxide (CO_2) and water (H_2O) once it has been discarded into the environment [7]. PHA is known to be produced from simple sugars such as glucose or fructose as well as from complex organic compounds such as plant oils. The most crucial factor which hampers the use of PHAs in a wide range of applications is its high production cost compared with that of petrochemical-based polymers. Various approaches are currently being undertaken to develop cheaper processes of producing high yields of PHA. Carbon substrate is the most significant factor which contributes to high production cost of PHA [7,12]. The use of cheap and renewable resources for PHA production will ensure a significant reduction in the overall production cost.

Plant oils such as palm oil are currently being investigated as potential carbon source because they are renewable and produce higher yields of polymer. Utilization of plant oils is estimated to reduce the overall production cost of PHA and energy consumption [13]. Almost two-folds higher poly(3-hydroxybutyrate) [P(3HB] yield is produced from plant oils (0.6 – 0.8 g/g) compared to that with glucose (0.3 – 0.4 g/g) [13]. The higher yield of the former is mainly due to the higher number of carbon atoms per gram of oil compared to sugar. Interest in the use of palm oil products for the production of PHA can be traced back to the early 1990s. Majid and co-workers reported the first attempts to use palm-oil based cooking oil (palm olein) as the sole carbon source to produce P(3HB) [14]. Saponified palm kernel oil was first reported as a suitable carbon substrate for the synthesis of PHA monomers with carbon number ranging from 6 – 14 using *Pseudomonas putida* [15]. Lately, several types of palm oil products were evaluated as carbon feedstock for PHA production and it was found that palm kernel oil produced the highest PHA yield [16,17,18,19,20]. Since palm oil is renewable and readily available in Malaysia, it has been identified as a viable feedstock for PHA production.

Today, Malaysia is one of the largest producer of palm oil in the world besides Indonesia [21]. In 2008, Malaysia ranked second with 17.73 million tons in overall palm oil production compared to 19.33 million tons produced by Indonesia [22]. Nevertheless, Malaysia alone exported a record breaking amount of 15.87 million tons of palm oil to the international palm oil market in 2009 [22]. The total export volume of oil palm products consisting palm oil, palm kernel oil, palm kernel cake, oleochemicals, biodiesel and finished products increased to 22.4 million tons in 2009 compared to 21.76 million tons in 2008 [22]. Development of large scale PHA production processes from palm oil products will enable the establishment of local PHA-based industries. The cost of carbon feedstock could be reduced significantly by integrating PHA production with oil palm plantation and palm oil processing mills. A similar strategy has been implemented in Brazil whereby PHA production was integrated successfully into sugar refineries [23].

2. Discovery of PHA and Development of Research Interest

2.1. Discovery of P(3HB) and other Hydroxyalkanoate (HA) Constituents in the Natural Environmental

The chronology of PHA began with the discovery of lipid-like inclusions in *Bacillus megaterium* in 1926 by a microbiologist, Lemoigne at Pasteur Institute in Paris [24]. These inclusion bodies (granules) were later identified as P(3HB) linked through ester bonds between the 3-hydroxyl group and the carboxylic group of the next monomer. Over the years, studies had revealed that the P(3HB) serves as energy and carbon storage material in microorganism [25]. The accumulation of P(3HB) is triggered when cells are exposed to excess carbon substrates while other essential nutrients such as nitrogen, magnesium, phosphate and/or oxygen are limited [26].

Initially, P(3HB) was thought to be the only type of storage polyesters synthesized and accumulated by bacteria. Most researchers accepted this as a fact until the discovery made by Wallen and Rohwedder in 1974, which proved that monomers other than 3HB were also present in the polyesters synthesized by various bacteria [27]. With the existing technology at that time, the presence of 3HB and 3-hydroxyvalerate (3HV) as the major constituents with 3-hydroxyhexanoate (3HHx) and 3-hydroxyheptanoate (3HHp) in smaller quantities were identified in the chloroform extract of activated sewage sludge [27]. Meanwhile, in another study, similar polymeric material extracted from marine sediments was found to contain 11 other short-chain monomers [28]. Since then, many researchers started to cultivate PHA producing bacteria using various carbon sources and culture conditions. This has led to the synthesis and characterization of a wide range of PHA.

2.2. Biosynthesis of Various PHA under Controlled Laboratory Conditions

Studies on PHA were mainly driven by commercial interest in its potential as a plastic material from renewable carbon sources. The properties of PHA are directly dependent on its monomer composition. Therefore, researchers were interested in determining all the different types of PHA that can be possibly produced by various bacteria. Many Gram positive and Gram negative bacteria have been identified to synthesize PHA with various monomer constituents. Many of the bacteria have been successfully cultivated under controlled laboratory conditions. Different kinds of carbon sources have been evaluated for the production of PHA polymers. To date, more than 150 types of monomers have been produced either from structurally related or unrelated carbon skeletons to the hydroxyalkanoic acids (HA) unit [29].

The type of carbon source fed to bacteria determines the type of PHA that is produced. In addition, the key enzyme of PHA synthesis, PHA synthase (PhaC), also determines the type of PHA that can be synthesized by a particular bacterium. To date, it is known that the substrate specificity of the PhaC is actually very versatile. Therefore, a diverse range of PHA having interesting properties has been synthesized. Bacterial production system is the most well established form of PHA production. To initiate PHA accumulation, the growth medium is prepared in such a way that one nutrient (generally nitrogen) limits growth while the carbon source is in excess. The depletion of the selected nutrient acts as a trigger for the metabolic shift to PHA biosynthesis.

PHA producing bacteria can be divided into two groups: growth associated PHA producers and non-growth associated PHA producers. Two different approaches have been developed for batch cultivations, which are one-stage cultivation and two-stage cultivation. One-stage cultivation approach is usually carried out for growth associated PHA producers. Here, growth of cells and PHA accumulation occur simultaneously. On the other hand, two-stage cultivation consists of a cell growth phase which is carried out in a separate nutrient enriched medium. The cells are then transferred into a nutrient (usually nitrogen) limited medium for PHA accumulation phase. This method of cultivation is preferred for non-growth associated PHA producers. The cultivation period for both these methods ranges from 24 to 96 h.

PHA accumulation in bacterial cells can be controlled by varying the ratio of carbon to nitrogen (C/N) [17,30,31]. This condition applies especially for growth-associated PHA-producing microorganisms. Nitrogen concentration in the culture medium affects the acetyl-CoA flux either to tricarboxylic acid (TCA) cycle or to the PHA polymerization process. In the presence of excess nitrogen, the biosynthesis of PHA is usually inhibited [32]. More acetyl-CoA will be channeled to the TCA cycle for cellular metabolism. Besides, the reaction of citrate synthase in TCA cycle also leads to the accumulation of high concentration of free CoASH which is known to inhibit the condensation activity of β-ketothiolase (PhaA) [33]. Therefore, it is necessary to determine the optimal C/N ratio in order to attain high PHA yield.

Normally, large-scale or industrial scale production systems use fed-batch cultivation method [34,35]. In this method, the cells are continuously fed with selected carbon source after it has entered the late exponential phase. The fed-batch method generally yields high cell densities which consequently reduces the overall production cost [36,37,38]. Besides this method, other modes of cultivation such as pH-stat based cultivation whereby carbon source is fed based on the fluctuation of pH [39] and chemo-stat method whereby culture medium is continuously exchanged with sterile growth medium [40] have also been evaluated.

The type of carbon substrate used to produce PHA varies among bacterial strains. This is because the metabolic pathways present and substrate specificity of PhaC is different and generally strain specific. *Cupriavidus necator* (formerly known as *Wautersia eutropha*, *Ralstonia eutropha*, *Alcaligenes eutrophus*) is a model strain which has been studied in detail for PHA production. *C. necator* is capable of accumulating P(3HB) efficiently from simple sugars such as glucose [13] and fructose [41]. Generally, *C. necator* has been investigated for the production of PHA containing 3-hydroxybutyrate (3HB), 3-hydroxyvalerate (3HV) and 4-hydroxybutyrate (4HB) monomers. On the other hand, substrates such as plant oils or free fatty acids were the favored carbon feedstock by microorganisms such as *Pseudomonas* sp. to produce PHA containing monomer units with carbon atom numbers ranging from 6 – 14 [42,43,44,45].

Nevertheless, other PHA production systems have also been evaluated. Attempts were also taken to synthesize PHA in transgenic plants and several studies are currently underway to produce PHA in a cost effective manner using plant systems [46,47,48,49,50]. PHA accumulation in plant cells are achieved by transforming the cells with PHA biosynthetic genes derived from native PHA producers. For example, *Arabidopsis thaliana* was transformed with the PhaC of *C. necator* to induce polymerization of the intermediates derived from the plant's endogenous 3-ketothiolase and acetoacetyl-CoA reductase. As a result, PHA accumulation was identified in the cytoplasm. Besides plants, other potential production systems in insect cells, yeast and cyanobacteria have been also evaluated [51]. Nevertheless, large-scale PHA production processes for commercialization purposes are currently derived from bacterial fermentations. Bacterial synthesis of PHA is normally carried out in batch or fed-batch cultures. The types of PHA synthesized by some bacterial strains are shown in Table 1.

Table 1. Biosynthesis of PHA containing short-chain-length (SCL) and medium-chain-length (MCL) monomers by different bacteria cultivated with various carbon sources

Bacterial strain	Types of PHA	Carbon source	Reference
Bacillus sp. JMa5	P(3HB)	Sucrose	[52]
Delftia acidovorans	P(3HB)	Glucose	[53]
	P(3HB-*co*-4HB)	1,4-butanediol and glucose	[53]
	P(3HB-*co*-3HV)	Sodium valerate and glucose	[54]
Cupriavidus necator	P(3HB)	Fructose	[55]
	P(3HB-*co*-4HB)	4-hydroxybutyric acid	[56]
	P(3HB-*co*-3HV)	Propionic acid and glucose	[57]
	P(3HB-*co*-MCL-PHA)	Sodium octanoate	[58]
Alcaligenes latus	P(3HB-*co*-3HP)	1,5-pentanediol	[59]
Aeromonas hydrophila	P(3HB-*co*-3HHx)	Lauric acid and oleic acid	[94]
Pseudomonas sp. 61-3	P(3HB-*co*-MCL-PHA)	Gluconate	[60]
P. putida	PHA copolymers with phenyl functional group	6-phenylhexanoic acid	[61]
	MCL-PHA	Saponified palm kernel oil	[16]
P. oleovorans	P(3HB) and MCL-PHA	Glucose, octanoic acid, oleic acid	[62]
P. aeruginosa 47T2	MCL-PHA	Oily substrates	[44]

2.3. Biodegradability of PHA

Intracellular and extracellular degradation of PHA have been investigated in detail in a number of microorganisms. PHAs are degraded internally by intracellular PHA depolymerases of PHA accumulating bacteria or extracellular PHA depolymerases of PHA degrading bacteria. Intracellular degradation occurs when the bacteria is stressed under carbon limitation conditions. Accumulated PHA granules are hydrolyzed by the cells as carbon and energy source [63,64]. P(3HB) is broken down to 3-hydroxybutyric acid by the PHA depolymerase and oligomer hydrolase [65]. The 3-hydroxybutyric acid is then oxidized by a dehydrogenase to acetylacetate. Acetyl-CoA is generated once β-ketothiolase reacts upon acetylacetate and is utilized for cell regeneration. In the natural environment, extracellular depolymerases are secreted by microorganisms to hydrolyze PHA polymer. The soluble fraction after hydrolysis normally consists of mixtures of oligomers or monomers. These products are then taken up for cell metabolism via absorption through the microorganism's cell wall [66]. The PHA polymers are degraded via surface erosion mechanism by the extracellular depolymerases. The end products of PHA degradation in an aerobic environment are CO_2 and H_2O [7]. In an anaerobic condition, methane is also produced.

The rate of biodegradation is influenced by a number of factors including the microbial population in a particular environment, temperature, pH, moisture, nutrient supply and the physical properties of the polymers itself [7,67,68]. Copolymers were found to degrade at a faster rate compared to P(3HB) homopolymer due to low crystallinity and porous surface

[69,70,71]. PHA film was found to degrade in less than 50 days when placed on soil surface under tropical conditions and the same films when buried in soil or mangrove sediments showed a faster rate of degradation [51,71,72]. Extracellular depolymerases have been purified from various microorganisms isolated from natural environments such as soil, activated sludge, lake water and sea water [67,69,73,74]. The most distinct character that differentiates the intracellular and extracellular depolymerases is that intracellular depolymerases are incapable of hydrolyzing extracellular PHA, and vice versa [67]. This phenomenon was believed to occur due to structural differences between the intracellular "native" (amorphous) granules and extracellular "denatured" (crystalline) PHA.

The biodegradation of various PHA-based biomaterials have been investigated *in vivo*. It is important to understand this mechanism because when PHA material is used as scaffold in tissue engineering application, the rate of degradation should be equal to the regenerative rate of tissue. Meanwhile, when it is used for as drug carrier, the rate of degradation will determine the amount of drug released [75]. Many successful studies of PHA degradation in living organisms have been demonstrated and no formation of toxic compounds was detected [76,77]. The main breakdown products of PHA are 3-hydroxyacids and they are naturally found in animals. Therefore, the biodegradation of PHA *in vivo* does not lead to formations of toxic compounds.

3. Types of PHA

3.1. Short-chain-length PHA (SCL-PHA)

3.1.1. Poly(3-hydroxybutyrate) [P(3HB]

P(3HB) is the most commonly found and naturally occurring PHA among microorganisms found in natural environment [24]. P(3HB) biosynthetic pathway is the simplest, which involves three enzymes and their encoding genes. This pathway is well established in *C. necator* through the metabolism of carbohydrates or triglycerides. Sugar such as glucose or triglycerides such as soybean oil and palm oil are metabolized via glycolysis and β-oxidation pathways, respectively to generate acetyl-CoA. These molecules of acetyl-CoA are either used for energy generation or PHA production. As for PHA production in *C. necator*, two acetyl-CoA molecules are condensed into acetoacetyl-CoA by PhaA (Figure 1). Then, this intermediate is reduced by NADPH-dependent reductase (PhaB) to produce (*R*)-3-hydroxybutyryl-CoA, which finally is polymerized into P(3HB) by PhaC [25].

P(3HB) homopolymer is a highly crystalline, brittle and stiff material [66]. It possesses a high Young's Modulus value of 3.5 GPa together with relatively high tensile strength of 43 MPa. Pure (no additives) P(3HB) exhibits an elongation to break of only 5% when compared with commercial polypropylene (PP) which is around 400 %. The melting temperature (T_m) of P(3HB) is about 179 °C which is very close to its degrading temperature of 185 °C, thus indicating that P(3HB) might not be suitable material for thermal processing at higher temperatures. In addition, its high glass transition temperature (T_g) of 4 °C indicates that P(3HB) is a crystalline and brittle material compared to other synthetic polymers such as low-density polyethylene (LDPE) or PP (Table 2). The average molecular weight (M_w) of P(3HB) is known to be in the range of 1×10^4 to 3×10^6 Da with a polydispersity of around two [66].

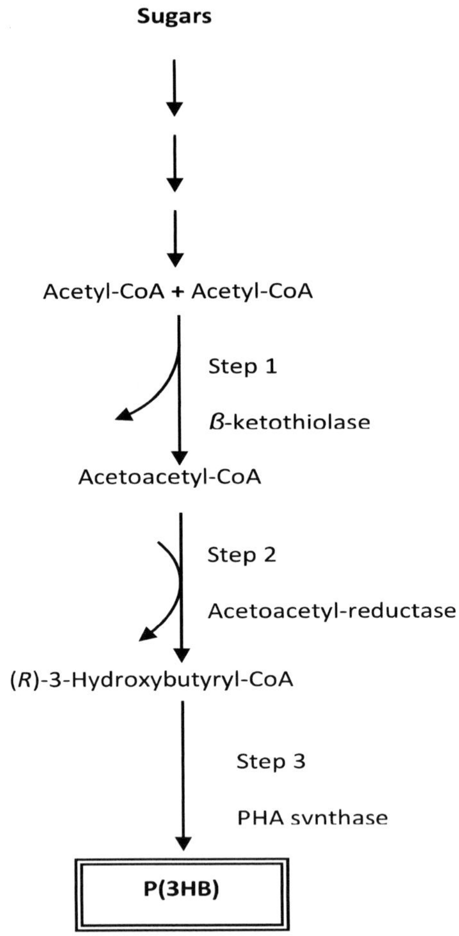

Figure 1. The most common P(3HB) biosynthesis pathway established in model strain *C. necator* [66].

Table 2. Comparison of the properties of P(3HB) with some common plastics [51]

Polymer	*Glass transition temperature, (T_g)*	*Melting temperature, (T_m)*	*Tensile strength (MPa)*	*Elongation to break (%)*
P(3HB)	4	179	43	5
Polypropylene	0	176	38	400
Low-density polyethylene	–30	130	10	620

Nevertheless, the quality of P(3HB) had been improved by genetically engineered transformant bacteria equipped with the ability to produce high molecular weight PHA under the direction of heterologous genes. Ultra-high-molecular weight P(3HB) homopolymer ranging from 3×10^6 to 11×10^6 Da could be synthesized by *Escherichia coli* transformant harboring the PHA synthase gene (*phaC*) of *C. necator* using glucose as the sole carbon source [78]. The mechanical properties of the ultra-high-molecular weight polymer were considerably improved compared to that of P(3HB) homopolymer produced by the wild-type microorganisms (Table 2). The elongation to break, Young's Modulus and tensile strength were 58%, 1.1 GPa and 62 MPa respectively [79]. The physical properties of the ultra-high-molecular weight P(3HB) were further improved by treating this polymer using cold-drawing procedures. The films produced in this manner were found to be more flexible yet strong as the elongation to break, Young's Modulus and tensile strength were 104%, 0.97 GPa and 175 MPa, respectively [80]. Beside altering the M_w of P(3HB) or developing different methods of polymer processing, incorporation of secondary monomers into the 3HB polymer backbone is also known to alter the overall property of resulting copolymers.

3.1.2. Poly(3-hydroxybutyrate-co-3-hydroxyvalerate) [P(3HB-co-3HV]

Poly(3-hydroxybutyrate-*co*-3-hydroxyvalerate) [P(3HB-*co*-3HV] is the first type of PHA to be commercialized by Imperial Chemical Industries (ICI), United Kingdom in 1980s [81]. This was the first example of P(3HB-*co*-3HV) copolymer production using axenic bacterial culture of *C. necator* fed with propionic acid as the additional carbon source to glucose [81,82,83]. Production of P(3HB-*co*-3HV) copolymer occurs when 3HV precursor substrates namely propionic or valeric acids or their salt form is fed to the microorganisms. Recently, efforts were taken to produce P(3HB-*co*-3HV) copolymer using *C. necator* grown on mixtures of various palm oil products and 3HV precursors such as sodium propionate and sodium valerate [17]. Polymerization of PHA containing 3HV monomers depends on the presence of 3-hydroxyvaleryl-CoA (3HV-CoA). Therefore, the addition of 3HV precursors is a prerequisite to promote the formation of 3HV-CoA. The metabolic pathways for P(3HB-*co*-3HV) biosynthesis from mixture of glucose and propionic acid is shown in Figure 2. Propionic acid is first converted to propionyl-CoA while acetyl-CoA is derived from glucose. Both acetyl-CoA and propionyl-CoA then undergo condensation by β-ketothiolase resulting in the formation of acetoacetyl-CoA and 3-ketovaleryl-CoA, respectively. Next, the condensation products are further reduced to (*R*)-3-hydroxybutyryl-CoA and (*R*)-3-hydroxyvaleryl-CoA, respectively by NADPH-dependent acetoacetyl-CoA dehydrogenase. Both 3HB and 3HV monomers are then polymerized into P(3HB-*co*-3HV) by the PHA synthase [84].

Other odd carbon numbered aliphatic fatty acids with higher carbon chain length are metabolized through the β-oxidation cycle and finally a pool of propionyl-CoA and acetyl-CoA are formed inside the bacterial cell. In addition, during the degradation of these odd carbon numbered fatty acids, the intermediates with five carbon atom could also be directly converted into 3HV monomers. Nevertheless, odd carbon numbered fatty acids are known to be toxic and they inhibit cell growth and PHA accumulation when present in higher concentrations [19,20,32]. Hence, various feeding methods have been implemented to reduce the inhibitory effects of these volatile fatty acids on cell cultures. Among the most effective feeding strategies are alternate feeding [85], pH-stat based feeding [86] and sequential feeding [87] of 3HV precursors.

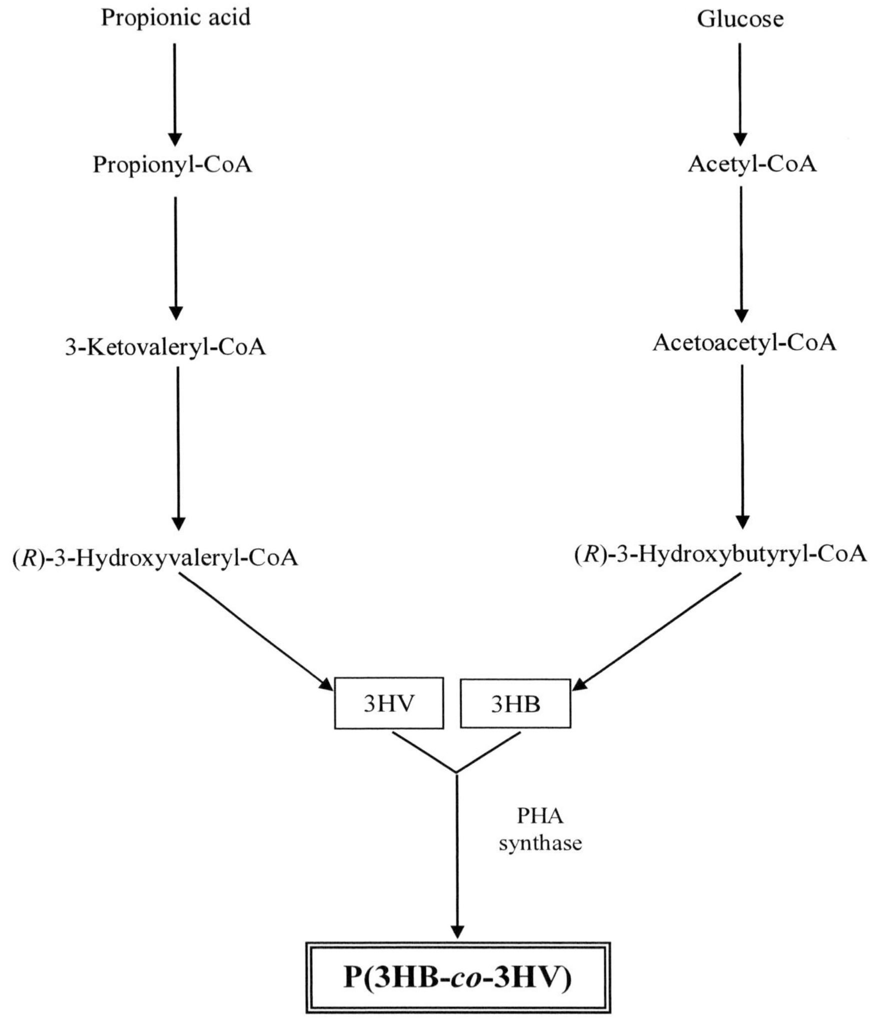

Figure 2. P(3HB-*co*-3HV) copolymer biosynthesis in *C. necator* from mixtures of glucose and propionic acid as the 3HV precursor [66].

The presence of 3HV units assists in the lowering of melting temperature, crystallinity as well as minimizing the polymer's stiffness [32]. Thus, P(3HB-*co*-3HV) could be categorized with better impact strength and processability. The physical properties exhibited by a copolymer are very much dependent on the ratio of the monomer fractions [88]. The thermal and mechanical properties of P(3HB-*co*-3HV) copolymers improve with increasing 3HV monomer composition (Table 3). T_m of P(3HB-*co*-25 mol% 3HV) was recorded at 137 °C which is much lower than 179 °C for P(3HB) homopolymer [82]. In addition, T_g also was found to decrease with an increase in 3HV monomer composition. A steady decrease in Young's Modulus and tensile strength indicates the improvement in material flexibility (Table 3). Nevetheless, the crystallinity of P(3HB-*co*-3HV) was maintained at 60% or higher even when different composition of 3HV monomer was incorporated into the 3HB polymer chain [82]. The high crystallinity of P(3HB-*co*-3HV) is attributed to the isodimorphism behavior of this copolymer, in which both the 3HB and 3HV monomers which are similar in size co-crystallize in the same lattice [66]. This phenomenon was only detected in copolymers of P(3HB-*co*-3HV).

Table 3. Physical properties of P(3HB-*co*-3HV) copolymers [82]

3HV monomer composition (mol%)	T_m [a] (°C)	T_g [b] (°C)	Young's Modulus (GPa)	Tensile strength (MPa)
0	*179*	*10*	*3.5*	*40*
9	*162*	*6*	*1.9*	*37*
14	*150*	*4*	*1.5*	*35*
20	*145*	*–1*	*1.2*	*32*
25	*137*	*–6*	*0.7*	*30*

[a] Melting temperature.
[b] Glass transition temperature.

3.2. Medium-chain-length PHA (MCL-PHA)

In general, SCL-PHA is highly crystalline and has poor tensile strength [66] whereas MCL-PHA is amorphous and very elastomeric. The former is too brittle and breaks easily while the latter is too sticky to be processed conveniently. MCL-PHA consists of monomers with carbon number ranging from 6 – 14. Bacteria from the genus *Pseudomonas* is generally associated with MCL-PHA production. However, an exception is the *Pseudomonas* sp. 61-3 which can incorporate both SCL- and MCL-PHA [89]. Polymer consisting mainly of 3-hydroxyoctanoate (3HO) is synthesized by *P. oleovorans* and was the first example for PHA consisting of MCL monomer [90]. In a separate study, *P. putida* was found to accumulate copolymer consisting of (3HO) and 3-hydroxyhexanoate (3HHx) as main and minor constituents respectively, when cultivated on octanoic acid as the carbon source. Beside fatty acids, sugar such as glucose was also identified as a potential carbon feedstock for *P. putida* to accumulate PHA containing MCL monomer. The monomer compositions were comprised of 3HHx, 3HO and 3-hydroxydecanoate (3HD) with other minor saturated and mono-unsaturated monomers of 12 carbon atoms [91].

PHA copolymer consisting of 3HB monomer as the major constituent and minor constituents of unsaturated MCL and saturated 16 carbon monomers were produced using *Pseudomonas* sp. A33 in the presence of 1,3-butanediol. The saturated and unsaturated 16 carbon monomers were 3-hydroxyhexadecanoate (3HHD), 3-hydroxydodecenoate (3HDDE) and 3-hydroxytetradecenoate (3HTDE) [92]. MCL-PHA had also been produced from saponified palm kernel oil using *P. putida* in shake-flask and fermenter scale experiments [15,16]. It was found that fluorescent *Pseudomonas* have good ability to accumulate a wide range of MCL-PHA from *n*-alkanoic acids.

This implies that the fatty acid metabolism of these bacteria, i.e. fatty acid *de novo* biosynthesis and fatty acid *β*-oxidation, are linked with PHA biosynthesis. Catabolism of fatty acids is represented by *β*-oxidation, a four-step process yielding acetyl-CoA (Figure 3). Fatty acid synthesis and *β*-oxidation display similar chemistries, however, different enzymes are used for these processes. Fatty acids will initially be converted by thiokinase into respective acyl-CoA thioesters before they enter the *β*-oxidation pathway. Acyl-CoA is then oxidized via *β*-oxidation pathway into acetyl-CoA by a recurring 4 step process (Figure 3). First, acyl-CoA is oxidized into enoyl-CoA by acyl-CoA dehydrogenase. Enoyl-CoA is then converted into (*S*)-3-hydroxyacyl-CoA by enoyl-CoA hydratase. Oxidation of *(S)*-3-hydroxyacyl-CoA by 3-hydroxyacyl-CoA dehydrogenase results in the formation of 3-ketoacyl-CoA. The 3-ketoacyl-CoA thiolase cleaves 3-ketoacyl-CoA that was formed into acetyl-CoA and acyl-CoA. However, the newly generated acyl-CoA lacks 2 carbon atoms, comparatively to those compounds that were present during the first cycle. In the case of fatty acids with even carbon number, further cycles go on until the original acyl-CoA is fully converted into acetyl-CoA [29].

The intermediates of *β*-oxidation cycle (namely (*S*)-3-hydroxyacyl-CoA) will be converted into (*R*)-3-hydroxyacyl-CoA by certain metabolic links since none of the intermediates are acknowledged as substrates by the PHA synthase [12,29]. These correlations are established by the activities of three different enzymes; epimerase, (*R*)-specific enoyl-CoA hydratase and 3-ketoacyl-CoA reductase. Epimerases catalyzes conversion of (*S*)-isomer of 3-hydroxyacyl-CoA into (*R*)-isomer. On the other hand, (*R*)-specific enoyl-CoA hydratase functions in converting enoyl-CoA into (*R*)-3-hydroxyacyl-CoA [93]. The 3-ketoacyl-CoA reductase reduces 3-ketoacyl-CoA to (*R*)-3-hydroxyacyl-CoA. The resulting (*R*)-3-hydroxyacyl-CoAs are polymerized by PHA synthase into corresponding MCL monomers. Meanwhile, the acetyl-CoA could be either channeled to TCA cycle, fatty acid synthesis or formation of P(3HB). The formation of P(3HB) is subjected to the availability of PHA synthase with substrate specificity towards 3HB-CoA.

3.3. Short-chain-length-medium-chain-length PHA (SCL-MCL PHA)

Advancements towards the production of hybrid polymers possessing the unique physical and mechanical properties of both SCL and MCL copolymers served as the stepping stone towards development of the SCL-MCL PHA. Bacteria from the genus *Aeromonas* are well known for the production of SCL-MCL PHA such as P(3HB-*co*-3HHx) copolymer [35,94,95]. Recently, P(3HB-*co*-3HHx) copolymer was produced from palm oil products using a *C. necator* PHB^-4 transformant harboring the PhaC of *A. caviae* [18]. It was found that the 3HHx monomer composition in the copolymer produced was maintained at 5 mol% at the end of 72 h fermentation regardless of the type of palm oil used. Production of SCL-MCL PHA terpolymers such as P(3HB-*co*-3HV-*co*-3HHx) [19,20,96,97,98] and P(3HB-*co*-3HV-*co*-3HHp) [99] have also been reported. Besides wild-type *Aeromonas* strains, transformant strains were also investigated for the production of P(3HB-*co*-3HV-*co*-3HHx). Park and co-workers synthesized P(3HB-*co*-3HV-*co*-3HHx) terpolymer from dodecanoic acid and odd numbered carbon fatty acids using an *E. coli* transformant harboring the PHA synthase gene of *Aeromonas* sp. [96]. After 90 h of fermentation, PHA content of 22 wt% with 24 mol% of 3HV and 5.9 mol% of 3HHx were obtained, respectively.

In another study, an *A. hydrophila* transformant harboring both the *β*-ketothiolase and acetoacetyl-CoA reductase encoding genes of *C. necator* was used to synthesize P(3HB-*co*-3HV-*co*-3HHx) from mixtures of dodecanoic acid and propionic acid [97]. The 3HV molar fraction ranged from 2.3 mol% to 7.1 mol% depending on the concentration of propionic acid in the culture medium. Recently, *C. necator* PHB$^-$4 transformants harboring the PhaC of either *A. caviae* or *Chromobacterium* sp. USM2 were investigated for the production of P(3HB-*co*-3HV-*co*-3HHx) terpolymers from mixtures of crude palm kernel oil and 3HV precursors such as sodium propionate and sodium valerate [19,20]. P(3HB-*co*-3HV-*co*-3HHx) with 3HV monomer composition ranging from 2 – 91 mol% were obtained by feeding the 3HV precursors at different intervals of cultivation. P(3HB-*co*-24 mol% 3HV-*co*-7 mol% 3HHx) exhibited mechanical properties similar to that of common LDPE. Generally, the terpolymers were found to be elastomeric material and their properties can be altered by changing the 3HV and 3HHx monomer compositions.

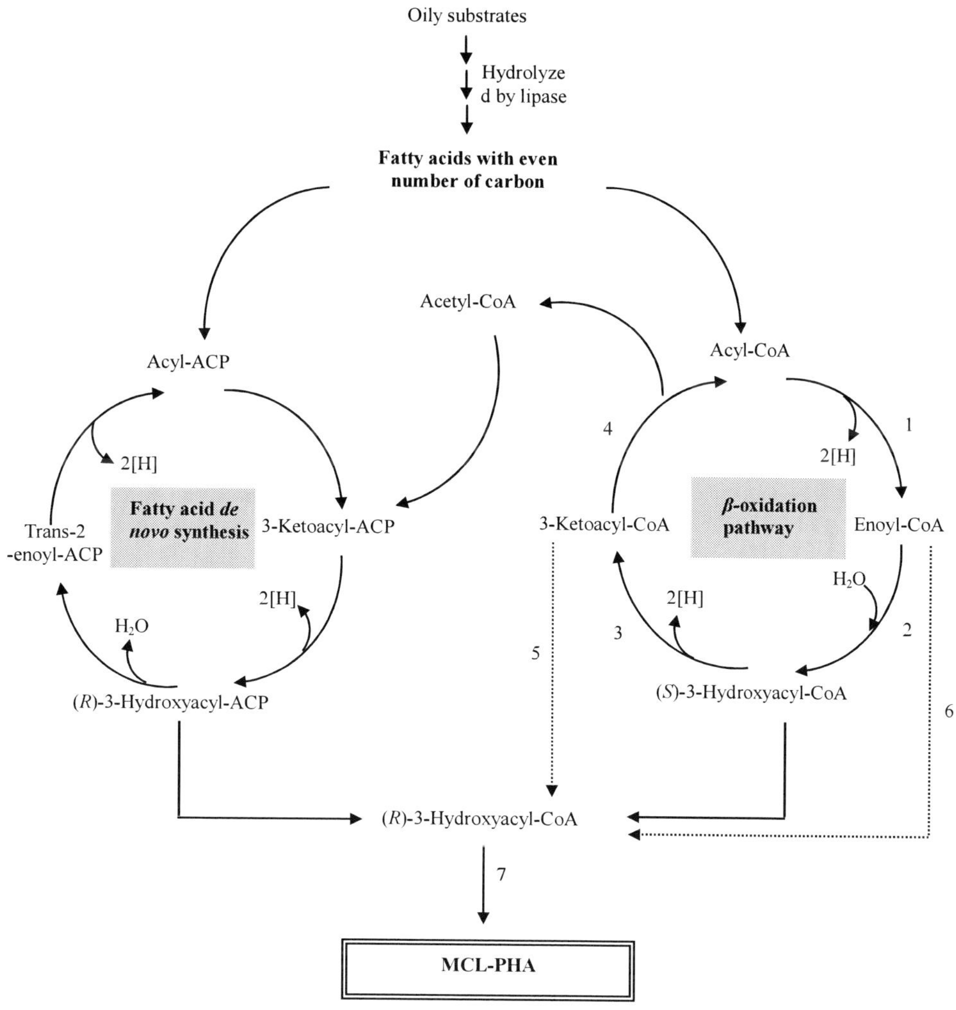

Figure 3. Proposed metabolic pathways for the biosynthesis of MCL-PHA. Oily substrates are first hydrolyzed by lipase into free fatty acids and then channeled via *β*-oxidation pathway for the generation of monomer intermediates. Enzymes: 1, acyl-CoA degydrogenase; 2, enoyl-CoA hydratase; 3, 3-hydroxyacyl-CoA dehydrogenase; 4, 3-ketoacyl-CoA thiolase; 5, 3-ketoacyl-CoA reductase; 6, (*R*)-specific enoyl-CoA hydratase; 7, PHA synthase [29].

4. Biosynthesis and Characterization of Various Types of PHA from Palm Oil Products

In the case of microbial fermentation, selection of cheap and sustainable starting material is a critical factor which determines the overall performance of the fermentation process as well as production cost of PHA. Therefore, the best approach is to choose readily available carbon substrates (preferably bio-based) which could be obtained in large quantities. This will ensure a lower substrate cost thus reducing the overall production cost. The most conventional and popular carbon feedstocks used for microbial growth and PHA accumulation are sugars such as glucose, fructose and other saccharides. Lately, plant oils are being investigated as suitable alternatives because they are renewable and produce higher yields of PHA. Soybean oil and palm oil are available in bulk quantities in several countries such The United States of America, Indonesia and Malaysia. Application of these oils as starting material for large-scale PHA production will be feasible as PHA could be marketed at reasonable price and this will boost the PHA industries. Abundant and uninterupted supply of palm oil in Malaysia enables it to be utilized as potential carbon feedstock for sustainable PHA production.

4.1. P(3HB)

During the early 1990s, Majid and co-workers reported their first attempt to use palm olein as the sole carbon source for the production of P(3HB) using *Alcaligenes* sp. AK201[14]. Approximately, 2 g/L of P(3HB) was obtained when 3 g/L of cooking oil was fed into the culture medium. This resulted in a yield coefficient of cooking oil to P(3HB) of 0.62 g/g. In comparison, fructose was found to yield only 0.35 g/g of P(3HB). Besides, relatively high M_w of more than 1×10^6 Da P(3HB) was obtained when cooking oil was fed to the bacteria [14]. In a subsequent study by Fukui and Doi, it was shown that up to 80 wt% of P(3HB) was synthesized by *C. necator* H16 from plant oils such as olive oil, corn oil and palm oil [100]. A scale-up fermentation (10 L capacity) with soybean oil had successfully led to the production of 76 wt% of P(3HB) from 126 g/L cell biomass by employing phosphate limitation strategy. The calculated yield of P(3HB) from soybean oil was 0.76 g/g which was considerably higher than that from sugars [34]. In the meantime, an attempt was also taken to isolate new microorganisms which grew and accumulated PHA from plant oils. Zazali and Tan had successfully isolated a Gram negative bacterium FLP1 from palm oil mill effluent with 80% similarity to *Burkholderia cepacia* which produced 50 wt% and 43.6 wt% of P(3HB) from crude plam oil (CPO) and palm kernel oil, respectively [101].

Recently, Lee and co-workers investigated the production of P(3HB) by growing *C. necator* H16 on various types of palm oils [17]. Different fractions of palm oil products from the kernel and mesocarp were fed to the cells grown in shake-flasks. GC analysis demonstrated that *C. necator* showed versatility in utilizing the various palm oils for the synthesis of P(3HB) (Table 4). P(3HB) content in the range of 67 – 78 wt% was obtained using these oils. Palm oil products derived from the kernel contributed to higher cell growth and PHA accumulation. Crude palm kernel oil (CPKO) is mainly used for non-edible purposes by the oleochemical industries as compared to palm oil and other plant oils.

Therefore, development of PHA production system using CPKO may not risk future sustainability of plant oils-based food products.

Table 4. Biosynthesis of P(3HB) using *C. necator* from different types of palm oils[a] [17]

Plant oils (5 g/L)	Total P(3HB) (g/L)	Cell dry weight (g/L)	P(3HB) content (wt%)
Crude palm kernel oil[b]	3.4	5.0	67
Crude palm oil[b]	3.5	4.6	75
Palm kernel oil[c]	4.1	5.5	75
Palm kernel oil[b]	4.3	5.6	77
Palm olein[b]	3.6	5.2	70
Cooking oil	4.2	5.4	78

[a] Incubated for 72 h at 30 °C at 200 rpm in mineral medium, initial pH 7.0.
[b] Source from Unitata Ltd.
[c] Source from Acidchem Int. Ltd.

A locally isolated *Chromobacterium* sp. USM2 also exhibited its ability to utilize CPKO for growth and P(3HB) accumulation [98]. Cell biomass of 3.0 ± 0.2 g/L with polymer accumulation of 23 ± 4 wt% was observed. By-products from the palm oil industries such as palm kernel acid oil (PKAO) and palm acid oil (PAO) have also been identified as potential carbon feedstock for P(3HB) production by *C. necator* [102]. More than 70 wt% of P(3HB) was accumulated when 5 g/L of PAO was supplemented into the culture medium. Meanwhile, up to 85 wt% of P(3HB) could be accumulated with similar concentration of PKAO. Traditionally, palm oil by-products are used for animal feed production. Now, it has been proven that these surplus materials could be converted into other value added product such as P(3HB).

4.2. P(3HB-*co*-3HV)

Besides P(3HB) homopolymer, palm oils were also investigated for the biosynthesis of P(3HB-*co*-3HV) copolymer. Lee and co-workers had demonstrated the production of P(3HB-*co*-3HV) from mixtures of various palm oils and 3HV precursors such as propionic acid and sodium valerate by *C. necator* at laboratory scale [17]. High copolymer accumulation of 90 wt% was obtained from 7.5 g/L of dry cell weight when CPKO and sodium propionate were fed to shaken flask cultures (Table 5). Sodium valerate or sodium propionate was fed at a concentration of 2.5 g/L during later hours of cultivation (48 and 60 h). It was found that early additions of these precursors resulted in low cell growth and impaired ability to accumulate PHA [17]. Copolymers with 3 – 10 mol% of 3HV were synthesized from a total of 5 g/L of precursor substrate and 5 g/L of palm oils.

Interestingly, the number-average molecular weight (M_n) of PHA obtained from the mixture of palm oils and sodium propionate was in the range of $5.7 - 7.7 \times 10^5$ Da (Table 5). The presence of glycerol in culture medium from hydrolysis of oil is of concern because it had been previously reported by Madden and co-workers that glycerol affected the molecular weight of PHA [103]. However, this phenomenon was not observed by Lee and co-workers [17]. The melting temperature (166 to 170 ºC) and glass transition temperature (0 to –0.9 ºC) of P(3HB-*co*-3HV) copolymers did not show clear differences. This is probably because the 3HV monomer composition was relatively low the copolymer is largely composed of 3HB. Besides, addition of 3HV precursor at later stage of cultivation might have resulted in uneven distribution of 3HV monomer. This was shown by the presence of two distinct T_m peaks which indicated that the PHA synthesized might be a blend of P(3HB) and P(3HB-*co*-3HV) [17]. Nevertheless, the distribution of 3HV could be controlled by adjusting the timing of precursor addition or employing different cultivation mode. Doi and co-workers reported the 3HV monomer was well regulated by two-stage cultivation using butyric acid and valeric acid as the properties of the copolymer are perfectly controlled [104]. It was clearly shown that mixture of palm oil and it products with 3HV precursors are suitable substrate for P(3HB-*co*-3HV) production.

Table 5. Biosynthesis of P(3HB-*co*-3HV) copolymer using *C. necator* from mixtures of palm oils and sodium propionate[a] [17]

Palm oil (5 g/L) + Sodium propionate (5 g/L)	Cell dry weight (g/L)	PHA content[b] (wt%)	PHA composition (mol%)		Molecular weight		Thermal properties		
			3HB	3HV	M_n[c] (x 10^5)	M_w/M_n[d]	T_g[e] (°C)	T_m[f] (°C)	ΔH_m[g] (J/g)
Crude palm kernel oil[h]	7.5	90	93	7	7.4	3.8	–0.8	166	64.7
Palm kernel oil[i]	7.3	65	93	7	7.7	3.5	–0.2	167	54.5
Cooking oil	6.3	85	95	5	5.7	4.1	–0.2	169	56.8
Palm olein[h]	6.7	88	96	4	7.1	3.9	–0.5	170	59.1
Crude palm oil[h]	3.6	74	97	3	6.7	3.9	–0.9	168	59.5

[a] Incubated for 72 h at 30 °C at 200 rpm in mineral medium, initial pH 7.0. Sodium propionate was added at 48 h (2.5 g/L) and 60 h (2.5 g/L) of cultivation.
[b] PHA content in freeze-dried cells.
[c] Number-average molecular weight.
[d] Polydispersity index.
[e] Glass transition temperature.
[f] Melting temperature.
[g] Enthalphy of fusion.
[h] Source from Unitata Ltd.
[i] Source from Acidchem Int. Ltd.
3HB, 3-hydroxybutyrate; 3HV, 3-hydroxyvalerate

4.3. P(3HB-co-3HHx)

The biosynthesis of random SCL-MCL PHA copolymer of P(3HB-*co*-3HHx) from alkanoic acid and oil was first found in *A. caviae* [105]. However, this bacterium could only accumulate less than 30 wt% PHA from cell biomass. In order to attain better PHA yield, *C. necator* PHB^{-}4 transformant harboring the PHA synthase of *A. caviae* was constructed [106]. This transformant was capable of producing P(3HB-*co*-3HHx) copolymer containing 4 – 5 mol% of 3HHx with contents above 80 wt% of the cell dry weight [34]. In 2005, Loo and co-workers demonstrated the excellent accumulation of P(3HB-*co*-3HHx) by the same transformant strain from palm oil products [18]. Up to 87 wt% of this copolymer was produced from 4.3 g/L of cell biomass using 5 g/L of CPKO (Table 6).

Figure 4 demonstrates the accumulation of P(3HB-*co*-3HHx) in the transformant grown on CPKO. The whole cells appear to be fully occupied with PHA granules. Besides CPKO, other palm oil products were also found to be useful for the biosynthesis of P(3HB-*co*-3HHx) with a constant composition of 3HHx monomer [18]. It is interesting to note that palm oil by-products such as PAO had also been identified as potential feedstock for cell growth and P(3HB-*co*-3HHx) copolymer accumulation (Table 6). Utilization of various palm oil products and by-products for PHA biosynthesis is important as reliance on a single source could be avoided.

Table 6. Biosynthesis of P(3HB-*co*-3HHx) copolymer using *C. necator* transformant harboring the PHA synthase gene of *A. caviae* from different types of palm oils[a] [18]

Carbon source (5 g/L)		*PHA content*[b] *(wt%)*	PHA composition (mol%)	
			3HB	3HHx
Crude palm kernel oil[c]	4.3	87 ± 2	95	5
Crude palm kernel oil[d]	3.2	70 ± 2	95	5
Palm olein[c]	3.7	64 ± 2	95	5
Crude palm oil[c]	3.1	54 ± 2	95	5
Palm acid oil[c]	3.8	40 ± 2	95	5

[a] Incubated for 72 h at 30 °C at 200 rpm in mineral medium, initial pH 7.0.
[b] PHA content in freeze-dried cells
[c] Source from Unitata Ltd.
[d] Source from Acidchem Int. Ltd.

The incorporation of small amount (5 mol%) of 3HHx monomers into the 3HB polymer chain had improved the thermal properties of the copolymer compared to P(3HB) homopolymer. The T_m had decreased from 179 °C [P(3HB] to approximately 125 to 155 °C (Table 7). The T_g were significantly lower as compared to P(3HB) homopolymer. These shows that P(3HB-*co*-3HHx) copolymer possesses improved properties. The presence of secondary monomers with longer side chain would increase the steric bulkiness and thus the crystallization of 3HB monomers could be reduced. Subsequently, the flexibility of the resulting copolymer could be improved compared to P(3HB) homopolymer [107]. The M_w of the copolymer were also relatively high in the range of $0.9 - 1.5 \times 10^6$ Da.

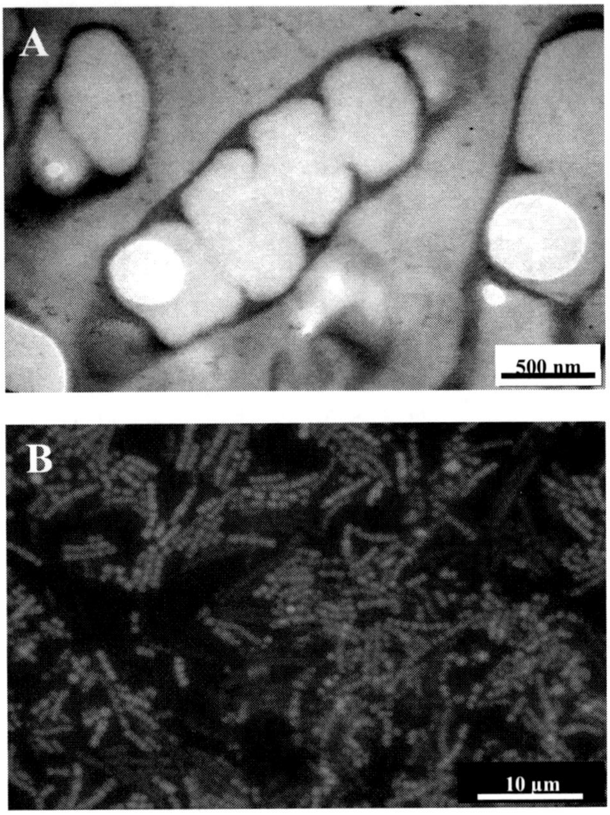

Figure 4. (A) Electron micrograph and (B) Fluoroscent micrograph showing high accumulation of PHA in transformant *C. necator* cells harboring the PHA synthase gene of *A. caviae* cultivated in mineral medium supplemented with CPKO for 72 h at 30 °C.

Table 7. Characteristics of P(3HB-*co*-5 mol% 3HHx) copolymer synthesized by *C. necator* transformant harboring the PHA synthase gene of *A. caviae* from different types of palm oils [18]

Carbon source[a]	Molecular weight			Thermal properties	
	M_w[b] (x 10^6)	M_n[c] (x 10^5)	M_w/M_n[d]	T_g[e] (°C)	T_m[f] (°C)
Crude palm kernel oil	1.5	4.6	3.3	–3	125, 143
Palm olein	1.2	3.9	3.0	–2	129, 147
Crude palm oil	1.4	3.5	3.9	–1	137, 154
Palm acid oil	0.9	2.8	3.3	–1	138, 155
Palm stearin	1.3	4.1	3.2	1	142, 153

[a] Source from Unitata Ltd.
[b] Molecular weight.
[c] Number-average molecular weight.
[d] Polydispersity index.
[e] Glass transition temperature.
[f] Melting temperature.

4.4. P(3HB-co-3HV-co-3HHx)

Production of P(3HB-*co*-3HV-*co*-3HHx) terpolymer had been studied in *Rhodospirillum rubrum* [108], *Rhodocyclus gelatinosus* [109], and *Rhodococcus* sp. [25]. *E. coli* transformant harboring *Aeromonas* PHA biosynthetic gene produced this terpolymer when grown on dodecanoic acid with the addition of different odd carbon number fatty acids [96]. In recent studies, Bhubalan and co-workers had successfully synthesized P(3HB-*co*-3HV-*co*-3HHx) terpolymers with various 3HV monomer compositions from mixtures of CPKO and sodium valerate or sodium propionate using *C. necator* PHB^{-}4 transformant harboring the PHA synthase gene of *A. caviae* or *Chromobacterium* sp. USM2 [19,20]. The composition of the 3HV monomer was found to be higher when sodium valerate was used as the 3HV precursor instead of sodium propionate. In addition, feeding time of precursor substrate was identified as an important factor in determining the monomer composition of 3HV.

When *C. necator* transformant harboring the PHA synthase gene of *A. caviae* was used, 3HV monomer compositions in the range of 2 to 60 mol% were obtained by feeding various concentrations of precursor at different stages of cultivation [19]. Nevertheless, two-fold higher concentration of sodium propionate was required to achieve the same amount of 3HV obtained using sodium valerate. The 3HHx monomer composition in the terpolymers was in the range of 2 to 7 mol%. In the case of the transformant harboring the PHA synthase gene of *Chromobacterium* sp. USM2, P(3HB-*co*-3HV-*co*-3HHx) terpolymers with higher 3HV monomer composition ranging from 2 to 91 mol% were synthesized [20]. The PHA synthase of *Chromobacterium* has an unusually high affinity towards 3HV monomer. The 3HHx monomer composition (7 to 9 mol%) of the terpolymers produced by this transformant was almost similar compared to the former.

The introduction of 3HV monomers into P(3HB-*co*-3HHx) polymer chain showed great changes in the thermal properties of the terpolymers (Tables 8 and 9). The melting temperatures of these terpolymers were in the range of 81 to 153 °C. Presence of two well separated melting peaks in the samples might have been caused by the rearrangement of the initial crystal morphology of copolymer [110]. During the primary crystallization process at given crystallization temperature, relatively thick crystalline lamellae may grow into spherulitic morphology. Subsequently, at room temperature thinner crystalline lamellae may be formed. Therefore, a difference in lamellae thickness would have accounted for the separate T_m values [110]. The T_g values of the terpolymers decreased with higher 3HV content (Tables 8 and 9) and this coincides with previous findings [32,111]. Terpolymer with 85 mol% of 3HV and 1 mol% of 3HHx possessed a T_m of 89 °C, which is much lower than that of P(3HB-*co*-3HHx) copolymers with 5 mol% of 3HHx (125 – 155 °C) [18].

The incorporation of both 3HV and 3HHx monomers greatly influenced the mechanical properties of the terpolymer in comparison to the homopolymer of P(3HB). This resulted in a material with more ductile properties. The M_w of the terpolymers produced were in the range of 3.3 – 25 × 10^5 Da with polydispersity values of 2.0 to 4.3 (Table 8 and 9). The results obtained was comparable to that reported by Fukui and Doi as P(3HB) with M_w of 1.0 × 10^6 could be synthesized in the presence of palm oil as the sole carbon source [100]. P(3HB-*co*-24 mol% 3HV-*co*-7 mol% 3HHx) was found to be a flexible yet strong material. It exhibited a high tensile strength of 20 MPa and recorded an elongation at break of 710% (Table 9). The mechanical property of this terpolymer is similar to that of common low-density polyethylene. Higher fraction of 3HHx in the terpolymer might have resulted in the interference of the crystallization process thus resulting in a polymer with increased flexibility. The solution-cast film of this terpolymer was very transparent and did not show any visible changes even after storage for more than 12 months under ambient conditions.

Table 8. Thermal and physical properties and molecular weights of P(3HB-*co*-3HV-*co*-3HHx) produced by *C. necator* transformant harboring the PHA synthase gene of *A. caviae*[a] [19]

3HV precursors (Time of feeding)	PHA composition (mol%)			T_g[b] (°C)	T_m[c] (°C)	Tensile strength (MPa)	Elongation at break (%)	M_w[d] ($\times 10^5$)	M_w/M_n[e]
	3HB	3HV	3HHx						
1 g/L Sodium valerate (0 h)	91	2	7	–3.4	144	22	312	8.7	3.9
4 g/L Sodium propionate (36 h)	87	9	4	–4.6	142,153	21	287	12	4.2
3 g/L Sodium valerate (60 h)	80	16	4	–3.5	140,151	20	321	11	3.4
7 g/L Sodium propionate (12 h)	71	26	3	–5.6	131,143	12	324	16	4.2
6 g/L Sodium valerate (12 h)	58	39	3	–7.9	N.D.	12	408	16	4.1
8 g/L Sodium valerate (12 h)	38	60	2	–13	81,87	14	421	25	4.3

[a] Incubated for 72 h at 30 °C at 200 rpm in mineral medium containing 5 g/L of CPKO, initial pH 7.0.
[b] Glass transition temperature.
[c] Melting temperature.
[d] Molecular weight.
[e] Polydispersity index.
3HB, 3-hydroxybutyrate; 3HV,3-hydroxyvalerate; 3HHx, 3-hydroxyhexanoate; N.D., not detected.

Table 9. Thermal and physical properties and molecular weights of P(3HB-*co*-3HV-*co*-3HHx) produced by *C. necator* transformant harboring the PHA synthase gene of *Chromobacterium* sp. USM2[a] [20]

3HV precursors (Time of feeding)	PHA composition (mol %)			T_g[b] (°C)	T_m[c] (°C)	Tensile strength (MPa)	Elongation at break (%)	M_w[d] (× 10^5)	M_w/ M_n[e]
	3HB	3HV	3HHx						
5 g/L Sodium propionate (48 h)	94	3	3	−3.8	129, 144	4	79	4.0	2.2
7 g/L Sodium valerate (48 h)	82	15	3	−2.3	147	6	84	3.3	2.6
1 g/L Sodium valerate (0 h)	69	24	7	−0.8	129, 139	20	710	4.0	2.0
9 g/L Sodium valerate (24 h)	66	32	2	−4.7	91, 148	5	282	4.6	2.2
7 g/L Sodium valerate (0 h)	14	85	1	−16.1	89	14.5	78	N.D.	N.D.

[a] Incubated for 72 h at 30 °C at 200 rpm in mineral medium containing 12 g/L of CPKO, initial pH 7.0.
[b] Glass transition temperature.
[c] Melting temperature.
[d] Molecular weight.
[e] Polydispersity index.
3HB, 3-hydroxybutyrate; 3HV,3-hydroxyvalerate; 3HHx, 3-hydroxyhexanoate; N.D., not detected.

4.5. MCL-PHA

Saponified palm kernel oil was first reported as a suitable substrate for the production of MCL-PHA using *P. putida* PGA1 by Tan and co-workers [15]. The MCL-PHA produced contained monomers with carbon numbers ranging from 6 to 14, with 3-hydroxyoctanoate (8C) as the prominent component. The PHA produced was compared with those produced by the major free fatty acids found in the palm kernel oil. The polymers were similar in property with those produced from lauric (C12:0), myristic (C14:0) and oleic (C18:1) acids. In a more recent study, production of MCL-PHA using the same strain and carbon source was investigated in fermenter experiment [16]. Ammonium was used as the growth limiting substrate. The amount of PHA accumulated and its specific production rate were found to be influenced by the residual ammonium concentration level in the culture medium. It was found that, by feeding low amount of ammonium to the culture, MCL-PHA yield and productivity were improved significantly. The MCL-PHA produced contained monomers with carbon number ranging from 8 to 16.

From these studies, *P. putida* PGA1 had been identified as a potential MCL-PHA producing microorganism from saponified palm kernel oil. The results obtained from both shaken-flasks and initial fermenter experiments suggest that this strain could be developed for large-scale MCL-PHA production systems.

5. Conclusion and Future Outlook

The increasing usage of finite resources in the past decades had led to their rapid exhaustion. Modernization and advancement in science and technology has increased demand for petroleum-based plastic materials. The durable nature of these plastics causes major concerns when these products enter the waste stream. The accumulation of plastic wastes in our environment is clearly visible and is worsening with time. Utilization of fossil carbon reserves will only increase the amount of green house gases in the environment. Ultimately, environmentally-friendly material derived from organic carbon sources are the more preferred and suitable alternative to fossil carbons. Even if we could not completely stop the utilization of fossil carbon supply, at least a reduction in its usage will help to reduce the carbon footprint of human activities.

In par with this, development of bio-based and biodegradable plastics has been given much attention lately. PHA had been identified as renewable and potential material for the production of biodegradable plastics. It is already being produced at industrial-scale and marketed in several countries. However, the cost of PHA is higher compared to conventional plastics mainly because of the carbon feedstocks and fermentation process used for the production of PHA. Efforts have been devoted to identify renewable and readily available raw material. In depth research in this aspect has led to the investigation of palm oil products for the production of various types of PHA. High yield production of PHA could be obtained from palm oil and its by-products. These bio-based raw materials are renewable and readily available in oil palm rich countries such as Malaysia. Through proper management of resources and sustainable usage, palm oil could be utilized for the production of value added materials such as PHA and biodiesel besides foodstuffs.

It is expected that the demand for palm oil will continue to rise as a result of increasing human consumption, thus, the expansion of oil palm plantation seems inevitable. Nevertheless, with proper land management and mitigation measures, deforestation and land wastage could be controlled and the biodiversity of ecosystems preserved. The future of PHA production from palm oil looks promising if an integrated system of PHA production with oil palm plantation and its milling industry could be designed and executed successfully. With all aims set for palm oil based PHA industry, sustainable production and utilization of palm oil will ensure its successful implementation.

Acknowledgments

Some of the results included in this chapter were obtained from research funded by Techno Fund provided by Ministry of Science, Technology and Innovation, Malaysia (MOSTI). Generous supply of palm oil products by Acidchem Int. Ltd. and Unitata Ltd. is gratefully acknowledged. K.B. acknowledges National Science Fellowship awarded by MOSTI for financial support. C.J.Y. acknowledges Malaysia Toray Science Foundation.

REFERENCES

[1] Sudesh, K. & Iwata, T. 2008. Sustainability of biobased and biodegradable plastics. *Clean-Soil, Air, Water, 36*, 433-442.

[2] Bohlmann, G. M. 2005. General Characteristics, Processability, Industrial Applications and Market Evolution of Biodegradable Polymers. In: *Handbook of Biodegradable Polymers* (edited by C. Bastioli). Pp. 183-218. Shawbury, Shrewsbury, Shropshire, UK: Rapra Technology Limited.

[3] Snell, K. D. & Peoples, O. P. 2009. PHA bioplastic: A value-added coproduct for biomass biorefineries. Biofuels, Bioproducts and Biorefining, *3(4)*, 456–467.

[4] Ningthoujam, D. S. 2010. Bioplastics: Environmental friendly biopolymers from microbes. http://www.kanglaonline.com. Accessed January 2010.

[5] Steinbüchel, A. 2001. Perspectives for biotechnological production and utilization of biopolymers: Metabolic engineering of polyhydroxyalkanoate biosynthesis pathways as a successful example. *Macromolecular Bioscience, 1(1)*, 1–24.

[6] Albuquerque, M. G. E., Eiroa, M., Torres, C., Nunes, B. R. & Reis, M. A. M. 2007. Strategies for the development of a side stream process for polyhydroxyalkanoate (PHA) production from sugar cane molasses. *Journal of Biotechnology, 130(4)*, 411–421.

[7] Khanna, S. & Srivastava, A. K. 2005. Recent advances in microbial polyhydroxyalkanoates. *Process Biochemistry, 40(2)*, 607–619.

[8] Lee, S. Y. 1996. Bacterial polyhydroxyalkanoates. *Biotechnology and Bioengineering, 49(1)*, 1–14.

[9] Lee, S. Y. & Choi, J. I. 1999. Production and degradation of polyhydroxyalkanoates in waste environment. *Waste Management, 19(2)*, 133–139.

[10] Kessler, B., Weusthuis, R., Witholt, B. & Eggink, G. 2001. Production of microbial polyesters: fermentation and downstream processes. *Advances in Biochemical Engineering/Biotechnology, 71*, 159–182.

[11] Kim, Y. & Lenz, R. 2001. Polyesters from Microorganisms. In: *Biopolyesters* (edited by Wolfgang Babel & A. Steinbüchel). Pp 51–79. New York, US: Springer.

[12] Tsuge, T. 2002. Metabolic improvements and use of inexpensive carbon sources in microbial production of polyhydroxyalkanoates. *Journal of Bioscience and Bioengineering, 94(6)*, 579–584.

[13] Akiyama, M., Tsuge, T. & Doi, Y. 2003. Environmental life cycle comparison of polyhydroxyalkanoates produced from renewable carbon resources by bacterial fermentation. *Polymer Degradation and Stability, 80(1)*, 183–194.

[14] Majid, M. I. A., Hori, K., Akiyama, M. & Doi, Y. 1994. Production of poly(3-hydroxybutyrate) from plant oils by *Alcaligenes* sp. *Studies in Polymer Science, 12*, 417–424.

[15] Tan, I. K. P., Sudesh Kumar, K., Theanmalar, M., Gan, S. N. & Gordon Iii, B. 1997. Saponified palm kernel oil and its major free fatty acids as carbon substrates for the production of polyhydroxyalkanoates in *Pseudomonas putida* PGA1. *Applied Microbiology and Biotechnology, 47(3)*, 207–211.

[16] Annuar, M. S. M., Tan, I. K. P., Ibrahim, S. & Ramachandran, K. B. 2007. Production of medium-chain-length poly(3-hydroxyalkanoates) from crude fatty acids mixture by *Pseudomonas putida*. *Food and Bioproducts Processing*, *85(2),* 104–119.

[17] Lee, W. H., Loo, C. Y., Nomura, C. T. & Sudesh, K. 2008. Biosynthesis of polyhydroxyalkanoate copolymers from mixtures of plant oils and 3-hydroxyvalerate precursors. *Bioresource Technology*, *99(15),* 6844–6851.

[18] Loo, C. Y., Lee, W. H., Tsuge, T., Doi, Y. & Sudesh, K. 2005. Biosynthesis and characterization of poly(3-hydroxybutyrate-*co*-3- hydroxyhexanoate) from palm oil products in a *Wautersia eutropha* mutant. *Biotechnology Letters*, *27(18),* 1405–1410.

[19] Bhubalan, K., Lee, W. H., Loo, C. Y., Yamamoto, T., Tsuge, T., Doi, Y. & Sudesh, K. 2008. Controlled biosynthesis and characterization of poly(3-hydroxybutyrate-*co*-3-hydroxyvalerate-*co*-3-hydroxyhexanoate) from mixtures of palm kernel oil and 3HV-precursors. *Polymer Degradation and Stability*, *93(1),* 17–23.

[20] Bhubalan, K., Rathi, D. N., Abe, H., Iwata, T. & Sudesh, K. 2010. Improved synthesis of P(3HB-*co*-3HV-*co*-3HHx) terpolymers by mutant *Cupriavidus necator* using the PHA synthase gene of *Chromobacterium* sp. USM2 with high affinity towards 3HV. *Polymer Degradation and Stability*, *95(8)*, 1436–1442.

[21] Yusoff, S. 2006. Renewable energy from palm oil - Innovation on effective utilization of waste. *Journal of Cleaner Production*, *14(1),* 87–93.

[22] MPOB. 2008. Malaysian Palm Oil Board. http://www.mpob.gov.my. Accessed January 2010.

[23] Nonato, R. V., Mantelatto, P. E. & Rossell, C. E. V. 2001. Integrated production of biodegradable plastic, sugar and ethanol. *Applied Microbiology and Biotechnology*, *57(1),* 1–5.

[24] Lemoigne, M. 1926. Products of dehydration and of polymerization of β-hydroxybutyric acid. *Bulletin des Societes Chimiques Biologie*, *8*, 770–782.

[25] Anderson, A. J. & Dawes, E. A. 1990. Occurrence, metabolism, metabolic role, and industrial uses of bacterial polyhydroxyalkanoates. *Microbiological Reviews*, *54(4),* 450–472.

[26] Steinbüchel, A. & Schlegel, H. G. 1989. Excretion of pyruvate by mutants of *Alcaligenes eutrophus*, which are impaired in the accumulation of poly(β-hydroxybutyric acid) (PHB), under conditions permitting synthesis of PHB. *Applied Microbiology and Biotechnology*, *31(2),* 168–175.

[27] Wallen, L. L. & Rohwedder, W. K. 1974. Poly-β-hydroxyalkanoate from activated sludge. *Environmental Science & Technology*, *8*, 576–579.

[28] Findlay, R. H. & White, D. C. 1983. Polymeric β-hydroxyalkanoates from environmental samples and *Bacillus megaterium*. *Applied Environmental Microbiology*, *45(1)*, 71–78.

[29] Steinbüchel, A. & Lütke-Eversloh, T. 2003. Metabolic engineering and pathway construction for biotechnological production of relevant polyhydroxyalkanoates in microorganisms. *Biochemical Engineering Journal*, *16(2),* 81–96.

[30] Tian, J., Song, H., Wu, B., Wang, Z. & Li, J. 2000. The construction of a recombined E. coli strain with stable and high production of poly(3-hydroxybutyrate-*co*-3-hydroxyvalerate). *Acta microbiologica Sinica*, *40(1)*, 26–31.

[31] Amirul, A. A., Syairah, S., Yahya, A. R. M., Azizan, M. N. M. & Majid, M. I. A. 2008. Synthesis of biodegradable polyesters by Gram negative bacterium isolated from Malaysian environment. *World Journal of Microbiology and Biotechnology*, *24(8)*, 1327–1332.

[32] Du, G., Si, Y. & Yu, J. 2001. Inhibitory effect of medium-chain-length fatty acids on synthesis of polyhydroxyalkanoates from volatile fatty acids by *Ralstonia eutropha*. *Biotechnology Letters*, *23(19)*, 1613–1617.

[33] Mansfield, D. A., Anderson, A. J. & Naylor, L. A. 1995. Regulation of PHB metabolism in Alcaligenes eutrophus. *Canadian Journal of Microbiology*, *41*, 44–49.

[34] Kahar, P., Tsuge, T., Taguchi, K. & Doi, Y. 2004. High yield production of polyhydroxyalkanoates from soybean oil by *Ralstonia eutropha* and its recombinant strain. *Polymer Degradation and Stability*, *83(1)*, 79–86.

[35] Chen, G. Q., Zhang, G., Park, S. J. & Lee, S. Y. 2001. Industrial scale production of poly(3-hydroxybutyrate-*co*-3-hydroxyhexanoate). *Applied Microbiology and Biotechnology*, *57(1-2)*, 50–55.

[36] Lee, S. Y. & Choi, J. I. 1998. Effect of fermentation performance on the economics of poly(3-hydroxybutyrate) production by *Alcaligenes latus*. *Polymer Degradation and Stability*, *59(1-3)*, 387–393.

[37] Kellerhals, M. B., Kessler, B. & Witholt, B. 1999. Development of a closed-loop control system for production of medium-chain-length poly(3-hydroxyalkanoates) (mcl-PHAs) from bacteria. *Macromolecular Symposia*, *144*, 385–389.

[38] Kellerhals, M. B., Kessler, B. & Witholt, B. 1999. Closed-loop control of bacterial high-cell-density fed-batch cultures: Production of mcl-PHAs by *Pseudomonas putida* KT2442 under single-substrate and cofeeding conditions. *Biotechnology and Bioengineering*, *65(3)*, 306–315.

[39] Choi, J. I. & Lee, S. Y. 1999. Factors affecting the economics of polyhydroxyalkanoate production by bacterial fermentation. *Applied Microbiology and Biotechnology*, *51(1)*, 13–21.

[40] Ren, Q., Ruth, K., Thony-Meyer, L. & Zinn, M. 2007. Process engineering for production of chiral hydroxycarboxylic acids from bacterial polyhydroxyalkanoates. *Macromolecular Rapid Communications*, *28(22)*, 2131–2136.

[41] Jung, Y. M. & Lee, Y. H. 2000. Utilization of oxidative pressure for enhanced production of poly-β-hydroxybutyrate and poly(3-hydroxybutyrate-3-hydroxyvalerate) in *Ralstonia eutropha*. *Journal of Bioscience and Bioengineering*, *90(3)*, 266–270.

[42] Simon-Colin, C., Raguénès, G., Costa, B. & Guezennec, J. 2008. Biosynthesis of medium chain length poly-3-hydroxyalkanoates by *Pseudomonas guezennei* from various carbon sources. *Reactive and Functional Polymers*, *68(11)*, 1534–1541.

[43] Tsuge, T., Yamamoto, T., Yano, K., Abe, H., Doi, Y. & Taguchi, S. 2009. Evaluating the ability of polyhydroxyalkanoate synthase mutants to produce P(3HB-*co*-3HA) from soybean oil. *Macromolecular Bioscience*, *9(1)*, 71–78.

[44] Haba, E., Vidal-Mas, J., Bassas, M., Espuny, M. J., Llorens, J. & Manresa, A. 2007. Poly 3-(hydroxyalkanoates) produced from oily substrates by *Pseudomonas aeruginosa* 47T2 (NCBIM 40044): Effect of nutrients and incubation temperature on polymer composition. *Biochemical Engineering Journal*, *35(2)*, 99–106.

[45] Bassas, M., Rodríguez, E., Llorens, J. & Manresa, A. 2006. Poly(3-hydroxyalkanoate) produced from *Pseudomonas aeruginosa* 42A2 (NCBIM 40045): Effect of fatty acid nature as nutrient. *Journal of Non-Crystalline Solids*, *352(21-22)*, 2259–2263.

[46] Poirier, Y., Dennis, D., Klomparens, K., Nawrath, C. & Sommerville, C. 1992. Perspectives on the production of polyhydroxyalkanoates in plants. *FEMS Microbiology Reviews*, *103(2-4)*, 237–246.

[47] Poirier, Y., Dennis, D. E., Klomparens, K. & Somerville, C. 1992. Polyhydroxybutyrate, a biodegradable thermoplastic, produced in transgenic plants. *Science*, *256(5056)*, 520–523.

[48] Matsumoto, K., Nagao, R., Murata, T., Arai, Y., Kichise, T., Nakashita, H., Taguchi, S., Shimada, H. & Doi, Y. 2005. Enhancement of poly(3-hydroxybutyrate-*co*-3-hydroxyvalerate) production in the transgenic *Arabidopsis thaliana* by the *in vitro* evolved highly active mutants of polyhydroxyalkanoate (PHA) synthase from *Aeromonas caviae*. *Biomacromolecules*, *6(4)*, 2126–2130.

[49] Matsumoto, K., Arai, Y., Nagao, R., Murata, T., Takase, K., Nakashita, H., Taguchi, S., Shimada, H. & Doi, Y. 2006. Synthesis of short-chain-length/medium-chain-length polyhydroxyalkanoate (PHA) copolymers in peroxisome of the transgenic *Arabidopsis thaliana* harboring the PHA synthase gene from *Pseudomonas* sp. 61-3. *Journal of Polymers and the Environment*, *14(4)*, 369–374.

[50] Nielsen, L. 2007. Polyhydroxyalkanoate production in sugarcane-recognizing temporospatial complexity. *Journal of Biotechnology*, *131(2)*, S28–S29.

[51] Sudesh, K., Abe, H. & Doi, Y. 2000. Synthesis, structure and properties of polyhydroxyalkanoates: biological polyesters. *Progress in Polymer Science*, *25*, 1503–1555.

[52] Wu, Q., Huang, H., Hu, G., Chen, J., Ho, K. P. & Chen, G. Q. 2001. Production of poly-3-hydroxybutrate by *Bacillus* sp. JMa5 cultivated in molasses media. *Antonie van Leeuwenhoek, International Journal of General and Molecular Microbiology*, *80(2)*, 111–118.

[53] Lee, W. H., Azizan, M. N. M. & Sudesh, K. 2004. Effects of culture conditions on the composition of poly(3-hydroxybutyrate-*co*-4-hydroxybutyrate) synthesized by *Comamonas acidovorans*. *Polymer Degradation and Stability*, *84(1)*, 129–134.

[54] Loo, C.Y. & Sudesh, K. 2007. Biosynthesis and native granule characteristics of poly(3-hydroxybutyrate-*co*-3-hydroxyvalerate) in *Delftia acidovorans*. *International Journal of Biological Macromolecules*, *40(5)*, 466–471.

[55] Jung, Y. M., Park, J. S. & Lee, Y. H. 2000. Metabolic engineering of *Alcaligenes eutrophus* through the transformation of cloned *phbCAB* genes for the investigation of the regulatory mechanism of polyhydroxyalkanoate biosynthesis. *Enzyme and Microbial Technology*, *26(2-4)*, 201–208.

[56] Sudesh, K., Fukui, T. & Doi, Y. 1998. Genetic analysis of *Comamonas acidovorans* polyhydroxyalkanoate synthase and factors affecting the incorporation of 4-hydroxybutyrate monomer. *Applied and Environmental Microbiology*, *64(9)*, 3437–3443.

[57] Holmes, P. A. 1985. Applications of PHB - A microbially produced biodegradable thermoplastic. *Physics in Technology*, *16(1)*, 32–36.

[58] Green, P. R., Kemper, J., Schechtman, L., Guo, L., Satkowski, M., Fiedler, S., Steinbüchel, A. & Rehm, B. H. 2002. Formation of short chain length/medium chain length polyhydroxyalkanoate copolymers by fatty acid beta-oxidation inhibited *Ralstonia eutropha*. *Biomacromolecules, 3(1)*, 208–213.

[59] Shimamura, E., Scandola, M. & Doi, Y. (1994). Microbial synthesis and characterization of poly(3-hydroxybutyrate-*co*-3-hydroxypropionate). *Macromolecules, 27*, 4429–4435.

[60] Abe, H., Doi, Y., Fukushima, T. & Eya, H. 1994. Biosynthesis from gluconate of a random copolyester consisting of 3-hydroxybutyrate and medium-chain-length 3-hydroxyalkanoates by *Pseudomonas* sp. 61-3. *International Journal of Biological Macromolecules, 16(3)*, 115–119.

[61] Sandoval, A., Arias-Barrau, E., Bermejo, F., Canedo, L., Naharro, G., Olivera, E. & Luengo, J. 2005. Production of 3-hydroxy-*n*-phenylalkanoic acids by a genetically engineered strain of *Pseudomonas putida*. *Applied Microbiology and Biotechnology, 67(1)*, 97–105.

[62] Ashby, R., Solaiman, D. & Foglia, T. 2002. The synthesis of short- and medium-chain-length poly(hydroxyalkanoate) mixtures from glucose- or alkanoic acid-grown *Pseudomonas oleovorans*. *Journal of Industrial Microbiology and Biotechnology, 28(3)*, 147–153.

[63] Madison, L. L. & Huisman, G. W. 1999. Metabolic engineering of poly(3-hydroxyalkanoates): From DNA to plastic. *Microbiology and Molecular Biology Reviews, 63(1)*, 21–53.

[64] Luengo, J. M., García, B., Sandoval, A., Naharro, G. & Olivera, E. R. 2003. Bioplastics from microorganisms. *Current Opinion in Microbiology, 6(3)*, 251–260.

[65] Kobayashi, T., Uchino, K., Abe, T., Yamazaki, Y. & Saito, T. 2005. Novel intracellular 3-hydroxybutyrate-oligomer hydrolase in *Wautersia eutropha* H16. *Journal of Bacteriology, 187(15)*, 5129–5135.

[66] Doi, Y. 1990. Microbial polyesters. Pp. 156. New York: VCH.

[67] Jendrossek, D., Schirmer, A. & Schlegel, H. G. 1996. Biodegradation of polyhydroxyalkanoic acids. *Appl Microbiol Biotechnol, 46(5-6)*, 451–463.

[68] Abou-Zeid, D. M., Müller, R. J. & Deckwer, W. D. 2001. Degradation of natural and synthetic polyesters under anaerobic conditions. *Journal of Biotechnology, 86(2)*, 113–126.

[69] Mergaert, J., Webb, A., Anderson, C., Wouters, A. & Swings, J. 1993. Microbial degradation of poly(3-hydroxybutyrate) and poly(3- hydroxybutyrate-*co*-3-hydroxyvalerate) in soils. *Applied and Environmental Microbiology, 59(10)*, 3233–3238.

[70] Wang, Y. W., Mo, W., Yao, H., Wu, Q., Chen, J. & Chen, G. Q. 2004. Biodegradation studies of poly(3-hydroxybutyrate-*co*-3-hydroxyhexanoate). *Polymer Degradation and Stability, 85(2)*, 815–821.

[71] Sridewi, N., Bhubalan, K. & Sudesh, K. 2006. Degradation of commercially important polyhydroxyalkanoates in tropical mangrove ecosystem. *Polymer Degradation and Stability, 91(12)*, 2931–2940.

[72] Yew, S. P., Tang, H. Y. & Sudesh, K. 2006. Photocatalytic activity and biodegradation of polyhydroxybutyrate films containing titanium dioxide. *Polymer Degradation and Stability, 91(8)*, 1800–1807.

[73] Kasuya, K., Doi, Y. & Yao, T. 1994. Enzymatic degradation of poly[R)-3-hydroxybutyrate] by *Comamonas testosteroni* ATSU of soil bacterium. *Polymer Degradation and Stability, 45(3)*, 379–386.
[74] Mukai, K., Yamada, K. & Doi, Y. 1994. Efficient hydrolysis of polyhydroxyalkanoates by *Pseudomonas stutzeri* YM1414 isolated from lake water. *Polymer Degradation and Stability, 43(3)*, 319–327.
[75] Pouton, C. W. & Akhtar, S. 1996. Biosynthetic polyhydroxyalkanoates and their potential in drug delivery. *Advanced Drug Delivery Reviews, 18(2)*, 133–162.
[76] Williams, S. F. & Martin, D. P. 2002. Applications of PHAs in medicine and pharmacy. *Biopolymers, 3*, 91–128.
[77] Zinn, M., Witholt, B. & Egli, T. 2001. Occurrence, synthesis and medical application of bacterial polyhydroxyalkanoate. *Advanced Drug Delivery Reviews, 53(1)*, 5–21.
[78] Kusaka, S., Iwata, T. & Doi, Y. 1998. Microbial synthesis and physical properties of ultra-high-molecular-weight poly[*R*)-3-hydroxybutyrate]. *Journal of Macromolecular Science - Pure and Applied Chemistry, 35(2)*, 319–335.
[79] Kusaka, S., Iwata, T. & Doi, Y. 1999. Properties and biodegradability of ultra-high-molecular-weight poly[*R*)-3-hydroxybutyrate] produced by a recombinant *Escherichia coli. International Journal of Biological Macromolecules, 25(1-3)*, 87–94.
[80] Aoyagi, Y., Doi, Y. & Iwata, T. 2003. Mechanical properties and highly ordered structure of ultra-high-molecular-weight poly[*R*)-3-hydroxybutyrate] films: Effects of annealing and two-step drawing. *Polymer Degradation and Stability, 79(2)*, 209–216.
[81] Byrom, D. 1992. Production of poly-β-hydroxybutyrate: poly-β-hydroxyvalerate copolymers. *FEMS Microbiology Reviews, 103*, 247–250.
[82] Holmes, P. A. 1988. Biologically produced (*R*)-3-hydroxyalkanoate polymers and copolymers In: *Developments in Crystalline Polymers-2 Polymer* (edited by D. C. Bassett). Pp. 1–65. London, UK: Elsevier.
[83] Holmes, P. A., Wright, L. F. & Collins, S. H. 1981. European Patent Application EP 052.459.
[84] Suriyamongkol, P., Weselake, R., Narine, S., Moloney, M. & Shah, S. 2007. Biotechnological approaches for the production of polyhydroxyalkanoates in microorganisms and plants - A review. *Biotechnology Advances, 25(2)*, 148–175.
[85] Madden, L. A., Anderson, A. J. & Asrar, J. 1998. Synthesis and characterization of poly(3-hydroxybutyrate) and poly(3-hydroxybutyrate-*co*-3-hydroxyvalerate) polymer mixtures produced in high-density fed-batch cultures of *Ralstonia eutropha* (*Alcaligenes eutrophus*). *Macromolecules, 31(17)*, 5660–5667.
[86] Choi, J. I. & Lee, S. Y. 1999. Production of poly(3-hydroxybutyrate) [P(3HB] with high P(3HB) content by recombinant *Escherichia coli* harboring the *Alcaligenes latus* P(3HB) biosynthesis genes and the *E. coli ftsZ* gene. *Journal of Microbiology and Biotechnology, 9(6)*, 722–725.
[87] Shang, L., Yim, S. C., Park, H. G. & Chang, H. N. 2004. Sequential feeding of glucose and valerate in a fed-batch culture of *Ralstonia eutropha* for production of poly(hydroxybutyrate-*co*-hydroxyvalerate) with high 3-hydroxyvalerate fraction. *Biotechnology Progress, 20(1)*, 140–144.
[88] Matsusaki, H., Abe, H. & Doi, Y. 2000. Biosynthesis and properties of poly(3-hydroxybutyrate-*co*-3-hydroxyalkanoates) by recombinant strains of *Pseudomonas* sp. 61-3. *Biomacromolecules, 1(1)*, 17–22.

[89] Matsusaki, H., Manji, S., Taguchi, K., Kato, M., Fukui, T. & Doi, Y. 1998. Cloning and molecular analysis of the poly(3-hydroxybutyrate) and poly(3-hydroxybutyrate-*co*-3-hydroxyalkanoate) biosynthesis genes in *Pseudomonas* sp. strain 61-3. *Journal of Bacteriology*, *180(24)*, 6459–6467.

[90] de Smet, M. J., Eggink, G., Witholt, B., Kingma, J. & Wynberg, H. 1983. Characterization of intracellular inclusions formed by *Pseudomonas oleovorans* during growth on octane. *Journal of Bacteriology*, *154(2)*, 870–878.

[91] Huijberts, G. N., Eggink, G., de Waard, P., Huisman, G. W. & Witholt, B. 1992. *Pseudomonas putida* KT2442 cultivated on glucose accumulates poly(3-hydroxyalkanoates) consisting of saturated and unsaturated monomers. *Applied and Environmental Microbiology*, *58(2)*, 536–544.

[92] Lee, E. Y. & Choi, C. Y. 1995. Gas chromatography-mass spectrometric analysis and its application to a screening procedure for novel bacterial polyhydroxyalkanoic acids containing long chain saturated and unsaturated monomers. *Journal of Fermentation and Bioengineering*, *80(4)*, 408–414.

[93] Fukui, T. & Doi, Y. 1997. Cloning and analysis of the poly(3-hydroxybutyrate-*co*-3-hydroxyhexanoate) biosynthesis genes of *Aeromonas caviae*. *Journal of Bacteriology*, *179(15)*, 4821–4830.

[94] Lee, S. H., Oh, D. H., Ahn, W. S., Lee, Y., Choi, J. I. & Lee, S. Y. 2000. Production of poly(3-hydroxybutyrate-*co*-3-hydroxyhexanoate) by high-cell-density cultivation of *Aeromonas hydrophila*. *Biotechnology and Bioengineering*, *67(2)*, 240–244.

[95] Kobayashi, G., Shiotani, T., Shima, Y. & Doi, Y. 1994. Biosynthesis and charcterization of poly(3-hydroxybutyrate-*co*-3-hydroxyalkanoate) from oils and fats by *Aeromonas* sp. OL-338 and *Aeromonas* sp. FA440. In: *Biodegradable Plastics and Polymers* (edited by Y. Doi & K. Fukuda). Pp. 410–416. Amsterdam, The Netherlands: Elsevier.

[96] Park, S. J., Ahn, W. S., Green, P. R. & Lee, S. Y. 2001. Biosynthesis of poly(3-hydroxybutyrate-*co*-3-hydroxyvalerate-*co*-3-hydroxyhexanoate) by metabolically engineered *Escherichia coli* strains. *Biotechnology and Bioengineering*, *74(1)*, 81–86.

[97] Zhao, W. & Chen, G. Q. 2007. Production and characterization of terpolyester poly(3-hydroxybutyrate-*co*-3-hydroxyvalerate-*co*-3-hydroxyhexanoate) by recombinant *Aeromonas hydrophila* 4AK4 harboring genes *phaAB*. *Process Biochemistry*, *42(9)*, 1342–1347.

[98] Bhubalan, K., Yong, K. H., Kam, Y. C. & K., S. 2010. Cloning and expression of the PHA synthase gene from a locally isolated *Chromobacterium* sp. USM2. *Malaysian Journal of Microbiology*, *6(1)*, 81–90.

[99] Fukui, T., Kichise, T., Yoshida, Y. & Doi, Y. 1997. Biosynthesis of poly(3-hydroxybutyrate-*co*-3-hydroxyvalerate-*co*-3-hydroxyheptanoate) terpolymers by recombinant *Alcaligenes eutrophus*. *Biotechnology Letters*, *19(11)*, 1093–1097.

[100] Fukui, T. & Doi, Y. 1998. Efficient production of polyhydroxyalkanoates from plant oils by *Alcaligenes eutrophus* and its recombinant strain. *Applied Microbiology and Biotechnology*, *49(3)*, 333–336.

[101] Zazali, A. & Tan, I. K. P. 2005. Isolation of palm oil-utilising, polyhydroxyalkanoate (PHA)-producing bacteria by an enrichment technique. *Bioresource Technology*, *96*, 1229–1234.

[102] Kek, Y. K., Lee, W. H. & Sudesh, K. 2008. Efficient bioconversion of palm acid oil and palm kernel acid oil to poly(3-hydroxybutyrate) by *Cupriavidus necator*. *Canadian Journal of Chemistry*, *86(6)*, 533–539.

[103] Madden, L. A., Anderson, A. J., Shah, D. T. & Asrar, J. 1999. Chain termination in polyhydroxyalkanoate synthesis: Involvement of exogenous hydroxy-compounds as chain transfer agents. *International Journal of Biological Macromolecules*, *25(1-3)*, 43–53.

[104] Doi, Y., Tamaki, A., Kunioka, M. & Soga, K. 1988. Production of copolyesters of 3-hydroxybutyrate and 3-hydroxyvalerate by *Alcaligenes eutrophus* from butyric and pentanoic acids. *Applied Microbiology and Biotechnology*, *28(4-5)*, 330–334.

[105] Doi, Y., Kitamura, S. & Abe, H. 1995. Microbial synthesis and characterization of poly(3-hydroxybutyrate-*co*-3-hydroxyhexanoate). *Macromolecules*, *28(14)*, 4822–4828.

[106] Tsuge, T., Saito, Y., Kikkawa, Y., Hiraishi, T. & Doi, Y. 2004. Biosynthesis and compositional regulation of poly[3-hydroxybutyrate)-*co*-(3-hydroxyhexanoate] in recombinant *Ralstonia eutropha* expressing mutated polyhydroxyalkanoate synthase genes. *Macromolecular Bioscience*, *4(3)*, 238–242.

[107] Taguchi, S. & Doi, Y. 2004. Evolution of polyhydroxyalkanoate (PHA) production system by "enzyme evolution": Successful case studies of directed evolution. *Macromolecular Bioscience*, *4(3)*, 145–156.

[108] Brandl, H., Knee, E. J., Fuller, R. C., Gross, R. A. & Renz, R. W. 1989. Ability of the phototrophic bacterium *Rhodospirillum rubum* to produce various poly(β-hydroxyalkanoates): potential carbon sources for biodegradable polyester. *International Journal of Biological Macromolecules*, *11*, 49–55.

[109] Liebergesell, M., Hustede, E., Timm, A., Steinbuchel, A., Fuller, R. C., Lenz, R. W. & Schlegel, H. G. 1991. Formation of poly(3-hydroxyalkanoates) by phototrophic and chemolithotrophic bacteria. *Archives of Microbiology*, *155(5)*, 415–421.

[110] Abe, H., Kikkawa, Y., Aoki, H., Akehata, T., Iwata, T. & Doi, Y. 1999. Crystallization behavior and thermal properties of melt-crystallized poly[*R*)-3-hydroxybutyric acid-*co*-6-hydroxyhexanoic acid] films. *International Journal of Biological Macromolecules*, *25(1-3)*, 177–183.

[111] Wang, Y., Yamada, S., Asakawa, N., Yamane, T., Yoshie, N. & Inoue, Y. 2001. Comonomer compositional distribution and thermal and morphological characteristics of bacterial poly(3-hydroxybutyrate-*co*-3-hydroxyvalerate)s with high 3-hydroxyvalerate content. *Biomacromolecules*, *2(4)*, 1315–1323.

PART 2. ALGAE AS A SOURCE OF RAW MATERIAL

In: Bioprocess Sciences and Technology
Editor: Min-Tze Liong

ISBN 978-1-61122-950-9

Chapter 4

MICROALGAE AS A FEEDSTOCK FOR THE PRODUCTION OF BIOFUELS: MICROALGAL BIOCHEMISTRY, ANALYTICAL TOOLS, AND TARGETED BIOPROSPECTING

L. G. Elliott[1], V. H. Work[1], P. Eduafo[2], R. Radakovits[2], R. E. Jinkerson[2], A. Darzins[3] and M. C. Posewitz[2,*]

[1]Division of Environmental Science and Engineering, Colorado School of Mines, 1500 Illinois St., Golden, Colorado 80401, USA

[2]Department of Chemistry and Geochemistry, Colorado School of Mines, 1500 Illinois St., Golden, Colorado 80401, USA

[3]National Renewable Energy Laboratory, 1617 Cole Blvd, Golden, Colorado 80401, USA

ABSTRACT

There are currently intensive research efforts aimed at increasing the accumulation of lipids, alcohols, hydrocarbons, H_2, polysaccharides, and other energy storage compounds in algae for use as renewable biofuels. Photosynthetic microorganisms (PSMs), more commonly known as cyanobacteria and algae, have emerged as promising platforms for the production of renewable biofuels. Compared to terrestrial crops, PSMs are typically more efficient solar collectors, use less or no land[1], can be converted into liquid fuels including JP-8 and other military grade fuels, are arguably more economically viable than lignocellulosic biomass, and offer other secondary uses that fossil fuels do not provide. Additionally, PSMs have evolved unique metabolic capabilities that allow the production of a greater diversity of biofuels. Their rapid generation times and extensive natural diversity, along with established genetic techniques, make water-oxidizing PSMs particularly well suited for the physiological and genetic manipulations required to optimize the conversion of sunlight, water, and CO_2 into forms of chemical energy that

* Corresponding author: *M. C. Posewitz* Email: mposewit@mines.edu

can be used as transportation fuels. Recently, many improvements have been realized, including increased lipid and carbohydrate production, improved H_2 yields, and the diversion of central metabolic intermediates into fungible biofuels. The application of genetic engineering to enhance biofuels production in PSMs is advancing rapidly with the development of more sophisticated algal genetic manipulation techniques. It is likely that many of these advances can be extended into industrially relevant organisms in order to improve algae and cyanobacteria as a feedstock for the production of biohydrogen, starch-derived alcohols, diesel fuel surrogates, and/or alkanes.

1. Introduction

As the use of petroleum-based fuels has grown, so have concerns over the environmental effects of green house gas (GHG) emissions, increased worldwide demand for energy, and international energy security. To ameliorate these concerns, there has been a recent surge in renewable energy research and technology development. One promising renewable energy scenario of considerable interest involves employing microalgae, which are an extraordinarily diverse group of photosynthetic microorganisms, as biological solar cells to capture and convert the photon energy of sunlight into an array of chemicals that can be readily harvested, transported, stored, and seamlessly integrated into the current fuel distribution infrastructure. This chapter is intended to (a) discuss the various microalgal chemical intermediates, especially lipids, which can be converted into biofuels, (b) present the underlying biochemistry behind the synthesis of these intermediates, and (c) highlight the analytical tools and targeted bioprospecting efforts used for the development of these organisms into a feedstock for the production of biofuels.

The recent surge of interest in developing microalgae for biotechnology follows many years of successful exploitation of a small number of well-studied microalgae as sources of bioactive compounds, nutraceuticals, human and animal food supplements, vitamins, polyunsaturated fatty acids, polysaccharides, proteins, and fluorescent pigments among others [1, 2]. Demonstrated to be more efficient than most terrestrial plants at oxygenic photosynthesis, in which photon energy is harvested to generate a chemical energy potential, microalgae fix inorganic carbon into renewable, energy-rich hydrocarbon stores [3, 4]. As such, microalgae are capable of producing a variety of chemical intermediates, many of which can be readily converted into biofuels: lipids, hydrocarbons, polyglucans, alcohols, terpenes, terpenoids, and H_2.

2. Microalgal Carbohydrates and Hydrocarbons

Microalgae predominantly produce two bioenergy carriers that are applicable for biofuel production: carbohydrates and hydrocarbons. Carbohydrates have long been used for the production of renewable biofuels through their conversion to alcohols and hydrocarbons, such as terpenes, terpenoids, and alkanes that can be used to produce diesel and gasoline fuel surrogates.

2.1. Starch Synthesis

Microalgae store glucans in a variety of distinct polyglucan macromolecules. The most common glucan found in nature, D-glucose, is polymerized in algae to serve as a transient energy store (e.g. glycogen and starch) or in structural polymers such as cellulose [5]. Chlorophyta (green algae), Dinophyta (dinoflagellates), Glaucophyta, and Rhodophyta (red algae) store glucans in α-1,4 and α-1,6 glycosidic linkages [6], but Phaeophyceae (brown algae) and Bacillariophyceae (diatoms) store glucans in β-1,3 and β-1,6 linkages [7]. Starch synthesis in Chlorophyta is the most well characterized algal polyglucan biosynthetic pathway due to the extensive study of the green alga *Chlamydomonas reinhardtii* [6, 8]. In green algae, glucans are stored in linear α-1,4 and branched α-1,6 glycosidic linkages that make up insoluble starch granules. These semi-crystalline granules are composed of two fractions: amylopectin (70-90%) and amylose (10-30%). Amylopectin is semi-crystalline and moderately branched (5% α-1,6 linkages), whereas amylase is a linear amorphous polyglucan with very few α-1,6 branches and is synthesized by granule-bound starch synthase I (GBSSI) [9]. Semi-crystalline amylopectin is a more complex molecule that consists of amorphous branching regions and crystalline regions of linear glucans arranged into double helixes. Amylopectin synthesis requires more enzymatic processing than amylose and is thought to follow the glucan trimming model proposed by Ball and coworkers [10, 11].

Starch synthesis (Figure 1) begins when fructose 6-phosphate from photosynthetic carbon fixation is converted to glucose 6-phosphate by a phosphoglucose isomerase (PGI1). Next, glucose 6-phosphate is converted to glucose 1-phosphate by a plastidial phosphoglucomutase (PGM). ADP-glucose pyrophosphorylase (AGPase) then catalyzes a major rate-controlling step of polysaccharide synthesis in plants: the reaction of glucose 1-phosphate with ATP, resulting in ADP-glucose and pyrophosphate. AGPase is a heterotetramer comprised of large regulatory and small catalytic subunits. It is allosterically activated by 3-phosphoglyceric acid (3-PGA), linking starch synthesis to photosynthesis [12]. This enzyme is inhibited by orthophosphate, preventing the storage of glucans during times of cellular energy debt.

The elongation of starch molecules proceeds when starch synthases (SS) transfer the glucose from the glycosylnucleotide ADP-glucose to the non-reducing end of an α-1,4 linked glucan, synthesizing a crystalline layer of the growing starch granule [13]. When enough α-1,4 linked glucans are assimilated, a branching enzyme (BE) hydrolyzes an α-1,4 linkage of a nearby donor glucan chain which is transferred to the accepter chain via a new branching α-1,6 glycosidic bond. These newly formed amylopectin branches are disordered and loosely packed. A proofreading step is needed to form more optimally packed starch structures. Isoforms of the debranching enzymes (DBE) pullulanase and isoamylase hydrolyze tightly and loosely packed α-1,6 linkages respectively, resulting in a tightly packed amorphous region of α-1,6 linkages that are ready to be elongated by starch synthases [11, 14]. This process is repeated, creating alternating layers of crystalline and amorphous regions, forming the semi-crystalline starch granule.

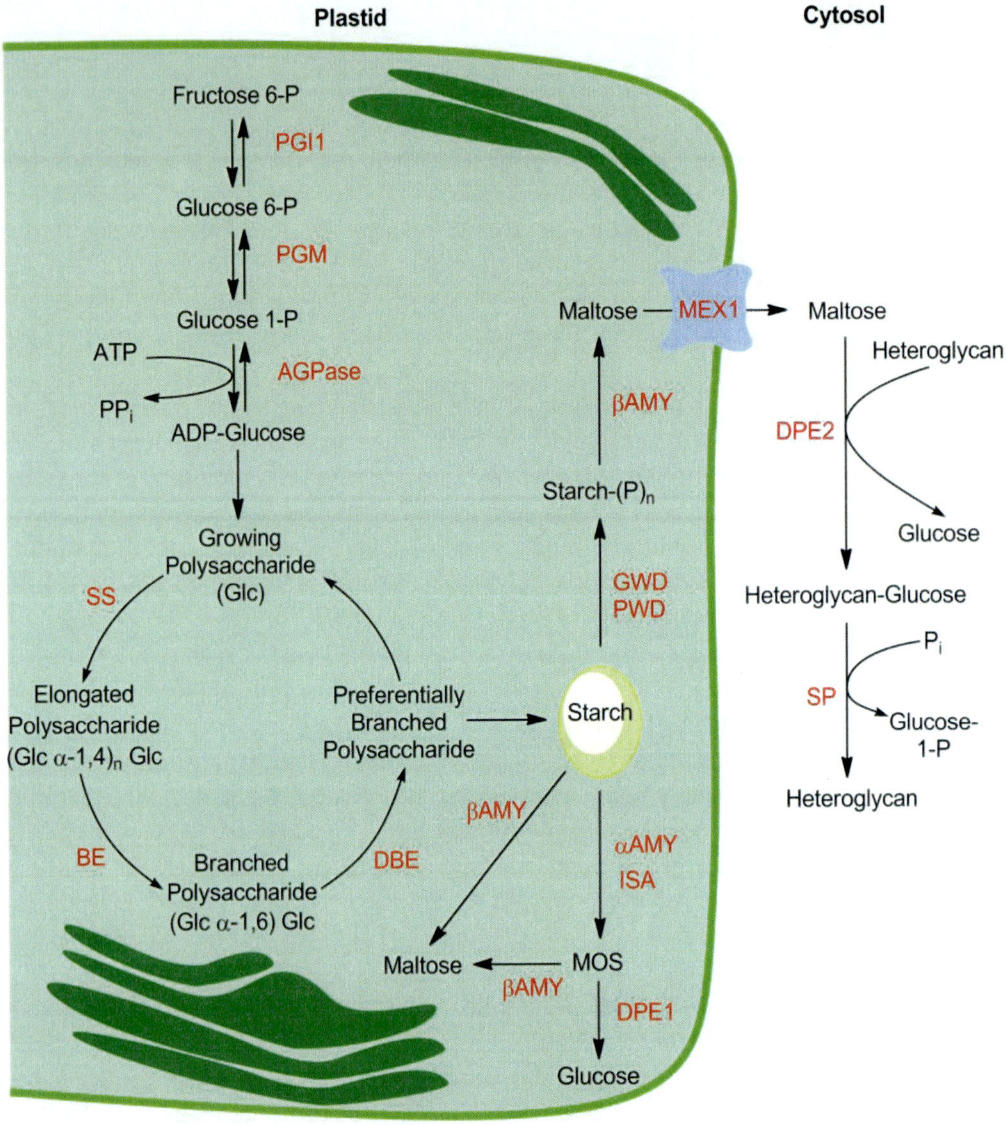

Figure 1. Overview of the metabolites and pathways involved in microalgal starch biosynthesis and degradation (black). Enzymes shown in red. During amylopectin synthesis glucans are added to the non-reducing end of the growing polysaccharide by α-1,4 glycosidic linkages. Once elongated, a branching enzyme targets the ends of the growing polysaccharides by forming new α-1,6 glycosidic linkages. Some of these branches are trimmed preferentially, and this process is repeated until a starch granule is formed. Hydrolytic and phosphorolytic [Starch-$(P)_n$] degradation pathways are shown. αAMY, α-amylase; AGPase, ADP-glucosepyrophosphorylase; βAMY, β-amylases; BE, branching enzymes; DBE, debranching enzymes; DEP, disproportionating enzyme (1 and 2) α-1,4glucanotransferase; Glc, glucose; GWD, glucan-water dikinases; ISA, isoamylases; MEX1, maltose transporter; MOS, malto-oligosaccharides; PGM, plastidial phosphoglucomutase; PGI1, phosphoglucose isomerase; P, phosphate; Pi, inorganic phosphate; PPi, pyrophosphate; PWD, phosphoglucan-water dikinases; SS, starch synthases.

2.2. Starch Degradation

Starch catabolism in algae is largely unknown [6], but putative degradation mechanisms have been inferred from known pathways in other plants such as *Arabidopsis thaliana* [15]. Hydrolytic and phosphorolytic starch degradation mechanisms are thought to exist in green algae. Hydrolytic starch degradation proceeds when the semi-crystalline glucans undergo enzymatic hydrolysis at the insoluble starch granule surface. In *Arabidopsis*, this is catalyzed by the chloroplast-targeted α-amylases (αAMY) [15]. Starch can also be degraded in *Arabidopsis* via phosphorolytic mechanisms even with all three α-amylases are knocked out [16]. Phosphorolytic degradation is initiated by glucan-water dikinases (GWD) which catalyze the transfer of β-phosphate in ATP to the C-6 position of a glucan in a crystalline amylopectin double helix [17]. Glucans can also be phosphorylated by the phosphoglucan-water dikinases (PWD) at the C-3 position. These phosphorylated glucans are thought to disrupt the crystalline structure of the amylopectin helices to allow access to glucan-metabolizing enzymes. Several approaches of attenuating starch degradation have been proposed to increase algal starch levels [18]. Knocking out key enzymes, such as GWD or PWD, may limit algal starch degradation and lead to a starch accumulation phenotype.

2.3. Terpenes and Terpenoids

Terpenes and terpenoids are a large class of compounds synthesized by algae that could be used for biofuel production. Terpenes are hydrocarbons, while terpenoids, also known as isoprenoids, are similar to terpenes except they contain functional groups. Two isoprene building blocks are used for terpene and terpenoid synthesis: isopentyldiphosphate (IPP) and dimethylallyl diphosphate (DMAPP). They are produced via the mevalonate or the non-mevalonate pathways from acetyl-CoA and pyruvate precursors, respectively. These activated isoprenes are then used to build terpene and terpenoids of various sizes and configurations. Terpenes follow the general molecular formula of $(C_5H_8)_n$, where n is the number of linked isoprenes [19, 20].

2.4. Alcohols

Traditionally, ethanol biofuels have been produced from the fermentation of food starches and sugars. Algal starches also have been shown to be a suitable feedstock for ethanol production via yeast fermentation [21]. Many algae have native fermentative metabolic pathways that produce ethanol and other alcohols; however, directly coupling ethanol production to photosynthetic carbon fixation would be preferred. This has been demonstrated in cyanobacteria where the introduction of pyruvate decarboxylase and alcohol dehydrogenase from *Zymomonas mobilis* has created a pathway for ethanol biosynthesis [22, 23]. Other pathways for the direct, non-fermentative production of higher alcohols has been demonstrated in *E. coli* and could be employed in algae. These approaches utilize native amino acid biosynthetic pathways to produce 1-propanol, 1-butanol, 2-methyl-1-butanol, isobutanol, and phenylethanol [24, 25].

2.5. *Botryococcus* Hydrocarbons

Botryococcus braunii is an oleaginous microalga that has been well studied and produces a variety of hydrocarbons including alkanes, fatty acids, terpenes, and terpenoids. Several strains of *B. braunii* are known and are classified by the type of hydrocarbons they produce. For example, the Race A strain produces dienes and trienes from C_{23} to C_{33}; the Race B strain produces very long chain triterpenoid hydrocarbons, C_{30}–C_{37} botryococcenes, and C_{31}–C_{34} methylated squalenes; and the Race L strain produces lycopadiene [26]. Race A has been observed to accumulate a hydrocarbon content of up to 61% of their dry weight, while Race B and L have been noted to accumulate hydrocarbons up to 40% and 8% dry weight, respectively [26].

3. Microalgal Hydrogen Production

Many eukaryotic microalgae and cyanobacteria have the ability produce H_2 either from dark, fermentative pathways coupled to saccharide catabolism or in the light from the electron transport chain at the level of ferredoxin [27-29]. It is now clear that many microalgae can withstand extended periods of anoxia and that hydrogenase activity is an integral component of this metabolism [30]. This anaerobic metabolism supports the secretion of alcohols and organic acids, as well as H_2 production during anaerobiosis [27, 31]. Currently, the most significant hurdles hindering H_2-production yields include (a) the sensitivity of hydrogenases to O_2, which prevents sustained H_2-photoproduction coupled to the photosynthetic electron transport chain [32], (b) the preferential use of competing metabolic pathways during aerobic metabolism, especially CO_2 fixation [33], and (c) the Thauer limit, which restricts dark, fermentative H_2 production to 4 moles of H_2 per mole of glucose [34]. To date, the use of sulfur deprivation remains the most effective means to sustain H_2 photoproduction [35]. This physiological stress attenuates photosynthesis to levels below respiration, allowing the establishment of anaerobiosis in sealed cultures and residual photosynthetic activity to provide reductant to the hydrogenase enzyme. Genetic techniques have been applied with the aim of increasing H_2-photoproduction activity by decreasing light harvesting antennae size, inhibiting state transitions, manipulating competing pathways, and engineering the hydrogenase [27-29]. Physiological, genetic, and biochemical approaches have rapidly advanced our understanding of H_2 metabolism and enzyme maturation in green microalgae, and several strategies are emerging to further advance our ability to optimize H_2 production in these eukaryotic organisms.

4. Oleaginous Microalgae

It is well known that microalgae are capable of producing various storage products that are of potential commercial value, and certain oleaginous species have been documented to produce high levels of non-polar triacylglycerols (TAGs) [3]. As a result, many research efforts have focused on utilizing these organisms as a biofuel feedstock and developing methods to readily harvest lipids [2, 36-38]. It is also well documented that some species of microalgae, particularly under environmental stress, can accumulate up to 50 percent of their dry weight as neutral storage lipids. The level of TAG accumulation achievable appears to be dependent on both environmental conditions and the metabolism of individual species [39]. Notably, *Chlamydomonas* has been shown to accumulate TAGs in the laboratory during nutrient deprivation and under high light [40].

Algal lipids can be harvested by various techniques and readily converted into diverse biofuels [41]. Many well-studied microalgae have been shown to produce high levels of lipids and theoretical calculations of maximum algal oil production from these organisms offer strong support that in best case scenarios they are capable of outcompeting traditional oilseed crops in terms of oil productivity [42]. However, as not all species are oleaginous, lipid productivity is a critical factor to consider for species selection, and maximizing lipid productivity is essential to the development of an economically viable algal biofuels industry [36]. Consequently, many current research efforts are directed towards genetically engineering well-characterized microalgae to optimize TAG production, which requires a significant investment of time and resources. Alternatively, some of the hurdles encountered in genetic optimization procedures can be bypassed, or at least lowered, by taking advantage of natural selection to identify native strains of microalgae that already possess more efficient metabolic traits desirable for commercial feedstock development.

4.1. Consideration of Carbon Source in Microalgal Lipid Accumulation

As evidenced by their broad distribution, microalgae have adapted biochemical mechanisms to withstand extended periods of unfavorable conditions such as nutrient deprivation, high light conditions, and drought. To a large extent, these adaptations manifest into a diverse composition of cellular lipids which are critical for the ability to survive a broad range of dynamic environmental conditions [3, 43]. As the environment becomes less favorable, many microalgae begin to accumulate lipids as a consequence of a limited supply of key nutrients. This likely occurs by two possible mechanisms, the first of which would entail *de novo* lipid synthesis by which extracellular carbon is actively fixed and stored in the form of lipids. The second mechanism would involve a regulatory pathway that when activated causes the termination of *de novo* carbon fixation and initiation of intracellular reorganization of previously fixed carbon. Here, carbon would be actively scavenged from various non-essential cellular components for biochemical conversion into energy-dense storage products such as TAGs. It is conceivable that a combination of the two proposed pathways occurs, although from the feedstock perspective, productivity would be greater if a constitutive lipid trigger could be found.

4.2. Algal Lipid Biochemistry

To narrow the range of biomolecules classified as lipids, this term will be used to reference fatty acid-containing or otherwise highly non-polar molecules that are associated with energy storage, cell membrane structure, and signaling. There have been relatively few recent investigations of microalgal lipid biochemistry. However, sequence homology comparisons of genes involved in lipid metabolism in *Arabidopsis* and putative genes in

Chlamydomonas indicate that lipid synthesis is likely very similar but may be less complex in microalgae [40, 44, 45]. Algal lipid biochemistry may eventually be found to deviate from the pathways detailed below, but these models give informed insight into the basic biochemistry involved.

4.2.1. Fatty Acids

The fundamental building-block of many lipids is the fatty acid, which is a carboxylic acid connected to a hydrocarbon with a methyl group at its terminus. The carboxyl carbon is in position C1 and subsequent carbons are defined as alpha (α), beta (β), and gamma (γ) carbons, respectively. Fatty acid chain length ranges from six carbons (hexanoic acid) to very-long-chain fatty acids (VLCFAs) containing over 22 carbons. Carboxylic acids containing five or fewer carbons are generally water-soluble and not classified as lipids. Carbon-carbon double bonds can be introduced within the hydrocarbon chain to confer structural diversity. Unsaturation is commonly found at positions toward the methyl terminus, or the omega (ω) carbon. Generally, sites of unsaturation are not seen before the γ carbon (C4), and multiple double bonds are usually found separated either by one or two single bonds. The majority of naturally derived unsaturated fatty acids are found in the *cis* conformation. However, some organisms produce fatty acids with *trans* unsaturation. This is most often found in conjugated fatty acids: those with two or more double bonds. Conjugated fatty acids with *trans* unsaturation are found in ruminants, some flower petals and seeds, some vegetables, and the red alga *Bossiella orbigniana* [46-48]. Also, partial hydrogenation, an industrial method of removing double bonds from unsaturated fatty acids, often results in byproducts with *trans* double bonds (trans fats).

4.2.2. Biosynthesis

Fatty acids (FAs) are a fundamental component of cellular structure and operation. The synthesis of algal lipids containing fatty acyl chains occurs via fatty acid synthase (FAS). In plants and photosynthetic algae, FA synthesis occurs predominately within the chloroplast and is catalyzed by a set of discrete enzymes collectively known as FAS II [40]. Many other eukaryotic organisms synthesize FAs in the cytosol [49], utilizing the multi-subunit enzyme complex FAS I. The discrete plastidial enzymes in plant and algal FA synthesis carry out analogous processes to their subunit counterparts in FAS I, but do not occur as single enzyme subunits.

Acetyl-CoA Carboxylase (ACCase)

The committed and highly regulated initial step in fatty acid synthesis is the formation of malonyl-CoA (3-C) via carboxylation of acetyl-CoA (2-C) by ACCase. This irreversible reaction takes place in the chloroplast and involves an ATP-dependent carboxylation of a biotin carrier followed by transcarboxylation of acetyl-CoA to form malonyl-CoA, which is the primary source of carbon in fatty acid synthesis. Six distinct enzymatic reactions follow, resulting in the formation of a fatty acid [50].

Malonyl-CoA/ACP Transacylase (MAT)

Malonyl-CoA is loaded onto an acyl carrier protein (ACP), resulting in the loss of coenzyme A (CoA-SH) and the formation of a thioester bond between malonate and the ACP, forming malonyl-ACP [51].

ß-ketoacyl-ACP Synthase (KAS)

To begin fatty acid synthesis, KAS III is required to condense malonyl-ACP with acetyl-CoA, forming acetoacetyl-ACP (a ketoacid) and releasing one CO_2. Acetoacetyl-ACP then undergoes a reduction-dehydration-reduction sequence (see below) to generate butyryl-ACP (4:0), which is then elongated. After the first condensation, malonyl-CoA is used as the sole carbon source. This requires a different type of ß-ketoacyl-ACP synthase, KAS I, which adds two carbons from malonyl-CoA to the elongating fatty acyl chain and releases the third as CO_2. The new acyl-ACP is recycled back to KAS I for condensation with malonyl-CoA and reduction-dehydration-reduction. This forms an acyl-ACP elongated by two carbons, and the cycle continues to 16:0. Plastidial fatty acid synthesis typically terminates at palmitic acid (16:0) [52]. Once the fatty acid has reached 16 carbons, a thioesterase cleaves the bond between the fatty acyl chain and its carrier protein. Fatty acyl-ACP thioesterase (FAT) catalyzes this hydrolysis, releasing water, ACP-SH, and the fatty acid [53, 54].

1. *ß-ketoacyl-ACP reductase (KAR)* reduces the ketoacid to a hydroxyacid. This reaction requires one NADPH molecule, which is oxidized to $NADP^+$. In the first round of reactions, this forms ß-hydroxybutyryl-ACP from acetoacetyl-ACP.
2. *ß-hydroxyacyl-ACP dehydratase (HD)* releases one water molecule as it removes the hydroxyl group and introduces an α-ß double bond to form a *trans*-2-Enoyl-ACP. This forms crotonyl-ACP in the first round.
3. *Enoyl-ACP reductase (ENR)* reduces the *trans*-2-Enoyl-ACP molecule's double bond to form a saturated acyl-ACP. One proton is used, and the reductant is again NADPH, which is oxidized to $NADP^+$. Crotonyl-ACP is reduced to butyryl-ACP (4:0) after the first cycle of reactions [55].

4.2.3. Plastidial Modification

Acyl chains both shorter and longer than 16 carbons are necessary components of many lipids. To form FAs shorter than 16 carbons, specialized thioesterases truncate fatty acid synthesis at a chain length specific to that thioesterase [52-54]. Plants and algae contain a wide range of FA chain lengths. 18-carbon chains are made within the chloroplast; however, further elongation and modification occurs within the endoplasmic reticulum (ER) [56]. Export of fatty acids from the chloroplast is done predominantly as FFAs [56].

The longest FA chain synthesized within the chloroplast contains 18 carbons; its elongation is catalyzed by a third type of ß-ketoacyl-ACP synthase, KAS II [52]. In this case, the thioesterase responsible for cleaving ACP from palmitate is regulated, allowing KAS II to condense malonyl-CoA with palmitoyl-ACP to form stearoyl-ACP (18:0) [57]. Both 16- and 18-carbon chains are highly prevalent in membrane and storage lipids [58]. Unsaturation can be introduced in the plastid both by soluble and integral membrane desaturases, or in the ER by integral membrane desaturases [40, 52]. They are named for the location of the double bond they form. A Δ^9-desaturase introduces a double bond at C9; whereas an ω-3 desaturase forms a double bond at the third carbon from the methyl end. Plastidial desaturation of stearoyl-ACP is performed by a soluble stearoyl-ACP Δ^9-desaturase, forming oleoyl-ACP (18:1). It has been found that a significant portion of the thioesterases present within the chloroplast are specific for 18:1-ACP, indicating that a large amount of 18:1 FAs are transported outside the plastid for modification. Similar fractions of 16:0 and 18:0 are also exported [52].

4.2.4. TAG Synthesis

A significant portion of lipids contain fatty acids esterified to a specialized head group. Because the hydrocarbon chains are relatively simple and not particularly differentiable, a head group allows the cell to diversify lipid function while using common building blocks to construct the basic form of the lipid. Head groups can serve to direct the lipid to specific cellular locations and engage in cellular communication and signal transduction [59]. In the synthesis of storage lipids, glycerol is the most important backbone for acyl chain attachment. Each carbon is hydroxylated, allowing derivatization at three locations. If glycerol is phosphorylated, the phosphoryl group can either be replaced by an acyl group or it can be bound by another functional group for further derivatization. TAGs contain three acyl groups linked by ester bonds to a glycerol backbone. Glycerol-3-phosphate (G3P), a product of glycolysis, is a primary substrate to begin constructing complex lipids.

The right-hand side of Figure 2 shows one common fate of activated acyl groups and glycerol-3-phosphate in an algal cell. During periods when the algal cell requires supplemental energy and can't photosynthesize, (e.g. dark conditions), this pathway would be inhibited, and lipids within the TAG droplet shown at the bottom right would be metabolized as a source of energy through ß-oxidation (see below).

Acyl-CoA synthetase. In order for a FFA to be available for attachment to a head group, it must first be activated. Once exported into the cytosol, FFAs headed to the ER for modification are derivatized by coenzyme A (CoA) via a thioester bond. Acyl-CoA synthetase catalyzes this two-step process, which involves an adenylate intermediate. The FFA displaces diphosphate from ATP, forming pyrophosphate (PP_i) and an acyl-AMP (acyl adenylate) intermediate. The immediate and highly exergonic hydrolysis of PP_i pulls this reaction toward the products. Once the acyl adenylate has been formed, CoA-SH displaces AMP, forming an acyl-CoA. This reaction requires a net of two ATPs, since ATP is taken to AMP [60].

Fatty acid elongase complex (FAE). ER-mediated elongation is facilitated by FAE, which recognizes acyl-CoAs and condenses them with malonyl-CoA. The FAE complex carries out functions similar to KAS, KAR, HD, and ENR, adding two carbons sequentially to the carboxyl end of the acyl chain [51].

Glycerol-3-phosphate acyltransferase (GPAT) links an activated acyl group (acyl-CoA) to G3P, forming lysophosphatidic acid (LPA). This monoacyl phospholipid is itself a signaling molecule and a precursor to membrane lipids, other signaling molecules, and nonpolar storage lipids [18, 61, 62].

Lysophosphatidic acid acyltransferase (LPAAT) is specific for LPA and adds another activated acyl group to form phosphatidic acid (PA), which is an intermediate for either polar or neutral lipids. As a polar lipid precursor, PA can be derivatized by one of several head groups including choline, inositol, ethanolamine, or serine, to become a glycerophospholipid. For neutral lipids, a third acyl group can replace PA's phosphate group to form a TAG [18].

Phosphatidic acid phosphatase (PAP) is the first step in forming a TAG from PA. This enzyme hydrolyzes the phosphoryl group from PA, forming diacylglycerol (DAG) [63].

Diacylglycerol acyltransferase (DAGAT) catalyzes the addition of the third acyl group to DAG, maximizing glycerol's capacity to carry acyl chains. This yields a highly nonpolar TAG, an efficient storage molecule that forms protein-coated lipid droplets (oleosomes) within the cytosol [18].

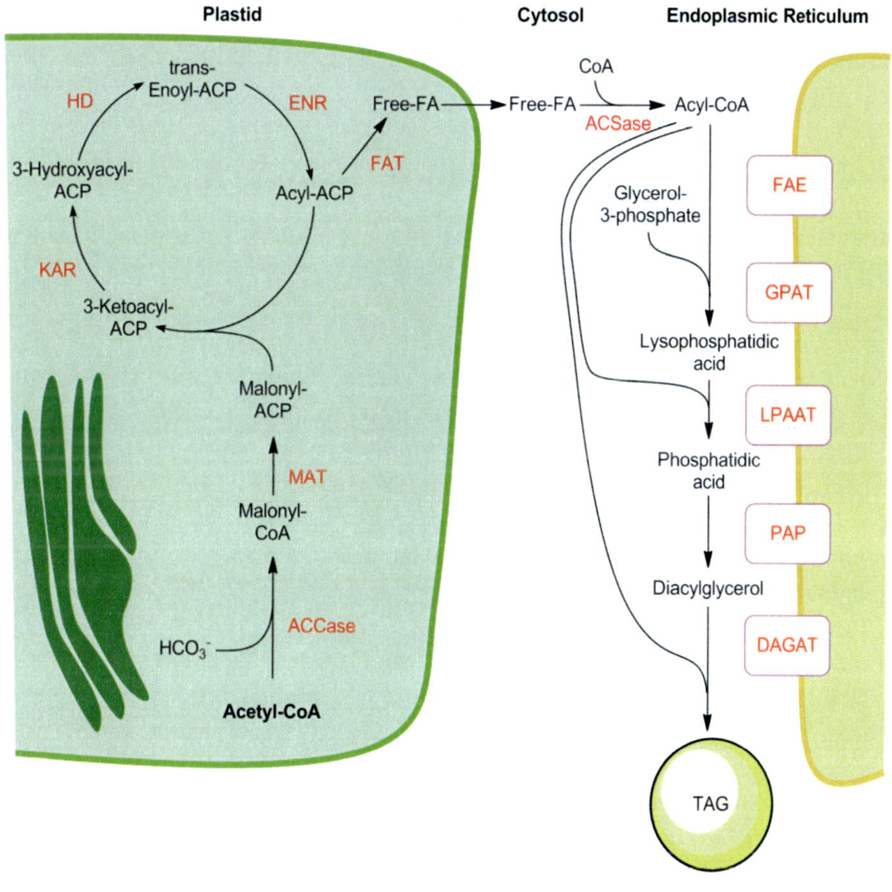

Figure 2. Overview of the metabolites and pathways involved in microalgal lipid biosynthesis (black). Enzymes shown in red. Free fatty acid synthesis occurs in the chloroplast, while TAG assembly may occur at the ER. ACCase, acetyl-CoA carboxylase; ACP, acylcarrier protein; ACSase, Acyl-CoA synthetase; CoA, coenzyme A; DAGAT, diacylglycerol acyltransferase; ENR, enoyl-ACP reductase; FAE, fatty acid elongase; FAT, fatty acyl-ACP thioesterase; GPAT, glycerol-3-phosphate acyltransferase; HD, 3-hydroxyacyl-ACP dehydratase; KAR, 3-ketoacyl-ACP reductase; KAS, 3-ketoacyl-ACP synthase; LPAAT, lyso-phosphatidic acid acyltransferase; MAT, malonyl-CoA:ACP transacylase; PAP, Phosphatidic acid Phosphatase; PDH, pyruvate dehydrogenase complex; TAG, triacylglycerols.

4.2.5. Fatty Acid Oxidation

Although the mobilization of TAGs is not yet well understood, it is believed that the interaction of lipases with the hydrophilic surface of lipid droplets exposes TAGs to lipase, an esterase that cleaves the acyl chain from its glycerol backbone. The products are FFAs that can be derivatized and shuttled into the mitochondrion for oxidation and energy generation [64]. Fatty acids are very energy dense. When completely oxidized under standard conditions, long chain (12C-20C) saturated FAs amount to an energy release of around 37 kJg^{-1} (about 9790 $kJmol^{-1}$) [65]. On a per weight basis, this is twice the energy content of carbohydrate or protein. The oxidation of one saturated palmitate (16:0) molecule requires 2 ATP for activation, and subsequently yields 8 acetyl-CoA, 7 $FADH_2$, and 7 NADH. Through the TCA cycle, 8 acetyl-CoAs yield 80 ATP. Each $FADH_2$ generates 1.5 ATP through the electron transport chain and oxidative phosphorylation, resulting in 10.5 ATP. Similarly, each NADH results in the formation of 2.5 ATP, producing 17.5 ATP. Overall, oxidation of palmitate yields 108 ATP, with a net of 106 ATP. However, there is not a perfect conversion of this energy into usable cellular energy. Under standard conditions, the energy stored in 106 ATP only amounts to 3233 $kJmol^{-1}$, but complete oxidation of palmitate releases almost 10,000 $kJmol^{-1}$ of energy, amounting to 33% conversion efficiency. However, under the conditions at which the cell operates, conversion efficiencies can be higher. Even with less-than-perfect conversion, the oxidation of fatty acids is a huge contributor to cellular energy.

4.2.7. Localization

FA oxidation generally occurs in the mitochondria, though it can also happen in the peroxisomes via an oxidase instead of a dehydrogenase [66]. However, there is no mechanism that can directly transport acyl-CoAs across the mitochondrial membranes; thus, a shuttling system regulated by malonyl-CoA is required to allow acyl-CoAs into the mitochondria for oxidation. Carnitine, a small carrier molecule, replaces CoA and facilitates transport of the acyl chain into the mitochondrial matrix. Once inside, the carnitine carrier is replaced by CoA and the acyl-CoA goes into ß-oxidation [67].

Carnitine acyltransferase I (CAT I), located within the outer mitochondrial membrane, facilitates the conversion of acyl-CoA to acyl-carnitine. CoA-SH is released back into the cytosol. Carnitine, CoA-SH, acyl-CoA, and acyl-carnitine move through the outer mitochondrial membrane without the aid of a transport protein. Thus, exchange of CoA with carnitine can occur either on the cytosolic or intermembrane side of the mitochondrion.

Carnitine-acylcarnitine translocase (CACT) passively facilitates the movement of acyl-carnitine and carnitine across the inner mitochondrial membrane. Generally, when the cell is undergoing oxidation and the generation of ATP, acyl-carnitine moves *into* the matrix and after delivering its acyl group, carnitine moves back into the cytosol.

Carnitine acyltransferase II (CAT II), located on the matrix side of the inner mitochondrial membrane, removes carnitine from the FA and re-derivatizes it with CoA from the mitochondrial CoA pool. The acyl-CoA is now localized and activated.

4.2.8. ß-Oxidation

The term ß-oxidation derives from the process of sequentially removing two-carbon units from the carboxyl end of the acyl chain. This results in a new FA minus 2C, where the ß-carbon has become the carboxyl carbon. It is essentially the reverse of FA synthesis: an oxidation-hydration-oxidation scheme followed by thiolysis to yield acetyl-CoA and an acyl-CoA minus 2C.

Acyl-CoA dehydrogenase (ADH) forms a double bond between the α and ß carbons via reduction of FAD to $FADH_2$. The product is a *trans*-Δ^2-Enoyl-CoA.

Enoyl-CoA hydratase (ECH) adds water across the double bond, hydroxylating the ß carbon and forming a methylene group at the α carbon. This forms ß-hydroxyacyl-CoA.

ß-hydroxyacyl-CoA dehydrogenase (HADH) converts the ß carbon's hydroxyl group to a keto group, reducing NAD^+ to NADH and releasing a proton to form ß-ketoacyl-CoA. This molecule has two carbonyl groups separated by one methylene group, an intrinsically unstable condition that readily leads to thiolysis.

Thiolase (Acyl-CoA acetyltransferase) facilitates the nucleophilic attack of a free CoA-SH to the ß carbon. This cleaves the α-ß bond and releases C1, the α carbon, and the attached CoA as acetyl-CoA. A new CoA-SH attaches to the shortened acyl chain, forming acyl-CoA minus two carbons. The new acyl-CoA returns to ADH and the cycle continues until the entire chain is converted to acetyl-CoA. In order to completely oxidize odd-chain or unsaturated fatty acids, other enzymes are required.

4.2.9. Unsaturated Fatty Acid Oxidation

Δ^3,Δ^2-Enoyl-CoA isomerase (ECI). In the instance of a Δ^9-monounsaturated FA, three cycles of ß-oxidation will occur before the double bond interferes. After three acetyl-CoAs have been released, the *cis* double bond comes into the ß-γ position, inhibiting ECH. ECI is responsible for repositioning the ß-γ double bond to form an α-ß double bond. In other words, ECI shifts Δ^3 unsaturation to the Δ^2 position. This provides ECH with a *trans* double bond in the correct position for ß-oxidation to continue.

2,4-dienoyl-CoA reductase (DECR). Oxidation of polyunsaturated fatty acids requires ECI and a second enzyme, DECR. For example, after three turns of ß-oxidation with a Δ^9,Δ^{12} diunsaturated FA, ECI shifts the Δ^9 double bond to allow for ECH, HADH, and thiolase to release one more acetyl-CoA. However, the Δ^{12} double bond has now moved into a position between C4 and C5 (Δ^4). ADH is able to oxidize this molecule to form a *trans*-Δ^2-Enoyl-CoA with *cis*-Δ^4 unsaturation. In order for ß-oxidation to continue, the Δ^4 double bond must be reduced by DECR, resulting in a Δ^3 double bond. This is isomerized by ECI and the newly formed *trans*-Δ^2-Enoyl-CoA reenters ß-oxidation [68, 69].

4.2.10. Odd-Chain Fatty Acid Oxidation

Although the majority of FAs contain an even number of carbons, some organisms synthesize odd-chain fatty acids. Breakdown of these FAs occurs in the same manner as even-chain FAs, except at the very end. The products at the end of odd-chain FA oxidation are acetyl-CoA and the three-carbon activated product propionyl-CoA. Unlike acetyl-CoA, propionyl-CoA cannot go through the TCA, and must be converted to succinyl-CoA, a four-carbon activated TCA intermediate.

Propionyl-CoA carboxylase (PCCase) requires bicarbonate (HCO_3^-) and one ATP to facilitate the binding of a single carboxyl group to the α carbon of propionyl-CoA, forming methylmalonyl-CoA. This reaction is facilitated by a biotin cofactor in a reaction similar to that of ACCase.

Methylmalonyl-CoA epimerase (MCE) rearranges methylmalonyl-CoA to its levorotary optical configuration, effectively forming a 3-carbon chain with a carboxyl group at its terminus and the middle carbon linked to carbonyl-CoA.

Methylmalonyl-CoA mutase (MCM) uses a vitamin B_{12} cofactor to move the carbonyl-CoA group and bind it to the terminal carbon [68, 69]. This forms succinyl-CoA, which is a TCA intermediate.

4.3. Algal Lipid Classification

The lipid composition of microalgae is not as well-studied as that of higher plants. However, there is substantial information about algal lipid structure, biochemistry, and metabolism. Microalgae contain a range of glycolipids and phospholipids, the major lipids of higher plants, including significant quantities of the glycosylglycerides monogalactosyldiacylglycerol (MGDG), digalactosyldiacylglycerol (DGDG), sulfoquinovosyldiacylglycerol (SQDG), and large amounts of the phosphoglycerides phosphatidylcholine [70], phosphatidylethanolamine (PE), and phosphatidylglycerol (PG). These comprise the main lipids found in the membranes of thylakoids, which dominate the intracellular membrane fraction in algae [71]. Microalgae have MGDG as their primary lipid, with smaller amounts of DGDG and two anionic lipids, SQDG and PG. MGDG is often present at fractions ranging from 40-55%, DGDG at 15-35%, SQDG at 10-20% and PG at 10-20% of the total lipid content [71]. The relative abundance of these lipids is highly dynamic and dependent on culturing conditions, but C16 and C18 fatty acids usually dominate. Algal lipids are often highly enriched in polyunsaturated fatty acids (PUFAs), especially in glycosylglycerides. Unusual lipid classes are present in some species of algae. For example, red algae contain sulfolipids such as phosphatidyl-sulfocholine and chlorosulfolipids, while some marine algae have the arsenic-containing lipids arsenoribosylphosphatidylglycerol. Most green alga have the betaine lipids diacylglyceryltrimethylhomoserine (DGTS) and diacylglyceryl hydroxymethyltrimethyl-β-alanine (DGTA) [71].

Betaine lipids belong to a class of complex polar lipids called glycerolipids. They can be found in lower plants (bryophytes), fungi, photosynthetic bacteria, and many algae [72].

Three primary betaine lipids are known: diacylglyceryltrimethylhomoserine (DGTS), diacylglyceryl hydroxymethyltrimethyl- β-alanine (DGTA), and diacylglyceryl carboxyhydroxymethylcholine (DGCC). DGTS and DGTA are the most common naturally-occurring betaine lipids and are widely distributed among bryophytes and algae. Labeling experiments with algae suggest that betaine lipids may play a metabolic role in the transfer of fatty acids between the cytoplasm and the chloroplast and that they may be involved in fatty acid biosynthesis by first accepting the fatty acids formed *de novo*, before they are processed and redistributed to other cellular lipids [72]. DGTS is a primary membrane lipid of some green algae (*C. reinhardtii*, *Volvox cateri* and *Dunaliella*) and algae that contain chlorophyll a and c [73]. Most algae with betaine lipids do not contain phosphatidylcholine; thus, it has been suggested that DGTS is a substitute for phosphatidylcholine. In green algae, the concentrations of DGTS and PC seem to be reciprocal, suggesting an apparent substitution of DGTS for PC as a membrane component. Interestingly, it has been found that *C. reinhardtii* accumulates high levels of DGTS and no PC [74]. In contrast, organisms such as *Ulva petusa* and *Chaetomorpha spirilis* have very low levels of PC and high amounts of DGTS [73]. It can be established from this inverse relationship that DGTS has progressively been replaced by PC during the evolution of green plants. DGTA is primarily found in the brown algae Fucales and Dictyotales [73]. DGTS is usually absent in algae that contain DGTA, and only *Crytomonas* and *Ochromonas* are known to produce both of these betaine lipids [75]. It has been demonstrated in *Ochromonas daninca* and *Crptomonas sp.* that DGTA can be uniquely synthesized from DGTS; therefore, species of algae that contain DGTA may contain low levels of DGTS as a precursor to DGTA [73].

4.4. Lipid Chemistry Affects Fuel Properties

One difference between higher plants and algae is that many algae are capable of producing long-chain polyunsaturated fatty acids (PUFAs), which are suggested to be beneficial for human health and nutrition, but can have an adverse impact on lipid-based biofuel performance. FA chemistry is important to consider for biofuel applications because chemical structure determines important biodiesel properties such as cold flow, oxidative stability, and cetane number. Increasing the PUFA content of the feedstock organism would improve fuel performance at the lower temperatures encountered in colder climates as a higher degree of unsaturation decreases the cloud point; however, this would also decrease the oxidative stability of the fuel, as unsaturated double bonds are prone to oxidation [3, 70]. With these effects in mind, the FA composition of the chosen feedstock organism(s) must be analytically determined, and will need to be tailored to achieve a reliable feedstock that can be readily incorporated into the current transportation fuel infrastructure.

Fatty acid synthesis can be manipulated by genetic modification; alternatively, variations in culture conditions have been shown to greatly affect microalgal FA composition, making it possible to direct FA synthesis towards more commercially valuable chemical structures within a controlled environment. Evidence also suggests that due to natural adaptations to particular environments, some algal species are better suited for producing commercially valuable FAs. For example, a *Parietochloris incisa* strain isolated from an alpine environment was found to contain the highest levels of long chain PUFAs than any other plant source described previously [76], alluding to the potential benefits of bioprospecting efforts directed towards discovering new, unique species of microalgae.

4.5. Qualitative Lipid Analysis of Microalgae

Upon microscopic analysis, some microalgae visibly produce oil bodies or vesicles that contain lipid in the form of TAG (Figure 3A). The extent of lipid accumulation can usually be assessed qualitatively *in situ* by using fluorescence microscopy and staining with neutral, lipophilic fluorophores such as Bodipy 493/503 (Molecular Probes, Invitrogen) (Figure 3B) or Nile Red. Notably, it is typical for cells to remain viable after exposure to Bodipy dye, a trait that can potentially be leveraged for high-throughput screening techniques to rapidly identify, and then culture, cells that produce neutral lipids.

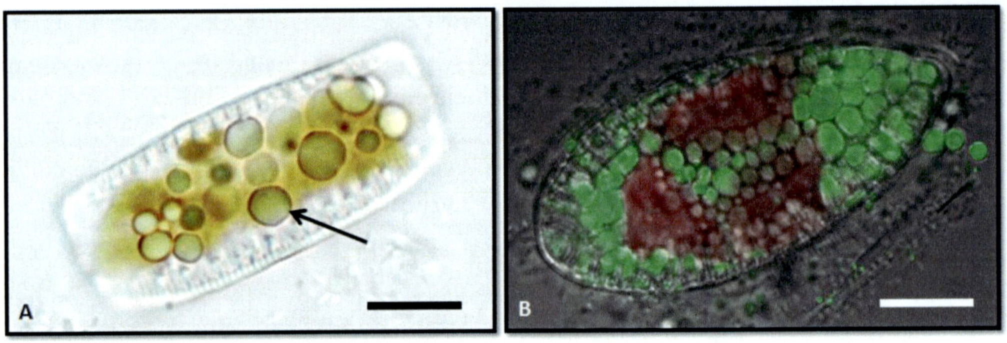

Figure 3. Image of a diatom species that visibly contains large oil bodies (arrow) (A). Fluorescence image of another diatom species where lipids were stained with Bodipy 493/503 (B). Chlorophyll is red and Bodipy is green. Scale bars are 10μM.

Unfortunately, some microalgae are recalcitrant to staining. This is likely due to the chemistry and composition of the cell wall and extracellular polysaccharide (EPS) matrices. Depending on the chemical makeup of these structures, dye uptake can be completely or partially occluded. However, protocols using heat shock, freeze fracturing, and/or treatment with chemicals like DMSO, methanol, ethanol, acetone, isopropanol, and glutaraldehyde have been shown to overcome this recalcitrance [77], but this is often at the cost of solubilizing chlorophyll. For example, DMSO treatment can enhance dye uptake but can make a structural analysis of the cell impossible as the solvent visibly disrupts lipid bodies and chloroplasts in some algae (Figure 4). Once solubilized, the chlorophyll phase-separates directly into the lipid bodies, which can be verified by fluorescence microscopy. Here, phase-separation is visualized as overlapping chlorophyll and Bodipy fluorescence signals, appearing as yellow/orange to orange/red droplets.

Electroporation can also be used to enhance dye uptake where the field strength of the electric pulse will determine the degree of dye penetration in recalcitrant cells. This method can preserve the fine structure of the cell and maintain viability. Unfortunately though, this technique is not well adapted for high-throughput analysis as the electroporation conditions require optimization for different microalgal strains and growth conditions.

An alternative to electroporation is high-pressure freezing-freeze substitution (HPF-FS), which fixes the cells but preserves the fine structure. Here, the biological sample is rapidly frozen at rates that approach −20,000°C/sec, wherein a fixative replaces the cellular water, preventing damaging ice crystals from forming and resulting in nearly perfect preservation of the cellular architecture. Frozen samples could be stained and imaged using electron microscopy. This has been used to elucidate a semi-crystalline secondary structure of TAG lipid bodies in *Schizochytrium sp.* and *Thraustochytrium sp.* that was not found in *Neochloris oleoabundans* and other oleaginous species [78].

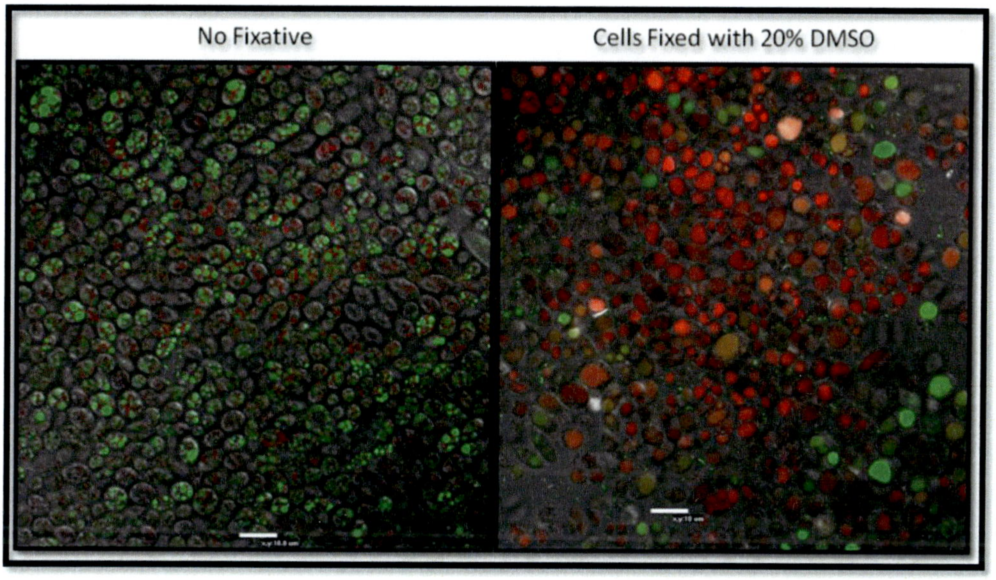

Figure 4. Comparison of Bodipy stained cells without and with the fixative DMSO. Fixative clearly disrupts intracellular structure and solubilizes chlorophyll. Chlorophyll is red and Bodipy is green, and overlapping signals are yellow/orange to orange/red. Scale bars are 10 μM.

4.6. Semi-Quantitative Lipid Analysis of Microalgae and FACS

Semi-quantitative lipid analysis can be performed using fluorophores such as Bodipy when coupled to flow cytometry. Using this technique, relative changes in fluorescence between populations of cells can be easily determined in addition to differences in light scattering properties. This is very useful for detecting both large and small changes within and between populations of cells. However, if cells in the population are recalcitrant to staining, this technique becomes much more challenging as fixation or electroporation must be carried out prior to analysis. Not only are these both time-consuming endeavors, but treatment with permeabilizing agents will also make the gathered data incomparable to analyses where such techniques were not employed. The true value of this type of analysis is realized when coupled to flow cytometry using a fluorescent activated cell sorter (FACS). This instrument facilitates the physical collection of individual cells or populations of cells into test tubes or the wells of microtiter plates. If for example, a desired group of traits, like intensity of Bodipy fluorescence on particles that also contain chlorophyll, are distinguishable by fluorescence, side scatter (a measure of intracellular complexity), and/or forward scatter (a measure of particle size), they can be used as gating criterion to physically sort cells with those combined traits [79-84].

4.7. Fully Quantitative Lipid Analysis of Microalgae

Qualitative information regarding lipid accumulation is useful for a rapid assessment of lipid productivity. However, a more detailed quantitative analysis is required to accurately assess the feedstock potential of the organisms. Quantitative determination of lipid composition can be revealed by fatty acid fingerprinting methods using gas chromatography [23] coupled to mass spectrometry (MS). Analysis by GC requires that the analyte be volatile, which is achieved most often by acid-catalyzed trans-esterification with methanol into the corresponding fatty acid methyl esters (FAMEs). Traditional analysis techniques involve lengthy extractions and long run times that result in a very low sample throughput. Fortunately, analysis can be accelerated significantly with the use of recently developed sample preparation and analysis techniques. For instance, methods have been developed that greatly decrease sample preparation time by performing *in situ* methylation or direct thermal desorption. Furthermore, an even greater sample throughput can be achieved if these methods are coupled to shortened GC run times, which is accomplished by manipulating the column chemistry, column length, carrier gas, and temperature ramp [85]. Despite the development of these sophisticated analytical methods, differences in TAG accumulation level and TAG chemistry remain largely uncharacterized for most microalgae.

5. Molecular Tools for Algal Feedstock Engineering

The development of tools for genetic engineering of algae has been steadily progressing for more than twenty years and many methods for transformation have been developed and tested in a wide range of algal model systems. These developments are likely to accelerate due to recent efforts to sequence algal genomes. Expressed sequence tag (EST) databases have been created; also, nuclear, mitochondrial, and chloroplast genomes from several microalgae have been sequenced and many more are reportedly being sequenced. While earlier efforts were focused mainly on genetic manipulation of the green alga *C. reinhardtii*, tools have also been developed for other algae that may be of greater interest for industrial applications. Transgene over-expression is becoming a routine procedure in many common algal model systems and several genetic selection markers have been proven to work well in algae. Some of the biggest challenges in algal genomics concern the knockout of endogenous genes, but recent progress with homologous recombination and RNA silencing may provide a way forward.

5.1. Methods for Transformation and Expression

To date, more than 30 different strains from several different phyla of microalgae have been successfully transformed. These include green algae (Chlorophyta) such as *C. reinhardtii*, *Chlorella ellipsoidea*, *Chlorella saccharophila*, *Chlorella vulgaris*, *V. carteri*, *Chlorella sorokiana*, *Chlorella kessleri*, *Ulva lactuca* and *Dunaliella viridis* [86-93]. Heterokontophyte algae that have been transformed include *Nannochloropsis oculata*, diatoms such as *Thalassiosira pseudonana* and *Navicula saprophila*, and brown algae (Phaeophyta) such as *Laminaria japonica* and *Undaria pinnatifada* [94-98]. Red algae, (Rhodophyta) including *Cyanidoschyzon merolae*, *Porphyra yezoensis*, *Kappaphycus alvarezii*, *Gracilaria changii*, and *Porphyridium sp.* have also been transformed [99-103]. Dinoflagellates that have been transformed include *Amphidinium sp.* and *Symbiodinium microadriaticum* [104]. The only euglenoid that has been transformed to date is *Euglena gracilis* [105].

Several methods for introducing transgenes into algae exist, including agitation in the presence of glass beads or silicon carbide whiskers [104], electroporation [86, 88, 94], biolistic microparticle bombardment [87, 91, 96], and *Agrobacterium tumefaciens* mediated gene transfer [100]. The efficiency of these transformation methods vary between different algal species, and the method of transformation must be carefully tested and optimized for each species of algae. In many cases, transformation has resulted in stable expression of transgenes, either from the nuclear or the chloroplast genome, but in some cases only transient expression has been achieved. Methods developed primarily in *C. reinhardtii* [106] demonstrate that the stability of expression can be improved through proper codon usage, the use of strong endogenous promoters, and inclusion of species specific 5', 3' and intronic sequences.

Many different genetic markers have been used to facilitate efficient isolation of genetic transformants. These markers comprise antibiotic resistance genes and fluorescent/biochemical markers. Many algae are resistant to a wide range of antibiotics and the choice of antibiotic will vary between different species of algae. Antibiotic resistance genes that have been used for algal transformant selection include bleomycin [107], spectinomycin and streptomycin [105], paromomycin [108], G418 [106] and hygromycin [110]. Fluorescent and biochemical markers have also been used to confirm genetic transformation in algae. These markers include luciferase [87], β-glucuronidase [91, 100], β-galactosidase [98, 102], and green fluorescent protein (GFP) [100].

Homologous recombination events are rare in algae; thus, transformation generally results in a random integration of transgenes into the nuclear genome. While this may be suitable for transgene expression or for random mutagenesis screens, it makes it difficult to delete specific target genes. Homologous recombination has been reported in the nuclear genomes of *C. reinhardtii* and *C. merolae* at a very low efficiency, and further optimization is needed to improve successful homologous recombination [99, 110]. RNA silencing is another option for gene inactivation that has been studied in algae as RNA silencing has been shown to be a feasible strategy in both *C. reinhardtii* and *P. tricornutum*. Recent improvements in gene knockdown strategies include the development of high throughput artificial micro-RNA (armiRNA) techniques in *C. reinhardtii* that are reportedly more target-specific and stable than traditional RNAi approaches [110].

Transgene expression and protein localization in the chloroplast are needed for the proper function of many genes involved in metabolic pathways. In *C. reinhardtii*, it is possible to achieve transformation of the chloroplast through homologous recombination [111]. Chloroplast transformation has not been demonstrated in diatoms. However, plastid targeting sequences in diatoms have been characterized, which allow for translocation of proteins into the chloroplast [112].

5.2. Microalgal Genomes

Algal genetic modifications are greatly facilitated by the availability of microalgal genome sequences, and the recent developments in rapid large-scale sequencing technology represent a revolution in microalgal research. Some of the past and current microalgal genome sequencing projects include *C. reinhardtii* [113], *Ostreococcus tauri* [114], *Fragilariopsis cylindrus*, *Pseudo-nitzschia*, *Thalassiosira rotula*, *Botryococcus braunii*, *C. vulgaris*, *D. salina*, *Micromonas pusilla*, *Galdieria sulphuraria*, *Porphyra purpurea*, *V. carteri*, and *Aureococcus annophageferrens* [115]. In addition, there are several completed and ongoing efforts to sequence plastid and mitochondrial genomes, as well as dynamic transcriptome analysis from many different microalgae [113].

6. BIOENERGY-FOCUSED BIOPROSPECTING FOR MICROALGAL DIVERSITY

At least 40,000 algal species have been described to date with some conservative and more generous estimates of actual diversity ranging from several hundred thousand to millions of extant microalgae and macroalgae species combined [116, 117]. Only a fraction of this enormous biological and genetic diversity has been examined to date, as it traditionally requires a significant investment of time and resources to perform conventional taxonomic and biological analyses. However, in conjunction with classical techniques and a recent surge in interest, the advent of molecular tools such as rDNA gene sequencing will help to rapidly expand the list of known and described species.

Microalgae thrive in a broad range of diverse natural habitats with many species being predominantly aquatic. With such an abundant diversity of species, it is feasible that there are undiscovered microalgae capable of producing and accumulating useful co-products. As such, metabolic diversity, especially with respect to differences in TAG metabolism, is likely to be enormous, and the need to assess this unknown potential has rapidly grown with the increasing interest of using these organisms for biofuels. This idea was extensively explored during the 1980's by the U.S. Department of Energy (DOE) during the Aquatic Species Program (ASP) directed from the Solar Energy Research Institute (SERI), later established as the National Renewable Energy Laboratory (NREL). Initial efforts were geared towards isolating strains from diverse environments throughout the U.S. Subsequent screening and characterization of these isolates revealed several strains of high interest for biofuels production. The program was eventually canceled as a result of low petroleum prices and federal budget cuts. Fortunately, information was gathered regarding all aspects of microalgal biology, process development, and the feasibility of using algae as a biofuel feedstock, which is presented in the ASP Close-Out report discussed below [118]. Most notably, this report provides compelling evidence that large-scale microalgal bioprospecting efforts targeting a broad spectrum of environments can uncover commercially valuable oleaginous species.

6.1. DOE's Former Aquatic Species Program: Establishment of a Microalgal Culture Collection for Biofuels Applications

From 1979 to 1996 the US Department of Energy (DOE) supported a relatively small algal biofuel research effort project known as the Aquatic Species Program (ASP). The major focus of this DOE program was the development of technologies for producing biodiesel from oil accumulating microalgae grown in large, open raceway ponds using the CO_2 derived from coal-fired power plants. The activities of the ASP included the isolation and characterization of microalgal strains, the study of the physiology and biochemistry of lipid accumulation, genetic engineering of microalgae for enhanced lipid production, and the demonstration of open raceway pond systems for the cultivation of microalgal biomass. Despite the significant advancements made in the biological and process engineering areas, the program was terminated in 1996. In 1998, a close-out report was prepared detailing the research accomplishments of ASP. One of the most important accomplishments was the establishment of microalgal culture collection which represented a large genetic resource with the potential for oil production [119].

At the beginning of the ASP, the information available in the general field of microalgal biology was rather limited. In addition, researchers exploring the feasibility of developing microalgal-derived biofuels did not have the benefit of accessing existing microalgal culture collections from which to identify microalgal strains that could serve as candidates for a commercial oil production process. Therefore, a major undertaking by ASP researchers from 1983-1987 was to build and characterize a large microalgal strain collection. Over the four-year period, microalgal strains were collected primarily from inland, shallow saline aquatic environments in the West, Northwest and Southeastern regions of the continental U.S. and from Hawaii. The rationale for collecting microalgal strains from these diverse locations was that these habitats, largely characterized by predominantly high solar irradiation and temperatures, would harbor strains that could naturally withstand substantial variations in temperature and salinity throughout the year. It was, therefore, hypothesized that these strains would be better able to withstand the fluctuating conditions that would be experienced at a commercial cultivation facility. The major goals of the strain collection work was to 1) isolate microalgal biodiversity from a variety of US aquatic environments; 2) characterize the isolated strains for their ability to grow rapidly under high light, high temperature, and saline conditions; 3) analyze the strains for lipid components; and 4) investigate the effects of various stress conditions on lipid composition.

The strain collection effort resulted in the identification of approximately 3,000 microalgal strains that were isolated from within the continental US (Alabama, Arizona, Califronia, Colorado, Mississippi, Nevada, New Mexico, Florida, Nebraska, Utah, and Washington) and Hawaii. Since most of the strains in the culture collection had not been characterized with respect to lipid content, a rapid screening method based on the lipophilic fluorescent dye Nile Red was developed [120]. In the presence of a neutral lipid environment (i.e., lipid droplets), Nile Red fluoresces yellow. Subsequent method development studies were able to demonstrate that an increase in Nile Red fluorescence (575nm) was correlated with an increase in total lipid content after nutrient deprivation. The Nile Red assay was then coupled with a screen for rapid growth of microalgal strains under high conductivity, high temperature (30°C), and high light conditions. Of the ten fastest growing strains, six were identified as cyanobacterial species belonging to the genus *Oscillatoria* and *Synechococcus*. The fastest growing eukaryotic algae identified belonged to the genus *Chlorococcum* (CHLOC4; 3.47 doublings/day), *Amphora* (AMPHO46; 2.84 doublings/day), *Nannochloris* (NANNO13; 2.78 doublings/day) and *Chlorella* (CHLOR23; 2.66 doublings/day) [119]. However, none of these fast-growing strains made the list of the ten highest Nile Red fluorescing strains. All ten of the strains identified as having the highest Nile Red fluorescence were diatoms comprising the genus *Nitzchia*, *Amphora* (AMPHOR45 and 27), and *Fragilaria*. The best microalgal strains that combined both high Nile Red fluorescence and rapid growth were *Chaetoceros muelleri* (CHAET9), *Navicula saprophila* (NAVIC2), and *Nitzchia pusilla* (NITZS12). Despite the identification of these superior strains through the ASP strain screening procedure, many of the 3000 strains in the collection were never characterized [119].

Between the years of 1983-1987 SERI published three Culture Collection Catalogs (Solar Energy Research Institute, SERI/SP-231-2486; SERI/SP-232-2863; SERI/SP-232-3079). The 1984-1985, 1985-1986, and 1986-1987 catalogs contained 11, 10, and 29 microalgal strains, respectively [119]. The 1987 catalog contained a total of 50 strains comprising a broad range of different microalgal classes including Chlorophyceae (26%), Bacillariophyceae (60%), Chrysophyceae (8%), and Eustigmatophyceae (6%) (121]. By the end of the ASP in 1996, approximately 37 of the 50 strains in the 1987 strain collection catalog were still viable. These strains, combined with an additional 260 strains that were not extensively characterized, brought the collection to approximately 300 strains [119] and were transferred to the Marine Bioproducts Engineering Center (MarBEC) headquartered at the University of Hawaii. MarBEC was established as a National Science Foundation (NSF) Engineering Research Center in 1998 through a partnership between the University of Hawaii at Monoa and the University of California at Berkeley. Since NSF funding for MarBEC ran out in 2004, the current status of the SERI culture collection is largely unknown. However, in the final report (http://mybiofuels.org/Pubs/MarBEC_Final_2004.pdf) a summary of the MarBEC Culture Collections stated that former NREL Marine Biofuels Collection is comprised of 183 microalgal strains. Based on this information, it is likely that only a fraction of the original 300 microalgal strains transferred to the University of Hawaii are viable today. The Executive Summary of the ASP close-out report emphasized the desire that future researchers should be able to make use of the culture collection; however, no additional publically available information regarding the current status of this largely untapped source of microalgal genetic diversity can be found.

6.2. Establishment of a Bioenergy-Focused Microalgal Strain Collection with FACS

The ASP strain screening effort was successful in that promising microalgae were identified, but was incredibly labor-intensive and time-consuming as it utilized traditional isolation and screening techniques. Approaches utilizing state-of-the-art technology to rapidly screen large communities of microalgae collected directly from the environment for TAG productivity and chemistry are greatly needed. One such approach employs FACS to rapidly screen large heterogeneous populations of cells for oleaginous microalgae and then subsequently sort those cells for cultivation. This technique utilizes the physical properties of various isomers of chlorophyll, which make algal cells highly autofluorescent at characteristic light wavelengths. Consequently, it is possible to identify and sort algae by coupling flow cytometry with a cell-sorting. Furthermore, by staining samples with fluorescent neutral lipid probes like Bodipy or Nile Red, it is possible to identify microalgal cells that accumulate TAGs. Together, these traits can be leveraged for bioprospecting using FACS to isolate oleaginous algae from large heterogeneous populations of microorganisms.

Recently, a new bioprospecting effort funded by the Colorado Center for Biorefining and Biofuels (C2B2) was undertaken using high-throughput FACS to rapidly re-establish a new bioenergy-focused microalgal strain collection at the National Renewable Energy Laboratory (NREL). Forty-five water samples were collected from a broad range of environments throughout the Southwestern states of Colorado, New Mexico, Arizona, and Utah. Differences in water chemistry were compared in terms of pH and conductivity, which is related to the total ions in solution. Conductivity measurements ranged from freshwater to brine, and pH ranged from near neutral to slightly basic.

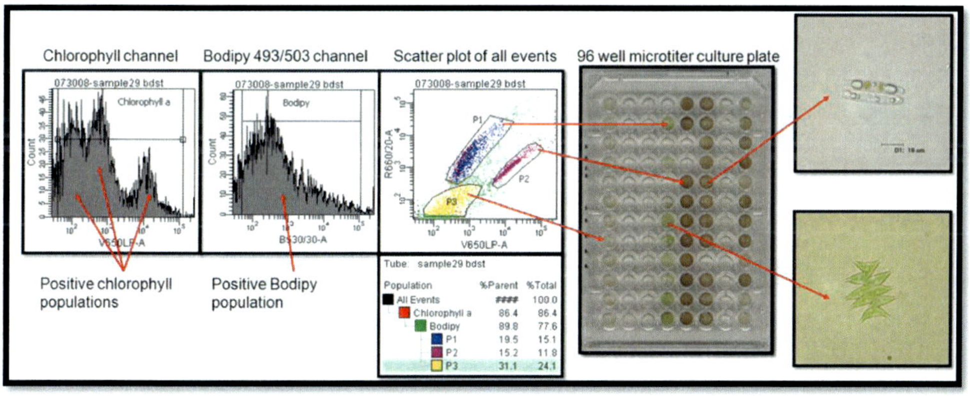

Figure 5. FACS process of algal isolation from a natural water sample depicting data gathering and analysis, gating, and growth after sorting, left to right respectively. Relative chlorophyll and Bodipy fluorescence of particles detected in stained water sample. Scatter plot of all events, where gate P1, P2, and P3 defines the sort criterion of particles with both positive chlorophyll and Bodipy fluorescence. Gated cells are sorted into a 96 well microtiter plate and growth is later characterized to identify unique isolates.

Stained water samples were directly analyzed on a BD FACSAria to identify populations with both positive chlorophyll and Bodipy 493/503 fluorescence (Figure 5). These populations were gated and single cells were sorted. The sort protocol was devised to isolate a representative number of the naturally-occurring microalgae from a particular water sample. This process resulted in the isolation of 245 robust, unialgal cultures followed by the successful cryopreservation of over 96% of the collection with methanol or dimethyl sulfoxide (DMSO) as a cryoprotectant. However, most species isolated using FACS were fresh water green algae despite the observation that the dominant species present in many of the water samples were diatoms. This selection bias likely occurred because diatoms, possessing a silica cell wall, are more fragile and may be less likely to survive the sorting process as there are considerable forces generated within the sample stream required to sort particles in a high throughput liquid platform. It is also possible that suboptimal cultivation conditions were responsible. Furthermore, dye recalcitrance, as discussed above, may have caused some oleaginous algal species to be overlooked during the sorting process.

A great deal of valuable information has been collected throughout this project resulting in the successful establishment of a bioenergy focused microalgal strain collection, which can be used to refine future high-throughput bioprospecting efforts. It can be concluded that FACS is very useful for analyses of algal populations and for subsequent physical separation of robust cells from such populations.

7. Conclusion

The development of microalgae into a feedstock for the production of biofuels is a promising renewable energy scenario that many believe capable of providing a substantial fraction of our transportation energy needs. However, viable commercial development of this technology will ultimately require the use of algal strains that have much higher and consistent production of chemical intermediates than any species described to date. Many efforts are currently focused on genetically optimizing metabolic pathways to increase lipid productivities in well-studied organisms, which are laborious and time consuming tasks. However, the difficulty could be lessened by taking advantage of natural selection and identifying native strains of microalgae that already possess the metabolic traits necessary for commercial feedstock development.

Acknowledgments

The authors acknowledge support from the Air Force Office of Scientific Research grant FA9550-05-1-0365 and the Office of Biological and Environmental Research, GTL program, Office of Science, U.S. Department of Energy, and support from the Colorado Center for Biorefining and Biofuels seed grant 07-2.

References

[1] León, R., Gaván, A., and Fernández, E. (2007) *Transgenic microalgae as green cell factories*, Springer Science+Business Media/Landes Bioscience, New York.

[2] Spolaore, P., Joannis-Cassan, C., Duran, E., and Isambert, A. (2006) Commercial applications of microalgae, *Journal of Bioscience and Bioengineering 101*, 87-96.

[3] Hu, Q., Sommerfeld, M., Jarvis, E., Ghirardi, M., Posewitz, M., Seibert, M., and Darzins, A. (2008) Microalgal triacylglycerols as feedstocks for biofuel production: perspectives and advances, *Plant Journal 54*, 621-639.

[4] Beer, L. L., Boyd, E. S., Peters, J. W., and Posewitz, M. C. (2009) Engineering algae for biohydrogen and biofuel production, *Curr. Opin. Biotechnol. 20*, 264-271.

[5] Robyt, J. F. (1998) *Essentials of Carbohydrate Chemistry*, Springer, New York.

[6] Ball, S. G., and Deschamps, P. (2009) Starch Metabolism, in *The Chlamydomonas sourcebook* (Harris, E. H., Stern, D. B., and Witman, G., Eds.) 2nd ed., pp 1-40, Academic Press, Amsterdam; Boston.

[7] Hirokawa, Y., Fujiwara, S., Suzuki, M., Akiyama, T., Sakamoto, M., Kobayashi, S., and Tsuzuki, M. (2008) Structural and physiological studies on the storage β-polyglucan of haptophyte *Pleurochrysis haptonemofera*, *Planta 227*, 589-599.

[8] Ball, S. G., and Morell, M. K. (2003) From bacterial glycogen to starch: understanding the biogenesis of the plant starch granule, *Annual Review of Plant Biology 54*, 207-233.

[9] van de Wal, M., D'Hulst, C., Vincken, J.-P., Buléon, A., Visser, R., and Ball, S. (1998) Amylose is synthesized in vitro by extension of and cleavage from amylopectin, *Journal of Biological Chemistry 273*, 22232-22240.

[10] Ball, S., Guan, H.-P., James, M., Myers, A., Keeling, P., Mouille, G., Buleon, A., Colonna, P., and Preiss, J. (1996) From glycogen to amylopectin: a model for the biogenesis of the plant starch granule, *Cell 86*, 349-352.

[11] Mouille, G., Maddelein, M. L., Libessart, N., Talaga, P., Decq, A., Delrue, B., and Ball, S. (1996) Preamylopectin processing: a mandatory step for starch biosynthesis in plants, *Plant Cell 8*, 1353-1366.

[12] Stark, D. M., Timmerman, K. P., Barry, G. F., Preiss, J., and Kishore, G. M. (1992) Regulation of the amount of starch in plant tissues by ADP glucose pyrophosphorylase, *Science 258*, 287-292.

[13] Ball, S. (1998) Regulation of Starch Biosynthesis, in *The Molecular Biology of Chloroplasts and Mitochondria in Chlamydomonas* (Rochaix, D., Goldschmidt-Clermont, M., and Merchant, S., Eds.), pp 549-567.

[14] Dauvillée, D., Colleoni, C., Mouille, G., Morell, M. K., d'Hulst, C., Wattebled, F., Liénard, L., Delvallé, D., Ral, J.-P., Myers, A. M., and Ball, S. G. (2001) Biochemical characterization of wild-type and mutant isoamylases of *Chlamydomonas reinhardtii* supports a function of the multimeric enzyme organization in amylopectin maturation, *Plant Physiology 125*, 1723-1731.

[15] Smith, A. M., Zeeman, S. C., and Smith, S. M. (2005) Starch degradation, *Annual Review of Plant Biology 56*, 73-98.

[16] Yu, T.-S., Zeeman, S. C., Thorneycroft, D., Fulton, D. C., Dunstan, H., Lue, W.-L., Hegemann, B. r., Tung, S.-Y., Umemoto, T., Chapple, A., Tsai, D.-L., Wang, S.-M., Smith, A. M., Chen, J., and Smith, S. M. (2005) α-Amylase is not required for breakdown of transitory starch in *Arabidopsis* leaves, *Journal of Biological Chemistry 280*, 9773-9779.

[17] Ritte, G., Heydenreich, M., Mahlow, S., Haebel, S., Kötting, O., and Steup, M. (2006) Phosphorylation of C6-and C3-positions of glucosyl residues in starch is catalysed by distinct dikinases, *FEBS letters 580*, 4872-4876.

[18] Radakovits, R., Jinkerson, R. E., Darzins, A., and Posewitz, M. C. (2010) Biofuels from eukaryotic microalgae, *Eukaryot. Cell*, doi:10.1128/EC.00364-00309.

[19] Fortman, J. L., Chhabra, S., Mukhopadhyay, A., Chou, H., Lee, T. S., Steen, E., and Keasling, J. D. (2008) Biofuel alternatives to ethanol: pumping the microbial well, *Trends in Biotechnology 26*, 375-381.

[20] Lee, S. K., Chou, H., Ham, T. S., Lee, T. S., and Keasling, J. D. (2008) Metabolic engineering of microorganisms for biofuels production: from bugs to synthetic biology to fuels, *Current Opinion in Biotechnology 19*, 556-563.

[21] Nguyen, M. T., Choi, S. P., Lee, J., Lee, J. H., and Sim, S. J. (2009) Hydrothermal acid pretreatment of *Chlamydomonas reinhardtii* biomass for ethanol production, *Journal of Microbiology and Biotechnology 19*, 161-166.

[22] Deng, M.-D., and Coleman, J. R. (1999) Ethanol synthesis by genetic engineering in cyanobacteria, *Applied and Environmental microbiology 65*, 523-528.

[23] Dexter, J., and Fu, P. (2009) Metabolic engineering of cyanobacteria for ethanol production, *Energy & Environmental Science 2*, 857-864.

[24] Atsumi, S., Hanai, T., and Liao, J. C. (2008) Non-fermentative pathways for synthesis of branched-chain higher alcohols as biofuels, *Nature 451*, 86-89.

[25] Connor, M. R., and Liao, J. C. (2009) Microbial production of advanced transportation fuels in non-natural hosts, *Current Opinion in Biotechnology 20*, 307-315.

[26] Metzger, P., and Largeau, C. (2005) *Botryococcus braunii*: a rich source for hydrocarbons and related ether lipids, *Applied microbiology and biotechnology 66*, 486-496.

[27] Hankamer, B., Lehr, F., Rupprecht, J., Mussgnug, J. H., Posten, C., and Kruse, O. (2007) Photosynthetic biomass and H2 production by green algae: from bioengineering to bioreactor scale-up, *Physiol Plant 131*, 10-21.

[28] Posewitz, M. C., Dubini, A., Meuser, J. E., Seibert, M., and Ghirardi, M. L. (2009) Hydrogenases, Hydrogen Production and Anoxia, in *The Chlamydomonas Sourcebook, Second Edition: Organellar and Metabolic Processes* (Stern, D., Ed.) 2 ed., pp 217-255, Academic Press, Oxford.

[29] Hemschemeier, A., Fouchard, S., Cournac, L., Peltier, G., and Happe, T. (2008) Hydrogen production by Chlamydomonas reinhardtii: an elaborate interplay of electron sources and sinks, *Planta 227*, 397-407.

[30] Grossman, A. R., Croft, M., Gladyshev, V. N., Merchant, S. S., Posewitz, M. C., Prochnik, S., and Spalding, M. H. (2007) Novel metabolism in Chlamydomonas through the lens of genomics, *Curr Opin Plant Biol 10*, 190-198.

[31] Dubini, A., Mus, F., Seibert, M., Grossman, A. R., and Posewitz, M. C. (2009) Flexibility in Anaerobic Metabolism as Revealed in a Mutant of Chlamydomonas reinhardtii Lacking Hydrogenase Activity, *J Biol Chem 284*, 7201-7213.

[32] Ghirardi, M. L., Togasaki, R. K., and Seibert, M. (1997) Oxygen sensitivity of algal H2-production, *Appl. Biochem. Biotech. 63*, 141-151.

[33] Cinco, R. M., Macinnis, J. M., and Greenbaum, E. (1993) The Role of Carbon-Dioxide in Light-Activated Hydrogen-Production by Chlamydomonas-Reinhardtii, *Photosynthesis Research 38*, 27-33.

[34] Thauer, R. K., Jungermann, K., and Decker, K. (1977) Energy conservation in chemotrophic anaerobic bacteria, *Bacteriol Rev 41*, 100-180.

[35] Melis, A., Zhang, L., Forestier, M., Ghirardi, M. L., and Seibert, M. (2000) Sustained photobiological hydrogen gas production upon reversible inactivation of oxygen evolution in the green alga Chlamydomonas reinhardtii, *Plant Physiol 122*, 127-136.

[36] Griffiths, M., and Harrison, S. (2009) Lipid productivity as a key characteristic for choosing algal species for biodiesel production, *Journal of Applied Phycology 21*, 493-507.

[37] Chisti, Y. (2008) Biodiesel from microalgae beats bioethanol, *Trends in Biotechnology 26*, 126-131.

[38] Ben-Amotz, A., Polle, J. E. W., and Subba Rao, D. V. (2009) *The alga Dunaliella: biodiversity, physiology, genomics and biotechnology*, Science Publishers, Enfield, NH.

[39] Huesemann, M. H., and Benemann, J. R. (2009) Biofuels from Microalgae: Review of Products, Processes and Potential, with Special Focus on *Dunaliella* sp., in *The Alga Dunaliella* (Ben-Amotz, A., Ed.), pp 445-474.

[40] Riekhof, W. R., and Benning, C. (2009) Glycerolipid biosynthesis, in *The Chlamydomonas Sourcebook* (Stern, D. B., Ed.) 2 ed., pp 41-68, Elsevier.

[41] Amin, S. (2009) Review on biofuel oil and gas production processes from microalgae, *Energy Conv. Manag. 50*, 1834-1840.

[42] Weyer, K., Bush, D., Darzins, A., and Willson, B. (2009) Theoretical Maximum Algal Oil Production, *BioEnergy Research*.

[43] Harwood, J. L., and Guschina, I. A. (2009) The versatility of algae and their lipid metabolism, *Biochimie 91*, 679-684.

[44] Hu, Q., Sommerfeld, M., Jarvis, E., Ghirardi, M., Posewitz, M., Seibert, M., and Darzins, A. (2008) Microalgal triacylglycerols as feedstocks for biofuel production: perspectives and advances, *Plant J. 54*, 621-639.

[45] Stern, D. B., and Harris, E. H., (Eds.) (2009) *The Chlamydomonas Sourcebook*, Vol. 2, 2 ed., Elsevier, Oxford.

[46] Willett, W., and Mozaffarian, D. (2008) Ruminant or industrial sources of trans fatty acids: public health issue or food label skirmish?, *Am J Clin Nutr 87*, 515-516.

[47] Cahoon, E. B., Carlson, T. J., Ripp, K. G., Schweiger, B. J., Cook, G. A., Hall, S. E., and Kinney, A. J. (1999) Biosynthetic origin of conjugated double bonds: production of fatty acid components of high-value drying oils in transgenic soybean embryos, *Proc. Natl. Acad. Sci. U.S.A. 96*, 12935-12940.

[48] Burgess, J., de la Rosa, R., Jacobs, R., and Butler, A. (1991) A new eicosapentaenoic acid formed from arachidonic acid in the coralline red algae *Bossiella orbigniana*, *Lipids 26*, 162-165.

[49] Smith, S. (1994) The animal fatty acid synthase: one gene, one polypeptide, seven enzymes, *The FASEB Journal 8*, 1248-1259.

[50] Sul, H. S., and Smith, S. (2008) Fatty acid synthesis in eukaryotes, in *Biochemistry of lipids, lipoproteins and membranes* (Vance, D. E., and Vance, J. E., Eds.) 5 ed., pp 155-190, Elsevier.

[51] Baud, S., and Lepiniec, L. (2009) Regulation of de novo fatty acid synthesis in maturing oilseeds of *Arabidopsis*, *Plant Physiol. Biochem. 47*, 448-455.

[52] Schmid, K. E., and Ohlroge, J. B. (2008) Lipid metabolism in plants, in *Biochemistry of Lipids, Lipoproteins and Membranes* (Vance, D. E., and Vance, J. E., Eds.) 5 ed., pp 97-130, Elsevier.

[53] Voelker, T., Worrell, A., Anderson, L., Bleibaum, J., Fan, C., Hawkins, D., Radke, S., and Davies, H. (1992) Fatty acid biosynthesis redirected to medium chains in transgenic oilseed plants, *Science 257*, 72-74.

[54] Voelker, T. A., and Davies, H. M. (1994) Alteration of the specificity and regulation of fatty acid synthesis of *Escherichia coli* by expression of a plant medium-chain acyl-acyl carrier protein thioesterase, *J. Bacteriol. 176*, 7320-7327.

[55] Heath, R. J., White, S. W., and Rock, C. O. (2001) Lipid biosynthesis as a target for antibacterial agents, *Progress in Lipid Research 40*, 467-497.

[56] Durrett, T. P., Benning, C., and Ohlrogge, J. (2008) Plant triacylglycerols as feedstocks for the production of biofuels, *Plant J. 54*, 593-607.

[57] Wilson, R. F., and Burton, J. W. (2002) New gene combinations governing saturated and unsaturated fatty acid compositionin soybean, in *Lipid Biotechnology* (Kuo, T. M., and Gardner, H. W., Eds.), pp 95-114, Marcel Dekker.

[58] Thompson, G. A. (1996) Lipids and membrane function in green algae, *Biochim. Biophys. Acta 1302*, 17-45.

[59] Carruthers, A., and Melchior, D. L. (1986) How bilayer lipids affect membrane protein activity, *11*, 331-335.

[60] Schneider, K., Kienow, L., Schmelzer, E., Colby, T., Bartsch, M., Miersch, O., Wasternack, C., Kombrink, E., and Stuible, H.-P. (2005) A new type of peroxisomal acyl-coenzyme A synthetase from *Arabidopsis thaliana* has the catalytic capacity to activate biosynthetic precursors of jasmonic acid, *Journal of Biological Chemistry 280*, 13962-13972.

[61] Moolenaar, W. H. (1995) Lysophosphatidic acid, a multifunctional phospholipid messenger, *J. Biol. Chem. 270*, 12949-12952.

[62] Ryu, S. B. (2004) Phospholipid-derived signaling mediated by phospholipase A in plants, *Trends Plant Sci. 9*, 229-235.

[63] Nakamura, Y., and Ohta, H. (2010) Phosphatidic acid phosphatases in seed plants, in *Lipid Signaling in Plants* (Munnik, T., Ed.), pp 131-141, Springer, Berlin.

[64] Huang, A. H. C. (1992) Oil bodies and oleosins in seeds, *Annu. Rev. Plant Physiol. 43*, 177-200.

[65] Turkish, A., and Sturley, S. L. (2006) Regulation of Triglyceride Metabolism. I. Eukaryotic neutral lipid synthesis: "Many ways to skin ACAT or a DGAT", *American Journal of Physiology: Gastrointestinal and Liver Physiology 292*, 953-957.

[66] Winkler, U., Säftel, W., and Stabenau, H. (1988) β-Oxidation of fatty acids in algae: localization of thiolase and acyl-CoA oxidizing enzymes in three different organisms, *Planta 175*, 91-98.

[67] Bourdin, B., Adenier, H., and Perrin, Y. (2007) Carnitine is associated with fatty acid metabolism in plants, *Plant Physiol. Biochem. 45*, 926-931.

[68] Nelson, D. L., and Cox, M. M., (Eds.) (2005) *Lehninger Principles of Biochemistry*, 4 ed., W.H. Freeman and Co., New York.

[69] Chensworth, J. M., Stuchbury, T., and Scaife, J. R. (1998) *An introduction to agricultural biochemistry*, Chapman & Hall, London.

[70] Drapcho, C. M., Nhuan, N. P., and Walker, T. H. (2008) *Biofuels engineering process technology*, McGraw-Hill, New York.

[71] Harwood, J. L. (1998) Membrane lipids in algae, in *Lipids in photosynthesis: structure, function and genetics* (Siegenthaler, P. A., Murata, N., Ed.), Kluwer Academic Publishers, Dordrecht.

[72] Christie, W. W. (2010) Betaine lipids: occurrence, biochemistry and analysis, in *The Lipid Library*, Dundee.

[73] Sato, N. (1992) Betaine lipids, *Journal of Plant Research 105*, 185-197.
[74] Giroud, C., Gerber, A., and Eichenberger, W. (1988) Lipids of *Chlamydomonas reinhardtii*. analysis of molecular species and intracellular site(s) of biosynthesis, *Plant Cell Physiol. 29*, 587-595.
[75] Eichenberger, W., Araki, S., and Müller, D. G. (1993) Betaine lipids and phospholipids in brown algae, *Phytochemistry 34*, 1323-1333.
[76] Bigogno, C., Khozin-Goldberg, I., Boussiba, S., Vonshak, A., and Cohen, Z. (2002) Lipid and fatty acid composition of the green oleaginous alga Parietochloris incisa, the richest plant source of arachidonic acid, *Phytochemistry 60*, 497-503.
[77] Chen, W., Zhang, C., Song, L., Sommerfeld, M., and Hu, Q. (2009) A high throughput Nile red method for quantitative measurement of neutral lipids in microalgae, *Journal of Microbiological Methods 77*, 41-47.
[78] Ashford, A., Barclay, W., Weaver, C., Giddings, T., and Zeller, S. (2000) Electron microscopy may reveal structure of docosahexaenoic acid-rich oil within Schizochytrium sp, *Lipids 35*, 1377-1387.
[79] Mattanovich, D., and Borth, N. (2006) Applications of cell sorting in biotechnology, *Microbial Cell Factories 5*, -.
[80] Crosbie, N. D., Pöckl, M., and Weisse, T. (2003) Rapid establishment of clonal isolates of freshwater autotrophic picoplankton by single-cell and single-colony sorting, *Journal of Microbiological Methods 55*, 361-370.
[81] Reckermann, M. (2000) Flow sorting in aquatic ecology, *Scientia Marina 64*, 235-246.
[82] Dubelaar, G. B. J., and Jonker, R. R. (2000) Flow cytometry as a tool for the study of phytoplankton, *Scientia Marina 64*, 135-156.
[83] Sensen, C. W., Heimann, K., and Melkonian, M. (1993) The production of clonal and axenic cultures of microalgae using fluorescence-activated cell sorting, *European Journal of Phycology 28*, 93 - 97.
[84] Sieracki, M., Poulton, N., and Crosbie, N. (2005) Automated Isolation Techniques for Microalgae, in *Algal Culturing Techniques* (Andersen, R. A., Ed.), pp 101-116.
[85] Laurens, L. (2008) Rapid Fatty Acid Fingerprinting in Microalgae by Gas Chromatography: Analytical Opportunities, *The Column 4*, 7-16.
[86] Shimogawara, K., Fujiwara, S., Grossman, A., and Usuda, H. (1998) High-efficiency transformation of *Chlamydomonas reinhardtii* by electroporation, *Genetics 148*, 1821-1828.
[87] Jarvis, E. E., and Brown, L. M. (1991) Transient expression of firefly luciferase in protoplasts of the green alga *Chlorella ellipsoidea*, *Current Genetics 19*, 317-321.
[88] Maruyama, M., Horáková, I., Honda, H., Xing, X., Shiragami, N., and Unno, H. (1994) Introduction of foreign DNA into *Chlorella saccharophila* by electroporation, *Biotechnology Techniques 8*, 821-826.
[89] Dawson, H. N., Burlingame, R., and Cannons, A. C. (1997) Stable transformation of *Chlorella*: rescue of nitrate reductase-deficient mutants with the nitrate reductase gene, *Current Microbiology 35*, 356-362.
[90] Schiedlmeier, B., Schmitt, R., Müller, W., Kirk, M. M., Gruber, H., Mages, W., and Kirk, D. L. (1994) Nuclear transformation of *Volvox carteri*, *Proceedings of the National Academy of Sciences, USA 91*, 5080-5084.
[91] El-Sheekh, M. M. (1999) Stable transformation of the intact cells of *Chlorella kessleri* with high velocity microprojectiles, *Biologia Plantarum 42*, 209-216.

[92] Huang, X., Weber, J. C., Hinson, T. K., Mathieson, A. C., and Minocha, S. C. (1996) Transient expression of the GUS reporter gene in the protoplasts and partially digested cells of *Ulva lactuca L.*(Chlorophyta), *Botanica marina 39*, 467-474.
[93] Sun, Y., Gao, X., Li, Q., Zhang, Q., and Xu, Z. (2006) Functional complementation of a nitrate reductase defective mutant of a green alga *Dunaliella viridis* by introducing the nitrate reductase gene, *Gene 377*, 140-149.
[94] Chen, H. L., Li, S. S., Huang, R., and Tsai, H. J. (2008) Conditional production of a functional fish growth hormone in the transgenic line of *Nannochloropsis oculata* (Eustigmatophyceae), *Journal of Phycology 44*, 768-776.
[95] Poulsen, N., Chesley, P. M., and Kröger, N. (2006) Molecular genetic manipulation of the diatom *Thalassiosira pseudonana* (Bacillariophyceae), *Journal of Phycology 42*, 1059-1065.
[96] Dunahay, T. G., Jarvis, E. E., and Roessler, P. G. (1995) Genetic transformation of the diatoms *Cyclotella cryptica* and *Navicula saprophila*, *Journal of Phycology 31*, 1004-1011.
[97] Qin, S., Sun, G. Q., Jiang, P., Zou, L. H., Wu, Y., and Tseng, C. K. (1999) Review of genetic engineering of *Laminaria japonica* (Laminariales, Phaeophyta) in China, *Hydrobiologia 398*, 469-472.
[98] Qin, S., Yu, D. Z., Jiang, P., Teng, C. Y., and Zeng, C. K. (2003) Stable expression of lacZ reporter gene in seaweed *Undaria pinnatifida*, *High Technol. Lett. 13*, 87-89.
[99] Minoda, A., Sakagami, R., Yagisawa, F., Kuroiwa, T., and Tanaka, K. (2004) Improvement of culture conditions and evidence for nuclear transformation by homologous recombination in a red alga, *Cyanidioschyzon merolae* 10D, *Plant and Cell Physiology 45*, 667-671.
[100] Cheney, D., Metz, B., and Stiller, J. (2001) *Agrobacterium*-mediated genetic transformation in the macroscopic marine red alga *Porphyra yezoensis*, *J Phycol (Suppl) 37*, 11.
[101] Kurtzman, A. M., and Cheney, D. P. (1991) Direct gene transfer and transient expression in a marine red alga using the biolistic method, *J Phycol (Suppl) 27*, 42.
[102] Gan, S. Y., Qin, S., Othman, R. Y., Yu, D., and Phang, S. M. (2003) Transient expression of lacZ in particle bombarded *Gracilaria changii* (Gracilariales, Rhodophyta), *Journal of Applied Phycology 15*, 345-349.
[103] Lapidot, M., Raveh, D., Sivan, A., Arad, S. M., and Shapira, M. (2002) Stable chloroplast transformation of the unicellular red alga *Porphyridium* species, *Plant Physiology 129*, 7-12.
[104] Michael, R., and Miller, D. J. (1998) Genetic transformation of dinoflagellates (*Amphidinium* and *Symbiodinium*): expression of GUS in microalgae using heterologous promoter constructs, *The Plant Journal 13*, 427-435.
[105] Doetsch, N. A., Favreau, M. R., Kuscuoglu, N., Thompson, M. D., and Hallick, R. B. (2001) Chloroplast transformation in *Euglena gracilis*: splicing of a group III twintron transcribed from a transgenic psbK operon, *Current Genetics 39*, 49-60.
[106] Eichler-Stahlberg, A., Weisheit, W., Ruecker, O., and Heitzer, M. (2009) Strategies to facilitate transgene expression in *Chlamydomonas reinhardtii*, *Planta 229*, 873-883.
[107] Lumbreras, V., Stevens, D. R., and Purton, S. (1998) Efficient foreign gene expression in *Chlamydomonas reinhardtii* mediated by an endogenous intron, *The Plant Journal 14*, 441-447.

[108] Sizova, I., Fuhrmann, M., and Hegemann, P. (2001) A *Streptomyces rimosus* aphVIII gene coding for a new type phosphotransferase provides stable antibiotic resistance to *Chlamydomonas reinhardtii*, *Gene 277*, 221-229.

[109] Berthold, P., Schmitt, R., and Mages, W. (2002) An engineered *Streptomyces hygroscopicus* aph 7 gene mediates dominant resistance against hygromycin B in *Chlamydomonas reinhardtii*, *Protist 153*, 401-412.

[110] Zhao, T., Wang, W., Bai, X., and Qi, Y. (2009) Gene silencing by artificial microRNAs in *Chlamydomonas*, *The Plant Journal 58*, 157-164.

[111] Marín-Navarro, J., Manuell, A. L., Wu, J., and P. Mayfield, S. (2007) Chloroplast translation regulation, *Photosynthesis Research 94*, 359-374.

[112] Gruber, A., Vugrinec, S., Hempel, F., Gould, S. B., Maier, U. G., and Kroth, P. G. (2007) Protein targeting into complex diatom plastids: functional characterisation of a specific targeting motif, *Plant molecular biology 64*, 519-530.

[113] Shrager, J., Hauser, C., Chang, C. W., Harris, E. H., Davies, J., McDermott, J., Tamse, R., Zhang, Z., and Grossman, A. R. (2003) *Chlamydomonas reinhardtii* genome project. A guide to the generation and use of the cDNA information, *Plant Physiology 131*, 401-408.

[114] Derelle, E., Ferraz, C., Rombauts, S., Rouze, P., Worden, A. Z., Robbens, S., Partensky, F., Degroeve, S., Echeynie, S., Cooke, R., Saeys, Y., Wuyts, J., Jabbari, K., Bowler, C., Panaud, O., Piegu, B., Ball, S. G., Ral, J. P., Bouget, F. Y., Piganeau, G., De Baets, B., Picard, A., Delseny, M., Demaille, J., Van de Peer, Y., and Moreau, H. (2006) Genome analysis of the smallest free-living eukaryote *Ostreococcus tauri* unveils many unique features, *Proceedings of the National Academy of Sciences, USA 103*, 11647-11652.

[115] Liolios, K., Mavromatis, K., Tavernarakis, N., and Kyrpides, N. C. (2008) The Genomes On Line Database (GOLD) in 2007: status of genomic and metagenomic projects and their associated metadata, *Nucleic acids research 36*, D475-479.

[116] Norton, T. A., Melkonian, M., and Andersen, R. A. (1996) Algal biodiversity, *Phycologia 35*, 308-326.

[117] Mann, D. G., and Droop, S. J. M. (1996) Biodiversity, biogeography and conservation of diatoms, Hydrobiologia 336, 19-32.

[118] Sheehan, J., Dunahay, T., Benemann, J., and Roessler, P. G. (1998) A Look back at the US Department of Energy's Acquatic Species Program - Biodiesel from Algae, Close Out Report TP-580-24190.

[119] Sheehan, J., Dunahay, T., Benemann, J., and Roessler, P. (1998) A look back at the US Department of Energy's Aquatic Species Program—Biodiesel from Algae, Report No. NREL/TP-580-24190, National Renewable Energy Laboratory, Golden, Colorado (http://govdocs.aquake.org/cgi/reprint/2004/915/9150010.pdf).

[120] Cooksey, K. E., Guckert, J. B., Williams, S. A., and Collis, P. R. (1987) Fluorometric determinations of neutral lipid content of microalgal cells using Nile Red., J. Microbiol. Methods 6, 333-345.

[121] Johansen, J., Lemke, P., Nagle, N., Chelf, P., Roessler, R., Galloway, R., and Toon, S. (1987) Addendum to Microalgae Culture Collection 1986-1987.

In: Bioprocess Sciences and Technology
Editor: Min-Tze Liong

ISBN 978-1-61122-950-9

Chapter 5

ALGAE FOR THE PRODUCTION OF PHARMACEUTICALS

***M. Frontasyeva*[1,*], *N. Tsibakhashvili*[2], *L. Mosulishvili*[3] *and E. Kirkesali*[3]**

[1]Frank Laboratory of Neutron Physics, Joint Institute for Nuclear Research, str. Joliot-Curie, 6, 141980 Dubna, Moscow Region, Russian Federation
[2]Ilya State University, av. Chavchavadze, 32, Tbilisi 0179, Georgia
[3]Andronikashvili Institute of Physics, str. Tamarashvili, 6, Tbilisi 0177, Georgia

ABSTRACT

The chapter reviews the implementation of biomass of different species of algae into the development of therapeutic and preventive pharmaceutical medications. The methods used for substantiating the biotechnology of their production as well as criteria for assessing quality of medications ba[1]sed on alga biomass are described. The role of modern analytical techniques to monitor both the quality of the original biomass of algae and ready substances intended for creation of new medications is considered. The widely employed blue-green microalga *Spirulina platensis* is used as an example of scientifically-based approach to develop new pharmaceutical medications. By means of instrumental neutron activation analysis, the possibility of targeted introduction of trace elements into *Spirulina* bio-complexes while maintaining its natural beneficial properties has been shown. The possibility of precise dosage of elements in Se-, Cr- and iodine-containing substances based on *Spirulina platensis* biomass is demonstrated. The advantages of biotechnological methods developed to ensure the quality of new pharmaceutical medications are discussed.

* Corresponding author: *M. Frontasyeva* E-mail: marina@nf.jinr.ru

1. Introduction

Algae are a heterogenic ecological group of predominantly phototrophic unicellular colonial or multicellular organisms normally living in aqueous media and also found in soil, tree bark, and on stones. They can be both prokaryotic (possessing no nucleus) and eukaryotic (possessing a nucleus). There are about 100 000 algal species which are split into 11 divisions according to the difference in a set of pigments, chromophore structure, morphological and biochemical features.

Algae have no roots, and they get all nutrients from the environment. The structure of multicellular algae resembles the structure of higher plants. The size of algae varies within a wide range from fractions of a micrometer (microalgae) to tens of meters (macroalgae, or macrophytes) [1–4].

Most species of algae live in marine water and are the main source of food for all living organisms in marine environment. Since ancient times, man has been using marine macroalgae for food and curing. The earliest evidence for this comes from China and dates back to 850 BC [5].

Now about 160 species of edible macrophytes are known, of which 25 are green, 54 are brown, and 81 are red [6]. Algae make up a substantial part of diets for the population of the coastal regions in China, Japan, and South- East Asia countries and are widely used in some regions of France, Ireland, Scotland, and Norway.

Algae are rich sources of chemicals and biologically active compounds used by humans in their activities. With technological progress, macroalgae have been more and more extensively used on a commercial scale in food, textile, paper-making, cosmetics, and pharmaceutical industries. As for microalgae, they can be used as extensively as macroalgae on cultivation in artificial ponds [7, 8].

Treatment of various diseases by medicines of algal origin, algotherapy, has found its appropriate place among other therapies. Modern pharmaceutical industry produces a lot of preparations made from algal biomass (powders, capsules, suspension), medicines from raw algae, and biologically active additives for treatment and prevention of various diseases.

The increasing demand for these preparations largely dictates interest in the researches aimed at developing new scientifically justified processes for the production.

This chapter deals with some typical features of algal biotechnologies and describes an example of elaborating and justifying the process for production of new medicines based on the *Spirulina platensis* blue-green microalga biomass enriched in some essential elements.

2. Marine Algae for Pharmaceutical

Marine algae are the oldest members of the Plant Kingdom, predecessors of the plant that left sea for land in the Paleozoic. They impressively vary in size, form, and color. The simplest of them, blue-green algae, form a bluish mucoid film on seashore stones and in shallow waters. Underwater forests are brown macrophytes: alaria, laminaria, macrocystis, nereocystis, eclonia, and lessonia. Very deep in the water live red algae.

The chemical composition of marine algae is diverse and depends on their habitat.

Algal products rank below land plant products in the quantitative and qualitative protein and carbohydrate composition, but they have a variety of valuable properties which land plants do not have. Among these properties there are an ability to take up great amounts of water and to increase in volume, a high content of specific colloid polymers (agar, alginic acids) and mannitol, and a great diversity of macro- and microelements, vitamins, and many other biologically active substances. A typical feature of algae is their ability to accumulate elements that are present in water in negligible amounts [9–16].

Green algae are the richest in proteins: 40–45% (bicarbonic acids, alanine, alginin, leupin). Their carbohydrate content is 30–35% and lipid content is 10%.

Brown algae contain 5–15% of proteins, 70% of carbohydrates (sugar mannitol, laminate, polyuronides—alginic and furinic acids, fuxidin, laminarin) and 1–3% of lipids.

Red algae contain 20% of proteins, up to 70% of carbohydrates (a simple sugar phloridzin, a disaccharide trehalose, sugar alcohols; a mucous sugar agar, the most valuable of polysaccharides), and about 3% of lipids [14].

Marine algae contain biologically active substances: polyunsaturated fatty acids, derivatives of chlorophyll, polysaccharides, fucoidans, glucans, galactins, pectins, alginic acid, plant sterols, and carotenoids. They feature the presence of vitamin A, which possesses antihemorrhagic properties, and other vitamins: B_1, B_2, B_{12}, C, D, K, PP (nicotinic acid), pantothenic acid, and folic acid [15].

Many algae were found to be effective against tumors (laminaria, fucus) and to show antimicrobic, antiviral, antimutagenic, radioprotective, and immunomodulatory effect [13, 17].

Carbohydrates of marine macrophytes have valuable properties. Mannitol is used as a sugar substitute in diabetes. Most widely used are polysaccharides: agar, carrageenans, and alginates. Of particular interest is alginic acid (brown algae, laminaria), a polysaccharide with sorption properties that compare well with any natural sorbent. Alginic acid is capable of sorbing water 300 times as much as its own weight. The salts of alginic acid (potassium, sodium, magnesium alginates) form viscous solutions and are used as thickeners, stabilizers, and binders in various industries. The pharmaceutical industry consumes 5% of all the alginates extracted from marine algae [16].

Marine macrophytes are currently harvested at an amount of 4,000,000 t/year, which is 4% of the world's marine resources. According to the data of the Food and Agriculture Organization (FAO), the allowable amount of harvest is up to 2,600,000 t/year of red algae and 14,500,000 t/year of brown algae. In addition, the so-called mariculture, farming of marine algae for industrial purposes, is developing and accounts for 1/4 of the total product in this industry [9, 18].

3. MICROALGAE

Microalgae are one of the most interesting groups of microorganisms, which are of interest for researchers from both the scientific and applied point of view. This abundant group includes both eukaryotes, microscopic algae, and prokaryotes, cyanobacteria, unicellular, filamentous, and colonial ones.

Out of more than 20,000 species of microalgae, only a few are practically used: *Chlorella, Dunaliella, Haematococcus, Scenedesmus, Porphyridium, Botryococcus, Spirulina spp., Aphazomenon flos aquae,* etc. [17–23].

Owing to their small size, relatively simple morphology, and reproduction modes, microalgae can be easily cultivated under laboratory conditions by microbiological methods in artificial mineral or organic solid or liquid nutrient media. Solid media are used to keep collections of strains, while the study of physiological and biological properties and cultivation of cells are carried out in liquid media.

As it is known, a large surface-to-volume ratio of a microorganism's cell results in a high intensity of exchange with environment in the course of metabolism. Plastic metabolism of microalgae allows the biomass quality to be controlled by targeted variation in cell cultivation conditions.

All microalgae used are photosynthetic organisms. A set of various pigments allows them to use the solar spectrum energy to the maximum. They are sometimes capable of adapting the composition of the photosynthetic pigments to the spectrum composition of light. In nature, some microalgae develop in sea water (*Chlorella, Dunaliella*), others in fresh water (*Spirulina*).

In the course of biosynthesis the cells accumulate a lot of useful compounds which have a nutritional value and probiotic effect. The term *probiotic* means an ability to produce an active effect on some physiological processes in a human. Among these compounds are proteins, carbohydrates, lipids, vitamins, etc. Table 1 presents the composition of some microalgae used in industry [18].

The first attempts of mass cultivation of microalgae date back to the 1940s. Subsequently, studies of their properties and search for new commercially beneficial strains began. By the 1990s, some microalgae, like *Chlorella, Spirulina, Scenedesmus,* etc., had been treated as commercially viable sources of high-quality protein [24, 25], vitamins [19, 21], lipids [21, 26, 27], and a variety of unique chemical compounds [21, 27–29] and as nutrient-dense foods.

It has become obvious that commercial cultivation of microalgae has advantages [18]:

1. Microalgae are simple to cultivate as they grow in natural environments (sunlight, water, pH = 9–11] without requiring any special agricultural techniques and may develop in both fresh and sea water [21, 30, 31].
2. Microalgae grow rapidly and produce more biomass per unit time than any other food source [32].
3. Microalgae show high productivity that ranges from 5.5 to 40 g/m^2 ◌̇◌̇◌̇◌̇ ? day (dry weight) [22, 30, 31, 33–36].
4. Cultivation of microalgae does not contribute to soil erosion, requires no herbicides and pesticides, and the energy consumption is minimal [21, 36].
5. In many developing countries the microalgal biomass is a substantial part of the population's diet.
6. In developed countries, microalgae are a source of valuable materials for the food, pharmaceutical, and chemical industries.

Table 1. Nutritional value of commercially produced food-grade microalgae [18]

Substances	*Aph. flos-aquae*	*Chlorella*	*Spirulina*
Moisture	6%	5%	5%
Protein	63%	58%	65%
Total lipid	3%	9%	5% •
Fatty acids	n/a	n/a	n/a
Carbohydrate	27%	23%	18%
Ash	7%	3%	7%
Calcium	140.0 mg	22.0 mg	100.0 mg
Phosphorus	51.0 mg	90.0 mg	90.0 mg
Iron	6.4 mg	13.0 mg	15.0 mg
Sodium	38.0 mg	n/a	60.0 mg
Chlorine	46.0 mg	n/a	44.0 mg
Chromium	40.0 mg	n/a	28.0 µg
Copper	60.0 µg	10.0 µg	120.0 µg
Magnesium	16.0 mg	32.0 mg	40.0 mg
Manganese	0.3 mg	n/a	0.5 mg
Zinc	0.3 mg	7.0 mg	0.3 mg
Potassium	100.0 mg	90.0 mg	120.0 mg
Others	n/a	n/a	n/a
Vitamins			
Ascorbic acid	5.0 mg	1.0 mg	0.5 mg
Carotene	2.000.0 RE	550 RE	2,300.0 RE
Biotin	3.6 µg	19 µg	0.5 µg
Cobalamin	8.0 µg	1.3 µg	3.2 µg
Choline	2.6 mg	n/a	n/a
Folic acid	1.0 µg	2.7 µg	1.0 µg
Pyridoxine	67.0 µg	140.0 µg	80.0 µg
Pantothenic acid	130.0 µg	130.0 µg	10.0 µg
Niacin	0.65 mg	2.38 mg	1.46 mg
Thiamin	0.03 mg	0.17 mg	0.31 mg
Inositol	n/a	13.2 mg	6.4 mg
Vitamin E	1.2 IU	0.1 IU	1.0 IU
Pigments and Other Compounds			
Carotenoids	n/a	n/a	37.0 mg
Phycocyanin	n/a	n/a	1,500.0 mg
Chlorophyll	300 mg	200 µg	115.0mg
Gamma Linolenic acid	n/a	n/a	135.0 mg
Glycolipids	n/a	n/a	200.0 mg
Sulfolipids	n/a	n/a	4—10.0 mg
Nucleic acids	n/a	30%	4.5%

Figures are given per 10 g dry weight.

Data iorAph. *flos-aquae* from Cell Tech. *Chlorella* sp. from Sun Chlorella, and *Spirulina* sp. from Earthrise Farms and SOSA Texcoco.

Yet, there are also some disadvantages and problems in commercial production of microalgae which should be taken into account [18]:

1. The price of 1 kg of microalgal biomass produced in developed countries is high.
2. Areas suitable for mass cultivation of microalgae are restricted.
3. It is difficult to ensure high quality and pureness of the biomass, and unified industry standards are absent.
4. Microalgae potentially may accumulate metals and other toxic elements from the nutrient medium and contain toxins [19, 21, 37, 38].

Systems for commercial cultivation of microalgae can be of the open and closed type. Outdoor cultivation systems first appeared in the United States (Carnegie Institute), Japan, and some other countries in the early 1950s. Essentially, this cultivation consists in growing algae in liquid nutrient media in tanks, raceways, and other reservoirs with various ways of mixing, supplying carbon dioxide, and controlling exposure to sunlight [39–45].

Outdoor cultivars are *Protococcophyceae* (*Chlorella, Scenedesmus*), *Volvocophyceae* (*Dunaliella*), the blue-green microalga *Spirulina*, and some others. The productivity averages between 15–20 g/m^2·day and 35–40 g/m^2·day.

In most outdoor systems microalgae are harvested by the draw-off/fill-up method: part of the suspension is periodically drawn off and an equivalent amount of the nutrient medium is added. However, continuous repetition of these operations leads to a decrease in the productivity due to disbalance of the medium composition, accumulation of metabolites, and culture dissynchronism. Therefore, research is continuously carried out to find strains of cultures stable under varying conditions [42–44].

Outdoor cultivation is characterized by a low reproduction coefficient; therefore, large artificial water areas are needed for getting large amounts of biomass. That is why it is called extensive cultivation. This method is widely used in Israel, Japan, and Australia.

The weather effect is smoothed by using artificial lighting, heating, and cooling and it appreciably affects the final cost of the biomass, and the decrease in the cost with decreasing the water areas and improvement of the product quality are the main problems to be solved in this production method [40, 41].

Another method—intensive cultivation—has been developing since the 1960s: growing of photosynthetic microalgae under precisely controlled conditions: stream flow of the biomass, automatic harvesting, supply of fresh nutrient medium, and stabilization of the optical density of the culture. Intensive cultivation is carried out in special apparatus, photobioreactors, which incorporate powerful light sources and a system of light guides to provide uniform illumination. This cultivation method requires substantial expenses but allows a considerably higher quality and purity of the biomass, which is particularly important for the raw material used in the pharmaceutical industry. Under intensive cultivation, the *Chlorella* harvest can be as high as 30–40 g of dry biomass per liter of suspension or 80–100 g per square meter of the illuminated surface. In all microalgae cultivation methods considered, the final stage is harvesting. The harvested biomass is supplied to pharmaceutical factories in dried and compressed form [45]. The largest part of the world's microalgal biomass resources (75%) is made into powders, tablets, capsules, or pastilles used as biologically active food additives (BAAs) and drugs [17]. Suspensions, aqueous emulsions, and raw pastes are also used for some particular purposes. The rest of the raw material is biochemically processed to produce valuable biologically active substances, many of which are unique and contained only in microalgae.

4. Properties of Some Microalgae Used in Pharmaceuticals

4.1. Chlorella

Chlorella is a unicellular green microalga using the red part of the solar spectrum for photosynthesis. It was first described as green fresh-water microalga by the Dutch scientist M. W. Beijerinck in 1890 [46].

There are 20 *Chlorella* species belonging to the class *Protococcophyceae* or *Chlorococcophyceae*. Most of them live in fresh water.

A cell 15 μm in size has a chloroplast and a nucleus. The cell wall contains not only cellulose but also sporopollenin, a substance very stable against attack of any enzymes. In human, cellulose adsorbs products of metabolism and they are excreted together with it. Due to this fact, *Chlorella* is capable of neutralizing the effect of toxic substances. Like *Spirulina* and unlike other algae, it has no companions (pathogenic bacteria and viruses) as it destroys them in the course of its life activity [47].

The most abundant and undemanding species is *Chlorella vulgaris*, which can be found in fresh and saline waters, humid soil, and tree bark. This species is characterized by high productivity and reproduction rate: the amount of biomass increases by a factor of 4 a day under normal conditions [48]. Other species, *Chlorella ellipsoidea, Chlorella pyronoidosa,* etc., are also in use.

Dried *Chlorella* matter is very nutritive. It contains 15% to 50–60% proteins, up to 40% carbohydrates, and 7–10% lipids, and it also contain vitamins A, B_2, K, and PP [18, 49]. A high protein content and oxygen regeneration ability are the main properties of *Chlorella* due to which it is recommended for use as food in space flights [21].

The chlorophyll content of *Chlorella* is 5–10 times higher than its content of spirulina. Since chlorophyll is identical in molecular structure to hemoglobin molecule, *Chlorella* helps to improve blood properties and shows good effect in treating anemia. It normalizes the glucose level in blood [18, 49–51].

Chlorophyll and fiber material of *Chlorella*'s cell walls favorably affect functioning of the digestive tract, assisting in treatment of dysbacteriosis and other dysfunctions.

Due to the presence of β-carotene in combination with vitamin E, chlorella is able to produce an anticarcinogenic effect [52].

It is used to treat cardiovascular diseases [63, 64] and exerts anti-inflammatory, antiviral, bactericidal, and detoxicating effect [48, 54, 55].

The most important property of chlorella for its biotechnological use is high metabolism plasticity, ability to change the biosynthesis direction depending on the cultivation conditions and various outside actions [55]. Some *Chlorella* cultures are capable of accumulating up to 85% of lipids in the cells under nitrogen deficiency (protein–lipid direction of metabolism). Other cultures show protein–carbohydrate direction of metabolism under the same conditions and can synthesize up to 60% of proteins.

Thus, *Chlorella* cells can be used as systems tending to synthesize proteins, carbohydrates, and lipids depending on the genetic properties. This allows one to manage not only the intensity but also the quality of microalgal biosynthesis.

4.2. Spirulina

Spirulina is one of the most interesting alga for biotechnology, which attracts attention of many researchers [25, 30, 31, 33, 35–28, 48, 56–64]. Its current taxonomic status is defined by both the International Code of Nomenclature of Bacteria and the International Code of Botanical Nomenclature. Therefore, it has a double name blue-green microalga–cyanobacterium [61-63]. It is similar to plankton in appearance and type of nutrition, a typical prokaryote in cell structure, and belongs to the class Cyanophyceae [31, 32, 48, 58].

In the 1970s, active studies of this microorganism began, and numerous interesting results led to an incredible rise in popularity of *Spirulina*. *Spirulina* is a filamentous plankton *cyanobacterium*, or a multicellular helical filamentous alga. Its unbranched helical trichomes (filaments) consist of cylindrical cells, are surrounded by a mucous sheath, and are capable of rotational and sliding movement. The filaments in a helix can be as long as 1 mm. Like all prokaryotes, *Spirulina* has a low level of cell differentiation. A cell lacks a nucleus and chromophores.

Spirulina is phototropic and grows in warm fresh-water or saline lakes with an alkaline medium. *Spirulina* includes several thousand various species. Class *Cyanophyceae*, to which spirulina belongs, includes two genera, *Spirulina* and *Arthrospira*. All food-grade strains, including those commercially grown, belong to *Arthrospira*, they are usually called *Spirulina* [57]. *Spirulina* contains a unique physiologically balanced combination of substances, vital for human: 60–70% protein, 10–2-% carbohydrates, 5% fat, 0.9% chlorophyll, and also vitamins, minerals, carotenoids, etc. (about 50 items in total) [62–64].

It should be mentioned that the data on the biochemical composition of *Spirulina* reported in various works are slightly different, attributed to the different strains used and are affected by cultivation conditions. Table 2 presents a sample composition of the *Spirulina* biomass [61].

The total set of proteins includes eight essential amino acids and ten others. Essential amino acids are needed for synthesis of hormones, enzymes, neuroregulators, immunoglobulins, etc. [30, 31, 36, 58, 65, 66]. *Spirulina* proteins require much less energy for their digestion than other products. Amino acid assimilability coefficient is 95.1% [63, 64], which explains their high biological activity. Carbohydrates (up to 20%) are mainly polysaccharides (glycogen, etc.) that almost do not require insulin for assimilation, which is very important for those having diabetes [60, 67]. Cell systems contain alginates—polysaccharide complexes capable of fixing and excreting radionuclides (Sr, Cs), heavy metals (Hg, Pb, As), nitrogen metabolism products, phenol compounds, and other toxic substances [60, 67–69].

Spirulina features optimum ratios of important minerals (Na, Cl, Ca, K, P, Fe, Mg, Mn, Zn, I, Cu, Cr, Se), vitamins (A, B_1, B_2, B_3, B_6, B_{12}, PP, E), biotin, folic acid, inositol, and pantothenic acid that are necessary for normal functioning of the immune, endocrine, nervous, cardiovascular, hematopoietic, and other systems (Table 3) [36, 60, 64, 70, 71].

Table 2. Biochemical composition of *Spirulina* [61]

Mass fraction, %		Minerals, %	
		Calcium	0.12
Proteins	60–70	Phosphorus	0.83
Carbohydrates	10–15	Potassium	1.4
Lipids	6.5–8.0	Sodium	0.03
Ash	7.5–8.0	Magnesium	0.37
Cellulose	2.0	Iron	0.05
Pigments, %		Zinc	0.003
Carotenoids	0.22–0.40	Vitamins, mg/kg	
Cholophyll	0.76–0.94	β-Carotene	1100–2400
Phycocyanin	0.8–1.0	B_1	55
Polyunsaturated fatty acids, %		B_2	35
		B_6	3–8
Linoleic	1.1–1.4	B_{12}	1.6–3.2
γ-Linolenic	0.9–1.2	E	190

Table 3. Vitamin and mineral content of *Spirulina platensis* biomass (per 10 g) [36, 64]

Vitamin content		Mineral content	
Vitamin A (retinol)	23000 IU	Ca	100 mg
Vitamin B_1 (thiamin)	31 mg	Fe	15 mg
Vitamin B_2 (riboflavin)	35 mg	Zn	300 mg
Vitamin B_3 (niacin)	1.46 mg	P	90 mg
Vitamin B_6 (pyridoxine)	80 μg	Mg	40 mg
Vitamin B_{12} (cyanocobalamin)	32 μg	Na	60 mg
Vitamin E (alpha tocopherol)	1 IU	K	160 mg
Folic acid	1 μg	Mn	500 mg
Pantothenic acid	10 μg	Cr	28 μg
Biotin	5 μg	Se	2 μg
Inositol	6.4 μg		

With a relatively low fat content (up to 5%), *Spirulina* is high in unsaturated acids (vitamin F), especially gamma linolenic, linolenic, and oleic acids that are necessary for synthesis of phospholipids, the main components of cell membranes and vascular walls, and prostaglandins, tissue hormones regulating blood pressure, condition of blood vessel walls, and metabolic processes in liver, and playing an important part in lipid metabolism [63, 72–74].

It has been found that *Spirulina* sulfolipids and glycolipids are able to suppress development of HIV [69–77] and other pathogenic viruses. Biochemical indicators of lipid and cholesterol metabolism in patients having brain, heart, and kidney blood vessel atherosclerosis show positive dynamics under the effect of spirulina. Out of all the substances produced by the pharmaceutical industry from the *Spirulina* biomass, the most valuable are pigments, unique natural dyes: chlorophyll, carotenoids, and phycobiliproteins. A high phycocyanin (blue pigment) content of *Spirulina* (about 10% of the biomass) has an immunostimulating and antitumoral effect. Therefore, the price of the pigments is very high, and some businessmen consider that it is most profitable to process *Spirulina* biomass predominantly into natural dyes [30, 60, 63, 78].

The *Spirulina* chlorophyll is almost identical to human hemoglobin molecule. Therefore, it is effective against anemia because it contributes to an increase in the hemoglobin level.

Cell walls of *Spirulina* consist of mucopolysaccharides which are 85–95% digested by digestive enzymes in the stomach, increasing assimilability of protein. This makes *Spirulina* favorably different from the other plants where cells containing cellulose that is not digested in the stomach. Therefore, *Spirulina* could positively benefit those suffering from gastric ulcer, gastritis, pancreatitis, enteritis, and other gastrointestinal disturbances.

Spirulina platensis is commercially cultivated in Mexico, the United States, India, Thailand, Japan, China, and Taiwan. Meanwhile, Bulgaria, Chile, Israel, Italy, Moldavia, and Ukraine are also developing its production. The growth of the world *Spirulina* biomass production is evident as seen from Figure 1 [61].

The continuous growth of the *Spirulina* production is also attributed to the ability of *Spirulina* to assimilate in varying environmental and climatic conditions. The biomass cultivation technology is relatively simple; *Spirulina* can be continuously grown in open ponds under natural light. Its growth rate and yield are 5–10 times higher than in traditional agricultural crops. Under certain conditions biomass can double in five hours [80–82].

There are projects for cultivation of sea-water-adapted spirulina in coastal tidal zones of seas and oceans with productivity of 7–12 g of dry biomass from 1 m^2 per day.

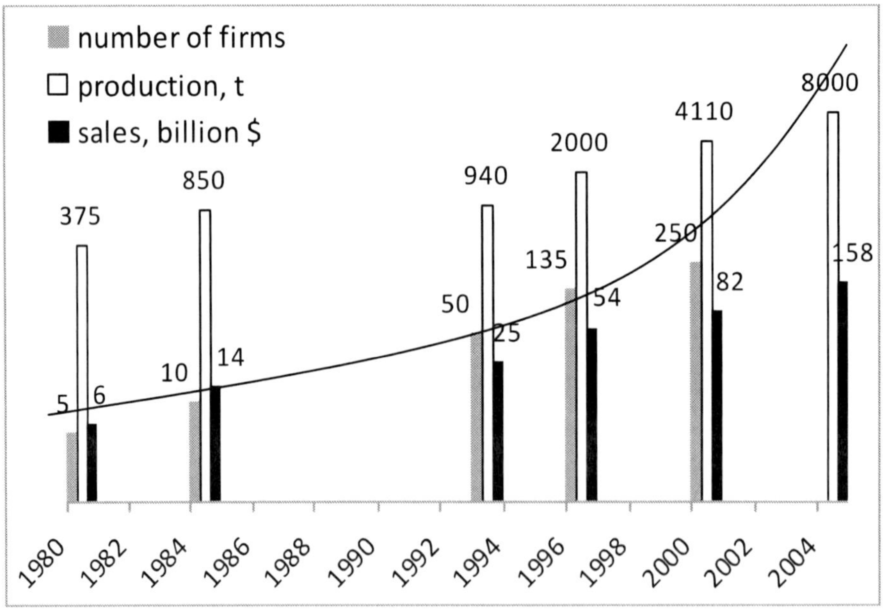

Figure 1. World production of *Spirulina* (1980–2004) [61].

4.3. Dunaliella

Dunaliella, a green unicellular microalga named after *Dunalle*, a French scientist who discovered it in 1905, belongs to the phylum *Chlorophyta.* The *Dunaliellaceae* includes 29 saline-water, fresh-water, and soil species. *Dunaliella salina* is of vast interest biotechnologically, attributed to its habitat in shallow lagoons of hypersaline lakes. It is found in Europe (Mediterranean, Black Sea, Azov Sea, and Caspian Sea coasts), Asia (West Siberia, Central Asia, India, Java), North America (California, Nevada, Utah), South America (Chile), North Africa (Algeria, Sahara), and South Australia.

Dunaliella cells are 2.8–40 μm in size, vary in shape, and move using flagella. Their chloroplast is most often green and sometimes yellow or brown changing to red. *Dunaliella* biomass is 10–57% protein, 10–32% carbohydrates, 7–30% lipids, 6% nucleic acids, 0.1–6% chlorophyll, and 0.03–10% carotenoids. Cell walls do not contain cellulose [83–85].

Saline lakes where *D. salina* lives are characterized by unstable salinity varying within a wide range of 30 to 300 g/l. Therefore, microalgae living there are plastic species capable of accommodating to extreme conditions and appropriately adapting their pigment system, respiratory and photosynthetic metabolism, nuclei acid metabolism, and other processes. To protect the cells, *D. salina* is able to synthesize substantial amounts of valuable biologically active substances such as β-carotene, glycerol, and unsaturated fatty acids [86–92]. Sometimes hypersynthesis of β-carotene results in up to 10% of the dry microalgal mass.

In Israel, experiments are carried out with the species of *D. bardawii*, which can grow in ocean and sea water with a wide range of salt concentration, even in the water of the Dead Sea almost saturated with salt. Under these extreme conditions, cells of *D. bardawi* are able to accumulate glycerol in amounts as large as 85% of dry biomass. The pharmaceutical industry employs the dunaliella biomass mainly for producing β-carotene and glycerol, and also for producing biologically active additives that have a lot of therapeutic and health-improving properties due to a rich combination of biologically active compounds.

4.4. Haematococcus

Haematococcus is a green microalga, *Chlorophyta* and genus *Haematococcaceae* known since the late 19th century. In 1899, the American Botanical Club published the first data on therapeutic properties of algae from the Hawaii and gave it the name *Haematococcus.* In 1938, American scientists isolated a reddish substance from *Haematococcus* and called it hematochrome, a term that was later changed to astaxanthin. At the same time, a similar substance was isolated from salmons, lobsters and shrimps, which were found to contribute a red color.

It was found that when the green alga *Haematococcus* is exposed to bright light, it changes color from green to red, producing astaxanthin that protects cells against free radicals arise from the effect of ultraviolet radiation [7, 17, 93–95].

Investigations have shown that astaxanthin is a unique carotenoid that is superior compared to β-carotene. Astaxanthin is 10 times more effective as an antioxidant than β-carotene and 100 times more effective than vitamin E. While interacting with other antioxidants, astaxanthin enhances their effectiveness.

Astaxanthin is able not only to neutralize the effect of free radicals but also to terminate chain reactions resulting in destruction of cells. Formed on both sides of bilipid cell membranes, astaxanthin molecules protect both outside and inside. Astaxanthin has been found to exhibit anticarcinogenic and radioprotective effects in human. Other therapeutic effects include improvement and protective properties of skin, stimulation of cell regeneration, protection of the nervous system and improvement of brain activity, prevention of cell membrane rupture, strengthening immunity, improvement of blood vessels, improvement of elasticity of ligaments and cartilages, inhibition of inflammatory processes, functional normalization of the gastrointestinal tract, vision improvement, decelerate the aging process, and improvement of working capacity.

One of the most cultivated *Haematococcus* strains is *Haematococcus pluvialis* [93]. However, production of astaxanthin from this microalgal biomass requires additional procedures that technologically often complicate commercial production. First the green mass of *Haematococcus pluvialis* is kept in a nutrient medium under daylight lamps for a week until the biomass doubles. Then the biomass is transferred to another tank with lamps submerged into it, kept for another week, and transferred again to tanks of larger volume (30 000 l) with deficient nutritional medium, where microalgae rapidly producing astaxanthin under extreme conditions. All these operations are bound up with a decrease in the growth rate and an increase biomass contamination. Therefore, most pharmaceuticals manufacturers prefer the use of expensive synthetic astaxanthin over naturally produced ones [17].

5. Substantiation of the Method for Making Substances for Pharmaceuticals on the Basis of the *Spirulina Platensis* Biomass

5.1. Formulation of the Problem

Development of substances for pharmaceuticals is the initial and decisive stage in their preparation. At this stage, apart from solving the formulated problem, one should take into account both preservation of useful natural properties of the biomass and purity of the resulting product. When any elements important for human are purposefully added to the biomass composition, the scientifically substantiated substance making technique should guarantee exact adherence to the doses corresponding to the purpose of the preparation. To this end, up-to-date biochemical and analytic methods are used.

The next stage is pharmacological studies aimed at testing substances, reaction to susceptibility, its effect on the immune system, enzymatic, genetic processes, and much other are tested. Equally important issues are estimation of the purity grade of the preparation, standardization, etc. [96].

This section deals with scientific substantiation of a possibility of developing substances for pharmaceuticals based on the *Spirulina platensis* biomass containing necessary doses of trace elements, such as selenium, iodine, chromium, while the natural composition and useful properties of the biomass are preserved.

The study of the role of microelements in living systems using modern biochemical and analytical techniques is a promising direction in Life Sciences.

According to Bertrand diagram [97] for each particular microelement, there is a certain concentration range of positive effect, with both excess and insufficient concentrations being harmful. Thus, it is evident that a precise choice of required doses is the most important task in designing their substances.

As a rule, in metabolism and exchange processes, microelements are best assimilated by the organism in a biologically accessible form, i.e. when they are included into a biological macromolecule. Hence it follows that it is desirable that for the substance base we should use biologically active biomass that is capable of assimilating required elements in prescribed quantities. The choice of biomass should be determined by its own beneficial therapeutic and preventative properties, as well as by the absence of harmful impurities in concentrations exceeding permissible levels.

Taking into account the requirements mentioned above, the Andronikashvili Institute of Physics (Tbilisi, Georgia) in cooperation with the Joint Institute for Nuclear Research (JINR, Dubna, Russia) developed a technique for production of substances for therapeutic and preventative preparations on the basis of blue-green algae *Spirulina platensis* [98–101].

Spirulina is a living microorganism, and in the process of cell cultivation it is capable of assimilating certain amounts of some microelements from a nutrient medium and incorporating them into the composition of its biological macromolecules. The analytical control of this process makes it possible to establish a unique dependence between the concentration of an element in the nutrient medium and its content in the resulting *S. platensis* biomass. This dependence serves as a basis for substantiation of the biotechnology for production of substances for pharmaceutical preparations with required doses of a given element. It is very important that concentrations of compounds added to the nutrient medium as loading have no influence on the conditions in which *Spirulina* cells normally grow and retain their beneficial natural properties. That is why, along with the analytical control, the protein composition of the obtained biomass was investigated by gel-electrophoresis technique.

The biotechnology of target-oriented incorporation of certain elements into *S. platensis* biomass composition in the process of cultivation was developed with such important elements as Se, Cr and I.

Selenium. The studies of the role of Se in the human over the last 20 years has shown it to be such an important element named the element of the century at the 7th International Symposium «Selenium-2000» (Venice, October 2000). Selenium is a normal component of some enzymes, proteins and amino acids. Low level raises the risk of diseases as cardiomyopathy, cancer, endemic osteoarhropathy, anaemia, etc. [102, 103]. The functions of Se are closely related to vitamin E and beta carotene (which contain in *S. platensis* biomass); therefore, in the treatment they are sometimes used in combination. Selenium contributes to the reduction of harmful effects of free radicals and also allows detoxification from elements such as As, Hg, Cd, Bi, etc. It participates in photochemical reactions related to vision, can influence the immune and endocrine systems, etc. Se added to the diet in particular doses slows down the ageing processes, favors treatment of cardiological patients and reduces the risk of cancer and AIDS [104, 105].

Iodine. Another equally important element incorporated in the composition of all living organisms and plants is iodine. It is important for their development, growth and functioning. Iodine intake strongly depends on the state of the environment, and its deficiency often has an endemic character. Iodine influences metabolism and enhances reduction-oxidation processes.

Thus, iodine deficiency can affect the physical, mental and emotional state of the organism. Among serious symptoms and results of iodine deficiency are cardiovascular diseases (atherosclerosis, vessel deformation, etc.), immunodeficiency (susceptibility to infections and colds), emotional disturbance (irritability, sleepiness, etc.), mental disorders (deterioration of memory, low level of intellectual development – low IQ, cretinism and others) [106].

Unfavorable ecological situation and lowering of living standards of the population in many countries, as well as studies on the assessment of iodine deficiency at the level of populations have put this problem among UN priorities in human health.

Chromium. Chromium interaction with *S. platensis* biomass is of special interest. Cr is known to be an important trace element, which at the same time possesses significant toxic properties. It exists in various valence states (from +2 to +6) and forms numerous complex compounds, most stable of which are Cr(III) and Cr(VI). Kinetically stable, non-toxic Cr(III) is most abundant in the environment. Toxic Cr(VI) penetrates cells easier than Cr(III), participates in the reduction-oxidation processes inside them, thus reducing to the stable Cr(III). These processes cause damage to cell genetic material, oxidative injuries, formation of cross-links and thus have carcinogenic, genotoxic and mutagenic effects [107].

On the other hand, Cr is a necessary chemical element without which normal functioning of the organism is impossible. The most important function of Cr is be the key constituent in the so-called glucose tolerance factor. Cr contributes to insulin participation in metabolism of carbohydrates (proteins, lipids, nucleic acids) and to transfer glucose into tissue cells. It also influences synthesis of macromolecules, activation of some enzymes and cholesterol metabolism [108, 109]. Deficiency of Cr in human results in derangement of lipid and fat metabolism, depressed physical growth and weight loss, decreased longevity and impaired coordination of movements. Sometimes Cr deficiency is associated with high cholesterol levels, tiredness and fatigue, alcoholic intolerance, etc. Cr deficiency most often becomes apparent in early childhood and young age through disturbed protein metabolism. In all the above-listed cases addition of Cr to the diet helps to treat the diseases. Thus, production of chromium-containing pharmaceuticals designed to eliminate Cr deficiency is an urgent matter.

It should be mentioned that the given concentrations is very important for chromium-containing preparations as well as for substances with other elements. However, additional measures should be taken to ensure predominant assimilation of Cr(III) rather than the toxic form Cr(VI).

5.2. Materials and Methods

The IPPAS B-256 strain of *S. platensis* from the algeological collection of the Timiryazev Institute of Plant Physiology, Russian Academy of Sciences, was used in the experiments. To stimulate growth, the *Spirulina* culture was treated in Potter homogenizer in order to partly shorten its threads. The cultivation of *S. platensis* cells was carried out in the Zaroukh standard water-salt nutrient medium at pH 8.5–11 and temperature +32–34 °C, with continuous stirring and illumination with a 5000 lx sodium lamp.

To study element accumulation dynamics in the *S. platensis* biomass, cultivation was carried out in a nutrient loading with the following concentrations:

- for selenium – selenic acid H_2SeO_3– from 0.5 to 15 mg/l
- for iodine – potassium iodine KI – from 10^{-8} to 10^{-4} g/l
- for chromium – (Cr(III)) – chromium acetate $Cr(CH_3COO)_3$ – from 0.5 to 15 mg/l.

In each experiment, after 5 days, *S. platensis* biomass was separated from the nutrient medium by filtering, rinsing, and centrifugation. The resulting substance was lyophilically dried in a special adsorption-condensation lyophilizer of an original design [110].

The protein content in the biomass determined by Lowry technique [111] was about 65%, corresponding to a normal level of natural *S. platensis* biomass. The quality of the biomass was also confirmed by comparison of electrophoregrams obtained by gel electrophoresis technique for samples grown with and without loading.

Samples for analysis were prepared in the form of small pellets using a special titanium mould.

Neutron activation analysis (NAA) using epithermal neutrons, a well-proven technique for determination of element composition of biological objects, was used as an analytical technique. Due to activation with resonance neutrons, the technique makes it possible to minimize matrix effects of biological samples and at the same time to determine concentrations of over 30 major, minor, and trace elements. Analytical research was conducted at the IBR-2 fast pulsed reactor of the Frank Laboratory of Neutron Physics, JINR (Dubna, Russia). Experimental techniques and analytical information processing procedure are described in [112, 113]. First, the multielement composition of the *S. platensis* biomass was studied by NAA technique and the concentrations of certain elements were compared with the corresponding permissible level values (Figure 2) [99]. In the course of cultivation *Spirulina* cells may assimilate some toxic elements, like Hg, As, Cr, Cd, Pb, etc., present in the nutrient as impurity. The background concentrations of these elements in the *Spirulina* biomass produced with reagents of the given grade are shown in Table 4. They do not exceed those permissible for human, according to the data at the website: [114].

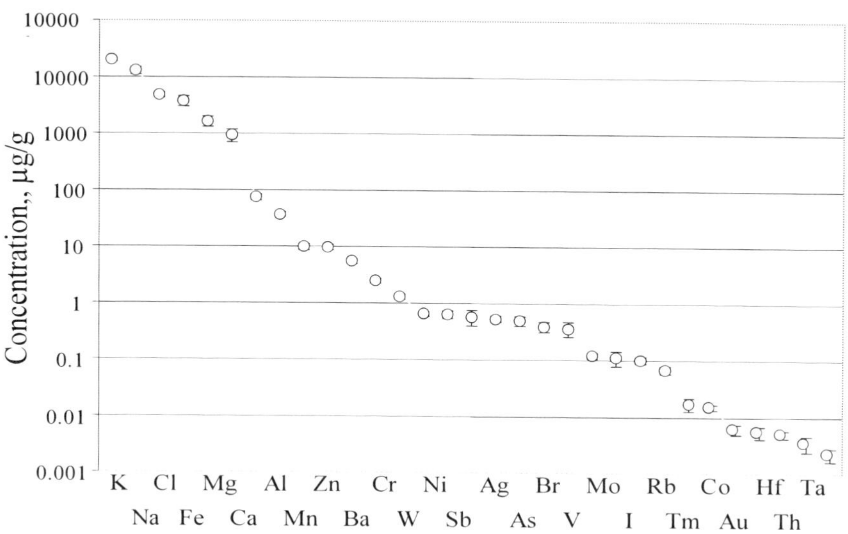

Figure 2. Background concentrations of macro- and microelements in *Spirulina* biomass.

Table 4. Background concentrations of some toxic elements in *Spirulina platensis* biomass

Element	Dry mass concentration, μg/g
Hg	3–5*
As	2–3*
Cr	3–4*
Pb	3**
Cd	0.2**

* NAA data obtained at the IBR-2 reactor in Dubna [115].

** data from [116].

5.3. Results and Discussion

Selenium

The results of determination of Se concentration by the NAA technique are shown in Figure 3. As can be seen from this figure, the curve of the Se content in the biomass versus Se concentration in the nutrient medium is well approximated by a polynomial of the 2nd order $y = -0.00008x^2 + 0.3x - 1$. Starting with the Se concentration of 100 μg/l, an intensive growth of Se accumulation by cells with a possible maximum in the range of 1100–1200 μg/l was observed. For pharmaceutical purposes, the range 100–1000 μg/l seems to be the most advantageous: at a high degree of Se assimilation the significant steepness of the curve allows rather exact determination of Se doses in a substance obtained [101].

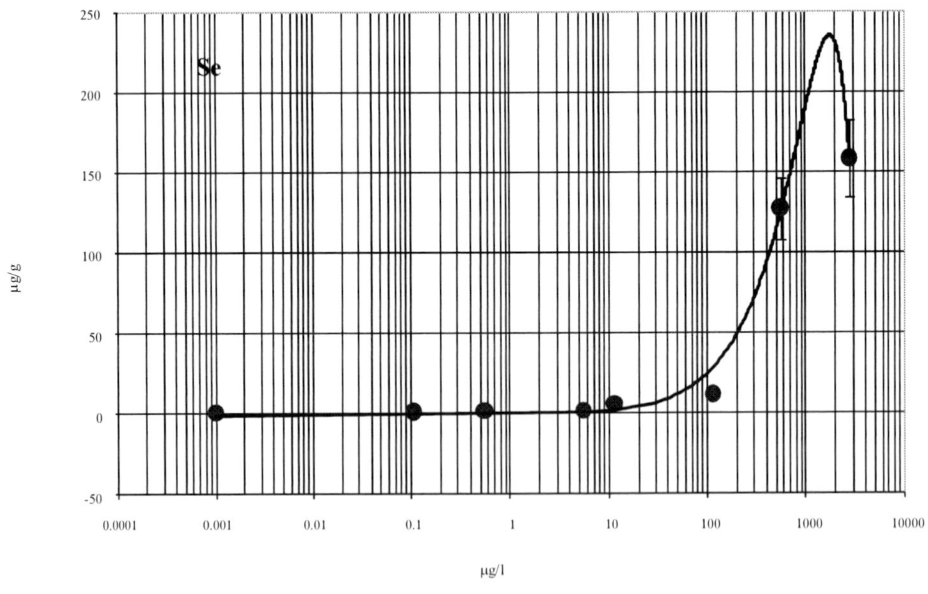

Figure 3. Selenium concentration in *Spirulina platensis* biomass *versus* its content in the nutrient medium.

Visual microscopic observation the state of culture, determination of the total protein content in the biomass as well as investigation of electrophoregrams revealed natural properties of the obtained selenium-containing biomass. Thus, by including Se in the composition of its biological macromolecules, *S. platensis* preserves its beneficial properties and compares favorably with other similar preparations, which are either a mechanical mixture of Se compounds with *Spirulina* powder [116] or are obtained at such high Se concentrations in a medium that cells grow while struggling for survival and no normal quality of biomass can be retained [117].

Chromium

From the curves obtained (Figure 4), *Spirulina* assimilates mainly Cr(III), while the degree of binding of Cr(VI) was approximately three times lower. As opposed to other microorganisms, e.g., *Arthrobacter oxydans, S. platensis* prefers the nontoxic form of Cr to the toxic one, even if both forms are present in the solution.

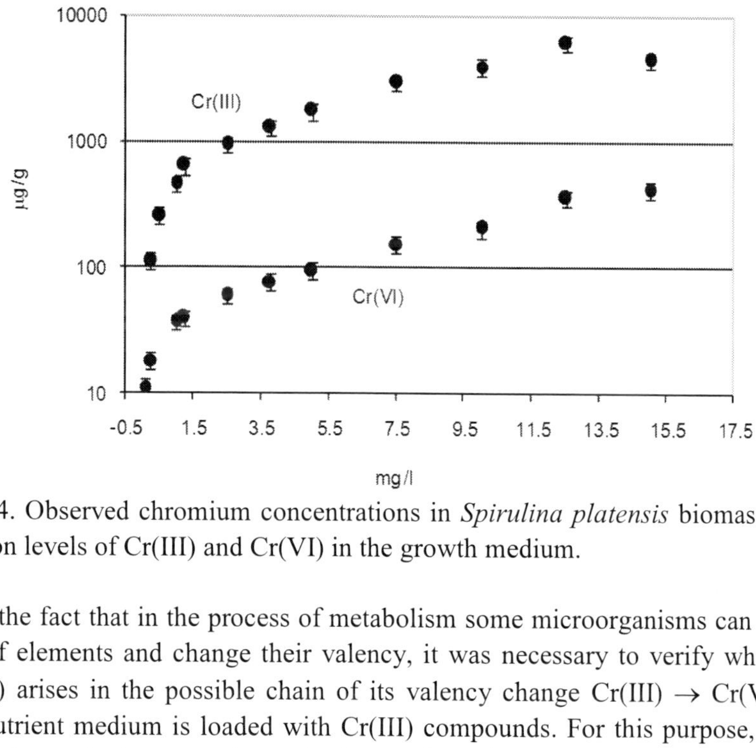

Figure 4. Observed chromium concentrations in *Spirulina platensis* biomass at different concentration levels of Cr(III) and Cr(VI) in the growth medium.

Due to the fact that in the process of metabolism some microorganisms can interact with a number of elements and change their valency, it was necessary to verify whether a toxic form Cr(VI) arises in the possible chain of its valency change Cr(III) → Cr(V) → Cr(VI) when the nutrient medium is loaded with Cr(III) compounds. For this purpose, we used the technique for colorimetric determination of Cr(VI) with diphenylcarbohydrazide $(C_6H_5NHNH)_2CO$, which reacts with Cr(VI) in concentrations of 0.1–10 µg/l to give a purple-red color and yields a photometric peak of 540 nm [118]. The investigations showed that Cr(VI) was absent in all cases of *S. platensis* cultivation with Cr(III) loading. The presence of the intermediate form Cr(VI) was checked by the electron paramagnetic resonance (EPR) technique with a sensitivity of the order of $5 \cdot 10^{-10}$g of Cr(V). The results showed the absence of a resonance signal, which is typical of Cr(V), in all samples under investigation. It is possible to obtain optimal Cr doses in the resulting biomass and to recommend 30–100 μg/l for food supplement and 200–250 μg/l for therapeutic and prevention purposes [100].

Iodine

The curve of the iodine concentration in the biomass versus the iodine concentration in the nutrient medium (Figure 5) can also be approximated by the 2nd order polynomial $y = 0.00001x^2 - 0.003x + 0.4$ [99].

The biomass enrichment coefficient can be defined as a ratio between the concentration of iodine (or other element) in the biomass and its concentration in the nutrient medium in accordance with the curve obtained. This coefficient may serve as an initial technological parameter governing the element dosage in treatment pills and the choice of the pill mass for a given substance.

The microscopic control of the cytological state of the culture as well as of the protein content of the obtained biomass demonstrated in all cases that the normal state of *S. platensis* and, consequently, its natural beneficial properties are preserved.

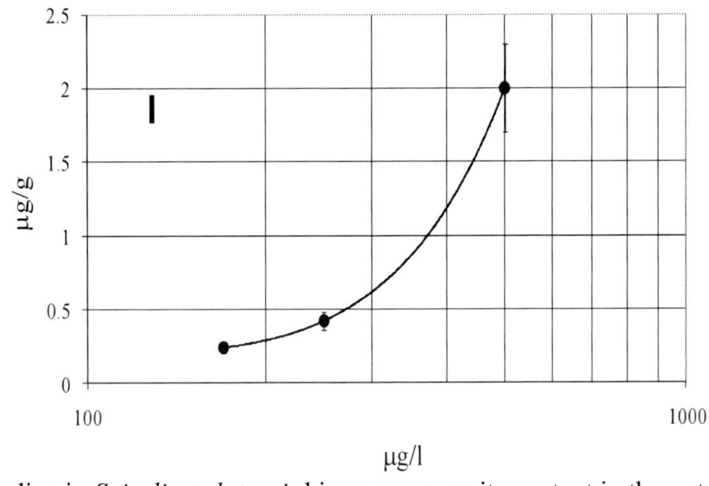

Figure 5. Iodine in *Spirulina platensis* biomass *versus* its content in the nutrient medium.

6. Conclusion

The performed investigations have demonstrated that cultivation of *S. platensis* cells under selected conditions allows target-oriented introduction of the required elements (Se, Cr, I, etc.) into the composition of biological macromolecules with preservation of their protein content and natural properties of biomass. The curves of concentrations of necessary elements in the resulting *S. platensis* substance versus concentrations of these elements in the nutrient medium were obtained with the NAA. It is demonstrated that therapeutic and preventative doses of each element can be accurately determined from the slopes of these curves.

The results of the neutron activation analysis of the multielement composition of the *S. platensis* biomass and their comparison with the permissible level demonstrated that the content of such toxic elements as Hg, Cd, As, etc. did not exceed the permissible level accepted in many countries of the world.

The developed technique can also be used for production of substances incorporating other vitally important elements such as Zn, Cu, Fe, etc.

REFERENCES

[1] http://en.wikipedia.org/wiki/algae

[2] Bold, H. C. & Wynne, M. J. 1985. *Introduction to the Algae.* Pp. 720. Engelwood Cliffs, New Jersey: Prentice Hall.

[3] Gorbunova, N. P. 1991. *Algology.* Moscow: Higer Scool (In Russian).

[4] Mazheika, I. S., Grube, E. T., Kamzolkina, O. V. & Dyakov. 2000. *Introduction to Algology and Mycology.* Pp. 190. Russian: Moscow University Pub.

[5] Waaland, J. R. 1981. Commercial utilization. In: *The Biology of Seaweeds* (edited by C. S. Lobban & M. J. Wynne). Pp. 736–741. Oxford: Blakwell Scientific Publications.

[6] Chapman, V. J. & Chapman, D. J. 1980. *Seaweeds and Their Uses.* Pp. 334. London: Chapman and Holl.

[7] Schwarz, R. E., Hirsch, C. F. et al. 1990. Pharmaceutical from cultured algae. *Journal of Industrial Microbiology and Biotechnology,* 5(2-3), 113–123.

[8] Andersen, R. A. 2005. *Algal Culturing Techniques.* Pp. 596. London: Elsevier, Academic Press.

[9] Vaskovsky, V. E. 1998. *Marine Macrophytes.* (In Russian). http://www.pereplet.ru/obrazovanie/stsos/588.html

[10] Titov, A. M. 2009. Marin Algae – Unique Materials for Pharmaceuticals and Supplements. (In Russian). http:// www.faros plus.ru

[11] Attaway, D. H. & Zaborsky, O. R. 1993. *Marin Biotechnology: Pharmaceutical and Bioactive Natural Products.* Vol. 1, Pp. 590. New York, London: Plenum Press, Springer.

[12] Moiseev, P. A. 1989. *Biological Resources of the Word Ocean.* Pp. 329. M. Agropromizdat. (In Russian).

[13] Hoppe, H. A., Levring, T. & Tanaka, R. 1979. *Marin Microalgae in Pharmaceuticals.* Vol. 1, Pp. 807. Berlin, New York: Walter de Gruyter.

[14] Saraxanakumar, D. E. et al. 2008. Antimicobacterial activity of the red alga Polysiphonia vigata. *Pharmaceutical Biology*, 46(4), 254 – 260.

[15] Node, H., Amano, H. et al. 1989. Antitumor activity of polysaccharides and lipids from marine algae. *Nippon Suisan Oakkaishi,* 55, 1265 – 1271.

[16] Lipkin, Y. 1985. Outdoor cultivation of sea vegetables. Plant and Soil. 89, 159 – 183.

[17] Raja, R., Hemaiswarya, S. et al. 2008. A perspective on the biotechnological potential of microalgae. *Critical Reviews in Microbiology,* 34, 77 – 88.

[18] Kay, R. A. 1991. Microalgae as food and supplement. Critical Reviews In *Food and Nutrition,* 30(6), 555 – 573.

[19] Richmond, A. 1986. *Handbook of Microalgal Mass Culture.* Pp. 528. Boca Raton, FL: CRC Press.

[20] Shelef, G. & Soeder, C. J. 1980. *Algal Biomass Production and Use.* Pp. 214. Amsterdam: Elsevier/ North Holland Press.
[21] Borowitzka, M. A. & Borovitzka, L. J. 1988. *Microalgal Biotechnology.* Pp. 466. Cambrige: Cambridge University Press.
[22] Stadler T. et al. eds. 1988. Algal biotechnology. Elsevier, *Applied Science,* New-York, Pp. 533.
[23] Norton, T. A., Melkonian, M. & Andersen, R. A. 1996. *Algal biodiversity.* Phycologia, 35, 308–326.
[24] Morimura, Y. & Tamiya, N. 1954. Preliminery experiments in use of Shlorella as human food. *Food Tecnology,* 8, 179.
[25] Krishnakumari, M. K., Ramesh, H. P. & Venkataraman, L. V. 1981. Food safety evaluation: acute oral and dermal effects of the algae Scenedesmus acutus and Spirulina platensis on albino rats. *Journal of Food Protection*, 44, 934.
[26] Klausner, A. 1986. *Alga culture: food for thought.* Bio/Technology, 4, 947.
[27] Voronkova, S. S., Sadkova, N. P. et al. 1980. Investigation of some physico- chemical properties of water-soluble protein from Spirulina platensis. *Applied Biochemistry and Microbiology,* 16, 266.
[28] Chen, B. H. & Chi, C. H. 1981. Process development and evaluation for algal glycerol production. *Biotechnology and Bioengineering,* 23, 1267.
[29] Fujimoto, S. 1979. Natural coloring agents. *Japan Examined Pat.,* 5, 437, 965.
[30] Ciferri, O. & Tiboni, O. 1985. The biochemistry and industrial potential of Spirulina. *Annual Review of Microbiology,* 39, 502.
[31] Ciferri, O. 1983. Spirulina, the edible microorganism. *Microbiological Reviews,* 47, 551.
[32] Vanossi, L. 1985. Chlorella as food. *Industrie Alimentari,* 21, 118.
[33] Guerin, O. & Moyse, A. 1975. Biological characteristics of Spirulina spp. *Annales Nutr. l'Aliment,* 29, 489.
[34] Payer, H. D., Pithakpol, B. et al. 1980. *Major resalts of Thai-German microalgae project at Bangkok.* Ergebnisse Limnologie, 11, 41.
[35] Richmond, A. & Vonshak, A. 1978. Spirulina culture in Israel. *Ergenisse Limonology,* 11, 274.
[36] Henrickson, R. 1989. *Earth food Spirulina.* Laguna Beach,CA: Ronore Enterprises.
[37] Shubert, L. E. 1988. The Use Spirulina (Cyanophyceae) and Chlorella (Chlorophyceae) as Food Sources for Animals and Humans. *Progress in Phycological Research.* (edited by C. Round & V. J. Chapman). Pp. 237. BioPress Ltd.
[38] Jonson, P. & Shubert, L. 1986. Accumulation of mercury and other elements by Spirulina (Cyanophyceae). *Nutrition Reports International,* 34, 1063.
[39] Le Gal, Y. & Miller- Feuga, A. 1998. *Marin Microorganisms for Industry.* Pp. 207. Editions Qual.
[40] Spolaore, P., Joannis- Cassan, C. et al. 2008. Commercial application of microalgae. *Critical Reviews in Microbiology,* 34(2), 77–88.
[41] Spolaore, P. et al. 2006. Commercial application of microalgae. *Journal of Bioscience and Bioengineering,* 101(2), 87–96.
[42] Borovitzka, M. A. 1999. Commercial production of microalgae: ponds, tanks, tubes and fermenters. *Journal of Biotechnology*, 70(1–3), 313 – 321.

[43] Borowitzka, L. J. & Borovitzka, M. A. 1989. Industrial production: methods and economics. In: *Algal and Cyanobacterial Biotechnology* (edited by R. C. Cresswell, Rees Tay & N. Shah.). Pp. 294-316. London: Longman Scientific.

[44] Richmond, A., Vonshak, A. & Arad, S. 1980. Enviromental limitation in outdoor production of algal biomass. In: *Algal Biomass* (edited by G. Shelef & C. Soeder). Pp. 65-72. Amsterdam: Elsevier/ North Holland Biomedical Press.

[45] Radmer, R. J. 1996. Algal diversity and commercial algal products. *BioScience,* 46(4), 253–270.

[46] http://www.healthywayproduction.com/encyclopedia.php

[47] Collins, M. 1978. Algal toxins. *Microbiological Reviews,* 42, 725.

[48] Trainor, F. R. 1978. Introductory Phycology. Pp. 525. New York: John Wiley & Sons.

[49] Lin, Y.C. 1969. The supplementary effect of algae on the nutritive value of soybean milk. *Journal of the Formosan Medical Association*, 68, 15.

[50] Fink, H., Herold, E. 1956. The protein value of unicellular green algae and their action in preventing liver necrosis. *Zeitscher Physiol Chem,* 305, 182.

[51] Lee, H. T., Lai, J. E. & Tung, Y. C. 1977. Hypoglycemic action of Chlorella. *Journal of the Formosan Medical Association,* 76, 272.

[52] Mares-Perlman, J. A., Millen, A. E. et al. 2002. The body of evidens to support a protective role for lutein and zeaxantin in delaying chronic disease: overview. *Journal of Nutrition,* 132, 5185–5245.

[53] Krinsky, M. I., Landrum, J. T. & Bone, R. A. 2003. Biologic mechanisms of the protective role of lutein and zeaxantin in the eye. *Annual Review of Nutrition,* 28, 171–201.

[54] Hills, C. & Nakamura, H. 1978. *Food from Sunlight.* Pp. 41. Bulder Creek, CA: University of the Trees Press.

[55] http://dic.academic.ru/dic.nsf/enc_biology/1488

[56] Lembi, C. A. & Waaland, I. R. 1988. *Algae in Human Affair*. Pp. 590. Cambrige: Cambrige Univer. Press.

[57] Vonshak, A. 1997. Spirulina platensis (Arthrospira): *Physiology, Cell-Biology, and Biotechnology*. Pp. 233. London, Bristol: Taylor and Francis.

[58] Richmond, A. Spirulina. 1988. In: *Microalgal Biotechnology* (edited by M. Borowitzka & J. Borowitzka). Pp. 83-121. Cambrige University Press.

[59] Fox, D. 1993. Health benefits of Spirulina. In: *Spirulina Algae of Life.* Bulletin 12, April 1993. Monaco: Institute of Oceanografy.

[60] Belay, A., Ota, Y. et al. 1993. Current knowledge of potential health benefits of Spirulina. *Journal of Applied Phycology,* 5, 235–241.

[61] Chernova, N. I., Korobkova, T. P. & Kiceleva, S. V. 2009. *Microalga Spirulina as the object of biotechnology*. (In Russian) http://bio.1september.ru/articlef.php

[62] http://biotex.ibss.org.ua/tech/8

[63] http://www.aryeda.info/497053038

[64] http://www.aquaria.com.ua/spirulina

[65] Becker, E. W. 1981. Nutritional properties of microalgae: potential and constraints. In: *Handbook of Microalgal Mass* Culture (edited by A. Richmond). Pp. 934. Boca Raton, FL: CRC Press.

[66] Pulz, O., Gross, W. 2004. Valuable products from biotechnology of microalgae. *Applied Microbiology Biotechnology,* 65, 635 – 648.

[67] Zang H.Q. et al. 2001. Chemo and radiao protective effects of polisaccharides of Spirulina platensis on hemopoietic system of mice and dogs. *Acta Pharmacol. Sin.* 22, 1121-1124.

[68] Evets, L. et al. 1994. Means to normalize the levels of immunoglobuline E using the supplements Spirulina. Grodnenski State Med. Univ. RF committee of Patent and Trade. *Patent* (19) RU (11) 2005486, Jan. 15, 1994, Russia.

[69] Hayashi, O. & Ishii, K. 2004. Enhancement of mucosal immune functions by dietary Spirulina platensis in human and animals. *Nutrition Science,* 7, 31–34.

[70] Herbert, V. 1982. Spirulina and vitamin B12. *Journal of the American Medical Association,* 248, 3096.

[71] Van den Berg, H., Dagnelie, P. C. et al. 1988. Vitamin B12 and seaweed. *Lancet,* 30, 242.

[72] Materassi, R., Paoletti, C. et al. 1980. Some considerations on the production of lipid substances by microalgae and cyanobacteria. In: *Algae Biomass* (edited by G. Shelef & C. J. Soeder). Pp. 619. New York: Elsevier/North-Holland Biomedical Press.

[73] Shifrin, N. S. & Chisholm, S. W. 1980. Phytoplankton lipids: environmental influences on production and possible commercial application. In: *Algal Biomass* (edited by G. J. Shelef & C. J. Soeder). Pp. 627. New York: Elsevier/North-Holland Biomedical Press.

[74] Kenyon, C. N. 1972. Fatty acid composition and physiological properties of some filamentos blue-green algae. *Archives of Microbiology,* 83, 216.

[75] Gustavson, K., Cardellinal, H. et al. 1989. AIDS-antiviral sulpholipids from cyanobacteria (blue–green algae). *Journal of the National Cancer Institute,* 81, 1254.

[76] Venkataraman, L. V. & Becker, E. W. 1985. *Biotechnology and Utilization of Algae.* Pp. 114-115. Mangalor, India: The Indian Experience Sharada Press.

[77] Hayashi, T., Hayashi, K. et al. 1996. Calcium spirulan, an ihibitor of enveloped virus replication from a blue-green alga Spirulina platensis. *Journal of Natural Products,* 59(1), 83 – 87.

[78] Li, B., Zhang, K. C. et al. 2004. Study on the anti-tumor immune activity of phycocyanin and polysaccharides from Spirulina platensis. *Journal of Ocean University of China,* 34, 396 – 402.

[79] Becker, E. W., Jacover, et al. 1986. Chemical and biochemical evaluations of the alga Spirulina with regard to its application in the treatment of obesity: a duble-blind cross-over study. *Nutrition Reports International,* 33, 565–574.

[80] Vonshak, A. 1987. Strain selection of Spirulina suitable for mass production. *Hydrobiologia*, 151/152, 75–77.

[81] Vonshak, A., Abeliovich, A. et al. 1982. Production of Spirulina platensis biomass: effect of environmental factors and population density. *Biomass,* 2, 175 – 185.

[82] Vonshak, A., Boussiba, S. et al. 1983. Production of Spirulina biomass: maintenance of microalgal culture outdoors. *Biotechnology and Bioengineering,* 25, 341–349.

[83] Borovkov, A. B. 2005. The green microalga Dunaliella salina. (In Russian). http: // repository. ibss. org. ua / dspace / handle / 99011/863

[84] Avron, M. & Ben-Amotz, A. (eds.) 1992. Dunaliella: *Physiology, Biochemistry and Biotechnology*. Pp. 239. Boca Raton FL: CRC Press.

[85] Borowitzka, M. A. & Borowitzka, L. J. 1988. Dundliella. In: *Microalgal Biotechnology* (edited by M. A. Borowitzka & L. J. Borovitzka). Pp. 27-58. New York: Cambrige University Press.

[86] Borowitzka, M. A., Borovitzka, L. J. & Moulton, T. P. 1984. The mass culture of Dunaliella salina for fine chemicals: from laboratory to pilot plant. *Hydrobiologia,* 116–117(1), 115–121.

[87] Fujii, Y., Sacamoto, S. et al. 1993. Effects of beta-caroten – rich algae Dunaliella Bardawiie on the dynamic changes of normal and neoplastic mammary cells and general metabolism in mice. *Anticancer Research,* 13(2), 389-393.

[88] Curtain, C. C., West, S. M. & Schlipalus, L. 1987. Manufacture of beta-caroten from the salt lake alga Dunaliella salina: the scientific and technical background. *Australian Journal of Biotechnology,* 1, 51–57.

[89] Borowitzka, L. J. & Borowitzka, M. A. 1990. Commercial production of β-caroten by Dunaliella salina in open ponds. *Bull. Mar, Sci.,* 47, 244-252.

[90] Raja, R., Hemaiswarya, S. & Rengasamy, R. 2007. Exploatation of Dunaliella for beta-carotene production. *Applied Microbiology and Biotechnology,* 74(3), 517–523.

[91] Gorbunova, S. Yu., Lelekov, A. S. & Borovkov, A. B. 2007. Dynamics of nitrogen and phosphorus in medium at intensive cultivation of microalga Dunaliella salina. *Marine Ecology,* 74, 21–24. (In Russian).

[92] Efremova, N., Rudic, V. et al. 2009. The obtening of green microalgae Dunaliella salina biomass with higth content of carotenoids. *Book of abstracts of Intern. conf. "Biologically active substances: fundamental and applied problems".* Pp. 37-38. May 25–30, Novy Svet, AR Crimea, Ukraine.

[93] Boussiba, S. 2000. Carotenogenesis in the green alga Haematococcus pluvialis: Cellular physiology and stress respons. *Physiologia Plantarum,* 108(2), 111–117.

[94] Guerin, M., Hantley, M. E. & Olaizola, M. 2003. Haematococcus astaxantin: application for human health and nutrition. *Trends in Biotechnology,* 21(5), 210–216.

[95] Lorenz, R. T. & Cysewski, G. R. 2000. Commercial potential of Haematococcus microalgae as a natural source of astaxantin. Trends in Biotechnology, 18(4), 160–167.

[96] Sambamurthy, K. 2006. *Pharmeceutical biotechnology.* Pp. 504. New Age International ISBN 8122417833.

[97] Mertz, W. 1981. *The essential trase elements.* 213, 1332.

[98] Mosulishvili, L. M., Belokobylsky, A. I., Kirkesali, E. I., Frontasyeva, M. V., Pavlov, S. S. & Aksenova, N. G. 2007. Neutron activation analysis for stadyng Cr uptake in the blue-green microalga Spirulina platensis. *Journal of Neutron Research,* 15(1), 49–54.

[99] Moshulishvili, L. M., Kirkesali, Ye. I., Beiokobylsky, A. I., Khizanishvili A. I., Frontasyeva, M. V., Pavlov, S. S. & Gundorina, S. F. 2002. Experimental substantiation of the possibility of developing selenium- and iodine- containing pharmaceuticals based on blue-green algae Spirulina platensis. *Journal of Pharmaceutical Biomedical Analysis,* 30(1), 87–97.

[100] Mosulishvili, L. M., Belokobilsky, A. I., Kirkesali, E. I., Frontasyeva, M. V. & Pavlov, S. S. 2002. *Patent of RF № 2230560*, priority of June 11, 2002.

[101] Mosulishvili, L. M., Belokobylsky, A. I., Khizanishvili, A. I., Kirkesali, E. I., Frontasyeva, M. V. & Pavlov S. S. 2001. *Patent of RF № 2209077*, priority of March 15, 2001.

[102] Combs, G. F. 1999. Chemopreventiv mechanisms of selenium. *Med. Klin., 94, Suppl.* 3, 18–24.

[103] Shumacher, K. 1999. Effect of selenium on the side effect profile chemotherapy/ radiotherapy in patient with breast carcinoma. *Design for a clinical stady* // Ibid. 45-48.

[104] Holben, D. H. & Smith, A. M. 1999. The divers role of selenium within selenoproteins: a review. *Journal of the American Dietetic Association,* 99(7), 836–843.

[105] Clark, L. C., Cantor, K. P. & Allaway, W. H. 1991. Selenium in forage crops and cancer mortality in US countries. *Archives of Environmental Health,* 46, 37–42.

[106] Voinar, A. I. 1962. *Trace elements in nature*. Moskow. (In Russian).

[107] Liu, K. J., Husler, J. et al. 2001. On the mechanism of Cr (VI)-induced carcinogenesis: dose dependence of uptake and cellular respons. *Molecular and Cell Biochemistry,* 222, 221.

[108] Andersen, R. A. 1988. Resent advances in the role of chromium in human health and diseases. In: *Essential and Toxic Trace Elements in Human Health and Diseases* (edited by A. S. Prasad.). Pp. 189-197. New York: Alan R. Liss.

[109] Mertz, W. et al. 1979. Chromium – an overview. In: *Chromium in Nutrition and Metabolism* (edited by D. Shapcott & Y. Hubert). Pp. 1-14. Amsterdam: Elsevier/North Holland Biomedical Press.

[110] Mosulishvili, L. M., Nadareishvili, V. S., Kharabadze, N. E. & Belokobylsky, A. I. 1980. *Patent URSS.* № 779765, Bul. 42.

[111] Meshkova, M. P. & Severin, E. E. 1979. *Practical Traning on Biological Chemistry.* Russian: MSU Publication.

[112] Frontasyeva, M. V. & Pavlov, S. S. 1999. REGATA Experimental Setup for Air Pollution Studies. In «Problems of Modern Physics». *On the 90th anniversary of Saratov State University and the 40 years of the JINR- SSU collaboration.* Editors: A.N. Sissakian, D.I. Trubetskov. Dubna, JINR, Pp. 152-158.

[113] Ostrovnaya T.M,. Nefedyeva L.S, Nazarov V.M., Borzakov S.V., Strelkova L.P. 1993. Software for INAA on the basis of relative and absolute methods using nuclear database. In «*Activation Analysis in Environment Protection*»,» D14-93-325, Dubna, Pp. 319-326.

[114] http://www.spirulina.com/SPBNutrition.html

[115] Ortega Calvo, J. J., Mazuelos, C., Hermosin, B. & Saiz- Jimenez, C. 1993. Chemical composition of Spirulina and eukaryotic algae products marketed in Spain. *Journal of Applied Physics.* 5, 425 – 435.

[116] Belokobilsky, A., Tsibakhashvili, N., Rcheulishvili, A. & Mosulishvili, L. 1999. Accumulation of Cd (II) in C- phycocyanin from Spirulina platensis. *Bul. Geor. AS,* Vol. 159, 1, 140-143.

[117] Hann, M. & Stemgel, H. 1985. *Diateticlie Zusammensetzung. Patent* DE3421644. 12. 12. 85.

[118] Tambiev, A. H., Kirikova, N. N. et al. 1997. *Patent RU* 2096037, 20. 11. 97.

[119] Uron, P. F. 1955. Stability of colorimetric reagent for chromium determination by s-diphenilcarbazide in various solvents. *Analytical Chemistry*, 27(13), 1355.

In: Bioprocess Sciences and Technology
Editor: Min-Tze Liong
ISBN 978-1-61122-950-9

Chapter 6

ALGAE: PROCESSES AND APPLICATIONS

G. Petkov[*]
Institute of Plant Physiology, Bulgarian Academy of Sciences,
Bl. 21, 1113 Sofia, Bulgaria

ABSTRACT

Algae and cyanobacteria, as photosynthetic organisms, are cultivated in mineral nutrition media bubbled with carbon dioxide. In large scale they are grown in naturally illuminated open sloped areas, covered ponds or photobioreactors. In order to enhance the yield of algal biomass attention is paid to the processes hydrodynamics, heat transfer and mass transfer in the algal device. Normally, daily light intensity is sufficient for intensive algal growth, but longer time of exposition ensures higher algal yield. Centrifugation or filtration is the main method of microalgae harvesting.

As a whole, the algal biomass is a valuable nutritional additive for humans and animals. The biochemical composition of algae comprises all essential amino acids and minerals in the right natural ratio. Their qualities as food and forage are completed by almost all vitamins, valuable oils and phytosterols. The eggs of algae-fed hens have much higher carotene content, harder shells [1]and the productivity as a whole is greater. Microalgae - zooplankton - fish is a nutritional chain, successfully applied in the industry. Dry algal biomass is a standardized product and can be kept in store for long.

Algae are important source of pharmaceuticals. Phycocyanin and phycoerythrin of cyanobacteria and red algae are used in the immunodiagnostic and radioactive protection. Algal biomass is rich of beta-carotene, tocopherol, vitamin K, coenzyme Q_{10} and other oil- and water soluble natural antioxidants. Highly unsaturated fatty acids with 4, 5, 6 double bonds, such as arachidonic, eicosapentaenoic, docosahexaenoic characterize the algal oils. Algal biomass, algal oils, hydrocarbons and hydrogen from microalgae are often pointed as proper renewable source of fuel.

[*] Corresponding author: *G. Petkov*. Email: gpetkov@bio.bas.bg

INTRODUCTION

When one speaks about algae most people have in mind the large green, red and brown fronds in the sea. They really are algae called macroalgae, but here we are going to discuss the microalgae, which are small, microscopic algae, and extremely wide spread in nature: water, soil, air, snow and ice, deserts, hot springs etc. and often called microalgae. The word microalgae is not strict taxonomic concepts but convenient for use in biotechnology. Cyanobacteria are prokaryotic organisms, without nucleus, and from taxonomic point of view are not algae. Both cyanobacteria and microalgae have similar environmental significance and, the methods of cultivation, separation of biomass, processing and commercial use are similar, too. Sometimes, the word "microalgae" is used for cyanobacteria and algae, bearing in mind both of them. Here is to mention that the words algology and phycology are in use to name the science of algae that from Latin and Greek mean literally "seaweed science".

Cyanobacterium *Spirulina* which naturally grows in the lakes Chad and Titikaka has been used as a food additive from times beyond memory. The alga *Chlorella* was one of the favourite subjects used to reveal the fundamental nature of photosynthesis and cell physiology. Microalgae were used at the beginnings of space exploration, and pointed as promising organisms at possible space travel in the Solar System. They convert water and CO_2 into valuable biomass and oxygen. Undoubtedly, great deeds as the space experiments in the 60-70 years have proved a driving force which supported the microalgal studies. Nowadays, they have plenty of applications though in the world trade with algae the demand is much higher than the supply. Algal biomass is source of valuable substances for the industry such as food, forage and pharmaceuticals; live algal cultures are useful modern implement in the fish breading, soil remediation of fertilizing.

The diversity of algal species is extremely high. Usually algal strains are isolated from nature or supplied from the algal culture collections and, abiding the common rules of microbiologic work algae are bred as laboratory cultures. In order to maintain the algal growth, the cultures are supplied with light, mineral nutrition media, carbon dioxide and stirring.

If one is going to deal with algae it is better to have some preliminary knowledge, which can be found in the books of basic plant physiology. The large scale algal cultivation requires some technical and technological skills, and application of the principles of mass transfer, heat transfer and hydrodynamics. The research and development in the area of large scale microalgae cultivation is an interdisciplinary pursue which requires a joint work of plant physiologist and chemical engineer.

PARAMETERS OF ALGAL PRODUCTIVITY

Having isolated an algal strain from nature, or received from an algal culture collection, one is able to cultivate it in laboratory. The algal culture collections are known under names in form of abbreviations and some of them can be found on http://www.phycol.ca/links. An algal strain is cultivated in an illuminated flask with proper mineral nutrition medium in sterile conditions. Intensive cultivation with stirring and bubbling CO_2 might begin when the algal cells are visibly grow. The transfer of cultivation on 1 m^2 is ready to get started if 2 - 3

dm^3 liquid culture with algal density 4 - 5 $g.dm^{-3}$ is at hand. It could be said that large scale cultivation of a cyanobacteria is wielded when achieve nearly 10 – 15 g. m^{-2} daily yield, and 20 – 25 g. m^{-2} for green algae at optimal temperature and 12 h light.

The large scale cultivation of algae is carried out on slopped areas [1, 2], open or covered ponds [3, 4], tubular devices [5, 6, 7], or in transparent plate columns [8, 9]. Sloped areas are predominantly used to cultivate green algae. The experience says that slopped areas, although being no universal, are more productive compared to horizontal ponds. Higher yield and better quality of biomass are achieved when use photobioreactors [6, 9, 10].

2.1. Light

Light is an important requirement of the photoautotrophic (self-feeding) growth and the algae, being photosynthetic organisms, need light. In a room, they are usually illuminated with luminescent lamps where the photon flux is 380-710 nm. A light intensity of 25 - 100 μmol m^{-2} s^{-1} (6 – 22 $W.m^{-2}$, 2 – 8 klx) does not saturate the photosynthetic abilities of the algae, even of the more sensitive ones. As a comparison, at 100 μmol m^{-2} s^{-1}, (8 klx), the green algae *Chlorella* or *Scenedesmus* binds about 15 μmol CO_2 m^{-2} s^{-1}. That is maximum photosynthetic rate for cucumber and near to the maximum for tomato which is achieved at about 1000 μmol m^{-2} s^{-1} or 54 klx [11, 12, 13]. The photosynthetic abilities of algae are saturated at about 30 klx, so that the statement: "alga needs light, the more the better" is a wrong one. At noon, when Sun is in zenith the illumination in the temperate zone is up to 90 klx. Therefore, the quantity of light from dawn to sunset is more than enough to satisfy the photosynthesis of algae. The really important thing for the algal yield is the continuance of the illumination. Both in laboratory or outdoors, when tend algae in large scale, the longer is the day the better. According to our experience, the yield comes to naught at 8 h daily illumination when breed intensively green algae. The values 8 h and 10000 lx solar light are tentative and useful in order to define the number of days of a cultivation period. That is economically important for the temperate zone and its significance grows with the latitude. For example, on N 42 – 44^0, the latitude of Bulgaria, one relying on sunlight could not achieve reasonable algal yield in the period from the middle of October to the middle of February. On latitude N 40 – 41^0, it is acceptable to the middle of November, according to the experience with *Spirulina* at Nigrita, Greece. Working on N 28^0, we have in mind Canary Islands, is possible to breed algae with a very good yield to the end of December that means round year.

In open ponds, when one breeds unicellular eukaryotic algae, a natural phenomenon arises, namely synchronization of algal cells. The natural diurnal light and temperature alterations provoke synchronized development of the cells. With a statistical probability equal to the stage of synchronization it may be accepted that all cells have similar behaviour. Then, from biochemical point of view, the whole culture behaves like one huge cell. That is why synchronous cultures of unicellular eukaryotic algae are suitable to study physiology and biochemical kinetics of cell ontogenesis [14, 15]. The large outdoor algal cultures in open ponds are oft synchronized on their own. The concentration of individual substances in a synchronous culture use to be very different with the time, compared to non-synchronous. Having known the kinetics of individual substances during ontogenesis, one can produce them in the right time, when their concentration is higher. That is an example for quick and direct implementation of research outcome in the algological practice.

2.2. Temperature

If the temperature holds sway over the growth of all terrestrial plants, it is even more important parameter for algae which are entirely in water. Increase of the temperature of a volume of algal suspension by 1^0C requires about 4000 time more energy compared to the same air volume. The algae remain too long in cold medium compared to the higher plants. The growth rate of an alga is maximal at the optimal temperature. Finding the optimal temperature is one of the first laboratory tasks when intensive cultivation of a new algal species begins (Table 1). Heating and cooling of the liquid are of importance in order to maintain optimal temperature when cultivate the algae on large scale. Heating or cooling, though being non-physiological concepts, contributes to the yield of algal biomass. As a tentative empiric rule, it could be generalized that algae which have higher temperature optimum are more productive [16, 17].

Table 1. Optimal cultivation temperature

Alga	**^0C**	**Source**
Arthronema africanum	36	[18]
Chlorella sp. R-06/2	39	[17]
Coelastrum sphaericum	27	[19]
Dunaliella salina	28	[20]
Isochrysis; *Porphyridium*	25	[21]
Porphyridium cruentum	25	[22]
Scenedesmus incrassatulus	33	[8]

The natural daily evaporation of nearly 10 $dm^3.m^{-2}$ water retains the temperature lower when algae are bred in open pond. In this case the algae will produce less than their genetic abilities allow. Additionally, the combined effect of low temperature and high light intensity is harmful: there is too much energy unleashed over thylakoids whose enzyme systems are not ready to fit photosynthetic response. Thereby, at least one advantage of the photobioreactors is plain: the temperature amplitude is smaller.

At night, without photosynthesis, the respiration remains the predominant physiological process, which leads to intensive protein synthesis, due to the metabolic degradation of carbohydrates. The increase of protein at night is presented on Table 2. That is an example for a quick and easy change of biomass to higher quality. Therefore, in the day when photosynthesis takes place, and at night when respiration dominates, it is important to maintain optimal temperature. Growing *Scenedesmus*, our experience says that the optimal temperature in the day is 33 ^{0}C, at night 23 ^{0}C [8].

The open ponds lose great amount of heat with the water evaporation and it is $\boldsymbol{Q} = \boldsymbol{i.m}$, where $\boldsymbol{Q}$ – heat waste; $\boldsymbol{i}$ – enthalpy of the water vapour, which is approximately 2250 $kJ.kg^{-1}$ at the range 25-35 ^{0}C; $\boldsymbol{m}$ – mass of evaporated water, which is about 5-10 $kg.m^{-2}$ according to our experience. The value of $\boldsymbol{Q}$ is so large that the heating up of opened ponds is pointless.

In order to maintain optimal temperature in a photobioreactor (PBR), it is necessary either to heat up the algal suspension or, blowing air, to evaporate certain amount of water and get rid of the extra heat. The evaluation of thermal waste is equal to quantification of heat transfer through the wall of PBR. The overall heat transfer, ***K*** can be evaluated according to the well known formulae [23]:

$$K^{-1} = \alpha_1^{-1} + \delta\,\lambda^{-1} + \alpha_2^{-1}; \qquad Q = K.F.\,\Delta t$$

where:

F – heat transfer area, m^2;
Δt – difference between temperatures of optimum and inner air;
α1 – heat transfer from inside to the wall, tentatively 1000-2000 $W.m^{-2}.K^{-1}$;
α2 – heat transfer from the wall to outside, tentatively 10-40 $W.m^{-2}.K^{-1}$;
δ – thickness of the wall, m;
λ – thermal conductivity of the wall material, $W.m^{-1}.K^{-1}$.

Table 2. Protein content (% m/m) of biomass at 18 h and at 8 h

Algal species	**18 h**	**8 h**	**Increase**	**Source**
Coelastrum sphaericum	34	52	53 %	[19]
Scenedesmus acutus	24	40	67 %	[16]
S. incrassatulus	36	51	42 %	[19]
S. incrassatulus	28	43	53 %	[16]
Spirulina platensis	50	57	14 %	[5]

Overall heat transfer was evaluated 11.2 ± 0.6 $W.m^{-2}.K^{-1}$ for covered pond, and 7.7 ± 0.9 $W.m^{-2}.K^{-1}$ for a plastic transparent plate absorber column [8, 24]. Both cases concern one-layer wall with 3 thermal resistances. In case of two-layer wall the thermal resistances are 7 and the heat loss is lower. The fact that 2 layers sink the illumination ought to be born in mind.

Sometimes, the algae are grown at low or high temperatures in order to characterize them and to study the influence of stress conditions. The main conclusion of some practical importance is that the percentage of unsaturated fatty acids is higher at lower temperature. This piece of knowledge is essential when study the algal membranes and thermal stress. Nevertheless, the yield of biomass is much smaller at lower temperature, so that the amount of unsaturated fatty acids will be smaller as an absolute value. Maintaining the temperature close to optimal is worth the effort.

2.3. Stirring

The intensive cultivation of algae requires stirring of the suspension. In order to increase the yield, every single cell of the algal suspension has to be illuminated and supplied with nutrients. At night, when algae do not photosynthesize, respiration is the most intensive process, and oxygen has to be supplied. The intensive algal breeding needs stirring around the clock.

The stirring rate depends on the morphological peculiarities of the alga. For example, speed 0.3 $m.s^{-1}$ may be recommended for *Spirulina*, *Porphyridium* and *Dunaliella*; about 2-3 times higher is good for *Chlorella*, *Chlorococcum*, and *Scenedesmus* [4, 25, 26]. An increase of the algal yield and increase of the optimal algal density are observed at higher stirring rate; not a few species have similar behaviour [27, 28, 29]. We share the same opinion reckoning, instead of "slow stirring" and "fast stirring", a part of dynamics of fluids ought to be applied.

There are several mechanical manners of stirring: pedal wheel stirring, circulation with centrifugal or plunger pump, bubbling of air, etc. Every one sort of stirring is bound with energy expense which is significant: the minimum of required power is about 6 $W.m^{-2}$ uninterrupted for 24 h. There are several really existing factors as viscosity of the algal suspension, friction, mechanical resistances which take much more energy than a theoretical calculation. The velocity of the algal suspension is a vector and every change of its direction requires energy. There are not a few phycological reports where the resistances have not been taken into consideration and, as an outcome there was underestimation of necessary power and the real energy expense. Similar errors keep on cropping up when airlift stirring is used. The whole energy expense shows that the latter is not more economic compared to mechanical stirring. As a matter of fact, an economically acceptable algal yield does really need electrical power of 10 $W.m^{-2}$ for stirring.

The proper choice of pumps and stirring manner in accordance with the morphology and requirement of cultivated alga or cyanobacteria, as well as energy expense, concerns questions of hydrodynamics. One of the basic concepts of hydrodynamics is namely turbulence [30, 31]. The quantitative characteristic of the turbulence is given with the number of Reynolds,

$$\boldsymbol{Re = v.l.\rho.\mu^{-1}}$$

where:

v – velocity of algal suspension, $\boldsymbol{m.s^{-1}}$
l – limiting size, $\boldsymbol{m}$
ρ – density of algal suspension, $\boldsymbol{kg.m^{-3}}$
μ - viscosity ***Pa.s***

The dimensionless Re-number is a criterion for the stage of turbulence. It is a ratio between two forces: the active force which drives the liquid and the passive force, the force of inner resistance of the liquid. Having this formula, there is no need to divide the velocity as "slow" and "fast". Moreover, not the velocity but Re-number is more proper characteristic of the current in algal device elements: pond, tube, pump, etc. (Table 3).

More often the physical density of the algal suspensions is 1000-1030 $kg.m^{-3}$, save the case of carotenogenesis with *Dunaliella*, where it is up to 1100 $kg.m^{-3}$. Data for the viscosity of algal cultures are given by Petkov and Bratkova, (1996) [32] and the impact of phytoplankton on the viscosity at natural conditions is described by Jenkinson and Biddanda, (1995) [33]. The viscosity of cyanobacteria and green algae depends on algal density and linearly on temperature and remains in the range of 0.9 – 1.2 mPa.s. The viscosity of *Porphyridium* is twice higher and strongly increases with the growth of algal density. The cultures of *Porphyridium* express behaviour of non-Newtonian fluid due to the extracellular polysaccharide. The energy expense for stirring of a pond with *Porphyridium* will be 2-3 times greater in comparison to *Spirulina*.

Table 3. Number of Reynolds at different parts of algal devices

Element of device	Limiting size, mm	*Stirring rate, m.s^{-1}*	Re
Sloped area	7 – 45	0.1 – 0.4	3000 – 5000
Tube	10 – 50	0.1 – 1.1	3000 – 2x10^4
Shallow pond	70 – 100	0.2 – 0.5	10^4 – 2x10^4
Centrifugal pump	25 – 50	1.0 – 2.5	5x10^4 – 10^5

Here is to mention that there is a difference between stirring and mass transfer. There will be a mass transfer process if a concentration gradient $\Delta C.\Delta n^{-1} \neq 0$ exists (C – concentration, n – distance). The salt concentration in an algal suspension is equal in all parts of the volume, so that there is stirring but there is not mass transfer. The stirring contributes to the intensification of the mass transfer of CO_2, as well as the increase of heat transfer when maintain optimal temperature.

As a recommendation, based on experience with the evaluation of various technological processes, the optimal Re number is between 10^4 – 1.5x 10^4 as a tentative value in the illuminated part of an algal device. This optimal interval emerges as a compromise between energy expense and sufficient intensity of the mass transfer process.

2.4. CO_2-Supply

The concentration of CO_2 in the air is 0.03 % (v/v) or 0.046 % (m/m). Looking to CO_2-supply only from the air one could achieve algal density up to 0.5 g.dm^{-3}; higher density and higher growth rate need bubbling of CO_2.

The mass transfer process concerns basically CO_2 supply and its distribution in the algal suspension. The solubility of CO_2 in the liquid solution is low and it is quickly consumed by the algal cells. Theoretically, production of 1 kg dry algal biomass requires 1.6 – 1.8 kg CO_2 depending on algal species. As a matter of fact, in an open pond, almost 10 kg CO_2 are usually expended for 1 kg algal biomass, really high level of desorption loss.

The minimization of desorption loss of CO_2 is important task, especially when its price is liable to calculation. The well known counter current absorption columns, which are applied at many chemical processes, ought to be a good example when make photobioreactor. The transparent plate absorber is the most effective photobioreactor, especially when green algae are bred [8, 9]. The performance of a column with size 1.5 x 0.5 x 0.3 m is more than 50 g daily. Having worked with 18 such columns we achieved mean daily yield 50 ± 1 g per a column, for 20 days in September, at 9 h illumination which the city terms allowed. Obviously, an optimistic opinion could be shared: extrapolating with the illumination to 12-14 hours in a country side with east and west opened horizons the algal yield will be greater. Besides, hydrodynamics, heat transfer and mass transfer could be very well visualized when work with transparent CO_2-absorption column, and it could be a useful model in the teaching of photoautotrophic biotechnology.

All C3 plants, including algae, suffer photorespiration: with light and low CO_2/O_2 ratio they release their CO_2 losing biomass. If the photorespiration is due to an evolutionary imperfection of RuBisCO historically bound to another CO_2/O_2 ratio or, an algobacterial symbiosis-bound mechanism, or a sort of way to release some extra-energy from the over-laden cell, are quests of fundamental plant biology. Anyway, the phenomenon is really important for the applied algology. It is known that scantiness of CO_2 supply under light conditions leads to release of glycolic acid in the medium [34, 35]. It happens often in the nature, for example, glycolic acid is a usual extracellular substance in the sea water [36]. In the laboratory at 8000 lx, glycolic acid was found by GC-MS in the medium of *Scenedesmus* 15 min after break off CO_2 supply [37]. The outcome of photorespiration is loss of biomass. Indeed, photorespiration is a negative process when the aim is high algal yield. The common experience says that a concentration of 10 – 15 $g.m^{-3}$ CO_2 is enough in order to avoid limitation of photosynthesis.

The solubility of CO_2 in water is about 25 fold higher then the solubility of O_2. Besides, as a large culture, algae are more often bred at pH > 7. As a matter of fact, the medium is a solution of $NaHCO_3$ with slowly increasing concentration. Therefore, even outdoors, the photorespiration of algae could strongly be thwarted in comparison to higher plants. The input of CO_2, supplied accordingly to the requirements of algal culture, eliminate thoroughly the sway of photorespiration.

Many sources are suitable for CO_2 supply: burning processes, fermentation, wellsprings. A scrubber system for CO_2 washing up is obligatory when use burning gases.

When one begins to grow algae in a nutrition medium at pH = 7, after a day or two, pH will be higher. The increase of pH is a consequence of accumulation of Na^+ in the medium. Even if the nutrition medium in the beginning of cultivation does not content $NaHCO_3$ at all, after some time it will be there. A lot of algae, being in possession of carbonic anhydrase, use $NaHCO_3$ as a source of carbon dioxide. That is why $NaHCO_3$ is often added in algal media. As a matter of fact, the chemical equilibrium $2NaHCO_3 = Na_2CO_3 + CO_2 + H_2O – Q$ depends on temperature and at 30 – 35 ^{0}C the partial pressure of CO_2 is significant. This means, desorption loss of CO_2 could thwart the economic success of an open pond if the price of CO_2 is not a negligible item of the prime cost, so that a second advantage of photobioreactors is plain: the CO_2 loss is much lower.

2.5. The Algal Suspension

2.5.1. Nutrition Medium

The algae have usually been bred in mineral nutrition media, and there are a lot of compositions and proportions. A well composed nutrition medium allows achieving a preliminary chosen algal density. The medium is almost fully exhausted of nutrients in the end of cultivation; the waste water after separation of biomass does not pollute the environment. When one is making up a new nutrition medium, it is good to bear in mind and to abide some tentative rules. The ash content and composition of the biomass of a strain is the reason for the right proportion of nutrients. The nitrogen source must be chosen accordingly to activity of the enzyme (urease, nitrate reductase) of the alga. Normally, as a substrate, $NaNO_3$, KNO_3, NH_4NO_3, NH_2CONH_2 or their combinations have been used.

Good results have been achieved with the nutrition medium suggested by Šetlík at al. (1965) for green algae, and modified by Georgiev et al. (1978) adding 2 g.dm^{-3} $NaHCO_3$ as a puffer [1, 38]. It has successfully been applied to the large scale breeding of *Chlorella* and *Scenedesmus* in opened ponds, for many years. Nevertheless, the high content of Na^+ and high pH value are disadvantages which do not allow long term use of the exhausted media for agricultural watering.

Economically considered and environmental friendly medium for breeding of green algae was described by Petkov (1995) with the next composition in mg.dm^{-3}: urea – 2100, KH_2PO_4 – 530, $MgSO_4$ – 370, $Fe_2(SO_4)_3$ – 20, H_3BO_3 – 1, $ZnSO_4.7H_2O$ – 1, $MnSO_4.4H_2O$ – 1, $CuSO_4.5H_2O$ – 0.2, $(NH_4)_2MoO_4$ – 0.02, $CoSO_4.7H_2O$ – 0.002 [39]. Very good outcome has been rendered by means of this medium, especially at experiments in photobioreactors. There are other advantages of this media, namely that there are no sediments, and no release of ammonia which happens oft at $pH > 7$, the most usual condition.

The most used medium for breeding of cyanobacterium *Spirulina* was composed by Zarrouk (1966) and concentration 0.2 M $NaHCO_3$ has been proved the most proper [40]. After that, almost all other media for *Spirulina* are small modifications of Zarrouk's medium. The concentration of total Na^+ in the other media is calculated as a sum from $NaHCO_3$ and Na_2CO_3 which does not matter, especially at bubbling CO_2.

The algae accumulate heavy metal ions, which means, the salts must be chosen carefully (Table 4). The content of Pb и Cd in nutrition medium, composed of distilled water and pure salts (p. a. grade) is lower than 5 mg.m^{-3}. In this case, theoretically, 1 g biomass is supposed to contain about 0.4 μg of these elements if all their ions are absorbed. Accumulation of heavy metals in higher concentrations takes place only at model experiments.

Sometimes, in order to increase the growth rate and to obtain qualitative biomass, successive heterotrophic and then autotrophic cultivation are applied. Usually, at the heterotrophic feeding, glucose or acetate are used as energy and carbon source. Having used organic substrates, sterile conditions are obligatory. If not, the bacterial biomass will predominate, due to its higher growth rate. However, the heterotrophic growth remains a subject of microbiology but photosynthesis is in the essence of algal biotechnology.

2.5.2. Nitrogen Starvation

The nitrogen starvation is a particular case of algae breeding. Growing algae in a medium with deliberately maintained scanty concentration of nitrogen source has been studied for several decades [41, 46, 47, 48, 49, 50]. The common outcome of this research is that the lipids have been enhanced, which was due to the triacylglycerols. At a lack of key substrate the algal culture can not grow intensively but to make storage of oils. What would one earn by increasing lipids nearly twice at nitrogen starvation, slanting down 2-3 folds the algal growth in the same time? The percentage of lipids in biomass will be higher nevertheless the whole amount of lipids will be lowered, compared to cultivation at optimal nitrogen feeding. Duplication of the total lipids of *Botryococcus* and twice lower growth at nitrogen deficiency for 20 days was described [51].

Table 4. Content of toxic metals (μg) in 1 g dry algal biomass

Alga	Pb	As	Cr	Hg	Cd	Source
Spirulina	5.1	-	2.7	-	1.5	[42]
Scenedesmus	7.0	4.0	3.5	0.6	0.4	[43]
Spirulina	<1	<1	-	0.05	0.05	[44]
Spirulina	2-5	-	9	0.1	-	[45]

Nitrogen deprivation remains a good manner when the aim is studying the algal behaviour. In this course it is good to mention that there are many algae whose total lipids contain high amount of fatty acids 20:5 or 22:6 in their triacylglycerols [49, 53, 54, 55]. High yield of biomass or total lipids are achieved at optimal conditions: the time and growth rate are of importance when bare in mind commercial use of algae.

2.5.3. Concentration of NaCl

Algae which usually dwell in fresh waters may successfully be grown in media based on sea water. The reason for doing this is clear: good husbandry of fresh water and utilization of some sea water salts. Here is to remember that one may use sea water when compose nutrition medium but the evaporation has to be compensated with addition of fresh water, otherwise the salt concentration increases.

There are some particular cases where algae can endure very high salt concentration. The green unicellular alga *Dunaliella* normally dwells in salt waters. When the concentration of NaCl increases due to the water evaporation, a process of glycerol accumulation begins in the cells of the alga as an answer to overcome the osmotic stress. At the same time the concentration of β-carotene increases and the effect is known as carotenogenesis [52, 56, 57, 58, 59]. Sustainable yield of β-carotene (0.6 $g.m^{-2}.d^{-1}$), glycerol (8 $g.m^{-2}.d^{-1}$), and daily growth of *Dunaliella* 20 $g.m^{-2}$ was achieved for a long period in outdoor ponds [52]. This sort of breeding is based on increase of NaCl from 31 $g.dm^{-3}$ to 150 $g.dm^{-3}$ which is deadly for other marine algae.

In order to provoke carotenogenesis in *D. salina*, Klyachko-Gurvich et al. (1997) after breeding it at optimal conditions, decrease the temperature for some days [60]. The percentage of β-carotene in dry biomass increased from 0.27 % at the end of exponential growth to 2.74 % in state of carotenogenesis. This is relatively quick and easy performable way to increase the quality of algal biomass.

It is worth mentioning that oleic acid appears to be the most variable fatty acid related with carotenogenesis in *Dunaliella salina* [61] and it shows the same behaviour at nitrogen deficiency of *Botryococcus* [61].

2.5.4. Extracellular Substances

At optimal conditions the algae excrete small amount of substances in the medium, save some of them as *Botryococcus* or *Porphyridium*, whose metabolism is naturally directed to abundant extracellular production. The concentration of total extracellular substances of green algae, for example, *Scenedesmus* that is subject of large scale breeding is 15-20 $mg.dm^{-3}$. At unfavourable conditions, for example at scanty amount of nutrients, the extracellular substances increase to 200 $mg.dm^{-3}$ [37]. The lipophilic excretions are predominantly hydrocarbons with 11-32 carbon atoms and small amount of free fatty acids with 16-18 carbon atoms. The predominant hydrophilic extracellular substances in the medium are succinic and lactic acid which originate from concomitant bacteria; others like hydroxyoctanoic acid are concerned by bacterial enzyme activity.

The methylated amino acid N-dimethylisopropenyl glycine was found amid the extracellular substances of *Scenedesmus*; N-dimethyl valine was found in the medium of its concomitant bacteria. Only one double bond makes the difference between these two substances. Probably, dimethylated amino acids are hardly or not at all assimilated by the concomitant bacteria. The presence of dimethylated amino acids can be imputed to extracellular function of algae, for example, chelating of metal ions [62].

2.5.5. Bacterial Contamination

Tending algae both in laboratory, in high volume photobioreactors, or in large scale ponds is carried out at non-sterile conditions. Nevertheless, as a rule the whole weight of concomitant microorganisms compared to the algal biomass is 1:100 due to the photoautotrophic manner of cultivation [63]. A carefully tended algal culture, being sun light illuminated and oversaturated with oxygen, contains insignificant amount of aerobic bacteria. On the other hand the algae, with a few exceptions, under normal conditions excrete in the medium very small amount of organic matter [37, 64, 65]. That is why the weight of total concomitant bacteria is also small. They are aerobic and non pathogenic, belong predominantly to the genera *Micrococcus*, *Pseudomonas*, *Bacillus*, *Flavobacterium* [66, 67].

SEPARATION OF BIOMASS

The microalgal biomass may be separated from the medium by means of filtration or centrifugation. Unfortunately, there are very few microalgae, filamentous only, which can successfully be filtrated even in laboratory. The performance of vacuum filtration of *Spirulina* is nearly 2 $kg.m^{-2}h^{-1}$ at algal density 3 $g.dm^{-3}$ which includes washing up the biomass and the filter after every portion and, of course, a lot of labour expense. A thin layer, no more than up to 5-6 mm, of filamentous microalgae is successfully yielding to filtration. The filtrated biomass is with 20 % dry weight.

The centrifugation, being a process of high performance and no requiring much labour, is the usual method for separation of algal biomass. The centrifuge must be sedimental one; filtration centrifuge is useless. Even a thin layer of microalgae does create very high hydrodynamic resistance for the water passing through. Generally speaking, the liquid phase, as an upper layer, must brim over but not pass through the bottom. The performance of the most often used centrifuges is a volume of 10^4 $m^3.h^{-1}$ microalgal suspension or up to 20 $kg.h^{-1}$ biomass. This includes washing up of the centrifuged biomass and second centrifugation as obligatory operation to achieve good quality. The bacterial number is much lower and practically no salts remain after washing up and centrifugation.

The most important characteristic of all types of centrifuges is how greater is the centrifugal acceleration, (a) compared to the acceleration of Earth

$$G = a.g^{-1} = v^2.g^{-1}.r^{-1}, \qquad v = 2\pi.n.r,$$

where:

v – velocity of the rotor, $m.s^{-1}$;
g – nominal acceleration, 9.81 $m.s^{-2}$;
r – centrifugal radius, m;
n – revolts per second, s^{-1}.
A value of ***G*** = 10 000 is acceptable for separation of microalgal biomass.

After the separation of biomass an equivalent amount of salts must be added in the nutrition medium in order to continue the breeding at steady state conditions. The separated biomass is immediately subjected to drying: in spray dryer, under vacuum freezing or smeared as a thin layer.

USE OF ALGAE AND ALGAL PRODUCTS

4.1. Algae as Foods, Forage and Pharmaceuticals

The cyanobacterium *Spirulina* and green microalgae *Chlorella* and *Scenedesmus* are widely used since long time and their advantage as nutritional additives is well proved [68, 69]. They are rich and valuable food due to high protein content, naturally balanced vitamins and minerals. All microelements of the algal biomass Fe, B, Zn, Mn, Cu, Mo, Co are naturally chelated and also are in a natural proportion. Microalgal biomass is one of the preferable sources for production of protein hydrolysates [70, 71, 72]. The protein content of large scale produced biomass is 45 ± 10 % as a mean value. All essential amino acids are presented in the proteins. The total lipid content of cyanobacteria is 7-8 % and it is about 15-25 % in the green algae and other eukaryotic that are in use. The lipid composition is more diverse in comparison to the traditional oils (Table 5). Higher lipid content, up to 38 %, was achieved with a starchless mutant of *Chlorella* after UV-provoked mutagenesis [73].

Table 5. Usual composition of algal lipids

Substances	**% m/m**
Hydrocarbons	2-3
Oils	2-12
Carotenoids	2-5
Sterols	0.5-5
Chlorophyll	8-12
Glycolipids	20-25
Sulpholipids	20-25
Phospholipids	15-20

The algal lipids are built of polyunsaturated fatty acids with 4, 5, 6 double bonds; such acids as arachidonic, eicosapentaenoic and docosahexaenoic characterize the algal oils. They are precursors in the biosynthesis of prostaglandins and thromboxanes in animal and human metabolism. Of importance for the human health is gamma-linolenic acid, γ-18:3 and *Spirulina* is the most proper source. Linoleic acid, 18:2 is known as vitamin F and it is abundantly presented in almost all cyanobacteria and algae. Synthesis of 18:2 does not take place in human body but is supplied with the food, meaning that it is really an essential fatty acid.

The green algae abundantly contain sterols which have hypocholesterolemic effect. Relatively high sterol content, 0.5 % from dry biomass, was found in *Scenedesmus*. In the sterol mixture chondrillasterol is 65 %. The main sterol of the green alga *Coelastrum sphaericum* is ergosterol, and usually it is about 90 % of the total sterols. Ergosterol is biological precursor to vitamin D_2. At the time being *Coelastrum sphaericum* is the most proper source for production of ergosterol on photoautotrophic way [74]. Strains of *Porphyridium cruentum* contain over 90 % 22-dehydrocholesterol.

Many cyanobacteria and algae, for example *Spirulina*, *Dunaliella*, are rich of β-carotene, the pro-vitamin A, whose role for the common health as antioxidant and particularly its role for the animal and human sight is well known [75]. Another useful antioxidant is astaxanthin which is produced from biomass of *Haematococcus* and *Chlamydomonas* [76, 77]. Plenty of reports confirm high amount of the antioxidants vitamin E, vitamin K, coenzyme Q_{10} in many cyanobacteria, red and green algae.

Dry microalgal biomass can be stored for several years without changes. That quality has to be taken into consideration by the authorities when the items of the emergency reserve are planed.

Microalgae have successfully been used to feed poultry, pigs and other livestock [78, 79]. Laying hens have been fed with 90 % combined forage and 10 % dry biomass of *Chlorella*. The outcome was an improvement of the bird productivity and the morphological characteristics of the eggs and the egg yolk pigmentation was more intensive by 2.5 units by the Roche's scale [80].

Microalgae, due to their high nutritional value combined with small size, are natural feed of many small animals: larvae, newly hatched fishes, garden snails, etc. Marine *Chlorella* is widely used in artificial nutrition chain phytoplankton – zooplankton – fish [81, 82]. The rotifer *Brachionus* has often been preferred in this chain [83, 84]. Fatty acid composition of phytoplankton lipids is important to the success of aquaculture, because the next members of this chain do not biosynthesize some fatty acids and also sterols [85]. It is evident that the algal species and its fatty acid composition are essential in these nutrition chains. Here ought to be mentioned that under the name "*Chlorella*" mixed cultures have often been studied or used [86].

The cyanobacteria contain phycocyanin which is applied in the immunodiagnostic and protection from ionizing radiation. Convenient methods for separation and purification of phycocyanin and allophycocyanin from *Arthronema africanum* and phycoerythrin from the red alga *Pophyridium cruentum* were described [18, 87, 88]. Antitumor effect of water-soluble extract from *Scenedesmus* has been reported, indicating that non-toxic agents can be used for therapeutic purposes [89].

The algae often are sources of metabolites unusual for terrestrial plants. Currently, there are more than 10 000 known species of microalgae and many of them seems to be promising sources of useful compounds. As well known, some cyanobacteria possess extremely strong toxins [90, 91].

The algal biomass, like the more standardized and high quality foodstuffs, is not a cheep product and there are no visible signs for a trend of slanting down its prime cost. By the same token, the processing of microalgal biomass for pharmaceutical aims is economically important to support at least several products.

4.2. Algae as Source of Fuel

The interest on the algae as a proper renewable source of fuel increases recently. Research and development activity is carried out worldwide even airline companies pay attention to algal fuels [92]. This topic, being extremely largely discussed in the World Wide Web is worth mentioning.

4.2.1. Hydrocarbons

The algae produce hydrocarbons which are located in their membranes. Their chain is build of 11 – 35 carbon atoms. They are saturated or with 1-3 double bonds and originate from fatty acids. Usually, the percentage of C_{17}, ΔC_{17}, C_{27} ΔC_{27} is significantly higher. Sometimes one of them is up to 90%. Hydrocarbons with uneven C-number predominate significantly. Iso- and anteiso- are present, too. Normally, the content of hydrocarbons is 2-3 % of the total lipids, or about 0.5 % of biomass. The algae usually excrete small amount of hydrocarbons and their concentration in the medium of *Scenedesmus* is nearly 10 $mg.dm^{-3}$ even at rather high algal density of 5-6 $g.dm^{-3}$ [37]. The level of extracellular substances of the green alga *Botryococcus* is significantly higher. This alga excretes terpenoid hydrocarbons which emerge on the surface [93, 94]. The energy required for breeding of *Botryococcus* is significantly higher than the energetic income of its hydrocarbons. Nevertheless, research and development work is promising.

4.2.2. Biogas

The only meaningful production of biogas from algae is when sea macroalgae stranded by waves are used, or if high amount of biomass is a result of eutrophication. Being of no other avail, they might be composted in methane tank together with all other cellulose containing waste [95]. Here has to be mentioned that the wrack of some macroalgae are important part of nutrition chains at the seaside [96]. Having at disposal so much cellulose material which rots freely in the environment, releasing green house gases, it would be pointless to deal with algae for biogas production. Biomass of cultivated macroalgae could be used as biogas source, for example, when they are a stage of devices for biologic treatment of water. Let the wild marine algae remain a part of natural nutrition chain in the sea.

4.2.3. Bio-hydrogen

The production of hydrogen by algae is worth mentioning though being a well known idea [97]. There are nowadays two manners for production of hydrogen from algae. The first of them concerns nitrogen fixing cyanobacteria. For example *Anabaena variabilis* produces hydrogen under condition of limitation of nitrogen. The production rate of 1 kg biomass is 80 $mg.h^{-1}$ hydrogen [98].

Hydrogen in the form of gas can be produced by green microalgae, *Chlamydomonas reinhardtii*, for example, at scantiness of sulfate [99]. The algae lose reversible the photosynthetic activity of PS-II at light and lack of sulfate. The achieved yield of hydrogen, though theoretically promising, is obviously very far from real practical use.

4.2.4. Biodiesel

From the beginning of the new century, microalgae have been supposed to be the proper renewable source for production of fatty acids, respectively turned into their methyl or ethyl esters, and used as biodiesel [100]. The idea has been fostered all over the world and often underpinned by wishful thinking proofs. Let us see the facts that the followers of microalgae-biodiesel-idea did not comment. The production of microalgae needs daily energy 0.15 $kWh.m^{-2}$ for stirring and centrifugation; the gain could be 0.05 $kWh.m^{-2}$ energy of biodiesel. Dry algal biomass can not be extruded like sunflower or other seeds. Algal lipids must be extracted with several times higher volume of organic solvent, or supercritical CO_2. The both processes are energy consuming; the first requires distillation of solvent, the second is going under high pressure.

The algal fatty acids are polyunsaturated that is a valuable quality when they are used as a food or forage, but it is a disadvantage if biodiesel is supposed to be produced. The double bonds of the unsaturated fatty acids should be partially saturated with H_2 under pressure, which means energy both for pressure and H_2 production. As a result, the energy expense is greater than the income. At the time being, production of biodiesel from microalgal oil is energetically unreasonable. The conclusion for the time being is plain: the biological value of cultivated microalgae as a food is much higher than as a fuel.

CONCLUSIONS

The microalgae render a large field for fundamental studies. Research and development bound with photoautotrophic technology of microalgae is worth the effort. The topic allows a quick transfer of knowledge from laboratory and workshop to the pilot plant and large scale production. A small enterprise for production of dry microalgal biomass may arise on the base of relatively low cost investment, having harnessed all available skills and knowledge. The product is of high biological value. There is an important advantage: every day is a day for harvest. A person who breeds microalgae could hardly be taken aback by the weather. It is good to remember that such an enterprise, having character of a farm for plant production, will always be classified as micro or small one. However, being no traditional farm, there is no well-trodden way for people who are up to breed microalgae; they ought to stick close to the algal science and knowledge. And a last reminder: nowadays microalgae are for eating, not for burning.

Acknowledgments

The author of this chapter thanks the NSF of Bulgaria, project D002-299/08.

References

[1] Šetlík, I., Prokeš B., Kubín Š., Přibil S. & Dittrt F. 1965. Analysis of technological principles of the production of algae. In: *Proceedings of III Counsel for Economic Collaboration*. Prague, Czech.

[2] Livansky, K. & Pilarski, P. 1993. Carbon dioxide supply to algal cultures. *Archiv für Hydrobiologie*, *97*(Algological Studies 69), 113-123.

[3] Richmond, A. & Becker, E. W. 1986. Technological aspects of mass cultivation, a general outline. In: *Handbook of microalgal mass culture* (edited by A. Richmond) Pp. 245-263. Florida, USA: CRC Press.

[4] Fournadzieva, S., Petkov, G., Pilarski, P. & Andreeva, R. 1999. Use of geothermal fluids and energy for mass microalgal cultivation. In: *Direct utilization of geothermal energy* (edited by K. Popovski et al.) Pp. 175-179. Oregon, USA: Oregon Institute of Technology.

[5] Torzillo, G., Carlozzi, P., Pushparaj, B., Montaini, E. & Materassi, R. 1993. A two-plane tubular photobioreactor for outdoor culture of *Spirulina*. *Biotechnology and Bioengineering*, *42*, 891-891.

[6] Pulz, O. 2001. Photobioreactors: production systems for phototrophic microorganisms. *Applied Microbiology and Biotechnology*, *57(3)*, 287-293.

[7] Carlozzi, P., Ena, A. & Carnevale, S. 2005. Hydrodynamic alterations during cyanobacteria (*Arthrospira platensis*) growth from low to high biomass concentration inside tubular photobioreactors. *Biotechnology Progress*, *21(2)*, 416-422.

[8] Petkov, G. D. & Muchtarova, M. G. 1995. Heat balance of a closed micro-algal photoreactor. *Archiv fur Hydrobiologie*, *111*(Algological Studies 79), 119-124.

[9] Petkov, G. D. 2000. Absorber tower as a photobioreactor for microalgae. *Russian Journal of Plant Physiology*, *47(6)*, 892-895.

[10] Del Campo, J. A., García-González, M. & Guerrero, M. G. 2007. Outdoor cultivation of microalgae for carotenoid production: current state and perspectives. *Applied Microbiology and Biotechnology*, *74(6)*, 1163-1174.

[11] Hoffmann, P. 1987. *Photosynthese*. Pp. 137-145. Berlin, Germany: Academie Verlag.

[12] Jones, H. G. 1992. Plants and microclimate: a quantitative approach to environmental plant physiology. Pp. 9-37. Cambridge, UK: University Press.

[13] Robertson, J. 2000. How to convert microeinsteins to lux. http://www.madsci.org/posts/archives/2000-08/965136947.Ph.r.html.

[14] Otsuka, H. & Morimura, Y. 1966. Change of fatty acid composition of *Chlorella ellipsoidea* during its cell cycle. *Plant and Cell Physiology*, *7*, 663-670.

[15] Reitz, R. C., Hamilton, J. G. & Cole, F. E. 1967. Fatty acid concentrations in synchronous cultures of *Chlorella pyrenoidosa*, grown in the presence and absence of glucose. *Lipids*, *2(5)*, 381-389.

[16] Bozkova, M., Pilarski, P. & Furnadzieva, S. 1987. Mass cultivation of strains from the genus *Scenedesmus Meyen. Khydrobiologiya*, *29*, 30-36.

[17] Gacheva, G. & Pilarski, P. 2008. The resistance of a new strain *Chlorella* sp. R-06/2, isolated from an extreme habitat to environmental stress factors. *General and Applied Plant Physiology*, *34*, 347-360.

[18] Chaneva, G., Furnadzhieva, S., Minkova, K. & Lukavsky, J. 2007. Effect of light and temperature on the cyanobacterium *Arthronema africanum*-a prospective phycobiliprotein-producing strain. *Journal of Applied Phycology, 19(5)*, 537-544.

[19] Dang, d. K., Jordanova, S., Dilov, H. & Bojadzieva, M. 1987. Some physiological-biochemical indices of *Coelastrum sphaericum* Kom 64/64 and *Coelastrum cambricum* A-370 under conditions of laboratory cultivation. *Khidrobiologiya*, *29*, 75-82.

[20] Petkov, G. D., Klyachko-Gurvich, G. L., Furnadzhieva, S. T., Pronina, N. A. & Ramasanov, Z. M. 1990. Genotypic differences and phenotypic changes of lipid fatty acid composition in strains of *Dunaliella salina. Soviet Plant Physiology*, *3*, 268-272.

[21] Kaplan, D., Cohen, Z. & Abelovich, A. 1986. Optimal growth conditions for *Isochrisis galbana. Biomass*, *9*, 37-48.

[22] Minkova, K. M., Georgiev, D. I. & Houbavenska, N. B. 1987. Light and temperature dependence of algal biomass and extracellular polysaccharide production from *Porphyridium cruentum. Comptes rendus de l'Academie bulgare des Sciences*, *40*, 87-89.

[23] Roberts, A. & Cass, F. Heat transfer by convection, In: *Chemical engineering practice.* (edited by H. W. Cremer). V.7. Pp. 63-140. London, UK: Butterworths Scientific Publications.

[24] Petkov, G. D. 1994. Estimation of heat waste in microalgal photoreactors. *Archiv fur Hydrobiologie, 104*(Algological Studies 74), 121-124.

[25] Gudin, C. & Chaumont, D. 1991. Cell fragility: the key problem of microalgae mass production in closed photobioreactors. *Bioresource Technology*, *38(2-3)*, 145-151.

[26] Silva, H. J., Cortinas, T. & Ertola, R. J. 1987. Effect of hydrodynamic stress on *Dunaliella* growth. *Journal of Chemical Technology and Biotechnology, 40(1)*, 41-49.

[27] Vonshak, A., Abelovich, A., Bossiba, S. & Richmond, A. 1982. Production of *Spirulina* biomass: effects of environmental factors and population density. *Biomass*, *2*, 175-186.

[28] Richmond, A. (1986) Outdoor mass cultures of microalgae. In: *Handbook of microalgal mass culture* (edited by A. Richmond) Pp. 285-329. Florida, USA: CRC Press.

[29] Richmond, A. 1988. A prerequisite for industrial microalgaculture: efficient utilization of solar irradiance. In: *Algal Biotechnology* (edited by T. Stadler et al.) Pp. 237-244. London, UK: Elsevier.

[30] Reynolds, O. 1883. An experimental investigation of the circumstances which determine whether the motion of water shall be direct or sinuous, and of the law of resistance in parallel channels. *Philosophical Transactions of the Royal Society of London*, *174*, 935-982.

[31] Franklin, N. L. & Cass, F. H. 1957. The flow of viscous fluids. Turbulence. In: *Chemical Engineering Practice*, V.4 (edited by H. W. Cremer) Pp. 303-379. London, UK: Butterworths Scientific Publications.

[32] Petkov, G. D. & Bratkova, S. G. 1996. Viscosity of algal cultures and estimation of turbulency in devices for the mass culture of microalgae. *Archiv für Hydrobiologie, 114*(Algological Studies 81), 99-104.

[33] Jenkinson, I. R. & Biddanda, B. A. 1995. Bulk-phase viscoelastic properties of seawater relationship with plankton components. *Journal of Plankton Research, 17(12)*, 2251-2274.

[34] Hellebust, J. A. 1965. Excretion of some organic compounds by marine phytoplankton. *Limnology and Oceanography, 10(2)*, 192-206.

[35] Stabenau, H. & Winkler, U. 2005. Glycolate metabolism in green algae. *Physiologia Plantarum, 123(3)*, 235-245.

[36] Leboulanger, C., Descolas-Gros, C. & Jupin, H. 1994. HPLC determination of glycolic acid in seawater. An estimation of phytoplankton photorespiration in the Gulf of Lions, western Mediterranean Sea. *Journal of Plankton Research*, *16(7)*, 897-903.

[37] Kambourova, R., Petkov, G. & Bankova, V. 2006. Extracellular polar substances in cultures of the green alga *Scenedesmus*. *Archiv für Hydrobiologie, 162*(Algological Studies 119), 155-162.

[38] Georgiev, D., Dilov, H. & Avramova, S. 1978. Milieu nutritif tamponnée methode de culture intensive des micro-algues vertes. *Khidrobiologiya, 7*, 14–23.

[39] Petkov, G. D. 1995. Nutrition medium for intensive cultivation of green microalgae in fresh and sea water. *Archiv für Hydrobiologie, 109*(Algological Studies 78), 81-85.

[40] Zarrouk, C. 1966. Contribution a l'étude d'une cyanophycée. Influence de diverses facteures physiques et chimiques sur la croissance et la photosynth`ese de *Spirulinamaxima (Setch. et Gardner) Geitler.* Paris, France: University of Paris.

[41] Piorreck, M., Baasch, K. H. & Pohl, P. 1984. Biomass production, total protein, chlorophylls, lipids and fatty acids of freshwater green and blue-green algae under different nitrogen regimes. *Phytochemistry*, *23(2)*, 207-216.

[42] Mannino, S. & Benelli, T. G. 1980. Costituenti minerali di biomasse di *Spirulina maxima*, in Prospettive della cultura di *Spirulina* in Italia, (edited by R. Materassi). Pp. 131-135. Rome, Italy: CNR.

[43] Becker, E. W. 1984. Biotechnology and exploitation of the green alga *Scenedesmus obliquus* in India. *Biomass*, *4*, 1-19.

[44] Henrikson, R. 2000. Earth food *Spirulina*. Five year testing of heavy metals in *Spirulina*, 1983-1987. Earthrise Farms. http://www.spirulinasource.com/earthfoodstan1.html

[45] Campanella, L., Crescentini, G., Avino, P. & Moauro, A. 1998. Determination of macrominerals and trace elements in the alga *Spirulina platensis*. *Analusis*, *26(5)*, 210-214.

[46] Iwamoto, H., Yonekawa, G. & Asai, T. 1955. Fat synthesis in unicellular algae. I. Culture conditions for fat accumulation in *Chlorella* cells. *Bulletin of Agricultural Chemical Society of Japan*, *19*, 240-246.

[47] Klyachko-Gurvich, G. L. & Zhukova, T. A. 1966. Changes in the biosynthesis of fatty acids due to nitrogen deficiency in *Chlorella pyrenoidosa*. *Russian Journal of Plant Physiology*, *13*, 15-24.

[48] Khozin-Goldberg, I., Bigogno, C., Shreshta, P. & Cohen, Z. 2002. Nitrogen starvation induces the accumulation of arachidonic acid in the freshwater green alga *Parietochloris incisa (Trebouxiophyceae)*. *Journal of Phycology*, *38*, 991-994.

[49] de la Jara, A., Mendoza, H., Martel, A., Molina, C., Nordströn, L., Rosa, V. De la. & Diaz, R. 2003. Flow cytometric determination of lipid content in a marine dinoflagellate, *Crypthecodinium cohnii. Journal of Applied Phycology, 15(5)*, 433-438.

[50] Mendoza, H., Molina-Cedres, C., de la Jara, A., Nordström, L., Freijanes, K. & Carmona, L. 2008. Variación cuantitativa y cualitativa de la composición en ácidos grasos de *Crypthecodinium cohnii* en condiciones de supresión de nitrógeno. *Grasas y Aceites, 59(1)*, 27-32.

[51] Zhila, N. O., Kalacheva, G. S. & Volova, T. G. 2005. Influence of nitrogen deficiency on biochemical composition of the green alga *Botryococcus. Journal of Applied Phycology, 17(4)*, 309-315.

[52] Ben-Amotz, A. & Avron, M. 1981. Glycerol and beta-carotene metabolism in the halotolerant alga *Dunaliella*: a model system for biosolar energy conversion. *Trends in Biochemical Science*, *6*, 297-299.

[53] Kyle, D. J., Sicotte, V. J., Singer, J. J. & Reeb, S. E. 1992. Bioproduction of docosahexaenoic acid (DHA) by microalgae. In: *Industrial Applications of Single Cell Oils* (edited by D. J. Kyle & C. Radledge) Pp. 287-300. Champaign, IL, USA: American Oil Chemists' Society.

[54] Kainz, M., Arts, M. T. & Mazumder, A. 2004. Essential fatty acids in the planktonic food web and their ecological role for higher trophic levels. *Limnology and Oceanography, 49*, 1784-1793.

[55] de Swaaf, M. E., Rijk, C., De, T., Van der Meer, P., Eggink, G. & Sijtsma, L. 2003. Analysis of docosahexaenoic acid biosynthesis in *Crypthecodinium cohnii* by ^{13}C labelling and desaturase inhibitor experiments. *Journal of Biotechnology, 103(1)*, 21-29.

[56] Masjuk, N. P. 1965. Effect of Na, Mg, Cl and SO_4 on the growth, reproduction and carotene formation of *Dunaliella salina Teod. Ukrainian Journal of Botany*, *22(5)*, 3-10.

[57] Al-Hasan, R. H., Ghanndoum, M. A. & Sallal, A. K. 1987. Correlative changes of growth, pigmentation and lipid composition of *Dunaliella salina* in response to halostress. *Journal of General Microbiology, 133(9)*, 2607-2616.

[58] Goyal, A. 2007. Osmoregulation in *Dunaliella*, Part II: Photosynthesis and starch contribute carbon for glycerol synthesis during a salt stress in *Dunaliella tertiolecta.* P*lant Physiology and Biochemistry, 45(9)*, 705-710.

[59] Mendoza, H., de la Jara, A., Freijanes, K., Carmona, L., Ramos, A. A., Duarte, V., De, S., Varela, J. C. S. 2008. Characterization of *Dunaliella salina* strains by flow cytometry: a new approach to select carotenoid hyperproducing strains. In: *Electronic Journal of Biotechnology 11(4)*, 13

[60] Klyachko-Gurvich, G. L., Pronina, N. A., Furnadzhieva, S., Ramazanov, Z. M. & Petkov, G. 1997. Influence of suboptimal temperature on the lipid composition and state of cell wall of *Dunaliella salina. Russian Journal of Plant Physiology*, *44*, 212-221.

[61] Mendoza, H., Martel, A., del Rio, J. M. & Garcia, G. R. 1999. Oleic acid is the main fatty acid related with carotenogenesis in *Dunaliella salina. Journal of Applied Phycology, 11(1)*, 15-19.

[62] Petkov, G., Kamborova, R. & Bankova, V. 2010. Extracellular substances of total bacterial flora from cultures of the green alga *Scenedesmus*. *Comptes rendus de l'Academie bulgare des Sciences, 63*, 235-240.

[63] Oron, G., Shelef, G., Levi, A., Meydan, A. & Azov, Y. 1979. Algae/bacteria ratio in high-rate ponds used for waste treatment. *Applied and Environmental Microbiology, 38(4)*, 570-576.

[64] Kambourova, R. & Petkov, G. 2004. Gravimetric quantification of total extracellular organic substances of microalgae. *Comptes rendus de l'Academie bulgare des Sciences, 57(8)*, 23-26.

[65] Kambourova, R., Bankova, V. & Petkov, G. 2004. Extracellular lipophilic substances of green alga *Scenedesmus*. *Comptes rendus de l'Academie bulgare des Sciences*, *57*, 77-82.

[66] Toncheva-Panova, T. 1978. The bacterial microflora of *Scenedesmus acutus Meyen* strain Tomaselli 8 and *Scenedesmus cf. acutus Meyen* Strain 1972/10-2 Ilkov. *Khidrobiologija*, *7*, 77-84.

[67] Toncheva-Panova, T. & Puneva, I. 1981. Cytochemical studies on some enzyme activities in bacterial contaminants of the genus *Scenedesmus*. II. Oxidoreductases. *Khidrobiologija*, *13*, 82-86.

[68] Soeder, C. J. 1980. *Massive cultivation of microalgae: results and prospects*. *Hydrobiologia*, *72*(1), 197-209.

[69] Pulz, O. & Gross, W. 2004. Valuable products from biotechnology of microalgae. *Applied Microbiology and Biotechnology*, *65(6)*, 635-648.

[70] Bozkova, M., Tchorbanov, B., Boyadyeva, M. & Pilarski, P. 1987. Production of enzymatic protein hydrolysates from green microalgae, cultivated under different seasonal conditions. *Biotechnology & Bioindustry*, *6*, 36-39.

[71] Tchorbanov, B. & Bozkova, M. 1988. Enzymatic hydrolysis of cell proteins in green algae Chlorella and Scenedesmus after extraction with organic solvents. *Enzyme and Microbial Technology, 10(4)*, 233-238.

[72] Morris, H. J., Almarales, A., Carrillo, O. & Bermurdez, R. C. 2008. Utilisation of *Chlorella vulgaris* cell biomass for the production of enzymatic protein hydrolysates. *Bioresource Technology*, *99(16)*, 7723-7729.

[73] Ramazanov, A. & Ramazanov, Z. 2006. Isolation and characterization of a starchless mutant of *Chlorella pyrenoidosa* STL-PI with a high growth rate, and high protein and polyunsaturated fatty acid content. *Phycological Research*, *54(4)*, 255-259.

[74] Petkov, G. D. & Kim D. D. 1999. Sterols of the green alga *Coelastrum*. *Archiv für Hydrobiologie*, *130*(Algological Studies 95), 89-92.

[75] McLaren, D. S. 2002. Vitamin A and its relatives: marvelous molecules in key life processes. *Sight and Life*, *3*, 3-17.

[76] Guerin, M., Huntley, M. E. & Olaizola, M. 2003. *Haematococcus astaxanthin:* applications for human health and nutrition. *Trends in Biotechnology*, *21(5)*, 210-216.

[77] Kang, C. D. & Sim, S. J. 2008. Direct extraction of astaxanthin from *Haematococcus* culture using vegetable oils. *Biotechnology Letters, 30(3)*, 441-444.

[78] Belay, A. 2002. The potential application of *Spirulina (Arthrospira)* as a nutritional and therapeutic supplement in health management. *The Journal of the American Nutraceutical Association*, *5(2)*, 27-48.

[79] Spolaore, P., Joannis-Cassan, C., Duran, E. & Isambert, A. 2006. Commercial applications of microalgae. *Journal of Bioscience and Bioengineering, 101(2)*, 87-96.

[80] Grigorova, S. 2006. Dry biomass of fresh water algae of Chlorella genus in the combined forages for laying hens. *Journal of Central European Agriculture, 6(4)*, 625-630.

[81] von Elert, E. & Wolffrom, T. 2001. Supplementation of cyanobacterial food with polyunsaturated fatty acids does not improve growth of *Daphnia. Limnology and Oceanography, 46*, 1552-1558.

[82] Wacker, A., Becher, P. & von Elert, E. 2002. Food quality effects of unsaturated fatty acids on larvae of the zebra mussel *Dreissena polymorpha. Limnology and Oceanography, 47(4)*, 1242-1248.

[83] Volkman, J. K., Jeffrey, S. W., Nichols, P. D., Rogers, G. I. & Garland, C. D. 1989. Fatty acid and lipid composition of 10 species of microalgae used in manculture. *Journal of Experimental Marine Biology and Ecology, 128*, 219-240.

[84] Martin-Creuzburg, D. & von Elert, E. 2004. Impact of 10 dietary sterols on growth and reproduction of *Daphnia galeata. Journal of Chemical Ecology, 30(3)*, 483-500.

[85] Khozin-Goldberg, I., Cohen, Z., Pimenta-Leibowitz, M., Nechev, J. & Zilberg, D. 2006. Feeding with arachidonic acid-rich triacylglycerols from the microalga *Parietochloris incisa* improved recovery of guppies from infection with *Tetrahymena sp. Aquaculture, 255(1-4)*, 142-150.

[86] Petkov, G. & Garcia, G. 2007. Which are fatty acids of the green alga *Chlorella*? *Biochemical Systematics and Ecology, 35*, 281-285.

[87] Minkova, K., Tchorbadjieva, M., Tchernov, A., Stojanova, M., Gigova, L. & Busheva, M. 2007. Improved procedure for separation and purification of *Arthronema africanum* phycobiliproteins. *Biotechnology Letters, 29(4)*, 647-651.

[88] Tchernov, A. A., Minkova, K. M., Georgiev, D. I., Houbavenska, N. B. 1993. Method for B-phycoerythrin purification from *Porphyridium cruentum. Biotechnology Techniques, 7(12)*, 853-858.

[89] Karaivanova, M., Gagov, I., Mihailov, Y., Dilov, H., Pilarski, P., Stoilov, I. & Furnadjieva, S. 1989. Antitumor and modulating properties of water-soluble extract from *Scenedesmus. Comptes rendus de l'Académie bulgare des Sciences, 42*, 103-106.

[90] Garson, M. J. 1993. The biosynthesis of marine natural products. *Chemical Reviews, 93(5)*, 1699-1733.

[91] Shimizu, Y. 2003. Microalgal metabolites. *Current Opinion in Microbiology, 6(3)*, 236-243.

[92] Rösch, C., Skarka, J. & Patyk, A. 2009. Microalgae – opportunities and challenges of an innovative energy source. In: *17th European Biomass Conference and Exhibition.* Hamburg, Germany.

[93] Metzger, P. & Largeau, C. 2005. *Botryococcus braunii:* a rich source for hydrocarbons and related ether lipids. *Applied Microbiology and Biotechnology, 66(5)*, 486-496.

[94] Eroglu, E. & Melis, A. 2010. Extracellular terpenoid hydrocarbon extraction and quantitation from the green microalgae *Botryococcus braunii var. Showa. Bioresource Technology, 101*, 2359-2366.

[95] Cuomo, V., Perretti, A., Palomba, I., Verde, A. & Cuomo, A. 1995. Utilisation of *Ulva rigida* biomass in the Venice Lagoon (Italy): biotransformation in compost. *Journal of Applied Phycology, 7(5)*, 479-485.

[96] Adin, R. & Riera, P. 2003. Preferential food source utilization among stranded macroalgae by *Talitrus saltator* (Amphipod, Talitridae): a stable isotopes study in the northern coast of Brittany (France). *Estuarine, Coastal and Shelf Science*, *56(1)*, 91-98.

[97] Gaffron, H. & Rubin, J. 1942. Fermentative and photochemical production of hydrogen in algae. *Journal of General Physiology*, *26(2)*, 219-240.

[98] Berberoglu, H., Jay, J. & Pilon, L. 2008. Effect of nutrient media on photobiological hydrogen production by *Anabaena variabilis* ATCC 29413. *International Journal of Hydrogen Energy*, *33(4)*, 1172-1184.

[99] Skjånes, K., Knutsen, G., Kallqvist, T. & Lindblad, P. 2008. H_2 production from marine and freshwater species of green algae during sulfur deprivation and considerations for bioreactor design. *International Journal of Hydrogen Energy, 33(2)*, 511-521.

[100] Chisti, Y. 2007. Biodiesel from microalgae. *Biotechnology Advances*, *25(3)*, 294-306.

In: Bioprocess Sciences and Technology
Editor: Min-Tze Liong

ISBN 978-1-61122-950-9

Chapter 7

ALGAE FOR THE PRODUCTION OF SCP

Y. Ghasemi**, *S. Rasoul-Amini and M. H. Morowvat

Department of Pharmaceutical Biotechnology and Pharmaceutical Sciences Research Center, Faculty of Pharmacy, Shiraz University of Medical Sciences, P.O. Box 71345-1583, Shiraz, Iran

ABSTRACT

The algae are a diverse collection of chlorophyll-a-containing organisms that includes many divisions of the plant kingdom, including seaweeds, and a number of single-celled and multicellular [1]microscopic forms. Broad assemblages of microalgae are grouped into major categories together with macroalgae (seaweeds) on the basis of pigmentation, cell wall composition, chemical constitution of food reserves, presence and kind of flagellation and features unique to different groups. Microalgae are important constituents of many ecosystems ranging from marine and fresh water environments to desert sands and from hot springs to snow and ice. They account for more than half total primary production at the base of the food chain worldwide.

Comprehensive analysis and nutritional studies have demonstrated that the algal proteins are of high quality and comparable to conventional vegetable proteins. However, due to high production costs as well as technical difficulties to incorporate the algal material into palatable food preparations, the propagation of algal proteins is still in its infancy. To date, the majority of microalgal preparations are marketed as health food, cosmetics or animal feed. Nutritional supplements produced from microalgae have been the primary focus of microalgal biotechnology for many years. Dried biomass or cell extracts produced from *Chlorella*, *Dunaliella* and *Spirulina* have dominated the commercial opportunities. These products are directed mainly at the nutraceutical or health food market and collectively are like worth many hundred of million dollars.

The microbial cell masses or microbial biomasses form a class of useful products. So that biomass production with the substantial exclusion of accompanying processes has been the subject of new development, the production of single-cell protein or microbial biomass. In the case of alga it has to be stresses that, due to technical and economical

*Corresponding author: *Y. Ghasemi* Email: ghasemiy@sums.ac.ir reasons, it is not the general intention to isolate and utilize the sole protein, but to propagate the whole algal biomass. Hence, the term SCP is not quite correct, because the microalgal material is definitely more than just proteins. The composition of an ideal biomass is based on components which are carbohydrates, proteins, vitamins, lipids and trace amount of mineral and salts. As the cells are capable of synthesizing all amino acids, they can provide the essential ones to humans and animals. Selected data on the amino acid profile of various algae are prepared and compared with some basic conventional food items and a reference pattern of a well-balanced protein recommended by WHO/FAO. The amino acid pattern of almost all algae compares favorably with that of the reference and the other food proteins. The existing commercial microalgae culture systems range in volume from about 102 L to > 1010 L. Types of culture systems predominantly are: (a) large open ponds, (b) circular ponds with a rotating arm to mix the cultures, (c) raceway ponds and (d) large bags. Other commercial large scale systems include tanks used in aquaculture, the cascade system, and heterotrophic fermenter systems.

This chapter will evaluate the properties, production systems and applications of SCP from algae for human consumption.

1. INTRODUCTION

The term single-cell protein (SCP) is used to describe protein derived from cells of microorganisms such as yeast, fungi, algae and bacteria which are grown on various carbon sources for synthesis [1]. In technical fermentation process, in addition to the final desired product of natural substances e. g., penicillin, vitamins, the multiplication and growth of culture of microbes itself also takes place. These microbial cell masses or microbial biomasses form a class of useful products. So that biomass production with the substantial exclusion of accompanying processes has been the subject of new development, the production of single-cell protein or microbial biomass [2]. In the case of alga it has to be stressed that, due to technical and economical reasons, it is not the general intention to isolate and utilize the sole protein, but to propagate the whole algal biomass. Hence, the term SCP is not quite correct, because the microalgal material is definitely more than just proteins [3].

Algae are important constituents of many ecosystems ranging from marine and fresh water environments to desert sands and from hot springs to snow and ice. They account for more than half of the total primary production at the base of the food chain worldwide [4]. Since the early 1950s intense efforts have been made to explore new alternate protein sources as food supplements, primarily in anticipation of a repeatedly predicted insufficient future protein supply called "protein gap". For these, i.e. yeasts, fungi, bacteria and microalgae, the name single cell protein, usually abbreviated to SCP, was coined to describe the protein production from biomass, originating from different microbial sources [4]. Comprehensive analysis and nutritional studies have demonstrated that the algal proteins are of high quality and comparable to conventional vegetable proteins. However, due to high production costs as well as technical difficulties to incorporate the algal material into palatable food preparations, the propagation of algal proteins is still in its infancy. To date, the majority of microalgal preparations are marketed as health food, as cosmetics or as animal feed [3]. Nutritional supplements produced from microalgae have been the primary focus of microalgal biotechnology for many years. Dried biomass or cell extracts produced from *Chlorella*, *Dunaliella* and *Spirulina* have dominated the commercial opportunities. These products are directed mainly at the nutraceutical or health food market and collectively are like worth many hundred of million dollars [5].

The production of microbial biomass is a manufacturing process of the cell mass of microbes from suitable organic raw materials in a fermentation process. Here, selected strains of microorganisms are multiplied on suitable raw materials in a technical cultivation process. Process development begins with microbial screening, in which suitable production strains are obtained from samples of soil, water, and air or from swabs of inorganic or biological materials and are subsequently optimized by selection, mutation or other genetic methods. Then the technical conditions of cultivation for the optimized strains are worked out and special metabolic pathways and cell structures are determined. In parallel to these biological investigations, process engineering and apparatus technology contribute to the technical performance of the process and the apparatus in which the production of bioprotein is to be carried out in order to make them ready for use on large technical scale. The need of research to adapt novel technological aspects on the large-scale production of SCP is very much to fulfill the demand of conventional food production [2].

Microalgae display a diversity of primary and secondary metabolites. These include primary and secondary amines, such as spermidine, 2-phenylethylamine and tyramine in *Scenedesmus acutus* and histidine and histamine in *Euglena,* Parsiguine in *Fischerella ambigua* [6], and Pars a in *Chroococcsus* [7]. *Chlorella*, *Scenedesmus,* and *Euglena* contain various di- and polyamines as well. The chlorophyte *Neospongiococcum saccatum* contains very high concentrations of the rare 1, 3-diaminopropane. Microalgae synthesize all of the amino acids necessary for protein synthesis and some unique ones too, including D-alanine, D-glutamine and β-amino acids are found in microcystin, a toxic principle in the cyanobacterium *Microcystis aeruginosa* [8].

2. History

Interest in pharmaceuticals from microalgae has benefited from the resurgent interest in ethnobotany. For example, in traditional medicine, *Cladophora glomerata* is used as an Asian burn remedy, *Pleurococcus naeglis* and *Trentepholia iolithus* as topical antibacterial ointments, and *Rhizoclonium rivulare* as a vermifuge [8]. Also stimulating interest were discoveries by ecologists in the 1930s and 1940s that suggested and then demonstrated the production of antibiotic and autotoxic substances from microalgae. Since then, it has become clear that algae release many kinds of substances to their surroundings, either actively or passively while living and upon death and subsequent decomposition. These substances influence other microorganisms. It is often difficult to judge whether interactions in nature or dual culture are due to competition for resources or production of interactive substances, or both. Ancillary to this is the difficulty of defining just what kinds of compounds are bioactive. Microalgae secrete vitamins, amino acids, fatty acids, siderophores, simple carbohydrates and other nutrilites that are essential or support growth of other microbes [8].

Germans are the pioneer in the production of SCP. During the First World War Germans faced the problem related to food demand at the time, a groups of scientists first established the culture of *Sacharomyces cerevisiae* for the production of SCP. The biomasses were utilized in the forms of soups and sausages. The group of same Germans could develop the culture of *Candida aroborea and C. utilis* during the Second World War as an alternative to foods [2].

In 1982, scientists of Kuwait Institute of Scientific Research developed the cultures of fungi, *Torulospsi* and bacteria *Methylomonas clara, Methylophilus methylotrophus* and *Alcaligenes* for the production of SCP by using carbohydrate source as methanol. A Belgium botanist J. Leonard harvested the biomass from *Spirulina platensis* in alkaline lakes as a feed stock. In 1982, the series of culture of green algae such as *Chlorella pyrenoidosa, Scenedesmus acutus,* and *Chlamydomonas reinhardtii* for the consumption of human beings was developed. In India, a research work is in progress at CFTRI (Central Food Technology Research Institute) Mysore, on *Spirulina* and other microalgae to develop some single cell proteins as a supplement to food [2].

3. Classification of the Algae

Algae are considered to be a loose group of organisms that have all or most of the following characteristics: aquatic, photosynthetic, simple vegetative structures without a vascular system, and reproductive bodies that lack a sterile layer of protecting cells. As such, algae are no longer regarded as a phylogenetic concept, but still represent an ecologically meaningful and important collection of organisms. Both prokaryotic and eukaryotic taxa are included. In addition, there is a wide range of vegetative morphologies, including unicells, colonies, pseudofilaments, pseudoparenchymatous structures, parenchymatous forms, and coenocytic or siphonous forms [9].

Algae do not represent a formal taxonomic group of organisms, but rather constitute a loose collection of divisions or phyla with representatives that have the characteristics noted previously. The divisions are distinguished from each other based on a combination of characteristics, including photosynthetic pigments, starch like reserve products, cell covering, and other aspects of cellular organization. There is little consensus among phycologists as to the exact number of algal divisions; 8–11 have been recognized in recent texts [10]:

A. Cyanobacteria

Cyanobacteria or blue–green algae are prokaryotes, that is, cells that have no membrane-bound organelles, including chloroplasts. Other characteristics of this division include unstacked thylakoids, phycobiliprotein pigments, cyanophycean starch, and peptidoglycan matrices or walls. Cyanobacteria inhabit the widest variety of freshwater habitats on Earth and can become important in surface blooms in nutrient-rich standing waters. Some of these blooms can be toxic to zooplankton and fish, as well as livestock that drink water containing these organisms. Some cyanobacteria also occur in extreme environments, such as hot springs, saline lakes, and endolithic desert soils and rocks.

B. Red Algae

Rhodophyta or red algae represent a division that is characterized by chloroplasts that have no external endoplasmic reticulum and unstacked thylakoids, phycobiliprotein pigments, floridean starch, and lack of flagella. They are predominantly marine in distribution; only approximately 3% of over 5000 species occur in truly freshwater habitats. Freshwater red algae are largely restricted to streams and rivers, but also can occur in other habitats, such as lakes, hot springs, soils, caves, and even sloth hair.

C. Green Algae

Chlorophyta or green algae constitute a division that has the following set of attributes: chloroplasts with no external endoplasmic reticulum, thylakoids typically in stacks of two to six, chlorophyll-*a* and –*b* as photosynthetic pigments, true starch, and cellulosic walls or scales. Some members of the green algae (Charophyeae) are part of a lineage that is thought to be ancestral to higher plants. Green algae are widespread in various habitats, but certain groups may have specific ecological requirements. For example, flagellated chlorophytes tend to be more abundant in standing waters that are nutrient rich. Coccoid unicells and colonies are common in the plankton of standing waters and slowly moving rivers when nutrients, light and temperature are reasonably high. The majority of filamentous and plantlike Chlorophyta are attached to hard surfaces in standing or flowing waters, but some can exist in the floating state or on soils or other subaerial habitats. Filamentous conjugating green algae are most frequent in stagnant waters of roadside ditches and ponds, and in the littoral zones of lakes, where they can form free-floating mats or intermingle with other algae in attached or floating masses. Desmids are more common in ponds and streams that have low conductance and moderate nutrient levels, and often intermingle with macrophytes.

D. Euglenoids

Photosynthetic Euglenophyta or euglenoids have chloroplasts surrounded by three membranes, thylakoids in stacks of three, chlorophyll-*a* and -*b* as photosynthetic pigments, paramylon, and a pellicle. Euglenoids are particularly abundant in the plankton of standing waters rich in nutrients and organic matter, and they can be associated with sediments, fringing higher plants, and leaf litter, although some may dominate in highly acidic environments.

E. Eustigmatophyte, Raphidiophyte, and Tribophyte Algae

Eustigmatophyte, raphidiophyte, and Tribophyte algae comprise a loose group of algae that share the following characteristics: chloroplasts with four surrounding membranes, thylakoids in stacks of three, chlorophyll-*a* and -*c* as the typical photosynthetic pigments, and chrysolaminarin as the photosynthetic reserve product (where known). Members of this group of algae have been collected from a wide variety of habitats.

F. Chrysophycean Algae

Chrysophyceae or chrysomonads are distinguished by chloroplasts that have four surrounding membranes, thylakoids in stacks of three, fucoxanthin that typically masks chlorophyll-*a* and -*c*, and chrysolaminarin as the photosynthetic reserve. Chrysophycean algae are typically associated with standing bodies of water that have low or moderate nutrients, alkalinity, and conductances, and a pH that is slightly acidic to neutral. In addition, the majority of genera tend to be planktonic; attached forms occur to a lesser extent.

G. Haptophyte Algae

Haptophyceae are characterized by chloroplasts that have four surrounding membranes, thylakoids in stacks of three, fucoxanthin that masks chlorophyll-*a* and -*c*, chrysolaminarin as the photosynthetic reserve, and a unique appendage associated with the flagellar apparatus, the haptonema. The two common genera are planktonic in lakes and ponds, and occasionally form predominant blooms, particularly in areas with low conductance. *Chrysochromulina breviturrita* has been used as an indicator of moderately acidic water.

H. Synurophyte Algae

Synurophyceae is characterized by chloroplasts that have four surrounding membranes, thylakoids in stacks of three, fucoxonthin that masks chlorophyll-*a* and -*c*, chrysolaminarin as the photosynthetic reserve product, and siliceous scales. Synurophytes are exclusively freshwater phytoplankters in lakes, ponds, and slowly flowing rivers. Habitats that support the largest flora are slightly acidic, low in conductance, alkalinity, and nutrients, and have moderate amounts of humic substances.

I. Diatoms

Bacillariophyceae or diatoms are distinguished by chloroplasts that have four surrounding membranes, thylakoids in stacks of three, fucoxanthin that masks chlorophyll-*a* and -*c*, chrysolaminarin as the photosynthetic reserve product, and a siliceous frustule that makes up the external covering. The diatoms are a complex and diverse group in terms of frustule morphology. Diatoms are found in all freshwater habitats, including standing and flowing waters, and planktonic and benthic habitats, and they can often dominate the microscopic flora. Because diatoms inhabit a broad array of habitats but many have specific habitat requirements, they have been used in freshwater environment assessment and to monitor long-term changes in ecological characteristics.

J. Dinoflagellates

Pyrrhophyta or dinoflagellates are characterized by chloroplasts that have three surrounding membranes, thylakoids in stacks of three, peridinin that masks chlorophyll-*a* and -*c*, true starch, a nucleus that has condensed chromosomes in cell cycle phases, a theca covering, and frequently a transverse and posterior flagellum. The dinoflagellates are typically minor components of the phytoplankton of lakes and ponds, but sometimes form dense blooms, particularly in the presence of high levels of nitrates and phosphates.

K. Cryptomonads

Cryptophyta, cryptomonads or cryptophyte algae, have chloroplasts that have four surrounding membranes in which a nucleomorph occurs between the outer and inner two membranes, thylakoids in loose pairs, phycocyanin or phycoerythrin that masks chlorophyll-*a* and -*c*, true starch as the photosynthetic reserve, a periplast, and two subapical flagella. Cryptomonads are typically planktonic in lakes and ponds, and are particularly diverse in temperate regions.

L. Brown Algae

Phaeophyceae or brown algae are distinguished by chloroplasts that have four surrounding membranes, thylakoids in stacks of three, fucoxanthin that masks chlorophyll-*a* and -*c*, laminarin as the photosynthetic reserve, and alginates commonly as the wall matrix component. There are six genera of freshwater brown algae. Brown algae are predominantly marine in distribution; less than 1% of the species are from fresh water. The inland species are benthic, either in lakes or streams, and distribution is quite scattered.

3.1. Cyanobacteria

Cyanobacteria or blue–green algae are prokaryotes, that is, cells that have no membrane-bound organelles, including chloroplasts. Other characteristics of this division include unstacked thylakoids, phycobiliprotein pigments, cyanophycean starch, and peptidoglycan matrices or walls. Cyanobacteria inhabit the widest variety of freshwater habitats on earth and can become important in surface blooms in nutrient-rich standing waters. Some of these blooms can be toxic to zooplankton and fish, as well as livestock that drink water containing these organisms. Cyanobacteria also occur in extreme environments, such as hot springs, saline lakes, and endolithic desert soils and rocks [10].

3.2. Green Algae

Chlorophyta or green algae constitute a division that has the following set of attributes: chloroplasts with no external endoplasmic reticulum, thylakoids typically in stacks of two to six, chlorophyll-*a* and-*b* as photosynthetic pigments, true starch, and cellulosic walls or scales. Some members of the green algae (Charophyeae) are part of a lineage that is thought to be ancestor to higher plants. Green algae are widespread in inland habitats, but certain groups may have specific ecological requirements. For example, flagellated chlorophytes tend to be more abundant in standing waters that are nutrient rich. Coccoid unicells and colonies are common in the plankton of standing waters and slowly moving rivers when nutrients, light and temperature are reasonably high. The majority of filamentous and plantlike Chlorophyta are attached to hard surfaces in standing or flowing waters, but some can exist in the floating state or on soils or other subaerial habitats. Filamentous conjugating green algae are most frequent in stagnant waters of roadside ditches and ponds, and in the littoral zones of lakes, where they can form free-floating mats or intermingle with other algae in attached or floating masses [10].

3.2.1. Chlorella

The genus *Chlorella* occupies a special position among the other genera in the order Chlorococales [10]. It is spherical in shape, about 2 to 10 μm in diameter, and is without flagella. *Chlorella* contains the green photosynthetic pigments chlorophyll-a and -b in its chloroplast. It depends on photosynthesis for growth and multiplies rapidly, requiring only carbon dioxide, water, sunlight, and a small amount of minerals. Its species possess spherical or ellipsoidal cells, exhibit a simple life cycle, and have simple nutritional requirements. In culture they grow more quickly than other microorganisms, rapidly overgrowing them [11]. The name *Chlorella* is taken from the Greek word *chloros* meaning green and the Latin diminutive suffix *-ella* meaning small The German biochemist Otto Heinrich Warburg received the Nobel Prize in physiology or medicine in 1931 for his study on photosynthesis in

Chlorella. In 1961 Melvin Calvin of the University of California received the Nobel Prize in chemistry for his research on the pathways of carbon dioxide assimilation in plants using *Chlorella*. The smallest *Chlorella* species are similar to bacteria and were the first algae to be isolated like bacteria and to be grown in pure cultures. In reproduction, which is exclusively asexual, each mature cell divides, producing 4, 8, or, more rarely, 16, autospores [11]. *Chlorella vulgaris* is a rich source of proteins, eight kinds of essential amino acids, vitamins (B-complex, ascorbic acid), minerals (potassium, sodium, magnesium, iron and calcium), β-carotene, chlorophyll, CGF (*Chlorella* growth factor) and other beneficial substances [12]. However, because it is a single-celled alga, harvest had posed practical difficulties for its large-scale use as a food source. Methods of mass production are now being used to cultivate it in large artificial circular ponds. It has been eaten in times of famine in areas such as China. *Chlorella* is important tool for physiological experiments. It has widely been used in the study of respiration and photosynthesis. It is also used for the purification of air in space capsule. An antibiotic, chlorellin is extracted from this alga. *Chlorella* is probably the first alga to have been grown extensively in axenic culture [11].

There are many reports of the pharmacological activity of *Chlorella* spp., for example, hypoglycemic effects of *Chlorella* in streptozocin induced diabetic mice [13], preventing dyslipidemia [14], antitumor immuno activity [15], lowering of blood pressure [16], prevention of stress induced ulcer [17], radio protective action [18], inhibition of interleukin 5 production and reduction of eosinophil infiltration [19], agglutinating activity [20], and stimulating effects on cytokines production [21]. *Chlorella* contains roughly 50 percent protein and amino acids, vitamin B6, minerals, chlorophyll, beta carotene, methyl cobalamine, the most absorbable form of vitamin B12, sporopollein which is effective in binding to neurotoxins and toxic metals and alpha and gamma linoleic acids. *Chlorella vulgaris* extract (CVE) helps the body detoxify also. The porphyrins in chlorophyll have their own strong metal binding effect. Chlorophyll also activates the receptor on the nucleus of the cell which is responsible for the formation of the peroxisomes, the cell organelles which are responsible for detoxification [22].

3.2.2. Dunaliella

Dunaliella is a unicellular, biflagellate, naked, green alga (chlorophyta, chlorophyceae). It is a halotolorant unicellular microalga and has no rigid cell wall under stress conditions (e. g. high light intensity, high salinity and nutrient deficiency), it can produce and accumulate high concentration of β-carotene in oil globules in the cell [11]. First sighted in 1838 in saltern evaporation ponds in the south of France by Michel Felix Dunal, it was named after its discoverer by Teodoresco in 1905 [23]. *D. salina* forms red blooms in the water because it accumulates about 14% of its dry weight as β-carotene, a valuable ingredient in the food and feed industries. *Dunaliella viridis*, in contrast, is not a β-carotene accumulative species, and has been used as live feed in marine aquaculture [24]. Algae of genus *Dunaliella* especially *D. salina* and *D. tertiolecta*, are among the microalgae most studied for mass culture. *Dunaliella* spp. is grown as a food source in aquaculture and *D. salina* is the richest algal source of β-carotene and glycerol. *Dunaliella salina* is the first microalga to be used commercially to produce fine chemicals [11]. *D. salina* is probably the most successful microalga for mass cultivation described so far, especially due to its high salinity requirement that minimizes the number of competitors and predators. Furthermore, due to lack of cell wall, dried *Dunaliella* is easily and fully digestible by animals and humans [25]. Commercial activity in the microalgae extractable chemical sector is currently limited to two main products: *Dunaliella*-derived carotenoid pigments as human nutritional; supplements and *Haematococcus*-derived pigment astaxanthin as a colouring agent [24]. The genetic characterization of strains with industrial potential has a great relevance in applied science because it permits partitioning of phenotypic variation into environmental (phenotypic plasticity) and genetic components [26]. Furthermore, genetic diversity studies permit exotic genotypes detection, which could be the source of genes with biotechnological potential. Most different molecular techniques can be used to reveal genetic polymorphisms [25]. One of them is random amplified polymorphic DNA (RAPD) or more specific direct sequencing of some conserved domains like ribosomal RNAs to find the relevance between genetic characterization of strains and potential of large scale cultivation of strains.

3.2.3. Scenedesmus

Scenedesmus belongs to the family Scenedesmaceae, division Chlorophycophyta. Species of *Scenedesmus* are widely distributed in freshwater and soil. The cylindrical cells, with rounded or pointed ends, are laterally jointed in groups of 4 or 8 or more rarely, 16. The terminal cells and some of the others in some species (e. g., *S. quadricauda*), have spines. Some species also have tufts of fine bristles which have buoyancy. The cells are uni-nucleate and have a laminate chloroplast that contains a pyrenoid. Reproduction of *Scenedesmus* is by autocolony formation in which each parental cell forms a miniature colony that is liberated thorough a tear in the parental wall. Sexual reproduction has been described only for *S. obliquus* [9]. It has been used for heavy metals removal [27], Mass culture of *Scenedesmus* species is widely studied [9,28].

4. Source of Single-Cell Proteins

Certain microbes which have high protein content are considered to be very much beneficial for the production of SCP. Such prepared biomasses can be utilized for the human consumption as protein rich food. Generally, these microbes can grow in an industrial bioreactor with the utilization of common wastes such as sewage, animal excreta, agricultural wastes, petroleum wastes, crude oil waste, paper and textile industry wastes, saw mill wastes, starchy waste from potato industry, beverage industrial wastes and distilleries waste [2]. The production of SCP from various microbes, particularly from fungi and bacteria has received considerable attention, in contrast, only a few studies have dealt with the feasibility of using SCP from microalgae [4]. Comprehensive analysis and nutritional studies have demonstrated that these algal proteins are of high quality and comparable to conventional vegetable proteins. However, due to high production costs as well as technical difficulties to incorporate the algal material into palatable food preparations, the propagation of algal proteins is still in its infancy [4].

5. Production of SCP

The production process of SCP usually consists of the following operational steps namely:

- Preparation of nutrient media
- Fermentation
- Separation and mechanical concentration of SCP
- Drying the SCP
- Final processing of the SCP [2].

The existing commercial microalgae culture systems range in volume from about 102 L to > 1010 L. Types of culture systems predominantly are: (a) large open ponds, (b) circular ponds with a rotating arm to mix the cultures, (c) raceway ponds and (d) large bags. Other commercial large scale systems include tanks used in aquaculture, the cascade system, and heterotrophic fermenter systems [29]. Table 1 summarizes the culture systems currently in use for commercial algal culture.

Table 1. Commercial microalgae culture systems currently in use and the algal species cultured [29]

Culture system	Algae	Approximate maximum volume (L)	Location
Tanks	Many species (for aquaculture)	1× 104	World Wide
Extensive open	*Dunaliella salina*	1× 109	Australia
Circular ponds with rotating arm	*Chlorella* spp.	1.5 × 104	Taiwan, Japan
Raceway ponds	*Chlorella* spp., *Spirulina* spp., *Dunaliella salina*	3× 104	Japan, Taiwan, USA, Thailand, China, India, Vietnam, Chile, Israel, China
Cascade system with baffles	*Chlorella* spp.	3× 104	Czech Republic, Bulgaria
Large bags	Many species (used for aqua culture)	1× 103	World Wide
Fermenters (heterotrophic)	*Chlorella* spp., *Crypthecodinium cohnii*	>103	Japan, Taiwan, Indonesia, USA
Two-stage system (indoors in closed reactor and then out-doors in paddlewheel ponds	*Hematococcus pluvialis*	?	USA

In the development of a high-yield food production process by microalgal cultures, optimization of medium components and environmental factors is vital because they can significantly affect the yield and volumetric productivity.

Methods employed for the optimization include one-at-a-time, statistical and mathematical, artificial neural networks, fuzzy logic, genetic algorithms, etc. Among them, the one-at-a-time and statistical methods are commonly used in the optimization of fermentation processes. The one-at-a-time method keeps the levels of all factors constant except one. The level of this factor is then changed within a desired range. This strategy is simple and easy, and hence has been widely employed for optimizing the fatty acid production from microalgae, however, the one-at-a-time method fails to take into account the interactions among factors and it often requires a relatively large number of experiments [30].

6. Microalgae as Human Food or Animal Feed

Microalgae for human nutrition are nowadays marketed in different forms such as tablets, capsules and liquids. They can also be incorporated into pastas, snack foods, candy bars or gums, and beverages. Owing to their diverse chemical properties, they can act as a nutritional supplement or represent a source of natural food colorants. The commercial applications are dominated by four strains: *Spirulina*, *Chlorella*, *D. salina* and *Aphanizomenon flos-aquae*. In addition to its use in human nutrition, microalgae can be incorporated into the feed for a wide variety of animals ranging from fish (aquaculture) to pets and farm animals. In fact, 30% of the current world algal production is sold for animal feed applications. Many nutritional and toxicological evaluations have proved the suitability of algal biomass as feed supplement. *Spirulina* is largely used in this domain and concerns many types of animal: cats, dogs, aquarium fish, ornamental birds, horses, cows and breeding bulls. Algae positively affect the physiology (by providing a large profile of natural vitamins, minerals, and essential fatty acids; improved immune response and fertility; and better weight control) and their external appearance (resulting in healthy skin and a lustrous coat) of animals [31].

Recently the utilization of algae and other forms of microorganisms as the source of SCP has gained increasing interest. This development prompted international organizations such as the *International Union of Pure and Applied Chemistry and the Protein-calorie Advisory Group of the United Nations Systems* to publish guidelines that stipulate various criteria of quality that should be fulfilled before the particular SCP can be declared as suitable for utilization as animal feed or human food. It should be stressed in this context that in the case of algae it is generally not the intention to use the biomass as the sole source of protein but as a supplement to the basic diet [11].

7. Chemical Composition of Microalgal SCP

In general, SCP has more nutritive value than the normal living cells [2]. Data on the chemical composition of algae give basic information on the nutritive potential of the algae biomass. However, it should be always kept in mind that algal cultivation basically represents a special form of agriculture, exposed to various environmental influences that alter the proportion of the individual cell constituents. In addition, this proportion can be modified by specific cultivation measures such as composition of the culture medium, and light intensity [11]. The composition of an ideal biomass is based on components which are carbohydrates, proteins, vitamins, lipids and trace amount of mineral and salts [2]. Various analyses of algal constituents have been published in the literature and a compilation of recent data on gross chemical composition of different algae is given in Table 2 [3].

Besides, these above mentioned nutritive components SCPs also contain nucleic acid, basically purine bases.

Table 2. General composition of different human food sources and algae (% of dry matter)* [3,31]

Commodity	Protein	Carbohydrate	Lipid
Baker's yeast	39	38	1
Meat	43	1	34
Milk	26	38	28
Rice	8	77	2
Soybean	37	30	20
Anabaena cylindrica	43-56	25-30	4-7
Aphanizomenon flos-aquae	62	23	3
Arthrospira maxima	60-71	13-16	6-7
Chlamydomonas reinhardtii	48	17	21
Chlorella pyrenoidosa	57	26	2
Chlorella vulgaris	51-58	12-17	14-22
Dunaliella salina	57	32	6
Euglena gracilis	39-61	14-18	14-20
Porphyridium cruentum	28-39	40-57	9-14
Scenedesmus obliquus	50-56	10-17	12-14
Spirogyra sp.	6-20	33-64	11-21
Spirulina maxima	60-71	13-16	6-7
Spirulina platensis	46-63	8-14	4-9
Synechococcus sp.	63	15	11

* It should be kept in mind that figures presented in this table are estimates, since the proportion of individual cell constituents largely depends on environmental parameters.

7.1. Protein/Amino Acids Content of Microalgae

The high protein content of various microalgal species is one of the main reasons to consider them as an unconventional source of protein [31]. Most of the figures published in the literature on concentration of algal proteins, dominantly enzymatic proteins, are based on estimates of so called crude proteins, commonly used in evaluating food and feed. These figures are the result of hydrolysis of the algal biomass and estimation of the total nitrogen [4]. Proteins are composed of different amino acids and hence the nutritional quality of a protein is determined basically by the content, proportion and availability of its amino acids [3]. As the cells are capable of synthesizing all amino acids, they can provide the essential ones to humans and animals [31]. Selected data on the amino acid profile of various algae are compiled in Table 3 and compared with some basic conventional food items and a reference pattern of a well-balanced protein recommended by WHO/FAO (1973). It can be seen that the amino acid pattern of almost all algae compares favorably with that of the reference and the other food proteins [3].

A more accurate method to evaluate the quality of proteins is the determination of the protein efficiency ratio (PER), expressed in terms of weight gain per unit of protein consumed by the test animal in short term feeding trial. However, still more specific nitrogen balance methods can be applied to evaluate the nutritive quality of a protein. One of these principles is the estimation of the biological value (BV), which is a measure of nitrogen retained for growth or maintenance. Another parameter, which reflects the quality of a protein, is the digestibility coefficient (DC). Finally, the net protein utilization (NPU) which is equivalent to the calculation BV $\times$ DC is a measure both of the digestibility of the protein and the biological value of the amino acids absorbed from the food [3]. Selected data of such metabolic studies are summarized in Table 4 [3].

7.2. Fat, Oil and Hydrocarbon Content of Microalgae

Fatty acids are primarily metabolites of acetyl CoA pathway which is generally determined, evolutionary very old, and therefore conservative [32]. Microalgae may contain significant quantities of fats and oils (lipids) with compositions similar to those of vegetable oils. Under certain conditions, microalgae have been reported to contain up to 85% of the dry weight as lipids. This exceeds the lipid content of most terrestrial plants. The range of potential applications for these algal fats and oils is very wide. The algal oils resemble fish and vegetable oils, and could therefore be considered as potential substitutes for petroleum products. Direct extraction and refinement of the microalgal oils also would seem to be a more efficient way of obtaining fuels from microalgae, compared to the alternative of fermenting the algal biomass to produce either methane or ethanol. Algal oils or fats can also be used as vegetable oil substitutes. The lipids of some algal species are also rich in essential oil fatty acids such as the C18 linoleic (18:2ω6) and γ-linoleic (18:3ω3) acids and their C20 derivatives, eicosapentaenoic acid (20:5ω3) and arachidonic acid (20:4ω6). These fatty acids are an essential component of the diet of humans and animals and are becoming important feed additives in aquaculture [11]. The major dietary sources of docosahexaenoic acid (DHA) are oils from marine fish and microalgae. Fish obtain most of their long-chain ω3-PUFAs (Poly unsaturated fatty acids), by consumption of marine microalgae, which are considered to be the primary producers of it [4,33]. Clinical studies have indicated that DHA is vital for proper visual and neurological development in infants. In addition, DHA consumption has been shown to benefit patients with chronic conditions, such as hypertension, coronary heart diseases, depression and diabetes [33]. The total oil and fat content of microalgae ranges from about 1% to 70% of dry weight [31]. The lipids of microalgae are generally esters of glycerol and fatty acids with a chain length of C14 to C22 [4].

Table 3. Amino acid profile of different algae as compared with conventional sources and the WHO/FAO (1973) reference pattern (per 100 g protein) [3]

Source	Ile	Leu	Val	Lys	Phe	Tyr	Met	Cys	Try	Thr	Ala	Arg	Asp	Glu	Gly	His	Pro	Ser
WHO/FAO	4.0	7.0	5.0	5.5	6.0		3.5		1.0									
Egg	6.6	8.8	7.2	5.3	5.8	4.2	3.2	2.3	1.7	5.0	-	6.2	11.0	12.6	4.2	2.4	4.2	6.9
Soybean	5.3	7.7	5.3	6.4	5.0	3.7	1.3	1.9	1.4	4.0	5.0	7.4	1.3	19.0	4.5	2.6	5.3	5.8
-*Chlorella vulgaris*	3.8	8.8	5.5	8.4	5.0	3.4	2.2	1.4	2.1	4.8	7.9	6.4	9.0	11.6	5.8	2.0	4.8	4.1
-*Dunaliella bardawil*	4.2	11.0	5.8	7.0	5.8	3.7	2.3	1.2	0.7	5.4	7.3	7.3	10.4	12.7	5.5	1.8	3.3	4.6
-*Scenedesmus obliquus*	3.6	7.3	6.0	5.6	4.8	3.2	1.5	0.6	0.3	5.1	9.0	7.1	8.4	10.7	7.1	2.1	3.9	3.8
-*Arthrospira maxima*	6.0	8.0	6.5	4.6	4.9	3.9	1.4	0.4	1.4	4.6	6.8	6.5	8.6	12.6	4.8	1.8	3.9	4.2
-*Spirulina platensis*	6.7	9.8	7.1	4.8	5.3	5.3	2.5	0.9	0.3	6.2	9.5	7.3	11.8	10.3	5.7	2.2	4.2	5.1
-*Aphanizomenon* sp.	2.9	5.2	3.2	3.5	2.5	-	0.7	0.2	0.7	3.3	4.7	3.8	4.7	7.8	2.9	0.9	2.9	2.9

Table 4. Comparative data on biological value (BV), digestibility coefficient (DC), net protein utilization (NPU) and protein efficiency ratio (PER), of differently processed algae [3]

Alga	Processing	BV	DC	NPU	PER
Casein	-	87.8	95.1	83.4	2.50
Egg	-	94.7	94.2	89.1	-
Scenedesmus obliquus	DD	75.0	88.0	67.3	1.99
Scenedesmus obliquus	SD	72.1	72.5	52.0	1.14
Scenedesmus obliquus	Cooked-SD	71.9	77.1	55.5	1.20
Chlorella sp.	AD	52.9	59.4	31.4	0.84
Chlorella sp.	DD	76.6	89.0	68.0	2.00
Coelastrum proboscideum	DD	76.0	88.0	68.0	2.10
Spirulina sp.	SD	77.6	83.9	65.0	1.78
Spirulina sp.	DD	68.0	75.5	52.7	2.10

AD: air dried, DD: drum dried, SD: sun dried.

Table 5. A summary of the identified fatty acids in the five naturally isolated microalgae: *Chlorella vulgaris* MCCS 013, *Dunaliella salina* MCCS 001, *Dunaliella salina* CCAP 19/18, *Scenedesmus obliquus* strain 019, and *Scenedesmus rubescens* MCCS 018 [4]

Fatty acid	C. Vulgaris MCCS 013	S. obliquus strain 019	S. rubescens MCCS 018	D. salina MCCS 001	D. salina CCAP 19/18	No. of carbons:No. of double bond(s)
2-Propenoic acid				+		3:1
Butanoic acid			+	+		4:0
2-methyl-2-propenoic acid	+					4:0
Pentanoic acid				+		5:0
Hexanoic acid	+	+	+	+		6:0
2-Hexenoic acid		+			+	6:1
3-Hexenoic acid				+		6:1
5-Hexenoic acid				+		6:1
2, 4-Hexadienedioic acid		+				6:2
Heptanedioic acid				+		7:0
2-Heptenoic acid				+		7:1
Octanoic acid	+		+	+		8:0
3-Octenoic acid				+		8:1
Nonanoic acid		+				9:0
Decanoic acid		+	+	+	+	10:0
Undecanoic acid	+	+	+	+	+	11:0
Dodecanoic acid		+		+	+	12:0

Table 5. Continued

Tridecanoic acid		+	+	+	+	13:0
Tetradecanoic acid		+	+	+	+	14:0
Pentadecanoic acid	+		+	+	+	15:0
2-pentenoic acid				+		15:1
4-Pentenoic acid				+		15:1
Hexadecanoic acid			+	+	+	16:0
7-Hexadecenoic acid		+				16:1
9-Hexadecenoic acid				+		16:1
9,12-Hexadecadienoic acid			+			16:2 ω4
7,10,13-Hexadecatrienoic acid	+					16:3 ω3
Heptadecanoic acid	+	+	+	+		17:0
Octadecanoic acid				+	+	18:0
6-Octadecenoic acid		+		+		18:1
8-Octadecenoic acid		+				18:1
9-Octadecenoic acid			+			18:1 ω9
10-Octadecenoic acid	+					18:1
15-Octadecenoic acid				+		18:1
16-Octadecanoic acid	+			+		18:1
9,12-Octadecadienoic acid				+		18:2
9,15-Octadecadienoic acid		+				18:2
10,13-Octadecadienoic acid		+				18:2

They may be either saturated or unsaturated. Some blue green algae, specially the filamentous species, tend to have large quantities of polyunsaturated fatty acids (25% to 60% of the total). Other blue green algae, namely those species that show facultative anoxygenic CO2 photoassimilation with sulphite as electron donor, lack polyunsaturated fatty acids in their lipids. The eukaryotic algae have a predominance of saturated and mono unsaturated fatty acids [11]. The total amount and relative proportion of fatty acids can be affected by nutritional and environmental factors, like nitrogen limitation [31]. Triglycerides are the most common storage lipids and may constitute up to 80% of the total lipid fraction. Aside from the triglycerides, the other major algal lipids are sulphoquinovosyl diglyceride (SL), monogalactosyl diglyceride (MGDG), digalactosyl diglyceride (DGDG), lecithin, phosphatidyl glycerol and phosphatidyl inositol. In addition to the mentioned lipids, microalgae can also synthesize some novel classes of lipids such as the chlorosulpholipids, which have been reported in the Chrysophyceae, Chlorophyceae, Xanthophyceae and Cyanophyceae. The hydrocarbon content of microalgae is generally less than 5% of dry weight. Some microalgae also produce methyl branched hydrocarbons and cyclic and acyclic triterpenes. To date, only the green alga *Botryococcus braunii* has been shown to produce large amounts of hydrocarbons, with levels of up to 90% of dry weight. This alga has been implicated as the source of many shale oil deposits and other oil deposits. Briefly, algal fats, oils and hydrocarbons have a wide range of existing and potential commercial applications [11].

Table 6. A summary of the identified fatty acids in the five naturally isolated microalgae: *Chlorella vulgaris* MCCS 013, *Dunaliella salina* MCCS 001, *Dunaliella salina* CCAP 19/18, *Scenedesmus obliquus* strain 019, and *Scenedesmus rubescens* MCCS 018 [4]

Fatty acid	C. Vulgaris MCCS 013	S. obliquus strain 019	S. rubescens MCCS 018	D. salina MCCS 001	D. salina CCAP 19/18	Type of FA
12,15-Octadecadienoic acid		+				18:2
9,11-Octadecadiynoic acid	+					18:2
9,12,15-Octadecatrienoic acid	+					18:3 ω3
6,9,12-Octadecatrienoic acid		+	+		+	18:3 ω6
Nonadecanoic acid		+		+	+	19:0
Eicosanoic acid	+	+	+		+	20:0
Eicosenoic acid					+	20:1 ω9
11,14,17-Eicosatrienoic acid	+					20:3 ω3
Heneicosanoic acid	+	+		+	+	21:0
Docosanoic acid	+					22:0
Tricosanoic acid	+			+		23:0
Tetracosanoic acid	+	+		+		24:0
15-Tetracosenoic acid				+		24:1
Heptacosanoic acid				+	+	27:0
Triacontanoic acid				+	+	30:0
Octatriacontanoic acid				+		38:0

7.3. Carbohydrate Content of Microalgae

Carbohydrates in microalgae can be found in the form of starch, glucose, sugars, and other polysaccharides [31].

7.4. Vitamin Content of Microalgae

Microalgae also represent a valuable source of nearly all essential vitamins (e. g., A, B1, B2, B6, B12, C, E, nicotinate, biotin, folic acid and pantothenic acid. Vitamins improve the nutritional value of algal cell but their quantity fluctuates with environmental factors, the harvesting treatment and the method of drying the cells [31].

8. Analytical Methods

The protein concentration could be measured by the Kochert method [4]. First a standard curve is generated by pipetting a range (10-100 μg) of protein concentration from the protein standard solution into a series of marked 12×100 mm test tubes, the volume of each tube is adjusted to 0.1 mL with distilled H2O. Then samples of the unknown protein (100 μL), after hydrolysis in 1 N NaOH for one h at 100°C, is dissolved into three separate test tubes and the volume of each tube is adjusted to 5 mL with the buffer in the presence of a blank (0.1 mL of the buffer solution) and they are mixed immediately by inversion. Absorbance at 595 nm is measured after 2 min. The weight of protein standard is plotted against the corresponding absorbance to generate a standard curve. The concentration of the unknown protein is determined graphically [34].

8.1. Solution Used for Protein Determination in the Kochert Method

(a) Protein reagent

Coomassie brilliant blue G-250 (100 mg) is dissolved in 50 mL of 95% ethanol. To this solution is added 100 mL of 85% (w/v) H3PO4. The resulting solution is diluted to a final volume of 1 L with H2O and stored at room temperature [34].

(b) Protein standard solution

Bovine serum albumin (100 mg) is dissolved in H2O to a final volume of 100 mL. The obtained solution is stored at 4°C [34].

Protein standard curve is obtained by plotting the absorbance against the concentration (μg/mL) for the protein standard solution.

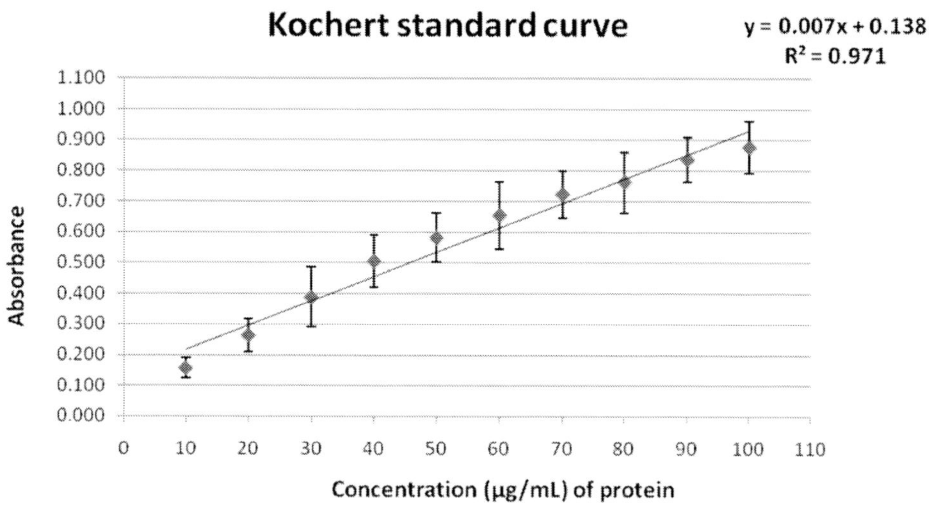

Figure 1. Standard curve for protein assay by Kochert method using Bovine serum albumin as standard. The equation and R-squared (R2) are given.

9. SCP Safety

The foreign protein in SCP can be unsuitable for humans and lead to skin reactions, allergies or gastrointestinal reactions resulting in nausea and vomiting. The SCP may even carry carcinogenic factors as contaminants derived from the substrates used. Hence, prior decontamination and purification of the final product is required before it is used as a food source [35].

10. Limitations for Use of SCP

Although algae are very good nutrition sources, there are some limitations for human consumption. The most important one is the presence of the algal cell wall. Humans lack the cellulose enzyme and hence they cannot digest the cellulose component of the algal wall. In order to be used as food for humans the algal walls must be digested before the final product is eaten. The cellulose digestion step is not required if the SCP is used as feed for cattle as they have cellulose-degrading symbiotic bacteria and protozoa in their rumen. Algal production is generally done outdoors and is dependent on the climatic conditions. Hence, productive algal species and favorable conditions are important. Elaborate methods and preparations are required to eliminate contamination [35].

11. Conclusion

SCP products should be evaluated extensively for safety purposes, to gain popularity among masses. The chemical composition of any SCP product must be characterized clearly in terms of percentage protein, type of amino acids, nucleic acid, lipids, fats, toxins and vitamins. Properties like density, particle size, texture, color and storage must be clearly indicated on the package for marketing. A microbiological description indicating species, strains and percentage of contaminants, if any, should be indicated. Final products for human consumption must be made to undergo rigorous testing during the pre-marketing stage. Possible toxic or carcinogenic compounds, heavy metals and polycyclic hydrocarbons must be assayed for and removed. It is of primary importance that SCP products are safe to eat and also inexpensive in order to be popular among masses. Further, genetically improved, high-yielding and nontoxic microbes can be grown for SCP production [35].

Briefly, it is suggested that microalgae, are a good and unstudied candidates to be used as SCP as human food or animal feed, because of their high content of protein, fatty acids and minerals. We can easily cultivate them, their growth rate is high, their productivity is high, there is no risk for pathogenicity, their culture media is simple and inexpensive, and finally there are different resources for finding and screening other strains of naturally isolated microalgae for SCP.

References

[1] Najafpour, G. D. 2007. Single Cell Protein. In: *Biochemical Engineering and Biotechnology*. Pp. 332-341. Amsterdam, The Netherland: Elsevier.

[2] Dixit, V. 1998. *Pharmaceutical Biotechnology*. New Delhi, India: CBS Publishers and Distributors.

[3] Becker, E.W. 2007. Microalgae as a source of protein. *Biotechnology Advances*, 25, 207-210.

[4] Rasoul-Amini, S., Ghasemi, Y., Morowvat, M. H. & Mohagheghzadeh, A. 2009. PCR amplification of 18S rRNA, Single cell protein production and fatty acid evaluation of some naturally isolated microalgae. *Food Chemistry*, *116 (1)*, 129-136.

[5] Apt, K. E. & Behrens, P. W. 1999. Commercial developments in microalgal biotechnology. *Journal of Phycology*, 35, 215-226.

[6] Ghasemi, Y., Tabatabaei Yazdi, M., Shafiee, A., Amini, M., Shokravi, Sh. & Zarrini, Gh. 2004. Parsiguine, a novel antimicrobial substance from *Fischerella ambigua*. *Pharmaceutical Biology*, *42(4-5)*, 318-322.

[7] Ghasemi, Y., Moradian, A., Mohagheghzadeh, A., Shokravi, Sh. & Morowvat, M. H. 2007a. Antifungal and antibacterial activity of the microalgae collected form paddy fields of Iran: characterization of antimicrobial activity of *Chroococcus dispersus*. *Journal of Biological Sciences*, *7 (6)*, 904-910.

[8] Metting, B. & Pyne, J. W. 1986. Biologically active compounds from microalgae. *Enzyme and Microbial Technology*, 8, 386-394.

[9] Richmond, A. 1986. *CRC Handbook of microalgal mass culture*. Boca Raton, Florida: CRC Press.

[10] Wehr, J. D. & Sheath, R.G. 2003. Introduction to fresh water algae. In: *Fresh water algae of North America* (edited by J. D. Wehr & R. G. Sheath) Pp 1-9. Amsterdam, The Netherland: Elsevier.

[11] Borowitzka, M. A. & Borowitzka, L. J. 1988. *Microalgal Biotechnology*. Cambridge, UK: Cambridge University Press.

[12] Rodriguez-Garcia, I. & Guil-Guerrero, J. L. 2008. Evaluation of the antioxidant activity of three microalgal species for use as dietary supplement and in the preservation of foods. *Food Chemistry*, 108, 1023-1026.

[13] Jong-Yuh, C. & Mei-Fen, S. 2005a. Potential hypoglycemic effects of *Chlorella in* streptozocin induced diabetic mice. *Life Sciences*, 77, 980-990.

[14] Jong-Yuh, C. & Mei-Fen, S. 2005b. Preventing dyslipidemia by *Chlorella pyrenoidosa* in rats and hamsters after chronic high fat diet treatment. *Life Sciences*, 76, 3001-3013.

[15] Noda, K., Tanaka, K., Yamada, A., Ogata, J., Tanaka, H. & Shoyama, Y. 2002. Simple assay for antitumor immuno active glycoprotein derived from *Chlorella vulgaris* strain CK22 using ELISA. *Phytotherapy Research*, 16, 581-585.

[16] Sansawa, H., Takahashi, M., Tsuchikura, S. & Endo, H. 2006. Effect of *Chlorella* and its fraction on blood pressure, cerebral stroke lesions and life span in stroke prone spontaneously hypertensive rats. *Journal of Nutritional Science and Vitaminology*, 52, 457-466.

[17] Tanaka, K., Yamada, A., Noda, K., Shyoama, Y., Kubo, C. & Nomoto, K. 1997. Oral administration of unicellular green algae, *Chlorella vulgaris,* prevents stress induced ulcer. *Planta Medica*, 63(5), 465-466.

[18] Sarma, L., Tiko, A. B., Kesavan, P. C. & Ogaki, M. 1993. Evaluation of radio protective action of a mutant (E-25) form of *Chlorella vulgaris* in mice. *Journal of Radiation Research*, 34(4), 277-284.

[19] Kralovec, J. A., Power, M. R., Liu, F., Maydanski, E., Ewart, H. S., Watson, L. V., Barrow, C. J. & Lin, T. J. 2005. An aqueous *Chlorella* extract inhibits IL-5 production by mast cells in vitro and reduces ovalbumin-induced eosinophil infiltration in the airway in mice *in vivo*. *International Immunopharmacology*, 5, 689-698.

[20] Chu, C. Y., Huang, R. & Lin, L. P. 2007. Analysis of the agglutinating activity from unicellular algae. *Journal of Applied Phycology*, 19, 401-408.

[21] Stephen Ewart, H., Bloch, O., Girouard, G. S., Kralovec, J., Barrow, C. J., Ben-Yehudah, G., Reyes Suárez, F. & Rapoport, M. J. 2007. Stimulation of cytokine production in human peripheral blood mononuclear cells by an aqueous *Chlorella* extract. *Planta Medica*, 73, 762-768.

[22] Kanno, T. & Klinghardt, D. 2007. *Chlorella vulgaris* and *Chlorella vulgaris* extract (CVE). Orem, Utah: Woodland publishing.

[23] Oren, A. 2005. A hundred years of *Dunaliella* research: 1905-2005, *Saline Systems*, 1:2.

[24] García, F., Freile-Pelegrín, Y. & Robledo, D. 2006. Physiological characterization of *Dunaliella* sp. (Chlorophyta, Volvocales) from Yucatan, Mexico. *Bioresource Technology*, 98, 1359-1365.

[25] Gómez, P. I. & González, M. A. 2004. Genetic variation among seven strains of *Dunaliella salina* (Chlorophyta) with industrial potential, based on RAPD banding pattern and on nuclear ITS rDNA sequences. *Aquaculture*, 233, 149-162.

[26] Ghasemi, Y., Morowvat, M. H., Rasoul-Amini, S., Mohagheghzadeh, A., Abolhassanzadeh, Z., Hamidi, M., Raee, M. J., Ghoshoon, M. B. & Shokravi, Sh. 2007b. PCR amplification of the 18S rRNA gene of *Dunaliella salina* MCCS 001 isolated from Maharlu Salt Lake of Iran. NCBI, EF682841.2, GI: 171920073.

[27] Peña-Castro, J. M., Martínez-Jerónimo, F., Esparza-García, F. & Cañizares-Villanueva, R. O. 2004. Heavy metals removal by the microalga *Scenedesmus incrassatulus* in continuous cultures. *Bioresource Technology*, 94, 219-222.

[28] Hodaifa, G., Eugenia Martínez, M. & Sánchez, S. 2008. Use of industrial wastewater from olive-oil extraction for biomass production of *Scenedesmus obliquus*. *Bioresource Technology*, 99, 1111-1117.

[29] Borowitzka, M. A. 1999. Commercial production of microalgae: ponds, tanks, tubes and fermenters. *Journal of Biotechnology*, 70, 313-321.

[30] Wen, Z. Y., & Chen, F. 2003. Heterotrophic production of eicosapentaenoic acid by microalgae. *Biotechnology Advances*, 21, 273-294.

[31] Spolaore, P., Joannis-Cassan, C., Duran, E. & Isambert, A. 2006. Commercial application of microalgae. *Journal of Bioscience and Bioengineering*, 101(2), 87-96.

[32] Petkov, G. & Garcia, G. 2007. Which are fatty acids of the green alga *Chlorella*?. *Biochemical Systematics and Ecology*, 35, 281-285.

[33] Pereira, S. L., Leonard, A. E., Huang, Y. S., Chuang, L. T. & Mukerji, P. 2004. Identification of two novel microalgal enzymes involved in the conversion of the ω3-fatty acids, eicosapentaenoic acid, into docosahexaenoic acid. *The Biochemical Journal*, 384, 357-366.

[34] Kochert, G. 1978. Protein determination by dye binding. In: *Handbook of phycological methods-physiochemical and biochemical methods* (edited by Hellbust, J. A. & Craigie, J. S.). Pp. 91-93. Cambridge, UK: Cambridge University Press.

[35] Anupama, & Ravindra, P. 2000. Value-added food: Single cell protein. *Biotechnology Advances*, 18, 459-479.

Part 3. Yeast as a Source of Raw Material

In: Bioprocess Sciences and Technology
Editor: Min-Tze Liong
ISBN 978-1-61122-950-9

Chapter 8

FERMENTING MICROORGANISMS FOR 2ND GENERATION BIOETHANOL PRODUCTION

*E. Tomás-Pejó**
CIEMAT, Renewable Energy Department Biomass Unit,
Av. Complutense 22, 28040 Madrid, Spain

ABSTRACT

The high demand for energy worldwide, coupled with the depletion of the world's petroleum supply and the concern about global climate change have resulted in a resurgence in the development of alternative fuels. Nowadays, bioethanol is one of the most important renewable fuels contributing to the mitigation of the negative environmental impacts generated by fossil fuels.

The current bioethanol production at industrial scale relies on ethanol from sugars and starch and it is called "first generation bioethanol". By contrast, bioethanol produced from lignocellulosic biom[1]ass is called "second generation bioethanol". It offers several advantages such as lower cost feedstocks, higher green house gases (GHG) reduction and non-competition with food-crops. One of the main configurations in second generation ethanol production is simultaneous saccharification and fermentation (SSF) where enzymatic hydrolysis and sugar fermentation are run in the same vessel at the same time. In SSF processes, owing to the different optimum temperature for fermentation and hydrolysis will gain importance the use of thermotolerant yeasts such as *Kluyveromyces marxianus* which could ferment at temperatures around 40 ºC closer to the optimum temperature for enzymes. When xylose fermenting microorganisms are available glucose and xylose are co-fermented to ethanol in a simultaneous saccharification and co-fermentation (SSCF) process. However, the highest level of process integration would be obtained when all steps (enzyme production, hydrolysis, xylose and glucose fermentation) are run in one reactor by cellulolytic microorganisms, which would be called consolidated bioprocessing (CBP).

* Corresponding author: *E. Tomás-Pejó* E-mail: mariaelia.tomas@ciemat.es

This chapter includes and updated revision about fermenting microorganism for second generation bioethanol production, emphasizing the importance of obtaining efficient xylose fermenting microorganisms. Moreover, recent advances in obtaining successful microorganisms for CBP are also reported.

1. Introduction

Burning fossil fuels for the generation of heat or energy represents the major source of antropogenic carbon emissions being the liquid fuels responsible for 30% of the emissions in industrialized countries [1].

Negative environmental consequences of fossil fuels have favoured the research for biomass conversion into biofuels. Liquid biofuels (bioethanol and biodiesel) are the only renewable products that can be easily integrated into current fuel distribution system and one of the few alternatives for short-term diversification in the transportation sector. Biodiesel is produced from animal or vegetable oil and fats, and bioethanol is produced by converting sugars into ethanol via fermentation. Both biofuels have a positive energetic balance because the energy contained in the product is higher to the energy required for its production and distribution [2].

The "first generation bioethanol" relies on ethanol from sugars and starch. However, sugar or starchy raw materials are also employed for animal or human feeding and its role on the reduction of GHG is not as high as desirable [3]. In this context the bioethanol produced from lignocelulosic biomass, so-called "second generation bioethanol", offers several advantages such as lower cost feedstocks, higher GHG reduction and non-competition with food-crops. Furthermore, lignocelulosic biomass is highly available because it can be found as forest or agricultural residues or even municipal solid wastes which could allow bioethanol fuel to play a much bigger role in the future. Life Cicle Analysis studies consider 30-50% reduction in CO_2 emissions to the biofuels produced with conventional technologies from cereals or vegetable oils with respect to fossil fuels. Notwithstanding, reductions as high as 80-90% are expected in case of biofuels produced from lignocellulose [4].

Lignocellulose is the main structural component of plants. It is mainly consisted of three components: cellulose, hemicellulose and lignin, beside other components less abundants such as extractives and ash which composition varies among different lignocellulosic materials. Cellulose is a linear polymer of D-glucose linked by β-1,4 bonds. By contrast, hemicellulose is a branched polymer consisting of more than one kind of sugar (D-xylose, L-arabinose, D-mannose, D-glucose and D-galactose) and related substances such as glucuronic acid, hexuronic acid and its by-products.

In order to convert the polysaccharides present in lignocellulose into potentially fermentable sugars, the lignocellulosic material needs to be subjected to a hydrolysis step. This hydrolysis can be performed by means of acid catalyst, a process known as "acid hydrolysis", or through enzymatic catalysts, known as "enzymatic hydrolysis". When employing concentrated acids in the acid hydrolysis, the huge amount of acid needed, the high cost of the acid recovery and the necessity of neutralization as well as corrosive problems in the tank make the process economically unfeasible. If diluted acids are used, higher temperatures are required, which provoke acusate corrosive problems and hemicellulosic sugar degradation. Due to these, ethanol production processes based on enzymatic hydrolysis are shown as more promising alternatives.

The ethanol production process from lignocellulose based on enzymatic hydrolysis includes the following main steps: pretreatment, hydrolysis of cellulose and hemicellulose, sugar fermentation, separation of lignin residue and recovery as well as purification of the ethanol to meet fuel specifications [5]. These steps can be carried out according to different process configurations which results in different levels of process integration.

The pretreatment of lignocellulose is a crucial step owing to the physical and chemical barriers caused by the close association of the main components (cellulose, hemicellulose and lignin). Its main purpose is to increase the enzyme accessibility improving digestibility of cellulose. However, when lignocellulosic materials are subjected to severe conditions during pretreatment (high temperature, high pressure, and acid environment) there could be certain lignin and sugar degradation which could lead in the presence of some inhibitory compounds [6, 7].

After pretreatment, the insoluble residue of the pretreated material consists mainly in cellulose and lignin since most of the hemicellulosic sugars are solubilized during the same. To convert cellulose into glucose at least three categories of enzymes, known as cellulases, are involved: endoglucanases hydrolyze internal β-1,4-glucosidic bonds in the cellulose chain, cellobiohydrolases cleave off cellobiose units from the end of the chain and β-glucosidases convert cellobiose into glucose. These enzymes work synergistically to hydrolyze cellulose by creating new accessible sites and relieving product inhibition. However, complex structure of lignocellulose requires not only cellulase and β-glucosidase activities for its complete degradation. Xylanases and other accessory enzymes act synergistically with cellulases and play an important role in an efficient hydrolysis process [8].

After hydrolysis the glucose is fermented to ethanol by the action of ethanologenic microorganisms with the consequent production of ATP and CO_2. The stoichometric conversion of glucose to ethanol is 0.51 g/g but in practice it is very difficult to reach such a high conversion because microorganisms divert certain part of the consumed energy into cellular metabolism and growth.

When performing processes based on enzymatic hydrolysis, different levels of process configuration are possible. Current strategies to produce bioethanol from lignocellulose are mainly based on simultaneous saccharification and fermentation (SSF) and simultaneous saccharification and co-fermentation (SSCF) processes. In the SSF/SSCF processes the enzymatic hydrolysis and sugar fermentation, mainly glucose, are run in the same vessel at the same time. It results in an important advantage when comparing with separate hydrolysis and fermentation (SHF) because glucose accumulation in the broth during hydrolysis have an inhibitory effect on enzyme activity and in the SSF/SSCF the glucose is continuously consumed by the yeast, which minimizes the end-product inhibition on enzymes. Notwithstanding, the optimum conditions for enzymatic hydrolysis and fermentation are quite different and as will be exposed next in the chapter the employment of thermotolerant yeasts such as *Kluyveromyces marxianus* capable of growing and ferment at temperatures above 40 °C is an interesting alternative when perfoming hydrolysis and fermentation simultaneously [9, 10].

The biomass composition in hemicellulosic or cellulosic sugars strongly influences the configuration process. Since pentoses forming the hemicellulosic fraction could be up to 30%

of the total carbohydrates the conversion of lignocellulose to bioethanol via fermentation is not economically viable unless hemicellulosic sugars are also fermented. Since xylose is the second most abundant carbohydrate in nature its commercial fermentation could provide an alternative for improving the global economy of the process.

The fourth configuration, known as consolidated bioprocessing (CBP), is an alternative in which enzyme production, substrate hydrolysis as well as glucose and xylose fermentation are accomplished in one single step. For this purpose it is needed a mixture or a single microorganism capable of hydrolyzing and fermenting the pretreated lignocellulosic biomass. In spite of all the advantages that would be expected when performing CBP there are no microorganisms commercially available with all desirable characteristics for an efficient CBP.

Among all improvements needed to move forward the implementation of the ethanol production process at industrial scale, the simplification of the process with the integration of all the stages offers the best prospects. Different levels of process integration of each configuration are illustrated in Figure 1. The integration of the steps implies to have an overall picture of pretreatment, hydrolysis and fermentation because each step has an important impact on the following. This integration could have benefits in terms of lower capital and operating costs.

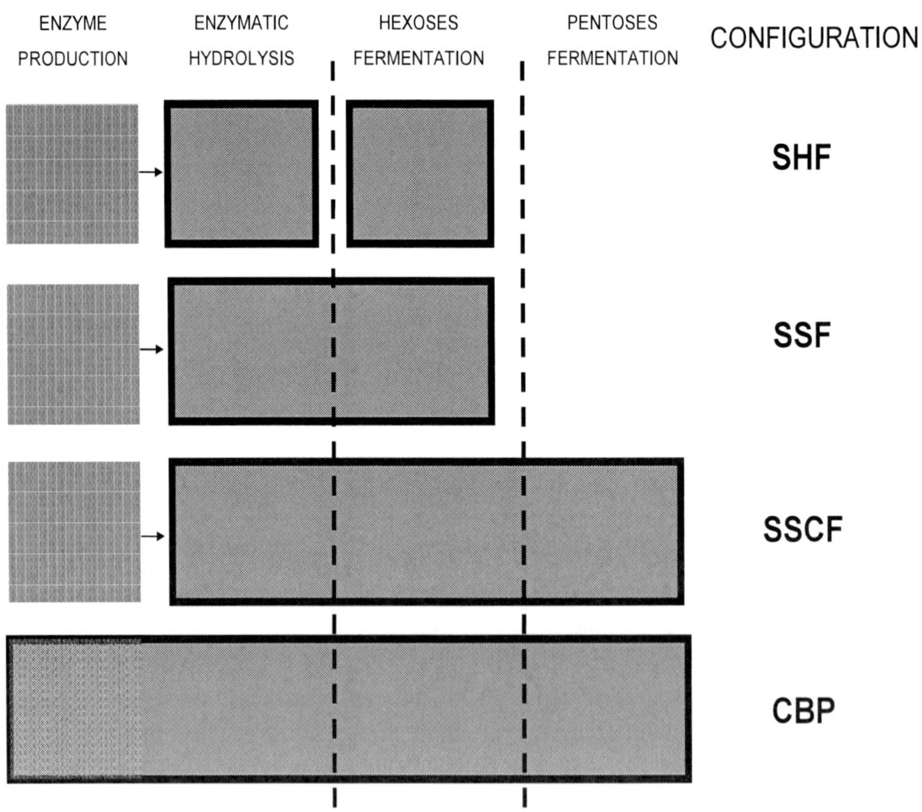

Figure 1. Different configuration with different levels of process integration for an ethanol production based on enzymatic hydrolysis.

2. SECOND GENERATION BIOETHANOL

2.1. Simultaneous Saccharification and Fermentation Processes

The alcoholic fermentation is a process widely used for the mankind since thousand years for brewery or wine production. When fermentation process is employed for second generation bioethanol production, sugars released during hydrolysis are fermented resulting in ethanol and CO_2 production.

There are several desired features to have into account for the choice of the fermenting microorganisms in an ethanol production process from lignocellulose such as high ethanol yields and productivities, minimal by-product formation, minimal nutrient supplementation, high ethanol tolerance or broad substrate utilization range [11]. Furthermore, it is remarkable the fact that some inhibitory compounds could be released during pretreatment owing to lignin, hemicellulose or cellulose degradation [7, 12]. These compounds could act as inhibitors on yeast and enzymes in the following SSF process, hence the necessity of using a robust microorganism with high tolerance to the toxic compounds.

The configuration with the lowest level of integration in the ethanol production implies a sequential procedure in which the enzymatic hydrolysis and fermentation are performed in two different reactors. In this process, so-called SHF, the cellulose is firstly hydrolyzed to glucose and afterwards glucose is fermented to ethanol. Its primary advantage is that each step can be conducted at optimal conditions and the major disadvantage is the end product inhibition because glucose accumulation in the broth inhibits the cellulolytic enzymes lowering enzymatic hydrolysis yields [13, 14].

The implementation and development of the process known as SSF has been one of the most interesting approaches related with the global ethanol production process from lignocellulose. Notwithstanding when using SSF configuration it is necessary a compromise in operational temperature since optimum temperature and pH for hydrolysis and fermentation are quite different.

During the SSF process glucose released by the action of cellulases is directly metabolized to ethanol by the yeasts. This continuous withdrawal of glucose from the medium minimizes the end-product inhibition on cellulase activity which is reflected in higher enzymatic hydrolysis yields and decreased enzyme loading requirements [15, 16]. Furthermore, in comparison to the sequential process, besides implying cost reduction owing to the employment of one single reactor, the SSF allows higher ethanol yields, shorter fermentation times and higher ethanol volumetric productivities [17, 18]. Following in the text are exposed the main features of the most promising microorganisms for ethanol production in SSF processes.

2.1.1. Saccharomyces cerevisiae

The yeast *S. cerevisiae* is the microorganism most commonly employed at industrial level in alcoholic fermentation processes. It can use all kind of hexoses present in lignocellulose, produces ethanol with a yield close to the maximum (0.51 g/g) and can produce ethanol concentrations as high as 18% [15]. Wild-type *S. cerevisiae* strains ferment glucose, mannose and fructose via the Embden-Meyerhof (EM) pathway of glycolysis, while galactose is fermented via Leloir pathway [19]. *Saccharomyces* is generally recognised as safe (GRAS) and in anaerobiosis it converts through the glycolysis 1 mol of glucose into 2 mol ethanol which results in the net production of 2 mol ATP plus CO_2.

S. cerevisiae incorporates glucose into the cell by a complex system for hexose transport [20, 21]. All the transporters transport glucose via facilitated diffusion, thus glucose uptake only requires a concentration gradient across de plasma membrane.

Apart from monosaccharides, *S. cerevisiae* can utilize the disaccharides sucrose and maltose, and some strains can also utilize mellobiose and the trisaccharides maltotriose and raffinose [22].

Beside all the advantages regarding its high ethanol yield and the wide range of consumed sugars, *S. cerevisiae* shows high tolerance to the toxic compounds released during pretreatment which is a crucial trait in an industrial ethanol production process from lignocellulose. Due to that, numerous researches have been focused on the employment of glucose fermenting *S. cerevisiae* strains in SSF processes from pretreated lignocellulosic biomass such as wheat straw [18], sunflower stalks [23], olive tree pruning [24] or even municipal solid waste [25].

In spite of all the mentioned interesting features, it presents an important drawback when fermenting the hemicellulosic fraction because is unable to ferment xylose. In this context, as will be stated below, huge efforts have been addressed to obtain recombinant *S. cerevisiae* strains capable of co-fermenting glucose and xylose.

2.1.2. Zymomonas Mobilis

Zymomonas mobilis is an industrial gram-negative microorganism that ferments glucose via Entner-Doudoroff (ED). The ED pathway yields only half as much ATP per mole glucose as the EM pathway. As a consequence, *Zymomonas* produces less biomass than yeast leading more carbon to ethanol [26]. It is also classified as GRAS and has simple nutritional needs.

It demonstrates some of the desirable traits that have to present an ideal fermenting microorganism such as high ethanol yield, high productivities as well as low pH and high ethanol tolerance [26]. Compared to yeast, *Z. mobilis* exhibits 5-10% higher ethanol yields and up to fivefold greater volumetric productivity [26] and in comparison to other bacterium such as *Escherichia coli*, it shows higher hexose metabolism [27].

Z. mobilis has been mainly used in processes for ethanol production from starch [28] or sugar biomass [29] but in the last years some approaches have been made on cellulosic materials such as paper sludge [30] or sugarcane bagasse [31].

One disadvantage when using *Z. mobilis* strains is that wild type can only utilize glucose, fructose and sucrose and it is unable to ferment pentose sugars. Furthermore, it has low tolerance to inorganic ions which are frequently present at inhibitory concentration in molasses and lignocellulosic hydrolysates [32].

2.1.3. Thermotolerant Yeasts

The employment of thermotolerant yeasts in SSF processes from lignocellulose is appearing as an interesting option since, as have been previously mentioned in the chapter, the optima conditions for enzymatic hydrolysis and fermentation in an SSF are quite different. Regarding optimum temperature, enzymes perform better at 50 °C while most of fermenting microorganisms have an optimum temperature at 30-37 °C. In this context, the employment of natural thermotolerant yeasts is glimpsed as an interesting alternative because they are able to grow and ferment at temperatures above 40 °C which implies, among other advantages, higher saccharification yields.

Thermotolerant strains of *Saccharomyces*, *Kluyveromyces* and *Fabospora* genera can grow at temperatures above 40 °C and ferment sugars at temperatures of 40 °C, 43 °C and 46 °C, respectively [33, 34]. Furthermore, some thermotolerant strains of *S. cerevisiae*, *K. marxianus* and *K. fragilis* have been successfully employed in SSF processes from lignocellulose [10, 34 and 35].

Besides improving the hydrolysis yields owing to the higher temperature of the process there are other advantages that could be exploited when using thermotolerant yeasts, such as, energy saving through a reduction in cooling costs, significant decreased risk of contamination and the possibility of continuous ethanol removal [36]. Furthermore, some studies concluded that only a 5 °C increase in the SSF temperature could greatly reduce the ethanol production cost [9].

In this context many efforts should be addressed towards indentifying thermostables, ethanol-resistant strains with wide spectrum of substrate utilization and with the capability of producing high ethanol concentrations at temperatures closer to the optimal for saccharification.

2.1.3.1. Kluyveromyces Marxianus

Yeasts belonging to the genus *Kluyveromyces* have been shown with potential for industrial applications. In particular *Kluyveromyces lactis* was the first yeast after *S. cerevisiae* to be given GRAS status, an important requisite for many industrial processes.

K. marxianus is closely related to *K. lactis* but has received much less attention for biotechnological applications in spite of utilizing wide range of sugars and growing faster than *K. lactis* even at temperatures above 40 °C.

Several strains belonging to the yeast species *K. marxianus* have been succesfully employed in SSF processes for bioethanol production from lignocellulose and, as can be observed in the Table 1, *K. marxianus* can efficiently ferment lignocellulosic materials to ethanol at temperatures between 37 °C to 45 °C. Some authors have even published xylose consumption by *K. marxianus* but in spite of consuming xylose the ethanol production is lower than other wild type or genetically modified yeasts [37]. Furthermore, the ethanol production in xylose utilizing strains is improved under micro aerobic conditions, which is very difficult to control in industrial fermentation.

Besides being successfully used for bioethanol production *K. marxianus* has been exploited for other wide range of applications such as enzyme production [46], cell protein synthesis [47] or aroma compounds production [48]. *K. marxianus* has also been proposed as a source of oligonucleotides used as flavour enhancers in food products; and oligopeptides added to dairy products [49].

2.2. Simultaneous Saccharification and Co-fermentation Processes

The use of lignocellulosic material for ethanol production requires the fermentation of both hexoses and pentoses which strongly influences the cost of the lignocellulosic fuel. Whereas the glucose fermentation can be carried out efficiently, the pentoses fermentation presents a challenge and currently there are no naturally ethanol producing strains capable of fermenting all the necessary sugars with high yields.

Table 1. Thermotolerant *K. marxianus* strains used in ethanol production processes from different lignocellulosic raw materials

Strain	Lignocellulosic raw material	Temperature of process (°C)	References
K. marxianus IMB1, IMB2, IMB3, IMB4 and IMB5	Switchgrass	45	[38]
K. marxianus IMB3	Barley straw	45	[39]
K. marxianus IMB4	Switchgrass	37, 41, and 45	[40]
K. marxianus DMKU 3-1042	Sugar cane juice	40 and 45	[36]
K. marxianus ATCC 12708	Orange-peel oil	37	[41]
K. marixianus ATCC 36907	Recycled paper sludge	38	[42]
K. marxianus Y01070	Industrial paper waste	40	[43]
K. marxianus CECT 10875	Wheat straw	42	[44]
	Eucalyptus, shorgum bagasse, poplar, *Brassica carinata* residue	42	[10]
	Recicled paper waste	42	[45]

Xylose fermentation to ethanol is crucial to meet desired overall yields of the process. In this context various strategies have been used to introduce the ethanol pathway in natural xylose consumers microorganisms or to modify the xylose-catabolizing pathway in natural ethanol producers [50].

In order to carry out the process known as simultaneous saccharification and co-fermentation (SSCF) is necessary to employ a microorganism able to ferment glucose as well as xylose. In the SSCF process the enzymatic hydrolysis is performed at the same time that glucose and xylose co-fermentation to ethanol. Besides the advantage that implies the higher amount of sugars available for fermentation, xylose consumption results in higher enzymatic hydrolysis yields because hemicellulosic sugars also have an inhibitory effect on cellulases [13, 51]. Thus, when employing a xylose utilizing microorganism the global efficiency of the process is increased even more.

In the recent years L-arabinose fermentation is gaining interest specially when employing raw materials such as grass or corn stover where the arabinan component is considerably high comparing to other lignocellulosic biomass. Several attempts have been made to genetically engineer *S. cerevisiae* for obtaining L-arabinose fermenting strains. Some strategies has been focused on the introduction of a bacterial arabinose utilization pathway [52] and others have been addressed to reconstruct in *S. cerevisiae* the fungal pathway [53]. However both approaches need thorough research for further optimization before being considered for industrial scale.

2.2.1. Natural Xylose Fermenting Microorganisms

Microorganisms that ferment xylose to ethanol are found among bacteria, fungi and yeasts. The desirable fermenting microorganism in an SSCF has to ferment xylose and it must do it in the presence of potential inhibitors released during pretreatment. It is worthy to mention that pentose fermenting yeasts are generally more inhibited by toxic compounds released during pretreatment than hexose fermenting yeast hence the importance of employing a robust microorganism.

Natural xylose fermenting microorganisms generally metabolize D-xylose to D-xylulose through two separate routes (Figure 2). In most fungi and xylose fermenting yeasts D-xylose is converted to D-xylulose by two oxidoreductases involving co-factors NADPH and $NADP^+$. It implies that D-xylose is firstly reduced to xylitol by NADPH-dependent xylose reductase (XR) and then xylitol is oxidized to D-xylulose by NAD^+ dependent xylitol dehydrogenase (XDH). Subsequently, D-xylulose is converted into D-xylulose-5-phosphate by the action of a xylulokinase (XK) and the D-xylulose-5-phosphate is further metabolized through the pentose phosphate pathway [54].

On the other hand, in most bacteria the xylose isomerase (XI) directly isomerizes D-xylose to D-xylulose and the D-xylulose is also phosphorylated to D-xylulose-5-phosphate by XK [55].

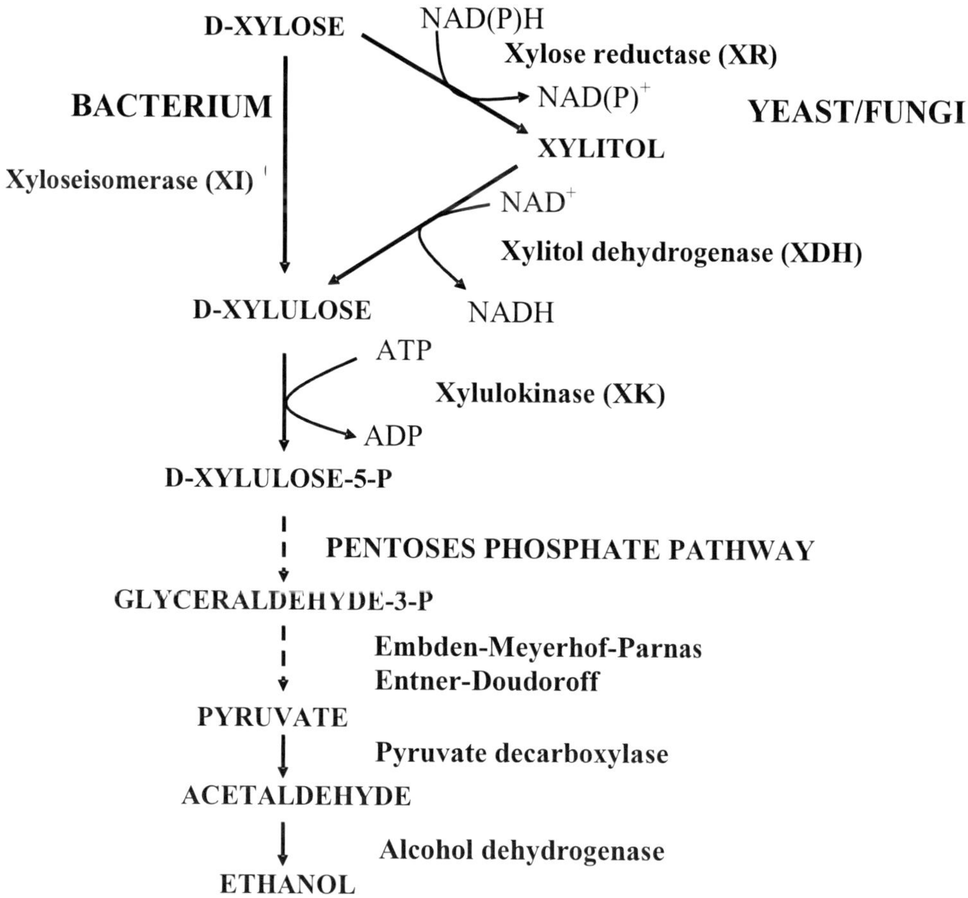

Figure 2. Both xylose fermenting pathways in bacteria, fungi and yeasts.

The problem when using natural xylose fermenting yeasts for ethanol production is that most of these yeasts excrete substantial amounts of xylitol as by-product. Xylitol excretion has been associated to the maintenance of the redox balance during xylose fermentation since XR and XDH have different co-factors preferences. The XR can use both NADPH and NADH as co-factor but XDH uses almost exclusively NAD^+ in the oxidation of xylitol to D-xylulose [20, 56]. As mainly NADPH is utilized in the first reaction, NAD^+ is therefore not made available for the XDH reaction and consequently this leads to substantial xylitol production and suboptimal ethanol yield on xylose [57].

The most commonly studied natural xylose fermenting yeast includes *Pichia stipitis, Candida shehatae* and *Pachysolen tannophilus*. The first has been extensively employed for ethanol production from lignocellulosic materials [58, 59]. However it ferments xylose directly to ethanol with very low rates and is very sensitive to pH changes [20]. Furthermore, many types of yeast that ferment xylose to ethanol, only do so with small amounts of oxygen. Accurate dosage of oxygen in large scale processes with viscous materials is not only economically undesired but also virtually impossible which entails interest in the ability for strict anaerobic growth such as exhibited by *S. cerevisiae*.

The first studies about xylose fermentation by fungi came from the 20s which reported xylose fermentation by *Fusarium lini*, *Monilia murmanica* and *Oidium laminarium.* Later on the direct conversion to xylose to ethanol was also studied in the molds *Fusarium oxysporum*, *Mucor*, *Rhizopus javanicus*, *Rhizopus oryzae* and several species of *Aspergillus.* In comparison to xylose fermenting yeasts, filamentous fungi are more tolerant to the inhibitors released during pretreatment but the main problem is that they show slow fermentation rates for a competitive industrial process [60, 61].

The ethanologenic xylose fermenting bacteria that show the most promising features for industrial exploitations in SSCF are *E. coli* and *Klebsiella oxytoca*. Both are naturally able to use a wide spectrum of sugars and great efforts have been concentrated on engineering these strains to selectively produce ethanol. However, anaerobic bacteria ferment pentoses but are inhibited at low sugar and ethanol concentration. In addition, sugar fermentation to ethanol occurs with considerable by-product formation.

2.2.2. Engineered Xylose Fermenting Microorganisms

2.2.2.1. Engineered Yeasts

It is not questionable that one of the most effective ethanol producers from hexoses is *S. cerevisiae*. Notwithstanding, as it has been previously indicated in the chapter, it presents an important hurdle since it is unable to ferment xylose to ethanol which is crucial for making the ethanol production process from lignocellulose cost-effective.

During the last years huge efforts have been addressed to modify *S. cerevisiae* strains for efficient xylose fermentation. Nevertheless, in most cases xylose fermentation by recombinant strains has been suboptimal. This limited xylose fermentation by recombinant *S. cerevisiae* has been ascribed to poor xylose uptake, co-factor imbalance owing to the different co-factor preference by XR and XDH, limitation in the pentose phosphate pathways and insufficient induction or activation of ethanologenic enzymes [62].

Wild *S. cerevisiae* is able to slowly metabolize the D-xylulose. Therefore, main strategies for constructing an efficient xylose fermenting *S. cerevisiae* strain have been focused on the conversion of D-xylose into D-xylulose. Recombinant *S. cerevisiae* strains have been obtained by introducing the *XYL1* and *XYL2* genes, mainly from *P. stipitis*, encoding XR and XDH, respectively, and the endogenous *XKS1* gene, encoding XK but as have been explained yeasts carrying XR/XDH genes excrete substantial amounts of xylitol owing to the co-factor redox imbalance. The over expression of the endogenous *XKS1* gene coding for XK enables *S. cerevisiae* to ferment xylose to ethanol more rapidly [62, 63]. Nevertheless, when *XKS1* is over expressed in a *S. cerevisiae* strain carrying *XYL1* and *XYL2*, in spite of the faster xylose uptake and higher ethanol yield, xylitol production is still a problem. In this context in the recent years one of the most effective approaches has been focused on altering the co-enzyme preference of the XR and XDH to diminish xylitol formation and enhance ethanol yields [64].

On the other hand, *xylA* gene encoding XI from bacterium has been also expressed in *S. cerevisiae*. Expression of XI, which directly converts D-xylose into D-xylulose, does not present this limitation regarding xylitol accumulation. Nevertheless, when using bacterial XI genes, such as XI from *Thermus thermophilus* [65], some difficulties are related with the functional expression of these genes in yeasts. This problem has been overcome with the discovery of a fungal XI gene in *Piromyces* sp [55] however, XI from this anaerobic fungus is strongly inhibited by xylitol. A recent study has successfully cloned in *S. cerevisiae* a XI from the anaerobic bacterium *Clostridium phytofermentans* less sensitive to inhibition by xylitol than the XI from *Piromyces* [66]. It is quite important because while xylitol formation is strongly reduced in yeasts expressing the XI in comparison with XR/XDH, other reaction in the cell still produce xylitol from xylose. Thus, it is interesting to express a XI with less xylitol inhibition.

As it is shown in the Table 2, different engineered *S. cerevisiae* strains have been successfully employed in ethanol production processes from lignocellulose.

Table 2. Engineered xylose fermenting *S. cerevisiae* strains with both xylose utilizing pathways employed for SSCF processes with lignocellulosic raw materials

Strain	Xylose pathway	Medium	References
S. cerevisiae TMB3001	XR-XDH-XK	Sugar cane baggasse	[62, 67]
S. cerevisiae TMB3400	XR-XDH-XK	Corn stover, wheat straw, barley straw, spruce hydrolysates	[62, 63]
S. cerevisiae TMB3066	XI-XK	Undetoxified spruce hydrolysates	[63]
S. cerevisiae F12	XR-XDH-XK	Wheat straw	[18, 68]
S. cerevisiae H158	XR-XDH-XK	Birch hydrolysates	[69]
S. cerevisiae CEN.PK	XR-XDH-XK	Birch hydrolysates	[69]
S. cerevisiae 424A	XR-XDH-XK	Corn stover	[56]

Another important limitation present in xylose fermenting *S. cerevisiae* strains is related with the lack of specific xylose transporters. *S. cerevisiae* takes up xylose by facilitated diffusion mainly through hexose transporters of high and low affinity encoded by genes belonging to *HXT* family [20, 21 and 70]. Xylose transportation through these transporters is significantly less efficient than for glucose thus, xylose uptake competes with glucose uptake [70]. Several strategies such as over expression of different genes encoding for different transporters in *S. cerevisiae* have been developed with the aim of enhancing xylose consumption [21]. Furthermore, different transporters with higher xylose affinity from different microorganisms have been successfully expressed in *S. cerevisiae* such as *Sut1* from *P. stipitis* [71], *Trxlt1* from *T. reseei* [70] and *Gxf1* from *Candida intermedia* [72]. The latter shows a xylose affinity about three times higher than hexose transporters for *S. cerevisiae*. The improvement on the efficiency of xylose uptake into the cell is still a huge challenge for optimizing xylose metabolism.

Besides the introduction of different genes necessary for the conversion of D-xylose into D-xylulose as well as the modification of the genes taking part in the transporters synthesis and redox balance, different strategies have been addressed to over express XK and to improve the pentose phosphate pathway.

Even with the introduction of pentose fermentation genes *S. cerevisiae* strains preferentially utilize glucose before the other monosaccharides. When glucose is present in the medium, alternative carbon sources are used only when glucose is depleted. It has to be taken into account that lignocellulosic hydrolysates are composed of sugars other than glucose which consumption is delayed owing to glucose repression. In this context deletion of the genes involved in glucose repression in xylose fermenting *S. cerevisiae* strains resulted in an increase of the xylose consumption rate as well as higher ethanol productivities [73]. Deregulation of the strong glucose repression effect in *S. cerevisiae* would be required to allow co-fermentation of sugars derived from lignocellulose for high ethanol productivity.

Beside *S. cerevisiae*, genetic modification of *P. stipitis* has received significant attention in the past few years. Since it needs microaerophilic conditions for growing, expression of *S. cerevisiae URA1* in *P. stipitis* has been an interesting approach. *URA1* gene encodes for dihydroorotate dehydrogenase that employs fumarate as an alternative electron acceptor which enables *P. stipitis* to grow in anaerobic conditions [74]. Further improvements have been also achieved by modification of its respiratory chain by deleting the *CYC1* gene encoding for cytochrome *c* [75].

2.2.2.2. Engineered Bacteria

The engineered ethanologenic bacteria that have shown more promising alternatives for fermenting xylose in industrial exploitation are *E. coli*, *Z. mobilis* and *K. oxytoca*.

E. coli is an attractive microorganisms for the conversion of lignocellulose to ethanol because it grows on a wide range of carbon sources including pentoses, it can sustain high anaerobic and aerobic glycolytic fluxes and presents quite good ethanol tolerance [76]. However, wild-type *E. coli* shows low ethanol yield because it converts sugar more efficiently to organic acids such as acetic or lactic acid instead of ethanol. Due to that, several approaches have been performed with the aim of redirecting glycolytic fluxes to ethanol. The most successful approach has been the transformation of *E. coli* with a plasmid containing genes from *Z. mobilis* encoding for PDC (pyruvate decarboxylase) and ADH (alcohol dehydrogenase) [77, 78] in an artificial operon (PET). In the recombinant strain *E. coli* KO11 both enzymes are over expressed at high level [79, 80]. Strain KO11 has been evaluated at laboratory scale for producing ethanol from many types of lignocellulosic biomass including barley hull [81], water energy crops [82], orange peel [83], corn cobs [84] or even waste house wood [85]. Recently *E. coli* LY01 has been derived from KO11 and it shows higher tolerance to the inhibitors present in lignocellulosic hydrolysates [86].

The major disadvantage when using *E. coli* is the public acceptance because of the existence of some patogenous strains as well as the neutral pH growth range. Due to that is not widely employed in SSCF approaches because its optimum pH (6.5) is not compatible with optimum pH for cellulolytic enzymes (4.8) [76].

Apart from *E. coli*, xylose is utilized by the engineered bacterium *Z. mobilis*. In *Zymomonas* glucose can easily cross the cell membrane by facilitated diffusion and is efficiently converted into ethanol by an overactive PDC-ADH system [26]. As have been previously mentioned, wild *Z. mobilis* strain can not produce ethanol from xylose but it has been successfully engineered by introducing a xylose-metabolizing pathway from *E. coli* [27, 87]. The newly strain *Z. mobilis* AX101 can use xylose after glucose consumption in lignocellulosic hydrolysates but showing high sensitivity to inhibitors present in the broth [88].

In addition to *E. coli* and *Z. mobilis*, the related bacteria *K. oxytoca* has been also transformed with the PET operon diverting the carbon flow to ethanol production (strain P2). It can grow on a wide range of sugars including hexoses and pentoses as well as cellobiose and cellotriose [89]. This latter trait makes the strain attractive for processes for ethanol production from lignocellulosic biomass presenting the chance of adding lower amount of extracellular ß-glucosidase for cellobiose hydrolysis. *K. oxytoca* has been employed successfully on various raw materials such as corn fiber [90] or sugar beet pulp [91].

2.3. Consolidated Bioprocessing (CBP)

The novel process known as consolidated bioprocessing (CBP) entails all the biologic reactions involved in the ethanol production process in one single step (Figure 1). It means that enzyme production, enzymatic hydrolysis as well as hexoses and pentoses fermentation are performed simultaneously. Since the enzyme cost could imply close to the 50% of the total cost in an SSF process [92], avoidance of the cost of the enzyme production is one of the largest potential saving associated with the implementation of CBP in comparison to current SSF/SSCF processes. For the achievement of this configuration is needed a single microorganism or microbial consortium able to hydrolyze and ferment the cellulose, and hemicellulose without exogenous enzyme addition.

Among all microorganisms capable of degrading cellulose and ferment glucose to ethanol, cellulolytic bacteria belonging to the genus *Clostridium* have been well studied and characterized. Those bacteria digest cellulose by an extracellular enzymatic complex called cellulosome which are stable enzyme complexes produced like protuberances on cell wall able to hydrolyze amorphous as well as crystalline cellulose. Among others, *Clostridium thermocellum* is an anaerobic thermophilic bacterium that employs cellulose as a sole carbon source and performs fermentation producing lactate, acetate, ethanol, H_2 and CO_2. Moreover, *C. thermocellum* expresses their cellulosome linked firmly to the cell wall and shows the highest known rate of enzymatic hydrolysis. Since this bacterium is unable to ferment pentoses, xylose fermentations could be performed in a co-culture by *Clostridium thermosaccharolyticum* which would allow simultaneous co-fermentation of glucose and xylose in a single reactor [93].

Direct cellulose to ethanol biotransformation can be also conducted, apart from *C. thermocellum,* by some fungi, including *Monilia* sp., *Aspergillus* sp., *Paecilomyces* sp., *Rhizopus* sp., *Neocallimastix* sp., *Trichoderma* sp. and *Neurospora crassa*. However, fungi have not been extensively studied as CBP microorganisms despite showing very interesting features because of the low ethanol yields.

In spite of all advantages that would imply the CBP, this potential cannot be carried out by microorganisms nowadays available and it requires further development of new and improved microorganisms. In this context, genetic modification of microorganisms for cellulose conversion to ethanol can be carried out according to two strategies. One is to improve the ethanol production and yields in naturally cellulolytic organisms such as *C. thermocellum*. Main objectives of this strategy include to redirect carbon fluxes toward ethanol and to insert new metabolic pathways for utilizing most of the lignocellulosic sugars. The most successful example to date is *Thermoanaerobacterium saccharolyticum,* an engineered thermophilic bacteria which produces cellulases and hemicellulases and able to ferment xylan polymers and primary sugars found in lignocellulosic biomass such as cellulose, glucose, xylose, mannose, galactose and arabinose [94].

The second strategy implies the introduction of the cellulolytic ability in microorganisms that exhibit high ethanol yields. Bigger efforts have been addressed to express a functional cellulase system in ethanol producers such as *Z. mobilis*, *S. cerevisiae* and enteric bacteria. In this context endoglucanase genes from *Pseudomonas fluorescens*, *Bacillus subtillis, Erwinia chrysanthemi* and *Acetobacter xylinum* have been introduced into *Z. mobilis*. Both endoglucanases from *E. chrysanthemi* encoded by *celY* and *celZ* and a cellulase gen from *A. xylinum* have been also successfully expressed in *E. coli* as well as *K. oxytoca* [95]. Furthermore recombinant *K. oxytoca* SZ21 containing genes for ethanol fermentation from *Z. mobilis* and endoglucanase genes from *E. chrysantherni* ferments efficiently amorphous cellulose to ethanol [96].

S. cerevisiae also have to be able to hydrolyze cellulose at high levels to sustain hydrolysis and fermentation of cellulose to ethanol in a CBP. In the last years *S. cerevisiae* has been engineered with genes encoding cellulases from various bacteria, filamentous fungi and plants. For example, β-glucosidase and endoglucanase from *Aspergillus aculeatus* and *Aspergillus oryzae* have been cloned and expressed on the cell surface of *S. cerevisiae* [97, 98] and two endoglucanases from *T. reseei* (*egI* or *Cel7B* and *egII* or *Cel5A)* and cellobiohydrolases from *T. reseei*, *Aspergillus niger* and *Phaenorachaete chrysosporium* have been also successfully cloned and expressed in *S. cerevisiae* Y294 [99, 100]. As indicated above, most of the cellulases that have been expressed in *S. cerevisiae* are of fungal origin but there have been also several examples of bacterial cellulases such as the expression of the endo-β-1,4-glucanase gene (*end1*) from *Butyvibrio fibrisolvens* [101].

Since the hydrolysis of hemicellulose and xylose utilization are also crucial for a cost-effective CBP process, several studies have been focused on the expression of hemicellulases in *S. cerevisiae* such as the co-expression of an endoxylanase (*xyl1*) from *Bacillus* sp. and an endoglucanase (*egl6*) from *Trichoderma* sp. [102].

3. Concluding Remarks

Currently, there is no doubt about the feasibility of using lignocellulosic biomass for producing ethanol fuel. Bioethanol from lignocellulose implies important advantages when comparing to ethanol from sugar or starch. As it has been indicated in the chapter it is remarkable the fact that lignocellulosic biomass composition in hemicellulosic or cellulosic sugars not only strongly influences the configuration process for ethanol production but also the required microorganisms.

Thus, there are several general desired features that should be present in microorganisms used in ethanol production from lignocellulose such as broad substrate utilization range, high ethanol yields and productivities, minimal by-product generation, minimal nutrient supplementation, high ethanol tolerance and high tolerance to the inhibitors generated during biomass pretreatment.

It is important to remark that successful fermentation of lignocellulose is strongly dependent on efficient utilization of D-xylose which is the second most common fermentable sugar in lignocellulosic pretreated materials. Furthermore, among all improvements needed to move forward the implementation of the ethanol production process an industrial scale, the simplification of the process with the integration of all the stages offers the best prospects. Depending on the fermenting microorganisms available, different level of process integration could be reached. In this context many attempts have been made to genetically engineer different microorganisms for efficient xylose fermentation to ethanol as well as an efficient cellulase production.

In spite of all the efforts, bacterial or yeast alternatives for industrial ethanol production in SSCF and CBP are not still widely used commercially and further research have still to be addressed to overcome the hurdles inherent in the process.

REFERENCES

[1] Gómez, L. D., Steele-King, C. G. & McQueen-Mason, S. J. 2008. Sustainable liquid biofuels from biomass: The writing's on the walls. *New Phytologist*, *178*, 473-485.

[2] Lechón, Y., Cabal, H., Lago, C., De la Rúa, C., Sáez, R. & Fernández, M. 2005. Análisis de ciclo de vida de combustibles alternativos para el transporte. Fase I. Análisis de ciclo de vida comparativo del etanol de cereales y la gasolina. Energía y cambio climático. Spanish Ministry of Environment and Ministry of Education and Science. ISBN.84-8320-312-X.

[3] Sims, R., Taylor, M., Saddler, J. & Mabee, W. E. 2008. From 1st to 2nd generation biofuel technologies: An overview of current industry and R&D activities. International Energy Agency. IEA.

[4] BioPlat. Spanish Technological Platform for Biomass. 2009. View document for 2030.

[5] Alvira, P., Tomás-Pejó, E., Ballesteros, M. & Negro, M. J. 2010. Pretreatment technologies for an efficient bioethanol production process based on enzymatic hydrolysis: A review. *Bioresource Technology*, *101*, 4851-4861.

[6] Almeida, J. R. M. & Hahn-Hägerdal, B. 2009. Developing *Saccharomyces cerevisiae* strains for second generation bioethanol: Improving xylose fermentation and inhibitor tolerance. *International Sugar Journal*, *111*, 172-180.

[7] Oliva, J. M., Sáez, F., Ballesteros, I., González, A., Negro, M. J., Manzanares, P. & Ballesteros, M. 2003. Effect of lignocellulosic degradation compounds from steam explosion pretreatment on ethanol fermentation by thermotolerant yeast *Kluyveromyces marxianus*. *Applied Biochemistry and Biotechnology - Part A Enzyme Engineering and Biotechnology*, *105,* 141-154.

[8] Merino, S. T. & Cherry, J. 2007. Progress and challenges in enzyme development for biomass utilization. *Advances in Biochemical Engineering Biotechnology*, *108,* 95-120.

[9] Abdel-Banat, B. M. A., Hoshida, H., Ano, A., Nonklang, S. & Akada, R. 2010. High-temperature fermentation: how can processes for ethanol production at high temperatures become superior to the traditional process using mesophilic yeast? *Applied Microbiology and Biotechnology*, *85,* 861-867.

[10] Ballesteros, M., Oliva, J. M., Negro, M. J., Manzanares, P. & Ballesteros, I. 2004. Ethanol from lignocellulosic materials by a simultaneous saccharification and fermentation process (SFS) with *Kluyveromyces marxianus* CECT 10875. *Process Biochemistry*, *39,* 1843-1848.

[11] Zaldivar, J., Nielsen, J. & Olsson, L. 2001. Fuel ethanol production from lignocellulose: a challenge for metabolic engineering and process integration. *Applied Microbiology and Biotechnology*, *56,* 17-34.

[12] Palmqvist, E. & Hahn-Hägerdal, B. 2000. Fermentation of lignocellulosic hydrolysates II: inhibitors and mechanism of inhibition. *Bioresource Technology*, *74,* 25-33

[13] Xiao, Z., Zhang, X., Greff, D. J. & Saddler, J. N. 2004. Effects of sugar inhibition on cellulases and β-glucosidase during enzymatic hydrolysis of softwood substrates. *Applied Biochemistry and Biotechnology*, *113-116,* 1115-1126

[14] Stenberg, K., Bollók, M., Réczey, K., Galbe, M., & Zacchi, G. 2000. Effect of substrate and cellulase concentration on simultaneous saccharification and fermentation of steam-pretreated softwood for ethanol production. *Biotechnology and Bioengineering*, *68,* 205-210.

[15] Lin, Y. & Tanaka, S. 2006. Ethanol fermentation from biomass resources: current state and prospects. *Applied Microbiology and Biotechnology*, *69,* 627-642.

[16] Olsson, L., Soerensen, H. R., Dam, B. P., Christensen, H., Krogh, K. M. & Meyer, A. S. 2006. Separate and simultaneous enzymatic hydrolysis and fermentation of wheat hemicellulose with recombinant xylose utilizing *Saccharomyces cerevisiae*. *Applied Microbiology and Biotechnology*, *129-132,* 117-129.

[17] Alfani, A. F., Gallifuoco, F., Saporosi, A., Spera, A. & Cantarella, M. 2000. Comparison of SHF and SSF process for the bioconversion of steam-exploded wheat straw. *Journal of Industrial Microbiology and Biotechnology*, *25,* 184-192.

[18] Tomás-Pejó E., Oliva J. M., Ballesteros M. & Olsson L. 2008. Comparison of SHF and SSF processes from steam-exploded wheat straw for ethanol production by xylose-fermenting and robust glucose-fermenting *Saccharomyces cerevisiae* strains. *Biotechnology and Bioengineering*, *100,* 1122-1131.

[19] Van Maris, A. J. A., Abbott, D., Bellissimi, E., Van den Brink, J., Kuyper, M., Luttik, M., Wisselink, H., Scheffers, W., van Dijken, J. & Pronk, J. 2006. Alcoholic fermentation of carbon sources in biomass hydrolysates by *Saccharomyces cerevisiae*: current status. *Antonie van Leeuwenhoek*, *90,* 391-418.

[20] Chu, B. C. H. & Lee, H. 2007. Genetic improvement of *Saccharomyces cerevisiae* for xylose fermentation. *Biotechnology Advances*, *25,* 425-441.

[21] Hamacher, T., Becker, J., Gardonyi, M., Hahn-Hägerdal, B. & Boles, E. 2002. Characterization of the xylose-transporting properties of yeast hexose transporters and their influence on xylose utilization. *Microbiology*, *148,* 2783-2788.

[22] Ostergaard, S., Olsson, L. & Nielsen, J. 2000. Metabolic engineering of *Saccharomyces cerevisiae*. *Microbiology and Molecular Biology Reviews*, *64,* 34-50.

[23] Ruíz, E., Cara, C., Ballesteros, M., Manzanaras, P., Ballesteros, I. & Castro, E. 2006. Ethanol production from pretreated olive tree wood and sunflower stalks by an SSF process. *Applied Biochemistry and Biotechnology*, *130,* 631-643.

[24] Cara, C., Ruíz, E., Ballesteros, M., Manzanares, P., Negro, M.J. & Castro, E. 2008. Production of fuerl ethanol from steam-explosion pretreated olive tree pruning. *Fuel*, *87,* 692-700.

[25] Ballesteros, M., Sáez, F., Ballesteros, I., Manzanares, P., Negro, M. J., Martínes, J. M., Castañeda, R. & Oliva, J. M. 2010. Ethanol production from organic fraction obtained after thermal pretreatment of municipal solid waste. *Applied Biochemistry and Biotechnology*, *101*, 423-431.

[26] Aristidou, A. & Penttila, M. 2000. Metabolic engineering application to renewable resource utilization. *Current Opinion in Biotechnology*, *11,* 187-198.

[27] Chen, J., Zhang, W., Tan, L., Wang, Y. & He, G. 2009. Optimization of metabolic pathways for bioconversion of lignocellulose to ethanol through genetic engineering. *Biotechnology Advances*, *27,* 593-598.

[28] Patle, S. & Lal, B. 2008. Investigation of the potential of agro-industrial material as low cost substrate for ethanol production by using *Candida tropicalis* and *Zymomonas mobilis*. *Biomass and Bioenergy*, *32,* 596-602.

[29] Cazetta, M. L., Celligoi, M. A. P. C., Buzato, J. B. & Scarmino, I. S. 2007. Fermentation of molasses by *Zymomonas mobilis*: Effects of temperature and sugar concentration on ethanol production. *Bioresource Technology*, *98,* 2824-2828.

[30] Yamashita, Y., Kurosumi, A., Sasaki, C. & Nakamura, Y. 2008. Ethanol production from paper sludge by immobilized *Zymomonas mobilis*. Biochemical Engineering Journal, 42, 314-319.

[31] Da Silveira dos Santos, D., Camelo, A. C., Rodrigues, K. C. P., Carlos, L. C. & Pereira, J. 2010. Ethanol production from sugarcane bagasse by *Zymomonas mobilis* using simultaneous saccharification and fermentation (SSF). *Applied Biochemistry and Biotechnology*, *161*, 1-8.

[32] Ranatunga, T. D., Jervis, J., Helm, R. F., McMillan, J. D. & Wooley, R. J. 2000. The effect of overliming on the toxicity of dilute acid pretreated lignocellulosics: The role of inorganics, uronic acids and ether-soluble organics. *Enzyme and Microbial Technology*, *27,* 240-247.

[33] Park, I., Kim, I., Kang, K., Sohn, H., Rhee, I., Jin, I. & Jang, H. 2010. Cellulose ethanol production from waste newsprint by simultaneous saccharification and fermentation using *Saccharomyces cerevisiae* KNU5377. *Process Biochemistry*, *38*, 2099-2110.

[34] Spindler, D. D., Wyman, C. E. & Grohmann, K. 1988. Evaluation of thermotolerant yeasts in controlled simultaneous saccharification and fermentation of cellulose to ethanol. *Biotechnology and Bioengineering*, *34,* 189-195.

[35] Araque, E., Parra, C., Rodriguez, M., Freer, J. & Baeza, J. 2008. Selection of thermotolerant yeast strains *Saccharomyces cerevisiae* for bioethanol production. *Enzyme and Microbial Technology*, *43,* 120-

[36] Limtong, S., Sringiew, C. & Yongmanitchai, W. 2007. Production of fuel ethanol at high temperature from sugar cane juice by a newly isolated *Kluyveromyces marxianus*. *Bioresource Technology*, *98,* 3367-3374.

[37] Wilkins, M. R., Mueller, M., Eichling, S. & Banat, I. M. 2008. Fermentation of xylose by the thermotolerant yeast strains *Kluyveromyces marxianus* IMB2, IMB4, and IMB5 under anaerobic conditions. *Process Biochemistry*, *43,* 346-350.

[38] Faga, B. A., Wilkins, M. R. & Banat, I. M. 2010. Ethanol production through simultaneous saccharification and fermentation of switchgrass using *Saccharomyces cerevisiae* D5A and thermotolerant *Kluyveromyces marxianus* IMB strains. *Bioresource Technology*, *101,* 2273-2279.

[39] Boyle, M., Barron, N. & McHale, A. P. 1997. Simultaneous saccharification and fermentation of straw to ethanol using the thermotolerant yeast strain *Kluyveromyces marxianus* IMB3. *Biotechnology Letters*, *19,* 49-51.

[40] Suryawati, L., Wilkins, M. R., Bellmer, D. D., Huhnke, R. L., Maness, N. O. & Banat, I. M. 2008. Simultaneous saccharification and fermentation of Kanlow switchgrass pretreated by hydrothermolysis using *Kluyveromyces marxianus* IMB4. *Biotechnology and Bioengineering*, *101,* 894-902.

[41] Wilkins, M. R., Suryawati, L., Maness, N. O. & Chrz, D. 2007. Ethanol production *Saccharomyces cerevisiae* and *Kluyveromyces marxianus* in the presence of orange-peel oil. *World Journal of Microbiology and Biotechnology*, *23,* 1161-1168.

[42] Lark, N., Xia, Y., Qin, C., Gong, C. S. & Tsao, G. T. 1997. Production of ethanol from recycled paper sludge using cellulase and yeast, *Kluyveromyces marxianus*. *Biomass and Bioenergy*, *12,* 135-143.

[43] Kádár, Z., Szengyel, Z. & Réczey, K. 2004. Simultaneous saccharification and fermentation (SSF) of industrial wastes for the production of ethanol. *Industrial Crops and Products*, *20,* 103-110.

[44] Tomás-Pejó, E., Oliva J. M., González, A., Ballesteros, I., & Ballesteros M. 2009. Bioethanol production from wheat straw by thermotolerant yeast *Kluyveromyces marxianus* CECT 10875 in a simultaneous saccharification and fermentation fed-batch process. *Fuel*, *88,* 2142-2147.

[45] Ballesteros, M., Oliva, J. M., Manzanares, P., Negro, M. J. & Ballesteros, I. 2002. Ethanol production from paper material using a simultaneous saccharification and fermentation system in a fed-batch basis. *World Journal of Microbiology and Biotechnology*, *18,* 559-561.

[46] Singh, R. S. & Bhermi, H.K. 2008. Production of extracellular exoinulinase from *Kluyveromyces marxianus* YS-1 using root tubers of *Asparagus officinalis*. *Bioresource Technology*, *99,* 7418-7423.

[47] Schultz, N., Chang, L., Hauck, A., Reuss, M. & Syldatk, C. 2006. Microbial production of single-cell protein from deproteinized whey concentrates. *Applied Microbiology and Biotechnology*, *69,* 515-520.

[48] Gao, F. & Daugulis, A. J. 2009. Bioproduction of the aroma compound 2-phenylethanol in a solid-liquid two-phase partitioning bioreactor system by *Kluyveromyces marxianus*. *Biotechnology and Bioengineering*, *104,* 332-339.

[49] Fonseca, G. G., Heinzle, E., Wittmann, C. & Gombert, A. K. 2008. The yeast *Kluyveromyces marxianus* and its biotechnological potential. *Applied Microbiology and Biotechnology*, *79,* 339-354.

[50] Stephanopoulos, G. 2007. Challenges in engineering microbes for biofuels production. *Science*, *315,* 801-804.

[51] Kumar, R. & Wyman, C. E. 2009. Effect of xylanase supplementation of cellulase on digestion of corn stover solids prepared by leading pretreatment technologies. *Bioresource Technology*, *100,* 4203-4213.

[52] Karhumaa, K., Wiedemann, B., Hahn-Hägerdal, B., Boles, E. & Gorwa-Grausland, M. F. 2006. Co-utilization of L-arabinose and D-xylose by laboratory and industrial *Saccharomyces cerevisiae* strains. *Microbial Cell Factories*, *5,* 18.

[53] Bettiga, M., Bengtsson, O., Hahn-Hägerdal, B. & Gorwa-Grauslund, M. F. 2009. Arabinose and xylose fermentation by recombinant *Saccharomyces cerevisiae* expressing a fungal pentose utilization pathway. *Microbial Cell Factories*, *8,* 40.

[54] Toivari, M. H., Salusjarvi, L., Ruohonen, L. & Penttila, M. 2004. Endogenous xylose pathway in *Saccharomyces cerevisiae*. *Applied and Environmental Microbiology*, *70,* 3681-3686.

[55] Van Maris, A. J. A., Winkler, A. A., Kuyper, M., De Laat, W. T. A. M., Van Dijken, J. P. & Pronk, J. T. Development of efficient xylose fermentation in *Saccharomyces cerevisiae*: Xylose isomerase as a key component. *Advances in Biochemical Engineering Biotechnology*, *108,* 179-204. 2007.

[56] Sedlak, M. & Ho, N. W. Y. 2004. Production of ethanol from cellulosic biomass hydrolysates using *Saccharomyces* yeast capable of cofermenting glucose and xylose. *Applied Biochemistry and Biotechnology*, *114,* 403-416.

[57] Wahlbom, C. F. & Hahn-Hägerdal, B. 2002. Furfural, 5-hydroxymethyl furfural, and acetoin act as external electron acceptors during anaerobic fermentation of xylose in recombinant *Saccharomyces cerevisiae*. *Biotechnology and Bioengineering*, *78,* 172-178.

[58] Gupta, R., Sharma, K. K. & Kuhad, R. C. 2009. Separate hydrolysis and fermentation (SHF) of *Prosopsis juriflora*, a woody substrate, for production of cellulosic ethanol by *Saccharomyces cerevisiae* and *Pichia stipitis*- NCIM 3498. *Bioresource Technology*, *100,* 1214-1220.

[59] Telli-Okur, M. & Eken-Saracoglu, N. 2008. Fermentation of sunflower seed hull hydrolysate to ethanol by *Pichia stipitis*. *Bioresource Technology*, *99,* 2162-2169.

[60] Jeffries, T. W. & Shi, N. Q. 1999. Genetic Engineering for improved xylose fermentation by yeasts. *Advances in Biochemical Engineering*, *65,* 118-161.

[61] Hahn-Hägerdal, B., Galbe, M., Gorwa-Grauslund, M. F., Lidén, G. & Zacchi, G. 2006. Bio-ethanol - the fuel of tomorrow from the residues of today. *Trends in Biotechnology*, *24,* 549-556.

[62] Eliasson, A., Christensson, C., Wahlbom, F. & Hahn-Hägerdal, B. 2000. Anaerobic xylose fermentation by recombinant *Saccharomyces cerevisiae* carrying *XYL1, XYL2* and *XKS1* in mineral medium chemostat cultures. *Applied and Environmental Microbiology*, *66,* 3381-3386.

[63] Karhumaa, K., Sánchez, R. G., Hahn-Hägerdal, B., & Gorwa-Grauslund, M. F. 2007. Comparison of the xylose reductase-xylitol dehydrogenase and the xylose isomerase pathways for xylose fermentation by recombinant *Saccharomyces cerevisiae*. *Microbial Cell Factories*, *6,* 5.

[64] Petschacher, B. & Nidetzky, B. 2008. Altering the coenzyme preference of xylose reductase to favor utilization of NADH enhances ethanol yield from xylose in a

metabolically engineered strain of *Saccharomyces cerevisiae. Microbial Cell Factories, 7,* 9.

[65] Walfridsson, M., Bao, X., Anderlund, M., Lilius, G., Bülow, L. & Hahn-Hägerdal, B. 1996. Ethanolic fermentation of xylose with *Saccharomyces cerevisiae* harboring the *Thermus thermophilus xylA* gene, which expresses an active xylose (glucose) isomerase. *Applied and Environmental Microbiology*, *62,* 4648-4651.

[66] Brat, D., Boles, E. & Wiedemann, B. 2009. Functional expression of a bacterial xylose isomerase in *Saccharomyces cerevisiae. Applied and Environmental Microbiology, 75,* 2304-2311.

[67] Martin, C., Galbe, M., Wahlbom, C. F., Hahn-Hägerdal, B. & Jonsson, L. J. 2002. Ethanol production from enzymatic hydrolysates of sugarcane bagasse using recombinant xylose-utilizing *Saccharomyces cerevisiae. Enzyme and Microbial Technology*, *31,* 274-282.

[68] Sonderegger, M., Jeppsson, H., Larsson, C., Gorwa-Grauslund, M. F., Boles, E., Olsson, L., Spencer-Martins, I., Hahn-Hägerdal, B. & Sauer, V. 2004. Fermentation performance of engineered and evolve xylose-fermenting *Saccharomyces cerevisiae* strains. *Biotechnology and Bioengineering*, *87,* 90-98.

[69] Johansson, B., Christensson, C., Hobley, T. & Hahn, H. 2001. Xylulokinase overexpression in two strains of *Saccharomyces cerevisiae* also expressing xylose reductase and xylitol dehydrogenase and its effect on fermentation of xylose and lignocellulosic hydrolysate. *Applied and Environmental Microbiology*, *67,* 4249-4255.

[70] Saloheimo, A., Rauta, J., Stasyk, O. V., Sibirny, A. A., Penttila, M. & Ruohonen, L. 2007. Xylose transport studies with xylose-utilizing *Saccharomyces cerevisiae* strains expressing heterologous and homologous permeases. *Applied Microbiology and Biotechnology*, *74,* 1041-1052.

[71] Katahira, S., Ito, M., Takema, H., Fujita, Y., Tanino, T., Tanaka, T., Fukuda, H. & Kondo, A. 2008. Improvement of ethanol productivity during xylose and glucose co-fermentation by xylose-assimilating *S. cerevisiae* via expression of glucose transporter *Sut1*. *Enzyme and Microbial Technology*, *43,* 115-119.

[72] Runquist, D., Fonseca, C., Radstrom, P., Spencer-Martins, I. & Hahn-Hägerdal, B. 2009. Expression of the *Gxf1* transporter from *Candida intermedia* improves fermentation performance in recombinant xylose-utilizing *Saccharomyces cerevisiae. Applied Microbiology and Biotechnology*, *82,* 123-130.

[73] Roca, C., Haack, M. B. & Olsson, L. 2004. Engineering of carbon catabolite repression in recombinant xylose fermenting *Saccharomyces cerevisiae. Applied Microbiology and Biotechnology*, *63,* 578-583.

[74] Shi, N. Q. & Jeffries, T. W. 1998. Anaerobic growth and improved fermentation of *Pichia stipitis* bearing a URA1 gene from *Saccharomyces cerevisiae. Applied Microbiology and Biotechnology*, *50,* 339-345.

[75] Shi, N. Q., Davis, B., Sherman, F., Cruz, J. & Jeffries, T. W. 1999. Disruption of the cytochrome *c* gene in xylose-utilizing yeast *Pichia stipitis* leads to higher ethanol production. *Yeast*, *15,* 1021-1030.

[76] Dien, B. S., Cotta, M. A. & Jeffries, T. W. 2003. Bacteria engineered for fuel ethanol production: Current status. *Applied Microbiology and Biotechnology*, *63,* 258-266.

[77] Ingram, L. O., Gomez, P. F., Moniruzzaman, M., Wood, B. E., Yomano, L. P. & York, S. W. 1998. Metabolic engineering of bacteria for ethanol production. *Biotechnology and Bioengineering, 58,* 205-214.

[78] Wang, Z., Chen, M., Xu, Y., Li, S., Lu, W., Ping, S., Zhang, W. & Lin, M. 2008. An ethanol-tolerant recombinant *Escherichia coli* expressing *Zymomonas mobilis pdc* and *adhB* genes for enhanced ethanol production from xylose. *Biotechnology Letters*, *30,* 657-663.

[79] Doran-Peterson, J., Cook, D. M. & Brandon, S. K. 2008. Microbial conversion of sugars from plant biomass to lactic acid or ethanol. *Plant Journal*, *54,* 582-592.

[80] Tao, H., Gonzalez, R., Martinez, A., Rodriguez, M., Ingram, L. O., Preston, J. F. & Shanmugam, K. T. 2001. Engineering a homo-ethanol pathway in *Escherichia coli*: Increased glycolytic flux and levels of expression of glycolytic genes during xylose fermentation. *Journal of Bacteriology*, *183,* 2979-2988.

[81] Kim, T. H., Taylor, F. & Hicks, K. B. 2008. Bioethanol production from barley hull using SAA (soaking in aqueous ammonia) pretreatment. *Bioresource Technology*, *99,* 5694-5702.

[82] Mishima, D., Kuniki, M., Sei, K., Soda, S., Ike, M. & Fujita, M. 2008. Ethanol production from candidate energy crops: Water hyacinth (*Eichhornia crassipes*) and water lettuce (*Pistia stratiotes L.*). *Bioresource Technology*, *99,* 2495-2500.

[83] Grohmann, K., Baldwin, E. A., Buslig, B. S. & Ingram, L. O. 1994. Fermentation of galacturonic acid and other sugars in orange peel hydrolysates by the ethanologenic strain of *Escherichia coli*. *Biotechnology Letters*, *16,* 281-286.

[84] Dien, B. S., Hespell, R. B., Ingram, L. O. & Bothast, R. J. 1997. Conversion of corn milling fibrous co-products into ethanol by recombinant *Escherichia coli* strains K011 and SL40. *World Journal of Microbiology and Biotechnology*, *13,* 619-625.

[85] Okuda, N., Ninomiya, K., Takao, M., Katakura, Y. & Shioya, S. 2007. Microaeration enhances productivity of bioethanol from hydrolysate of waste house wood using ethanologenic *Escherichia coli* KO11. *Journal of Bioscience and Bioengineering*, *103,* 350-357.

[86] Brandon, S. K., Eiteman, M. A., Patel, K., Richbourg, M. M., Miller, D. J., Anderson, W. F. & Peterson, J. D. 2008. Hydrolysis of Tifton 85 bermudagrass in a pressurized batch hot water reactor. *Journal of Chemical Technology and Biotechnology*, *83,* 505-512.

[87] Zhang, M., Eddy, C., Deanda, K., Finkelstein, M. & Picataggio, S. 1995. Metabolic engineering of a pentose metabolisms pathway in ethanologenic *Zymomonas mobilis*. *Science*, *267,* 240-243.

[88] Mohagheghi, A., Evans, K., Chou, Y. C. & Zhang, M. 2002. Co-fermentation of glucose, xylose, and arabinose by genomic dna-integrated xylose/arabinose fermenting strain of *Zymomonas mobilis* AX101. *Applied Biochemistry and Biotechnology - Part A Enzyme Engineering and Biotechnology*, *98-100,* 885-898.

[89] Golias, H., Dumsday, G. J., Stanley, G. A. & Pamment, N. B. 2002. Evaluation of a recombinant *Klebsiella oxytoca* strain for ethanol production from cellulose by simultaneous saccharification and fermentation: comparison with native cellobiose-utilising yeast strains and performance in co-culture with thermotolerant yeast and *Zymomonas mobilis*. *Journal of Biotechnology*, *96,* 155-168.

[90] Moniruzzaman, M., Dien, B. S., Ferrer, B., Hespell, R. B., Dale, B. E., Ingram, L. O. & Bothast, R. J. 1996. Ethanol production from AFEX pretreated corn fiber by recombinant bacteria. *Biotechnology Letters*, *18,* 985-990.

[91] Doran, J. B. & Foster, B. 2000. Ethanol production from sugar beet pulp using engineered bacteria. *International sugar journal*, *102,* 336-340.

[92] Galbe, M. & Zacchi, G. 2002. A review of the production of ethanol from softwood. *Applied Microbiology and Biotechnology*, *59,* 609-617.

[93] Lynd, L. R., Van Zyl, W. H., Mcbride, J. E. & Laser, M. 2005. Consolidated bioprocessing of cellulosic biomass: An update. *Current Opinion in Biotechnology*, *16,* 577-583.

[94] Joe Shaw, A., Jenney, J., Adams, M. W. W. & Lynd, L. R. 2008. End-product pathways in the xylose fermenting bacterium *Thermoanaerobacterium saccharolyticum*. *Enzyme and Microbial Technology*, *42,* 453-458.

[95] Zhou, S., Davis, F. C. & Ingram, L. O. 2001. Gene integration and expression and extracellular secretion of *Erwinia chrysanthemi* endoglucanase CelY (*celY*) and CelZ (*celZ*) in ethanologenic *Klebsiella oxytoca* P2. *Applied and Environmental Microbiology*, *67,* 6-14.

[96] Zhou, S. & Ingram, L. O. 2001. Simultaneous saccharification and fermentation of amorphous cellulose to ethanol by recombinant *Klebsiella oxytoca* SZ21 without supplemental cellulase. *Biotechnology Letters*, *23,* 1455-1462.

[97] Kotaka, A., Bando, H., Kaya, M., Kato-Murai, M., Kuroda, K., Sahara, H., Hata, Y., Kondo, A. & Ueda, M. 2008. Direct ethanol production from barley β-glucan by sake yeast displaying *Aspergillus oryzae* β-glucosidase and endoglucanase. *Journal of Bioscience and Bioengineering*, *105,* 622-627.

[98] Ueda, M. & Tanaka, A. 2000. Cell surface engineering of yeast: Construction of arming yeast with biocatalyst. *Journal of Bioscience and Bioengineering*, *90,* 125-136.

[99] Den Haan, R., Mcbride, J. E., Grange, D. C. L., Lynd, L. R. & Van Zyl, W. H. 2007. Functional expression of cellobiohydrolases in *Saccharomyces cerevisiae* towards one-step conversion of cellulose to ethanol. *Enzyme and Microbial Technology*, *40,* 1291-1299.

[100] Du Plessis, L., Rose, S. H. & Van Zyl, W. H. 2010. Exploring improved endoglucanase expression in *Saccharomyces cerevisiae* strains. *Applied Microbiology and Biotechnology*, 86,1503-1511.

[101] Van Rensburg, P., Van Zyl, W. H. & Pretorius, I. S. 1996. Co-expression of a *Phanerochaete chrysosporium* cellobiohydrolase gene and a *Butyrivibrio fibrisolvens* endo-β-1,4-glucanase gene in *Saccharomyces cerevisiae*. *Current Genetics*, *30,* 246-250.

[102] Hyung, L. J., Lim, M. Y., Kim, M. J., Heo, S. Y., Seo, J. H., Kim, Y. H. & Nam, S. W. 2007. Constitutive coexpression of *Bacillus* endoxylanase and *Trichoderma* endoglucanase genes in *Saccharomyces cerevisiae*. *Journal of Microbiology and Biotechnology*, *17,* 2076-2080.

In: Bioprocess Sciences and Technology
Editor: Min-Tze Liong
ISBN 978-1-61122-950-9

Chapter 9

SINGLE CELL PROTEIN FROM YEAST

P. Ravindra*, Z. H. Yim, P. P. Lee, S. H. Phan and E. S. Chan
School of Engineering and Information Technology,
Universiti Malaysia Sabah, Kota Kinabalu

ABSTRACT

This chapter discusses on the science and technology in the production of SCP from yeast. The contents begin with a brief introduction on the yeast and fundamental knowledge that[1] related to physiology, i.e. reproduction, growth of yeast and etc. were included. Comparison among different SCP microorganism in terms of various aspects was included in the following sub-chapter. The resources for production of yeast-based SCP that include those conventional and non-conventional as well as renewable and non-renewable sources will be covered. Issues related to the toxicity and safety of yeast-based SCP which related to nucleic acid, digestibility, toxic and possible carcinogenic compounds are also highlighted. Next, discussion on the conventional processes for production of yeast based SCP, optimal parameters for SCP production, the growth kinetics of yeast as well as process design and control were also discussed.

1. INTRODUCTION

Protein of microbial origin is named as single-cell protein (SCP). SCP can be derived from a variety of microorganisms including bacteria, yeast, fungi and microscopic algae. Humans have been using microorganism as biomass in production of various food products such as alcoholic drinks, cheese, yogurt, soybean paste and etc. Since centuries ago, people from Africa and Mexico have been consuming filamentous blue green alga, Spirulina as their staple food.

Production of SCP from yeast at industrial scale was firstly started since World War I when 'Torula yeast' (*Candida utilises*) was produced in Germany and used in soups and

* Corresponding author: *Pogaku Ravindra* E-mail: r:*dr_ravindra@hotmail.com*

sausages. Popularity of SCP increased tremendously again during World War II as well as during mid 1950s. By 1967, British Petroleum was producing SCP at an industrial scale. The interest in SCP was mainly generated in the wake of protein deficiency especially in the developing countries. It was reported that the deficit of protein in these countries was at 12 million tons in 1985 and rose to 22 million tons by 2000. SCP is expected to relieve this deficiency since it may be used directly as human food supplement and it may be used as animal feed to reduce the demand pressure on the currently used feeds from protein rich soybean and fish proteins and even cereals, which can be diverted for human consumption.

To date, there are still many works focus on the food and feed production by using microorganisms. Value of SCP is determined by the parameters related to nutritional content including vitamins, nitrogen, carbohydrates, fats, cell wall, nucleic acids, and profile of protein and amino acids.

It is noteworthy that SCP is not pure protein but rather, dehydrated cells consisting of mixtures of proteins, lipids, carbohydrates, nucleic acids and a variety of other non-protein nitrogenous compounds, vitamins, and inorganic compounds. Microbial protein is a nontraditional protein; it is not a palatable, desirable food and must be incorporated directly or indirectly into other foods. In recent years SCP has being produced from many species of microorganisms. Particularly, yeasts are widely used in practical. Uses of yeasts for SCP production are economical since the microbe is able to grow in a wide range of medium including those easily available wastes.

2. Fundamental of Yeast

Yeasts are unicellular fungi, a large, heterogeneous group of eukaryotic microorganisms. However, some species with yeast forms may become multicellular through the formation of a string of connected budding cells known as pseudohyphae or false hyphae as seen in most molds [1]. Yeasts do not form an exact taxonomic or phylogenetic grouping. Currently, about 1,500 species (only 1% of all yeasts) were already described and they dominate fungal diversity in the oceans [1, 2, 3]. Often, 'yeast' is perceived as a synonym for *Saccharomyces cerevisiae*. However, the phylogenetic diversity of yeasts is shown by their placement in both divisions Ascomycota and Basidiomycota [4]. Meanwhile, depending on the species the sizes of yeast cells are ranging from typically 3 to more than 40 μm [5].

Yeast reproduces almost as rapidly as bacteria. Yeasts have asexual and sexual reproductive cycles and they reproduce asexually by budding except a few other yeasts reproduce by binary fission (Figure 1). However, the most common mode of vegetative growth in yeast is asexual reproduction by budding [6]. At this stage, a small bud, or daughter cell, is formed on the parent cell. The budding yeasts are classified in the order Saccharomycetales [7]. The nucleus of the parent cell splits into a daughter nucleus and migrates into the daughter cell. The bud continues to grow until it separates from the parent cell and forms a new cell [8]. Meanwhile, yeast, such as *Schizosaccharomyces pombe*, reproduces by binary fission instead of budding [6].

S. cerevisiae and *S. pombe* have similar life cycles. Both yeasts can proliferate in either a diploid or a haploid state: the diploid cells, as an alternative to dividing in the ordinary way, can go through meiosis to form haploid cells; the haploid cells, as an alternative to dividing, can mate (conjugate) with one another to form diploid cells (Figure 1). The haploid phase makes it easy to isolate and study on mutations that inactivate a gene without the complication of having a second gene copy in the cell. In both species of yeasts, food and sex play important parts in controlling the cell-division cycle, so that these organisms can be used to investigate the general question of how the division cycle is regulated by environmental factors and by cell-cell interactions.

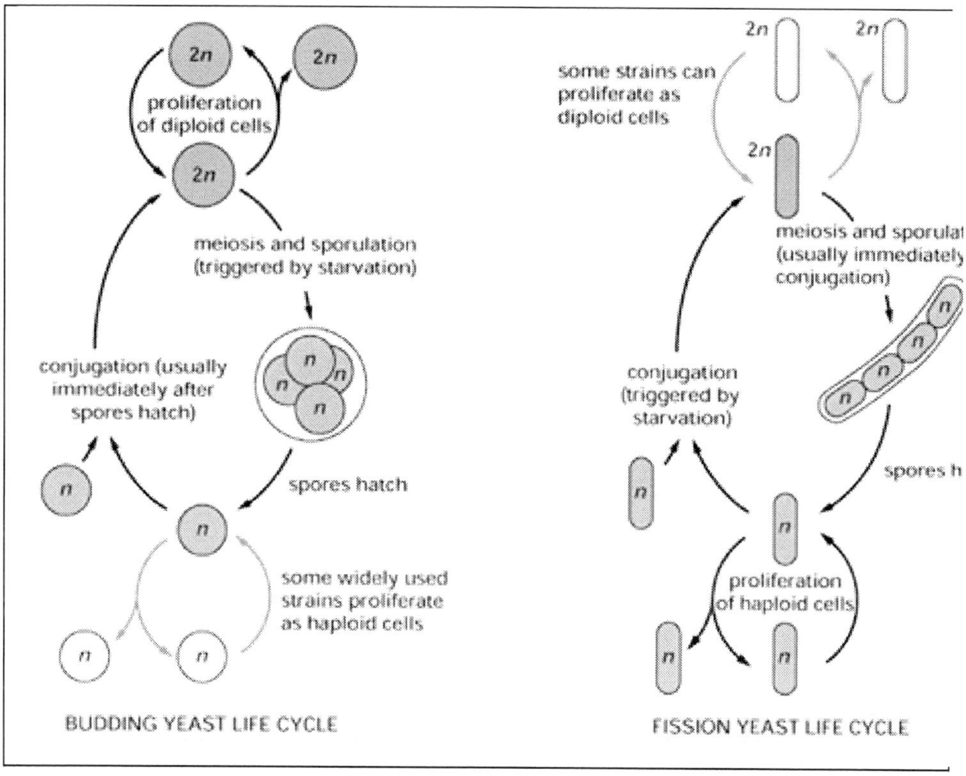

Figure 1. Reproduction processes of yeasts, (a) *Saccharomyces cerevisiae* and (b) *Schizosaccharomyces pombe* [9, 10].

The proportion of the life cycle spent in the diploid or haploid state varies with the species and according to the environment. When nutrients are plentiful, normal wild-type varieties of budding yeast proliferate as diploid cells, with a cell-cycle time of about 2 hours [11]. However, under high stress conditions (such as starvation, heating, low pH and etc.) haploid cells will generally die. Meanwhile, diploid cells can undergo sporulation, entering sexual reproduction (meiosis) and producing a variety of haploid spores, which can go on to mate (conjugate), reforming the diploid [12], depending on the environment and other factors. On the other hand, fission yeasts typically proliferate as haploid cells, which fuse in response to starvation to form diploid cells that promptly go through meiosis and sporulation to regenerate haploid cells. The most widely used laboratory strains of budding yeasts are mutants that can proliferate, like fission yeasts, as haploids [12].

Yeasts are chemoorganotrophs as they use organic compounds as a source of energy and do not require sunlight to grow. Carbon is obtained mostly from hexose sugars such as glucose and fructose, or disaccharides such as sucrose and maltose. Some species can metabolize pentose sugars like ribose, alcohols, and organic acids [13]. Yeast species either require oxygen for aerobic cellular respiration (obligate aerobes), or are anaerobic but also have aerobic methods of energy production (facultative anaerobes). Unlike bacteria, there are no known yeast species that grow only anaerobically (obligate anaerobes). Yeasts grow best in a neutral or slightly acidic pH environment.

Yeasts vary in what temperature range they grow best. For example, *Leucosporidium frigidum* grows at −2 to 20 °C, *Saccharomyces telluris* at 5 to 35 °C and *Candida slooffi* at 28 to 45 °C [14]. The cells can survive freezing under certain conditions, with viability decreasing over time.

In terms of application, *S. cerevisiae* is one of the yeast species that has been widely used in baking and fermenting alcoholic beverages [15]. On the other hand, other species of yeast, such as *Candida albicans*, is opportunistic pathogen and can cause infections in humans. Recently, yeasts have been used to generate electricity in microbial fuel cells, and produce ethanol for the biofuel industry [16].

3. Yeast as SCP

Members of *Saccharomyces*, *Candida* and *Torulopsis* have been widely produced as SCP. Particularly, the first two genera are used for some commercial processes using various substrates. The SCP produced from these genera has 55-60% crude protein which has very rich vitamin B and has good amino acid balance except for a deficiency in S-containing amino acids.

The SCP is used both for human food and animal feed supplementation. The risk of bacterial contamination is low and recovery by continuous centrifugation is easy.

The difficulties in use of yeasts as SCP are:

(i) Marginally slower growth rates than fastest growing bacteria,
(ii) High nucleic acid content (up to 15%), which needs to be reduced, and
(iii) Methionine supplementation may be necessary to overcome S-containing amino acid deficiency of its proteins.

Table 1 shows the doubling time for living beings. Bacteria and yeasts have the shortest doubling time among living organism. Thus, the superior characteristic provides the microorganisms exceptional advantages to be an alternative protein source for human and animal.

The comparison of features among the SCP microorganisms is shown in Table 2. Overall, bacterial and yeast based SCP are more preferable at the industrial scale of production. For example, the microorganisms are able to grow faster than others in wide range of substrates. Bacterial SCP is marginally superior to yeast in terms of its protein content and quality, but yeast based SCP still preferable at larger scale production due to its easy handling (biomass recovery), higher vitamin B content and relatively less toxin produced at the end product.

Table 1. The doubling period of mass for some living beings [15]

Organisms	Time for doubling mass
Bacteria and yeasts	20-120 minutes
Filamentous fungi and algae	2-6 hours
Grass and some plants	1-2 weeks
Chicken	2-4 weeks
Pig	4-6 weeks
Cattle	1-2 months
Human being	3-6 months

Table 2. Features and the SCP produced from different microorganisms

Feature	Algae	Bacteria	Yeast	Filamentous fungi
Growth Rate	Low	Highest	Quite high	Lower than bacteria and yeast
Substrate	Light, inorganic carbon sources, e.g., CO_2 (preferably)	A wide range of substrates	Most substrates, except, hydro carbons and CO_2	Limited substrates (mostly starchy and cellulosic materials)
pH range	Up to 2	5 to 7	5 to 7	3 to 8
Cultivated in	Open ponds, tanks; in sunlight	Bioreactors	Bioreactors	Bioreactors
Risk of contamination	Serious	High; precautions necessary	-	Low if grown below pH 5
Protein content (crude)	Up to 60%	80% or more	55 - 60%	50 - 55%
Amino acid profile	Generally good; low in S - containing amino acids	Generally good; a small deficit in S - containing acids	Generally good; deficit in S - containing amino acids	Low in S-containing amino acids
Nucleic acid content	-	Very high (20% RNA)	High (15% RNA)	High (15% RNA)
Removal of nucleic acids	-	Necessary	Necessary	Necessary
Toxins	-	Geam bacteria may produce endotoxins	-	Many species produce mycotoxins
Other features	1. Low yield (1-2 g dry wt/1)	-	High B vitamin content	Chitin may contain a significant proportion of N content, which is unavailable
	2. High chlorophyll content; unsuitable for humans			

Reference: [17].

4. Resources for Production of Yeast-based SCP

Many fungal species are used as protein-rich food. Most popular among them are the yeast species *Candida, Hansenula, Pitchia, Torulopsis* and *Saccharomyces*. Many other filamentous fungi also serve as sources of SCP (Table 3). A recent trend in SCP production is the exploitation of fungal species for bioconversion of lignocellulosic wastes [18].

The design and strategy of SCP processes is dependent on the choice of substrates that are normally abundant and in proximity to the production plant. The most widely used substrates for SCP production are the carbon and energy source that is derived from carbohydrates because the mono and disaccharides are natural microbial substrates, and that carbohydrates are a renewable and abundant resources.

Table 3. Yeasts used for SCP production

Microorganism	**Substrate**	**Used as**	**Used commercially**
C. utilis	1. Confectionery	-	Yes (U.K.); Symba process
	2. Ethanol	Feed	Yes (USA)
	3. Sulphite liquor	-	Yes (Europe, USA, Russia)
C. intermedia	Whey	-	Yes; Vienna process
C. krusei (+ *Lactobacillus bulgarius*)	Whey	-	Yes; Kiel process
C. krusei SO1& *Saccharomyces* species LK3G	Sorghum hydrolysate [19]	-	-
C. lipolytica	n-alkanes (C10 - C23) + ammonia	-	Yes (Russia)
C. tropicalis ceppo 571	Sulfite waste liquor [20]	-	-
Kluyveromyces fragilis	Whey	Food	Yes (France); Le Bel process
Marine yeast	Prawn shell wastes [21]	-	-
Mixed cultures of yeasts	Dairy wastes [22]	-	-
Pichia pastoris	Methanol [23]		
Saccharomyces cerevisiae	Molasses	Food	Yes
Schwanniomyces occidentalis	Starch [24]		
Yeast	Plant origin liquid waste [25]		

Reference: [26].

Molasses is a by-product of the sugar manufacturing process. There are 3.5 to 4.5 kg of molasses may be obtained from every 100 kg of plant [27]. Molasses also contains minerals, organic compounds and vitamins that are valuable nutrients in fermentation processes. These compounds are including protein, potassium, calcium, magnesium, phosphorus, biotin, pantothenic acid, inositol and thiamine. In fact, about 9% of the dry matter in yeast grown on molasses has been estimated to originate from substances other than sucrose [28]. Nevertheless, biomass production from molasses requires supplementation with a suitable nitrogen source, as well as phosphorus. The traditional nitrogen sources used are ammonia or ammonium salts, and phosphorus can be added in the form of salts.

A cheaper and abundantly available SCP substrate of carbohydrate origin is starch. This carbohydrate may be obtained from bulb plants of tropical and temperate regions, or from rice, maize and cereals. In tropical countries, cassava has been proposed as a good source of starch for SCP processes [29].

The Symba process developed in Sweden utilized starchy wastes combining two yeasts in sequential mixed culture: The amylase producing *Endomycopsis fibuligira*, and the fast growing *C. utilis* [27, 30]. The process consists of three phases: The incoming starch waste from potato tubers is fed through heat exchangers and sterilised. The medium is then fed to a first bioreactor where the starch hydrolysing yeast grows and hydrolyses starch. The hydrolysed solution is then fed to a second reactor where culture conditions favour the proliferation of *C. utilis*.

The Quorn myco-protein production process is currently supported on glucose, nearly all of which is obtained from maize, but it has been reported earlier to use wheat-starch, a by-product of the production of wheat gluten (the protein fraction) and wheat flour [31, 32]. This means that the process may be applied with various sources of starch as the carbon source [33, 34].

Whey is a by-product from the removal of protein and fat from milk. Whey can be obtained from the curding process in cheese production and after ultrafiltration procedures for the production of spreading cheeses. At these processes, the protein fractions corresponding to lactalbumins and lactoglobulins are incorporated to the casein fraction and all the proteins are in native form. About 9 kg of whey may be obtained for the production of every kg of cheese [35]. The major component in whey is lactose (4-6%, w/v) and other nutrients in trace amount such as protein, mineral salts and lactic acid [36].

Whey was found to be an exceptionally suitable substrate for the production of SCP. In production of yeast from whey by using lactose assimilating *Kluyveromyces marxianus* (formerly *K. fragilis*) was started in 1956 by a French dairy company, Fromageries Bel. In 1983, the company was processing 8000 tons of yeast in continuous culture [37]. Although similar operations were later installed in other western European countries and the US, the Bel process was the largest and longest running operation using whey substrate in the world. The yeast was mostly designed for animal feeding.

Alkanes were used as a promising substrate for SCP production, especially in the former Soviet Union, where the structural deficit in feed protein was compensated by the surplus of oil. Yeasts are able to assimilate n-alkanes and 1-alkenes in liquid culture [38].

There are some technical challenges in the SCP production from hydrocarbons due to the low water solubility of the substrate and the high degree of aeration required for its metabolization (the substrate is essentially oxygen free and therefore presents a high oxidation potential). As a result, the cost of aerating the culture is increased. In addition, the oxidation processes are highly exothermic and cooling costs are correspondingly higher [39].

The toxicity of the substrate has raised suspicions on the safety of continuous feeding with SCP containing trace amounts of alkanes. Therefore, the SCP production plants using alkanes are no longer in operation.

Methanol is a residual liquid of the petrochemical industry and has been used as a substrate for a number of SCP production systems. Methanol tolerance (up to 6 g/L) and assimilation is a specialised prerequisite which may be found in yeast species such as *Hansenula, Pichia, Candida* and *Torulopsis*. Methanol has more advantages over other petrochemical by-products, but its volatility nature results in loss during drying process. A number of industrial examples of SCP production systems using methylotrophic yeasts were implemented in the 70's [40].

Methylotrophic yeasts have enormous capacity to synthesize proteins which participate in methanol assimilation. This capability has been exploited for the heterologous expression of proteins at the pilot and industrial scales [41]. However, the production of heterologous proteins is a field with a separate orientation to SCP production and is not covered in detail in this chapter.

Cellulose can be considered as the most abundant renewable resource in the world. SCP production from these substrates requires their breakdown into assimilable forms in order to be rapidly taken up in solution by the growing organism. Therefore, the substrate must be subjected to pretreatments in a number of steps which include milling and chemical or enzymatic hydrolysis. The cost of these steps has prevented the generalised production of SCP from cellulose [42]. The production of industrial cellulose for paper and tissue production includes a number of steps designed to remove the lignin and hemicelluloses from wood where they act as cell wall constituents and as cementing agents joining fibers to larger aggregates. Wood is cooked in a medium containing calcium sulfite with excess free sulfur dioxide. Lignin is thus converted to lignosulfonates and hemicellulose is hydrolysed to monosaccharides. These in turn, may be further broken down to furfurols. The amount of free sugars in the spent liquor is variable with the type of procedure chosen since various cellulose fibers may be obtained with different degrees of degradation, as well as the differences in the type of wood used. However, up to half of the matter contained in wood may be extracted out of the fibers, and converted to biodegradable material [43].

Spent sulfite liquor has been used as a substrate for fermentations since 1909 in Sweden, and later in many other parts of the world. The first organism to be used was *Saccharomyces cerevisiae,* although this organism is unable to metabolise pentoses which are found in considerable amounts in this waste product. Later, other organisms better suited for the assimilation of all the sugar monomers were chosen, namely *C. tropicalis* and *C. utilis*. Nevertheless, microorganisms are susceptible to sulfite, which is removed previous to the fermentation process [44]. Yeast produced from sulfite liquor has been used for feeding at war periods, but lost favour in peace time, being destined for fodder in most instances.

However, experiences of baker's yeast produced from sulfite liquor exist in Finland [27]. It is in Finland that an innovating process consuming sulfite liquor was put into operation in 1975. The Pekilo process is a continuous fermentation consuming pulp mill effluent using the filamentous fungus *Paecilomyces variotii*. It transforms sulfite liquor containing 32 g/L assimilable sugars with a yield of 55% (weight in biomass over unit weight of carbon substrate). The protein content of the fungus exceeds 55% (w/w), and has been officially approved as a food in Finland. In 1983, the projected biomass production of the process was estimated to be around 7000 tons per year [27]. The process is not currently in operation.

5. Toxicity and Safety

Among the criteria to be fulfilled in the final products of SCP are including satisfactory nutritious values and pass all toxicity tests, before they are readily used as food or feed. Nucleic acid content, several toxins and unwanted compounds accumulated during growth are the major concern in the issue of toxicity and safety. Removal of these compounds is important to ensure a quality and safe SCP based product.

Nucleic acids are a necessary component of all cells. Nucleic acids are formed during rapid proliferation of microbial cells RNA [45]. Thus, the nucleic acid content of yeast (around 10% of dry weight) is approximately five-fold higher than in the average mammalian organ. Degradation of nucleic acid in body leads to the production of uric acid. Lacking of the uricase enzyme in body would results in accumulation of uric acid. Consumption of more than 2 g nucleic acid equivalent per day may lead to kidney stone formation and gout in human bodies [46]. Meanwhile, pyrimidines are degraded to orotic acid, the accumulation of which results in liver damage. Thus, removal or reduction of nucleic acids in SCP to acceptable and safe range is indispensible.

Reduction of nucleic acid can be achieved in few ways, such as activation of endogenous RNAase by brief heat treatment up to 60–70°C for 20 min, alkaline hydrolysis of nucleic acids, modifications of cultural conditions with respect to nitrogen, carbon, phosphorous and zinc content or chemical extraction.

Alkaline hydrolysis wipes out RNA but also reduces the nutritional values of the protein components. A compromise method which is effective consists of incubation at pH 9.5 and followed by a heat shock which precipitates the protein and ended with sodium chloride extraction. In some yeast products, thermal shock at 60°C is used and followed by pancreatic ribonuclease, to reduce the nucleic acids from 9 to 2%. Similar results have been achieved by a series of short heat bursts which activate intracellular ribonucleases in yeast [47].

Reduction of nucleic acids can also be achieved by the endogenous polynucleotide phosphorylase and RNase in Brevibacterium [48]. Two derivatives of pancreatic RNase and an endonuclease of *Staphylococcus aureus*, immobilized on corn cobs, have been used to reduce the percentage of nucleic acids in SCP concentrates of yeasts, from 5–15% to 0.5% with a protein loss of only 6% after treatment [49]. Immobilized nucleases like benzonases on corncobs were also used to reduce the nucleic acid content in protein concentrates. The percentage of DNA was reported to be reduced to 3–6% and RNA to 50% with loss of protein in the process being only 1% [50]. An immobilized pancreatic RNase was also investigated for the degradation of yeast ribonucleic acid. The rapid reaction rates obtainable at relatively low temperatures offer a potential alternative method for purifying yeast SCP with minimal loss of derived protein [51]. Methods for reduction of nucleic acid content in SCP obtained from gas oil are also reported [52].

Toxicological tests taking into account aspects other than RNA content, such as allergenicity and mutagenicity, show that SCP is an acceptable product in the case of yeasts [53].

In conclusion, SCP from yeast origin may be used as a source of protein but its nucleic acid content, as well as its deficiencies in sulfur containing amino acids render it as a food factor which must be formulated along with other compensating sources of protein. In most cases, yeast based SCP has been included in animal feeds with excellent results, and it is normally accepted that a 10% contribution of this product in mixtures with other sources providing carbohydrates, lipids and vitamins, is acceptable. Yeast protein is most commonly included in poultry food formulations. However, the advent of aquaculture has recently seen the emergence of ever more sophisticated feeds for fish cultivation, and SCP of fungal origin has proved to be well digested by fish. In addition, the ingestion of yeast biomass appears to increase resistance to mycoses which often decimate fish farms, particularly at water temperatures above 10 °C [54].

6. Technologies in Production of Yeast-based SCP

6.1. Production of SCP

Production of SCP is ordinarily carried out in the continuous mode to maximize yields and economic returns. The process of SCP production from any microorganisms or substrates would have the following basic steps.

1. Provision of a carbon source; it may need physical and/or chemical pretreatments.
2. Addition to the carbon source, of sources of nitrogen, phosphorus and other nutrients needed to support optimal growth of the selected microorganisms.
3. Prevention of contamination by maintaining sterile or hygienic conditions. The medium components may be heated or sterilized by filtration and fermentation equipments may be sterilized.
4. The selected microorganism is inoculated in a pure state.
5. SCP processes are highly aerobic (except those using algae). Therefore, adequate aeration must be provided. In addition, cooling is necessary as considerable heat is generated.
6. The microbial biomass is recovered from the medium.
7. Processing of the biomass for enhancing its usefulness and/or storability.

6.2. Optimal Parameters for SCP Production

The cost of the final product relies on the type of substrates, yield and nutritional value of the SCP. The yield is controlled by the organism and the substrate used for fermentation. Mixed cultures of *Kluyveromyces marxianus* and *Trichoderma reesei* gave a SCP yield of 51% and the efficiency of conversion of beet-pulp into protein was reported to be 39.4%. The yield was 54% when yeast extract was substituted in the basal medium and beet-pulp level was increased from 2 to 4%. The conversion efficiency in the latter case was increased to 41.8% [55]. Among 67 potential yeasts analyzed, *C. tropicalis* and *Yarrowia lipolytica* were found to be the most efficient for SCP production on diesel oil as the sole carbon source.

Maximum yield was reported at 168 h for diesel oil concentrations in the range of 40–60 mL/L [56].

Culture conditions, pretreatment of substrates, nutrient supplementation, types of fermentation processes, and strain improvement can all alter the final SCP composition. Several fungal cultures produced double the amount of crude protein on alkali-treated rice straw when compared to untreated straw [57]. When *Kluyveromyces fragilis* was grown on whole whey as the substrate, the chemical score of the protein produced was 91%. Whey protein product, a protein isolate could be recovered with a yield of 80%. The protein content of the isolate was reported to be 75% and the nucleic acids were reduced by 90.8%. Cell wall debris was also considerably reduced in the final product [58]. Production of mycoprotein shows greater efficiency in conversion of substrate to protein when compared to the conversion of feed by farm animals.

Cloning techniques are also used for improvement of SCP production. The use of genetic engineering is not restricted to the design and production of microbes with high protein yield alone, but also finds application in the improvement of downstream processing in the post-fermentation stage. It is used in the production of 'designer' or 'recombinant' proteins of high nutritive value and their subsequent purification. Novel techniques like specifically designed tags or modification of sequences with target-gene product are used for efficient recovery of native and modified proteins [59, 60].

Improvement in protein recovery can also be achieved by suppression or elimination of protease activity in the microbial cells. Mutants with a modified protease gene sequence can be isolated and used for this purpose. For example, the LDHP1 strain of *S. cerevisiae*, a protease-deficient mutant produced by osmotic shock, showed an improved beta-galactosidaseyield [61].

Recombinant DNA technology can be used to isolate mutant genes that can produce high amounts of specific amino acids like glutamate, tryptophan, and phenylalanine along with high protein yields. Such genes were combined by recombination techniques and incorporated into an organism with a wide substrate range. A mutant variety of *S. cerevisiae* produced a high concentration of intracellular glutamate [62]. Mutants of *S. cerevisiae*, which were resistant to toxic ethionine, showed accumulations of methionine as high as 163 times that of parental strain. These high levels were achieved due to enhanced specific activity and reduced feedback inhibition by threonine of the enzymes aspartokinase and homoserine dehydrogenase [63]. About 180 000 tons of lysine is needed annually to improve the nutritive value of food and SCP. The yield of expensive amino acids like phenylalanine can be improved by using a genetically manipulated strain of *Rhodotorula rubra* [64].

High cost restricts the use of tryptophan by humans. Use of genetic recombination to combine four mutations into one strain of *S. cerevisiae* was reported to help increase the production of tryptophan to 12 g/L [65]. Strain improvement is done by complementing the auxotrophic mutants with wild-type genes [66, 67]. The substrate range of such strains can be extensively widened when compared to the auxotrophic variety. This would allow the utilization of a variety of unused substrates for the growth of microbes for SCP production.

Unstructured logistic models were also applied to stimulate the kinetic processes of SCP batch cultures with two species of yeast, in which glucose and xylose or a mixture of the two were used as a carbon source. The aim was to increase the productivity of the final product i.e. SCP [68, 69].

Solid state fermentation (SSF) involves the growth of microbes on a chiefly insoluble substrate where there is no free liquid. Research in industrial fermentation technology has largely concentrated on submerged fermentation and until recently SSF did not generate much interest. A renewal of interest in SSF over the past 10 years stems from the need to reduce manufacturing costs from wastes by taking advantage of engineering principles and techniques [70]. These include simplicity in operating the Koji fermenter, simultaneous saccharification, fermentation of lignocellulosic substrates, and release of concentrated hydrolytic enzymes in the medium during growth. Generally, under conditions of low water activity and an intractable solid substrate, fungi show luxuriant growth when compared to bacteria and algae. Mycelial tips of fungi have immense turgor pressure, which assists in their penetration of hard substrates. Hence better growth of fungi in SSF gives much higher biomass when compared to submerged fermentation. Further, the SSF process is simple and has many advantages over the submerged fermentation. However, substrate particle size, moisture level and C/N ratio are critical factors for SSF [70, 71, 72, 73, 74].

A study regarding the cost analysis of SCP reported that the cost of SCP produced by submerged fermentation was 2–5-fold the cost of fishmeal [75]. Perhaps SSF processes should be used to reduce the cost of SCP production. Currently, SSF is being used for the production of protein-enriched feed [72, 73, 76]. Industrial scale application of SSF for production of SCP would help increase yields and improve conversion efficiency, which would reduce the overall cost of the final product.

6.3. Growth Kinetics

The production of SCP by fermentation has initiated a substantial effort to understand the basis of microbial growth quantitatively. The main parameters which characterize the performance of a process for SCP production are Specific growth rate (μ), Growth Yield (Y_x/s) and Biomass Productivity (P_x).

Under conditions favorable to microbial growth, a given amount of biomass (X, gh^{-1}) is expected to experiment an increase (dX) for a given period of time (dt) according to the following expression:

$$dX = \mu\ X\ dt\ (g/L)\ \text{or,}$$
$$dX/dt = \mu\ X\ (g/L \cdot h)$$

where m is the specific growth rate (h^{-1}), characteristic for the organism and the conditions applied. The specific growth rate m has the dimension of reciprocal time (h^{-1}). It is analogous to the compound interest rate on an investment, thus a specific growth rate of 0.1 h^{-1} is equivalent to a compound interest rate of 10% per hour [69].

For a batch culture, under conditions of exponential growth in which the specific growth rate can be considered constant, $\mu = \mu_{max}$, integration of this expression results in the following equation which describes the amount of biomass produced at any one time during exponential phase.

$$X = X_o\ e^{\mu max\ (t-to)}\ (g/L)$$

where X_o is the amount of biomass at initial reference time (t_o), X is the amount of biomass at time t, and μ_{max} the maximum specific growth rate for an experiment under determined conditions. It has an upper limit for every organism. The maximum absolute values obtained for m experimentally (μ_{max}) may reach values as high as 0.55 h^{-1} in yeast and filamentous fungi. However, in batch culture, growth at exponential phase ($\mu = \mu_{max}$) occurs during a limited period of time. In fact, m has a variable value, being affected by different growth parameters. Thus, the substrate concentration has a notable influence on it, according to the Monod equation [9]:

$$\mu = \mu_{max}\, S / (S + K_s)\ (h^{-1}).$$

where μ_{max} is the maximum specific growth rate of the organism and K_s (mol/L) the saturation constant, a measure of the organism's affinity for the limiting substrate. It can be considered that K_s are the substrate concentration at which the growth rate is half μ_{max}. Control of substrate concentration has been used to control the value of μ in culture.

The rate at which microbial growth takes place is independent from the efficiency of the bioconversion process from substrate to biomass. Growth Yield is defined by the quotient

$$Y_{x/s} = Dx/Ds\ [\text{g biomass (g substrate)}^{-1}]$$

where Dx is the increase in biomass consequent on utilization of the amount of substrate Ds. The maximum growth yield obtained for yeast and fungal culture growing on carbohydrates under aerobic conditions ranges between 0.4 and 0.5 g biomass per g of substrate. This is also considered the theoretical maximum in thermodynamic terms [27]. In physiological terms, these yields are achieved under oxidative metabolic patterns, with conditions of no limiting oxygen transfer. Maximum growth yields are normally achieved under conditions of carbon substrate limitation, with m values around 0.1-0.25 h^{-1}.

The biomass productivity can be defined as the amount of biomass produced per volume and per unit of time for a given fermentation process.

$$P_x = \mu X\ (g/L \cdot h)$$

This parameter has important repercussions in process design, since it combines two parameters which can be estimated in economic terms: amount of product and process time.

6.4. Process Design and Control

The physiological characteristics of yeast recommend the control of the carbon source concentrations, as a limiting substrate, as well as an adequate supply of oxygen for the maintenance of balanced growth under an oxidative metabolic pattern. However, since microbial growth is a time dependent process, it exerts continuous modifications on all process parameters which influence physiology and substrate concentration. Therefore, an adequate technology which maintains appropriate growth conditions for a prolonged period of time must be implemented specifically for the purpose of obtaining high yield and productivity values.

Batch fermentations are clearly inadequate for the purpose of biomass production, since the conditions in the reaction medium change with time [77]. Fed-batch fermentations are better suited for the purpose of biomass production, since they involve the control of the carbon source supply through feeding rates. However, as the biomass concentration increases, the oxygen demand of the culture reaches a level which cannot be met in engineering or economic terms.

Fed-batch culture is still in use for bakers' yeast production using well established and proven models [78, 79]. However, they have not been favored for the production of SCP at a large industrial scale.

Prolonging a microbial culture by continuous addition of fresh medium with the simultaneous harvesting of product has been implemented successfully in industrial fermentations destined to biomass production. The most commonly used principle has been the chemostat: a perfectly mixed suspension of biomass into which medium is fed at a constant rate, and the culture is harvested at the same rate so that the culture volume remains constant.

The technical implications of chemostat culture are various and extremely relevant. They will be explained briefly below, although the mathematical demonstrations may be found in classic reviews [80, 81, 82, 83]. Since the rates of incoming and out flowing medium to and from the reactor are identical, the volume remains constant. When these conditions are maintained, a steady-state is attained where the specific growth rate and all parameters remain constant. If the conditions are carefully controlled, the process may be maintained at production settings which are optimal for long periods. Production periods as long as six weeks have been implemented in many fungal and yeast SCP production processes based on carbohydrate carbon sources [32, 35]. This practical time limit has not been surpassed due to the increasing risk of contamination with time, as well as the appearance of undesirable genetic variations in the culture after a critical number of generations, as will be discussed below.

Another very important implication is that in chemostat culture, the specific growth rate μ may be set by the dilution rate (D, h^{-1}) of the process (D = F/V = flow rate/reactor volume). Other technical measures may be implemented to control substrate and oxygen concentrations under steady-state conditions for any set dilution rate. Thirdly, the continuous removal of biomass relieves the limitations on oxygen supply that apply for batch and fed-batch culture systems. Finally, a continuously operating installation requires smaller operating volumes than batch operated installations. The reaction volumes also determine the dimensions of all surrounding facilities, with important consequences on capital investment, as will be discussed in section 5.

Under chemostat culture, high yielding (Y = 0.45-0.5) SCP processes have been operated at μ values ranging between 0.2 and 0.3 h^{-1} in yeast cultures [77, 84] and between 0.15 and 0.2 h^{-1} in cultures using filamentous fungi [85].

The limitations imposed by oxygen transfer through the gas-liquid interface have provoked an important scientific and technological effort directed at optimizing this bottleneck process.

Massive aeration is not recommended on economic grounds, but on the technical side also, as the higher the proportion of gas pumped through the solution, the greater is the partial volume occupied by the gas, and therefore the reactor volume. Evaporation and cooling of the medium is another undesired consequence of over aeration. In addition, a common problem of industrial fermentations is the profuse appearance of foam on the head space of the reactor, causing reactor pressurization, spillages and contamination hazard.

Among the various designs which have been put to effect, the deep-jet fermenter and the air-lift fermenter have been the most successfully applied [86]. But it is the air-lift in its different variations that has enjoyed the greatest success as the configuration of choice for continuous SCP production. This configuration is presently used in the production of myco-protein which is the basis for some commercial products.

The control of key process variables is a critical element of SCP production, from oxygen transfer, substrate and product concentration, to the appearance of minimal amounts of toxic compounds through undesired metabolic processes, which may compromise the quality of the final product. Many of these are carried out automatically as a result of a rapid development in all aspects of control, from sensor design to the computer algorithms which modulate the control responses [87, 88, 89, 90]. However, since the cost effectiveness of SCP production is under constant scrutiny due to the price competition with plant proteins, simplified control devices are preferred. These often take advantage of existing conventional technology such as oxygen and pH control.

Experimental evidence of extracellular medium acidification by cell cultures has pointed towards the existence of a relationship between proton production and cell growth, and the formal relationship between these two parameters has been recently demonstrated [91]. This has enabled for the on-line estimation of biomass and growth-linked product synthesis through pH control analysis using formal relationships which are applicable for a wide range of organisms [92].

Oxygen concentration is a key parameter which must be subject to monitoring and control. Specific electrodes placed at different reactor locations give an account of the levels of dissolved oxygen in the medium. In SCP fermentation processes, the oxygen concentration must never fall below the critical point. The various control devices used to maintain oxygen levels within the oxidative physiological range, span from increased agitation and aeration, to more sophisticated control of carbon source dosage and oxygen gas injection [93, 94].

The biomass from yeast fermentation processes is harvested normally by continuous centrifugation. This process results in biomass concentrations around 30% (w/v). Filamentous fungi are harvested by filtration [85]. The biomass is then treated for RNA reduction and dried in steam drums of spray driers. Drying is expensive, but results in stabilized product with shelf lives of years. This is a key feature in the animal feed and fodder business. For a more detailed revision of the different processes, flow charts and reaction configurations used for commercial SCP production the reader can be referred to the work of [95].

7. Conclusion

Yeast based SCP has a proven record as a source of protein which may be obtained with large productivities in compact installations. The daunting prospects of world famine in the 60's lead to great expectations about the social and economic relevance of microbial proteins as a food source. Sustainable food security implies physical and economic access to food for all. This requires greater efforts directed at food production and distribution as well as improving living standards of the people. However, in the recent decades, there were newspaper reports regarding destruction of genetically modified cloned crops in Europe. People fear that microbial genes cloned into crops would enter the human system and lead to diseases, allergies or even cancer. Precautions and control measures must be taken to alleviate these fears. SCP products should be evaluated extensively for safety purposes, to gain popularity among masses. Nevertheless, there are success stories which paved the way for a new market of specialized foods of microbial origin. The compactness and high degree of control achieved yeast based SCP production process provide a high degree of safety to the consumer, against the uncertainties which regularly surround meat products. Other such cases with lower international resonance have constantly arisen, and it is likely that more such cases will continue to surface in the future, requiring a higher degree of control on food quality for animal and human consumption. Substitution of animal products by foods of plant origin is not totally exempt from these dangers, as the use of pesticides or the appearance of carcinogens such as nitrosamines has already been reported in the past. SCP products for animal and human nutrition are safer in this respect, since the components from which they are produced are easily controlled, and their genetic background is well known. They represent a line of non-conventional substitutes which will continue to have a market for these reasons.

References

[1] Kurtzman, C. P. & Fell, J. W. 2005. Biodiversity and Ecophysiology of Yeasts. In: *The Yeast Handbook* (edited by P. Gábor & C. L. de la Rosa). Pp. 11-30. Berlin: Springer.

[2] Bass, D., Howe, A., Brown, N., Barton, H., Demidova, M., Michelle, H., Li, L., Sanders, H., Watkinson, S. C., Willcock, S. & Richards, T. A. 2007. Yeast forms dominate fungal diversity in the deep oceans. *Proceedings. Biological Sciences/The Royal Society*, *274*, 3069–3077

[3] Kurtzman, C. P. & Piškur, J. 2006. Taxonomy and phylogenetic diversity among the yeasts. In: *Comparative Genomics: Using Fungi as Models* (edited by P. Sunnerhagen & J. Piškur) Pp. 29-46. Berlin: Springer.

[4] Kurtzman, C. P. 1994. Molecular taxonomy of the yeasts. *Yeast*, *10*, 1727–1740.

[5] Walker, K., Skelton, H. & Smith, K. 2002. Cutaneous lesions showing giant yeast forms of Blastomyces dermatitidis. *Journal of Cutaneous Pathology*, *29*, 616–618.

[6] Balasubramanian, M. K., Bi, E. & Glotzer, M. 2004. Comparative analysis of cytokinesis in budding yeast, fission yeast and animal cells. *Current Biology*, *14*, R806–818.

[7] Yeast Virtual Library. 2010. What are yeasts? http://wiki.yeastgenome.org/index.php/General_Topics. Accessed 04/01/2010.

[8] Yeong, F. M. 2005. Severing all ties between mother and daughter: Cell separation in budding yeast. *Molecular Microbiology*, *55*, 1325–1331.

[9] Forsburg, S. L. & Nurse, P. 1991. Cell cycle regulation in the yeasts *Saccharomyces cerevisiae* and *Schizosaccharomyces pombe*. *Annual Review of Cell and Developmental Biology*, *7*, 227–256

[10] Hartwell, L. H. 1991. Twenty-five years of cell cycle genetics. *Genetics*, *129*, 975–980.

[11] Pringle, J. R. & Hartwell, L. H. 1981. The *Saccharomyces cerevisiae* cell cycle. In: *Molecular Biology of the Yeast; Saccharomyces, Life Cycle and Inheritance* (edited by J. N. Strathern, E. W. Jones & J. R. Broach) Pp. 97-142. NY: Cold Spring Harbor Laboratory.

[12] Watson, J. D., Hopkins, N. H., Roberts, J. W., Steitz, J. A. & Weiner, A. M. 1987. *Molecular Biology of the Gene, 4th Ed*. Pp. 550-618. Menlo Park, CA: Benjamin-Cummings.

[13] Barnett, J. A. 1975. The entry of D-ribose into some yeasts of the genus *Pichia*. *Journal of General Microbiology*, *90*, 1–12.

[14] Arthur, H. & Watson, K. 1976. Thermal adaptation in yeast: growth temperatures, membrane lipid, and cytochrome composition of psychrophilic, mesophilic, and thermophilic yeasts. *The Journal of Bacteriology*, *128*, 56–68.

[15] Legras, J.-L., Merdinoglu, D., Cornuet, J.-M. & Karst, F. 2007. Bread, beer and wine: *Saccharomyces cerevisiae* diversity reflects human history. *Molecular Ecology*, *16*, 2091–2102.

[16] Biofuelcell. 2007. Helsinki University of Technology. http://automation.tkk.fi/Biofuelcell. Accessed 04/01/2010.

[17] Features of Different Microorganisms and the SCP Produced From Them http://www.molecular-plant-biotechnology.info/foods-and-beverages/features-of-different-microorganisms-and-the-SCP-produced-from-them.htm Accessed 01/04/2010

[18] Anupama & Ravindra, P. 2000. Value-added food: Single cell protein. *Biotechnology Advances*, *18*, 459–479.

[19] Kolani, S., Delgenes, J. P., Moletta, R., Traore, A. & Doh, A. 1996. Optimization of cell yield of *Candida krusei* SO1 and *Saccharomyces* sp. LK3G cultured in sorghum hydrolysate. *Bioresource Technology*, *57*, 275–281.

[20] Guven, M. & Cansunar, E. 1989. Single cell protein production and reduction of pollutants in waste sulphite liquor. *Mikrobiyoloji bulteni*, *23*, 329–335.

[21] Rhishipal, R. & Philip, R. 1998. Selection of marine yeast for generation of single cell protein from prawn shell wastes. *Bioresource Technology*, *65*, 255–256.

[22] Saliceti-Piazza, L., Dale, M. C., Moelhman, M., Ooks, M. R. & Wankat, P. C. 1992. Free and immobilized yeasts for BOD reduction in dairy wastes: Growth on low levels of lactose, lactic acid and glycerol. Paper No. 165k. AICHE Annual Meeting, Miami, FL.

[23] Wagner, G. H. 1990. Emerging applications of methylotrophic yeasts. *FEMS Microbiology Reviews*, *7*, 279–83.

[24] Deibel, M. R., Hiebsch, R. R. & Klein, R. D. 1988. Secreted amylolytic enzymes from *Schwanniomyces occidentalis*: purification by fast protein liquid chromatography (FPLC) and preliminary characterization. *Preparative Biochemistry and Biotechnology*, *18*, 77–120.

[25] Chanda, S. & Chakrabarti, S. 1996. Plant origin liquid waste, a source for single cell protein production by yeast. *Bioresearch Technology*, *57*, 51–54.

[26] Microorganisms Used for SCP Production. http://www.molecular-plant-biotechnology.info/foods-and-beverages/microoraganisms-used-for-SCP-production.htm Accessed 01/04/2010

[27] Oura, E. 1983. Biomass from Carbohydrates. In: *Biotechnology, Vol. 3.* (edited by H.-J. Rehm & G. Reed) Pp.3. Weinheim: Verlag Chemie.

[28] Olbrich, H. 1973. Biotin activity of molasses. *Branntweinwirtschaft, 113*, 270.

[29] Forage, A. J. & Righelato, R. C. 1979. Biomass from Carbohydrates. In: *Microbial Biomass: Economic Microbiology. Vol. 4* (edited by A. H. Rose) Pp. 289. London, UK: Academic Press.

[30] Jarl, K. 1969. Symba yeast process. *Food Technology*, *23*, 1009.

[31] Trinci, A. P. J. 1992. Myco-protein: A twenty-year overnight success story. *Mycological Research*, *96*, 1.

[32] Trinci, A. P. J. 1994. Evolution of the Quorn myco-protein fungus *Fusarium graminearum* A3/5. *Microbiology*, *140*, 2181.

[33] Anderson, C., Longton, J., Maddix, C., Scammell, G. W. & Solomons, G. L. 1975. The growth of microfungi on carbohydrates. In: *Single Cell Protein* (edited by Tannenbaum & D. I. C. Wang). Pp. 314. Cambridge MA: MIT Press.

[34] Steinkraus, K. H. 1986. Microbial biomass protein grown on edible substrates: The indigenous fermented foods. In: *Microbial Biomass Protein, Section I.* (edited by M. Moo-Young & K. F. Gregory) Pp. 33. London: Elsevier Applied Science.

[35] Muller, L. L. 1969. Yeast products from whey. *Process Biochemistry, 4*, 21.

[36] Moebus, O. & Teuber, M. 1983. General aspects of production of biomass by yeast fermentation from whey and permeate. In: *Production and Feeding of Single Cell Protein* (edited by M. P. Ferranti and A. Fiechter) Pp. 153. London, UK: Applied Science.

[37] Moulin, G., Malige, B. & Galzy, P. 1983. Balanced flora of an industrial fermenter. Production of yeast from whey. *Journal of Dairy Science*, *66*, 21.

[38] Rivière, J. 1977. Microbial proteins. In: *Industrial Applications of Microbiology. Chapter 4.* Pp. 105. UK: Surrey University Press.

[39] Einsele, A. 1983. Biomass from higher alkanes. In: *Biotechnology, Vol. 3* (edited by H.-J. Rehmand & G. Reed) Pp. 43. Weinheim: Verlag Chemie.

[40] Faust, U. & Präve, P. 1983. Biomass from methane and methanol. In: *Biotechnology, Vol. 3* (edited by H.-J. Rehm & G. Reed) Pp. 83. Weinheim: Verlag Chemie.

[41] Gellissen, G. 2000. Heterologous protein production in methylotrophic yeasts. *Applied Microbiology and Biotechnology, 54*, 741.

[42] Ferranti, M. P. & Fiechter, A. 1983. Production and feeding of single cell protein. London, UK: Applied Science Publishers.

[43] Butshek, G. & Zellstoffablaugen, G. K. 1962. Zellstoffablaugen. *In: Die Hefen, Vol 2.* (edited by F. Rieff, R. Kautzmann, H. Lüers & M. Lindemann) Pp. 121. Nuremberg, Germany: Verlag Hans Carl.

[44] Webb, F. C. 1964. *Biochemical Engineering*. London, UK: D. Van Nostrand Ltd.

[45] Singh, B. D. 1998. *Biotechnology*. Pp. 498–510. New Delhi, India: Kalyani Publishers.

[46] Calloway, D. H. 1974. The place of single cell protein in man's diet. In: *Single cell protein.* (edited by Davis P.) Pp. 129–146. New York, USA: Academic Press

[47] Solomons, G. L. & Litchfield J. H. 1983. Single Cell Protein. *Critical Reviews in Biotechnology*, *1*, 21-58, Boca Raton, USA: CRC Press.
[48] Yang, H. H, Thayer, D. W. & Yang, S. P. 1979. Reduction of endogenous nucleic acid in single cell protein. *Applied and Environmental Microbiology*, *38*, 143–147.
[49] Martinez, M. C., Sanchez-Montero, J. M., Sinisterra, J.V. & Ballesteros, A. 1990. New insolubilized derivatives of ribonuclease and endonuclease for elimination of nucleic acids in single cell protein concentrates. *Biotechnology and Applied Biochemistry*, *12*, 643–652.
[50] Moreno, J. M., Sanchez-Montero, J. M., Ballesteros, A. & Sinesterra, J. V. 1991. Hydrolysis of nucleic acids in single cell protein concentrates using immobilized benzonases. *Biotechnology and Applied Biochemistry, 31*, 43–51.
[51] Dale, B. E. & White, D. H. 1979. Degradation of ribonucleic acids by immobolized ribonuclease. *Biotechnology and Bioengineering*, *21*, 1639–1648.
[52] Abou-Zeid, A. A., Khan, J. A. & Abulnaja, K. O. On methods for reduction of nucleic acid content in single cell protein from gas oil. *Bioresource Technology*, *52*, 21–24.
[53] Schoch, U. & Schlatter, C. H. 1983. Toxicological evaluation of SCP produced from whey. In: *Production and feeding of single cell protein* (edited by M.P. Ferranti & A. Fiechter). London, UK: Applied Science Publishers.
[54] FAO, 2010. FAO Aquaculture Development programme, Fish feeds and feeding in developing countries. ADCP/REP/83/18. http://www.fao.org. Accessed 28/03/2010.
[55] Ghanem, K. M. 1992. Single cell protein production from beet-pulp by mixed culture. *Microbiologia*, *8*, 39–43.
[56] Ashy, M. A. & Abou-Zeid, A. 1982. Potentialities of yeasts in production of single cell proteins (SCP). *Zentralblatt Fur Mikrobiologie*, *137*, 387–394.
[57] Darmwal, N. S. & Gaur, A. C. 1991. Isolation of cellulolytic fungi for cellulase and protein production from rice straw. *Zentralblatt Fur Mikrobiologie*, *146*, 467–470.
[58] Galvez, A., Ramirez, M. J. & Garcia-Garibay, M. 1990. Chemical composition of a mixture of single cell protein obtained from *Kluyveromyces fragilis* and whey proteins. *Archivos Latinoamericanos de Nutrición*, *40*, 252–262.
[59] Nygren, P. A., Stahl, S. & Uhlen, M. 1994. Engineering proteins to facilitate bioprocessing. *Trends in Biotechnology*, *12*, 184–188.
[60] Murby, M., Uhlen, M. & Stahl, S. 1996. Upstream strategies to minimize proteolytic degradation upon recombinant production in *Escherichia coli. Protein Expression and Purification*, *7*, 129–136.
[61] Becerra, M., Cerdan, E. & Gonzalez-Siso, M. I. 1997. Heterologous *Kluyveromyces lacti* beta-galactosidase production and release by *Saccharomyces cerevisiae* osmotic remedial thermosensitive autolytic mutants. *Biochimica et Biophysica Acta*, *1335*, 235–241.
[62] Hill, F. 1994. *Yeast with high natural glutamic acid content.* Patent EP592785.
[63] Martinez-Force, E. & Benitez, T. 1993. Regulation of aspartate derived amino acid biosynthesis in yeast *Saccharomyces cerevisiae. Current Microbiology*, *26*, 313–322.
[64] De Boer, L. & Dijkhuizen L. 1990. Microbial and enzymatic process for L-phenylalanine production. *Advances in Biochemical Engineering Biotechnology*, *41*, 1–27.

[65] Moller, A. 1994. L-Tryptophan production from anthranilic acid by amino acid auxotrophic mutants of *Candida utilis*. *Process Biochemistry*, *29*, 521–527.

[66] Jungehulsing, U., Arntz, C., Smit, R. & Tudzynski, P. 1994. The *Claviceps purpurea* glyceraldehyde-3-phosphate dehydrogenase gene cloning, characterization and use for improvement of a dominant selection system. *Current Genetics*, *25*, 101–106.

[67] Niederberger, P. 1989. Amino acid production in microbial eukaryotes and prokaryotes other than coryneforms. In: *Society for General Microbiology Symposium, 44* (edited by I. Hunter & M. Rhodes) Pp. 2-24. Cambridge, UK: University Press.

[68] Liu, G., Zheng, Z., Song, D. & Cen, P. 1995. Kinetic models of batch and fed-batch culture of single cell protein with double carbon sources. *Chinese Journal of Biotechnology*, *11(3)*, 163–170.

[69] Ahmad, M. N. & Holland, C. R. 1995. Growth kinetics of single cell protein in batch fermenter. *Journal of Food Engineering*, *26*, 443–452.

[70] Pandey, A. & Soccol, C. R. 1998. Bioconversion of biomass: A case study of lignocellulosic bioconversion in solid state fermentation. *Brazilian Archives of Biology and Technology*, *41*, 379–390.

[71] Tengerdy, R.P. 1985. Solid state fermentation. *Trends in Biotechnology*, *3(4)*, 96–99.

[72] Rodriguez, J., Ferraz, A., Nogueira, R. F., Ferrer, I., Esposito, E. & Duran, N. 1997. Lignin biodegradation by the ascomycete *Chrysonilia sitophila*. *Biotechnology and Applied Biochemistry*, *62(2–3)*, 233–242.

[73] Zadrazil, F. & Puniya, A. K. 1995. Studies on effect of particle size on solid state fermentation of sugarcane bagasse into animal feed using white rot fungi. *Bioresearch Technology*, *54*, 85–87.

[74] Nigam, P. & Singh, D. 1994. Solid state (substrate) fermentation systems and their applications in biotechnology. *Journal of Basic Microbiology*, *34*, 405–414.

[75] Perera, W. M., Carter, C. G. & Houlihan, D. F. 1995. Feed consumption, growth and growth efficiency of rainbow trout (*Onchorhyncus mykiss* (Walbaum) fed on diet containing bacterial single cell protein. *British Journal of Nutrition*, *73(4)*, 591–603.

[76] Chaudhari, L. C., Singh, R. & Kamra, D. N. 1994. Biodelignification of sugar cane bagasse by *Pleurotus* and *Pleurotus cornucopiae*. *Indian Journal of Microbiology*, *34*, 55–57.

[77] Fiechter, A., Käppeli, O. & Meussdoerffer, F. 1987. Batch and continuous culture. In: *The Yeasts, 2nd Ed., Vol. 2.* (edited by A. H. Rose & J. S. Harrison) Pp. 99. London, UK: Academic Press.

[78] Chen, S. L. & Chinger, M. 1985. Production of Baker's Yeast. In: *Comprehensive Biotechnology, Vol. 3* (edited by H. W. Blanch, S. Drew & D. I. C. Wang) Pp. 429. Oxford, UK: Pergamon.

[79] Peppler, H. J. 1979. Production of yeast and yeast products. In: *Microbial Technology, 2nd Ed., Vol. 1* (edited by D. Perlman & H. J. Peppler) Pp. 157. New York, USA: Academic Press.

[80] Kubitschek, H. E. 1970. Introduction to research with continuous cultures. *Prentice Hall. Biological Techniques Series*. New Jersey, USA: Englewood.

[81] Fiechter, A. 1975. Continuous cultivation of yeasts. In: *Methods in Cell Biology. Vol. XI. (Yeast cells)* (edited by D. M. Prescott) Pp. 97. London, UK: Academic Press.

[82] Calcott, P. H. 1981. *Continuous culture of cells. Vol. I* (edited by P. H. Calcott) Boca Raton, USA: CRC Press.

[83] Cooney, C. L. 1986. Continuous culture: A tool for research, development and production. In: *Perspectives in Biotechnology and Applied Microbiology* (edited by D. I. Alanl & M. Moo-Young) Pp. 271. London, UK: Elsevier Applied Science Publishers.

[84] Castrillo, J. I. & Ugalde, U. O. 1993. Patterns of energy metabolism and growth kinetics of *Kluyveromyces marxianus* in whey chemostat culture. *Applied Microbiology and Biotechnology, 40(2-3)*, 386.

[85] Solomons, G. L. 1985. Production of biomass by filamentous fungi. In: *Comprehensive Biotechnology, Vol. 3* (edited by: H.W. Blanch, S. Drew & D.I.C. Wang) Pp. 429. Oxford, UK: Pergamon.

[86] Stanbury, P. F., Whitaker, A. & Hall, S. J. 2000. *Principles of fermentation technology, 2nd Ed.* Oxford, UK: Butterworth-Heinemann.

[87] Wang, H. Y. 1986. Bioinstrumentation & computer control of fermentation processes. In: *Manual of industrial microbiology and biotechnology* (edited by A. L. Demain & N. A. Solomon) Pp. 308. Washington D.C., USA: American Society for Microbiology.

[88] Ritzka, A., Sosnitza, P., Ulber, R. & Sheper, T. 1997. Fermentation monitoring and process control. *Current Opinion in Biotechnology, 8(2)*, 160.

[89] Royce, P. N. 1993. A discussion of recent developments in fermentation monitoring and control from a practical perspective. *Critical Reviews in Biotechnology, 13,* 117.

[90] Sonnleitner, B. 1996. New concepts for quantitative bioprocess research and development. In: *Advances in Biochemical Engineering and Biotechnology 54* (edited by T. Sheper) Pp. 155. Berlin, Germany: Springer-Verlag.

[91] Castrillo, J. I., De Miguel, I. & Ugalde, U. O. 1995. Proton production and consumption pathways in yeast metabolism. A chemostat culture analysis. *Yeast, 11(14)*, 1353.

[92] Vicente, A., Castrillo, J. I., Teixeira, J. A. & Ugalde, U. 1998. On-line estimation of biomass through pH control analysis in aerobic yeast fermentation systems. *Biotechnology and Bioengineering, 58(4)*, 445.

[93] Zhang, X.-C., Visala, A., Halme, A. & Linko, P. 1994. Functional state modeling and fuzzy control of fed-batch aerobic baker's yeast process. *Journal of Biotechnology, 37*, 1.

[94] von Schalien, R., Fagervik, K., Saxen, B., Ringbom, K. & Rydstroem, M. 1995. Adaptative on-line model for aerobic *Saccharomyces cerevisiae* fermentation. *Biotechnology and Bioengineering, 48(6)*, 631.

[95] Ward, O. P. 1992. Biomass production. In: *Fermentation Biotechnology, Chapter 6.* Pp. 91. Chichester, UK: John Wiley & Sons.

In: Bioprocess Sciences and Technology
Editor: Min-Tze Liong
ISBN 978-1-61122-950-9

Chapter 10

GANODERMA LUCIDUM PRODUCTION OF PHARMACEUTICALS

B. Boh*[*] *and M. Berovic
Faculty of Natural Sciences and Engineering, University of Ljubljana,
Vegova 4, 1000 Ljubljana, Slovenia
Faculty of Chemistry and Chemical Technology, University of Ljubljana,
Askerceva 9, 1000 Ljubljana, Slovenia

ABSTRACT

The chapter Higher Fungi for the production of pharmaceuticals provides an overview of wood degrading basidiomycetes with pharmacological effects. Since these mushrooms are rare in nature, artificial cultivation of fruit bodies has been established, as well as a biotechnological cultivation of fungal biomass in bioreactors, both on solid substrates and in liquid media. Cultivation technologies are presented, including traditional cultivation of fruit bodies on wood logs and on sawdust-based substrates, and modern biotechnological cultivation of mycelia in bioreactors by submerged and solid-state cultivation technologies. Bioreactor design and process parameters are presented and discussed. An industrially important medicinal mushroom *Ganoderma lucidum* - Ling zhi or Reishi is described as an outstanding example, with emphasis on the cultivation possibilities, structure and properties of pharmacologically active compounds, pharmacological effects, and medicinal applications.

1. CULTIVATION OF HIGHER FUNGI WITH MEDICINAL EFFECTS

The number of mushrooms on Earth is estimated at 140,000, yet maybe only 10% of species are known and named [1]. Basidiomycetes mushrooms comprise a vast but largely untapped source of new pharmaceutical products in fruit bodies, cultured mycelium and culture broth. They are of different chemical composition, such as polysaccharides, glycopeptide-protein complexes, proteoglycans, proteins and triterpenoids, with most scientific attention focussed to the group of non-cellulosic β-glucans [2]. Immunomodulating and antitumor activities of these metabolites are related to immune cells, resulting in the production of biologic response modifiers [3]. Clinical evidence for antitumor and other medicinal activities come primarily from some commercialised purified polysaccharides, such as lentinan from Shiitake - *Lentinula edodes,* krestin from *Coriolus versicolor*, grifolan from *Grifola frondosa*, and schizophyllan from *Schizophyllum commune* [1,2], but polysaccharide preparations of some other medicinal mushrooms also show promissing results. This chapter reviews cultivation possibilities, and focusses on biochemical and medicinal studies of *Ganoderma lucidum* – a basidiomycete with promissing pharmacological effects.

1.1. Wood Degrading Basidiomycetes with Pharmacological Effects

Almost unknown in western scientific research only three decades ago, some of the wood degrading basidiomycetes became intensely and systematically studied due to their promising pharmacological effects. Among them, *Ganoderma lucidum* – Ling zhi or Reishi (Figure 1), *Grifola frondosa* – Maitake (Figure 2), and *Lentinula edodes* – Shiitake (Figure 3) have been known from the traditional Asian medicine, while *Pleurotus ostreatus* – Oyster mushroom (Figure 4) and *Agaricus brasiliensis* – Brasilian agaricus (Figure 5) have been previously used primarily as culinary mushrooms.

Technological possibilities for commercial cultivation gave rise to the number of scientific papers and patents. *Ganoderma lucidum* has been identified as a medicinal mushroom with the largest number of patented inventions, followed by *Lentinula edodes* and *Pleurotus ostreatus* (Figure 6).

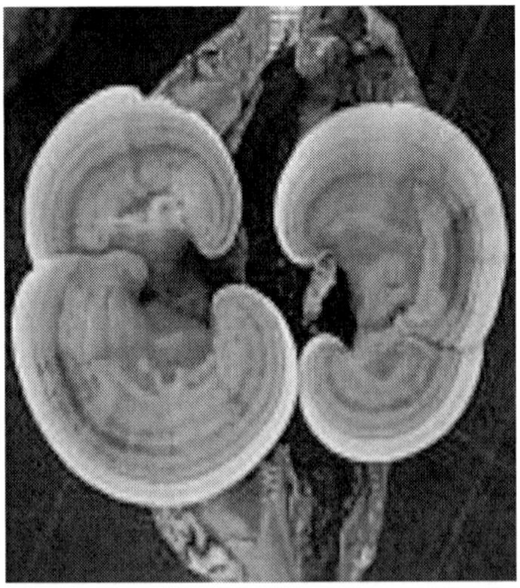

Figure 1. *Ganoderma lucidum* – Ling zhi or Reishi.

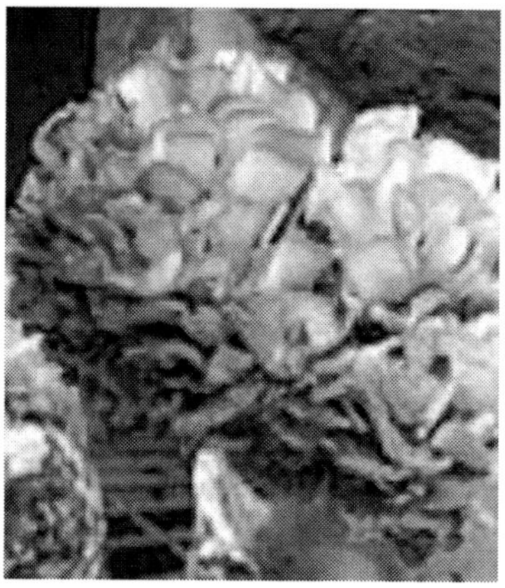

Figure 2. *Grifola frondosa* – Maitake.

Figure 3. *Lentinula edodes* – Shiitake.

Figure 4. *Pleurotus ostreatus* – Oyster mushroom.

Figure 5. *Agaricus brasiliensis* –Brasilian agaricus.

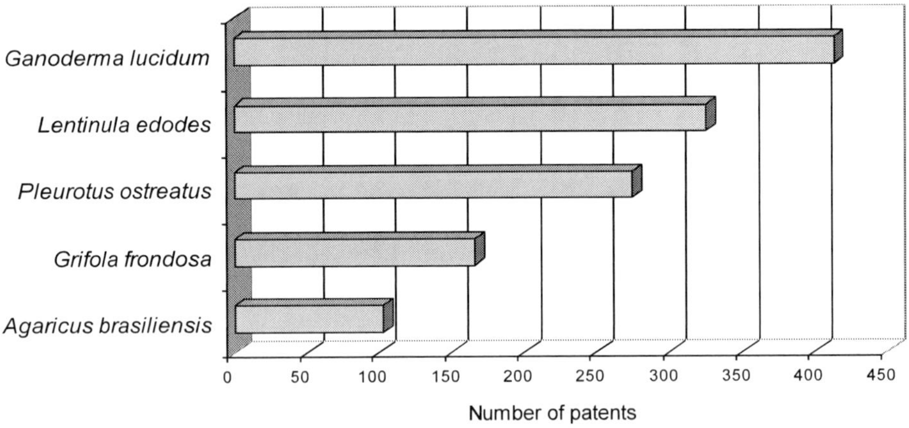

Figure 6. Comparison of the number of patents on five important wood degrading basidiomycetes with pharmacological effects (Free patents Online, patent claims search for US, EP, JP and PCT WO patents).

1.2. Substrates for Fruit Bodies Cultivation

In the wild, wood degrading mushrooms grow primarily on hardwood of trees. However, under artificial cultivation conditions, they thrive on various other substrates containing lignin and cellulose, and therefore have a high potential for recycling organic waste materials of different types. Based on an extensive literature survey, the following groups of lignocellulosic substrates have been identified for artificial cultivation of edible and medicinal wood-degrading mushrooms (Figure 7).

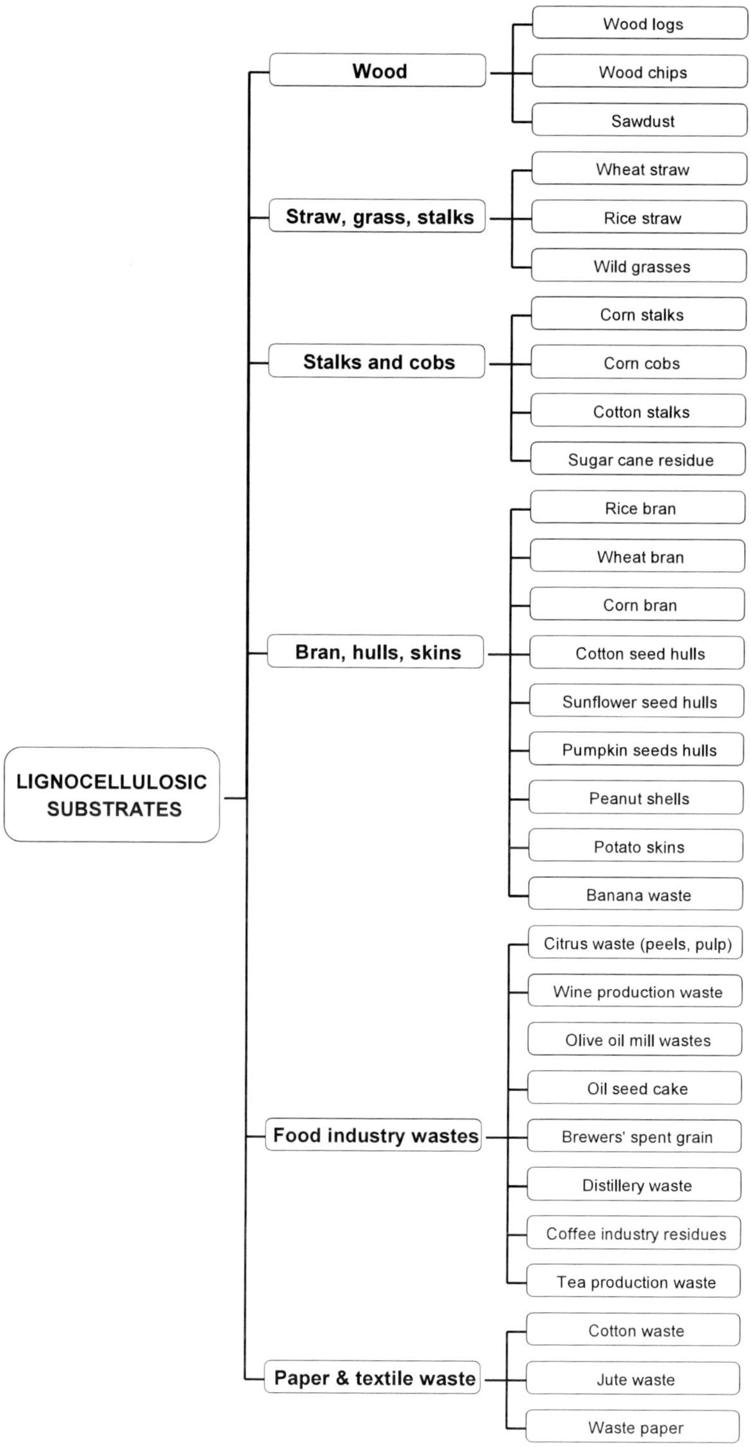

Figure 7. Lignocellulosic substrates for artificial cultivation of edible and medicinal mushrooms.

1.3. Cultivation Technologies

1.3.1. Overview of Cultivation Technologies

Since medicinal mushrooms are scarce in nature, cultivation of fruit bodies on artificial media has been introduced. Traditional cultivation of fruit bodies on wood logs has been known for centuries. With time, cultivation methods have diversified, modified and developed (Figure 8). Besides on wood logs, fruit bodies are being produced on sawdust substrates in trays or beds, and in sterilised plastic bags or in bottles. In addition, production of fungal mycelia has been developed in bioreactors, utilizing submerged cultivation technologies in liquid media, or solid state substrates.

1.3.2. Cultivation of Fungal Biomas in Bioreactors

Traditional cultivation of fruit bodies takes several months, and it is difficult to control the quality of products. In order to decrease the cultivation time and to improve the quality, great efforts have been invested into the controlled cultivation in bioreactors, both submerged and solid state processes. Types and definitions of bioreactors have been described elsewhere, e.g. by Berovic [12].

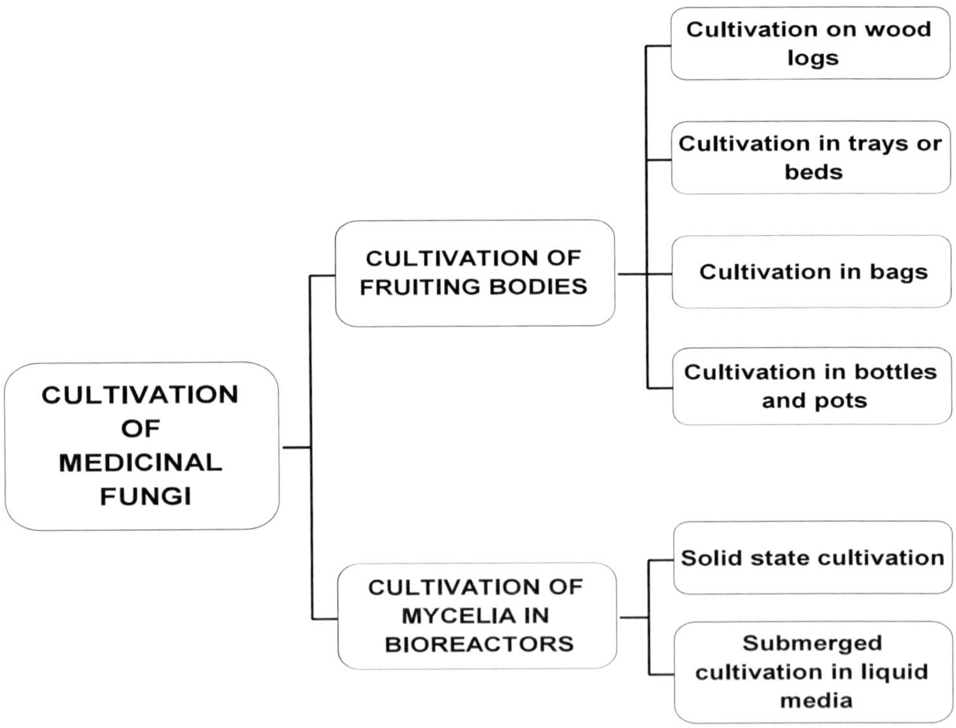

Figure 8. Technological possibilities of cultivating medicinal fungi fruit bodies or fungal biomass on a commercial scale.

1.3.2.1. Submerged Cultivation

Mushroom submerged cultivation variables have complex interactions, therefore various technical problems need to be solved for the production to be successful at an industrial scale. The problems in submerged cultivation of fungal biomass increase with an increased broth viscosity during the cultivation due to the changes in the morphology, fungal biomass and the production of extracellular oilysaccharides (ECP). Therefore, the most important factors in a large-scale submerged fungal culture in a bioreactor are heat and mass transfer, particularly the liquid phase oxygen supply.

Growth of medicinal mushrooms mycelia in submerged cultures is similar to that of lower fungi in most cases, and the formation of large pellets is not unusual [5]. Apart from medium composition, the agitation intensity is another major influence on pellet formation and structure. There seems to be an inverse relationship between agitation intensity and the pellet size. Vigorous agitation appears to prevent pellet formation, and an increased agitation results in small and more compact pellets [6, 7, 8]. Freely dispersed mycelia and its entanglement to form clumps can cause the cultivation media to become highly viscous and non-Newtonian, and to exhibit pseudoplastic or shear thinning behavior. The pelleted type of growth consists of compact discrete masses of hyphae that generally lead to Newtonian behavior of the broth. In many cases, as the size of the pellets increases, the diffusion of nutrients, particularly oxygen, into the center of the pellets becomes limited [6, 9, 10]. Frequently, the central region of larger pellets undergoes autolysis as a result of nutrient limitations. This autolysis can have a significant effect on both cellular metabolism and product synthesis. Pellets up to a size of 200–400 mm are optimum, because the viscosity still remains low, and there are no serious transport limitations into the pellets, as it can be observed for larger pellets [11].

Stirred-tank Bioreactor

The most common type of aerobic bioreactors, used in various higher fungi cultivations, is the stirred-tank reactor, which may feature a specific internal configuration design to provide a specific circulation pattern.

The operating principles of a stirred-tank bioreactor are relatively simple. For optimal mixing, the tank features not only an agitator system, but also baffles, typically from four to eight, which help to prevent a whirlpool effect that could impede proper mixing. As the bioreaction progresses, the bubbles produced by the air supply are broken up by the agitator as they travel upward. Many types of agitators are currently used, with the most common one being the four bladed disk turbine. Newer designs, featuring 12 or 18 blades, or concave ones, have also been shown to improve the hydrodynamics. Power consumption or power input, needed for the agitation of a cultivation broth in a stirred tank reactor, is one of the most outstanding parameters. Large energy consumption at Rusthon turbines from 5 up to 10 KW/m^3 makes this types of reactors high energy consumption machines [12].

Air-lift Bioreactor

Air-lift reactor can be described as a bubble column containing a draught tube. The reactor typically consists of the riser - the central tube with the gas-liquid up-flow, and the downcomer - the region containing the down flowing fluid. Air is typically fed through a sparger ring into the bottom of a central draught tube that controls the circulation of air and the medium. Air flows up the tube, forming bubbles, and exhaust gas disengages at the top of the column. The degassed liquid then flows downward, and the product is drained from the tank. The tube can be designed to serve as an internal heat exchanger, or a heat exchanger can be added to an internal circulation loop [12].

Airlift systems provide some advantages, such as simple design with no moving parts or agitator shaft seals, for less maintenance, less risk of defects and easier sterilization, and lower shear rate, for greater flexibility, especially suitable for higher fungi cultivation. They also provide large, specific interfacial contact-areas with low energy input.

Practical applications depend on the ability to achieve the required rates of momentum, heat and mass transfer at acceptable capital and operating costs. The technical and economic feasibility of using an air-lift reactor has been conclusively established for a number of processes, and these reactors find an increasing use in aerobic cultivations. The simplicity of their design and construction, better defined flow patterns, and comparatively low power inputs for required transport rates, make them very attractive. Air lifts can be scaled up to very large sizes in a single stream. A disadvantage is that oxygen transfer may not be as efficient [13].

1.3.2.2. Solid-state Cultivation

Solid-state cultivation involves the growth and metabolism of fungi in beds of moist solid materials in which the interparticle spaces contain a continuous gas phase and little or no free water. The water activity (a_w) of substrates is an important aspect of fungal cultivation, because the growth of fungi is controlled by the level of a_w. The water activity of a substrate is not the same parameter as its water content, but refers to free, unbound, or active water that can support the growth of fungi [14].

Most bioreactors used in the solid state cultivation can be classified as belonging to one of the five bioreactor types: trays, packed beds, rotating drums, stirred bioreactors and air-solid fluidized beds.

Packed Bed Bioreactors

The characteristics of packed-bed bioreactors are: a static substrate mass packed into a column resting on a perforated plate, and a forced aeration with controlled temperature and humidity air through the substrate bed, usually coming from below the bed, but sometimes from above [15]. Very often the design of a packed bed bioreactor consists of a glass or plastic column with entries at both top and bottom for aeration, while the temperature is controlled either by putting the column in a temperature-controlled room, or by passing water through double jackets, fitted around the column. If the air has low humidity, then the evaporative cooling becomes signficant, although the bottom of the bed tends to dry out. Humidity in the bioreactor can be controlled by using moist air, which can be additionally manipulated to aid in the regulation of the incubation temperature. Several researchers have been successfully using packed bed bioreactors for a variety of applications [16].

Rotating Drum Bioreactors

Rotating drum bioreactors are characterized as cylindrical reactors lying horizontally or slightly inclined, with low pressure air fed into the reactor headspace. They are partially filled with substrate, and rotate around the central axis to promote mixing of the substrate, which in some cases might be promoted by baffles.

Mixing provided in a rotating drum bioreactor is gentle, and among all methods of automated mixing causes the least damage to fungal hyphae and to the agglomeration of substance particles. However, fungi which are highly sensitive to abrasive damage may still be affected. In addition, temperature control is quite difficult, since it is difficult to water-jacket the moving body of the drum. These problems increase significantly with scale [17].

Stirred Aerated Beds Bioreactors

Stirred aerated beds bioreactors fall into the categories of horizontal or vertical stirred bioreactors, depending on the orientation (Figures 9, 10). Horizontal stirred tank bioreactors are similar to rotating drum bioreactors, exept that mixing is provided by the internal scapers or paddles attached to a central shaft. Agitation may be continuous or intermittent. Low pressure air is introduced into the bioreactor headspace [9, 10].

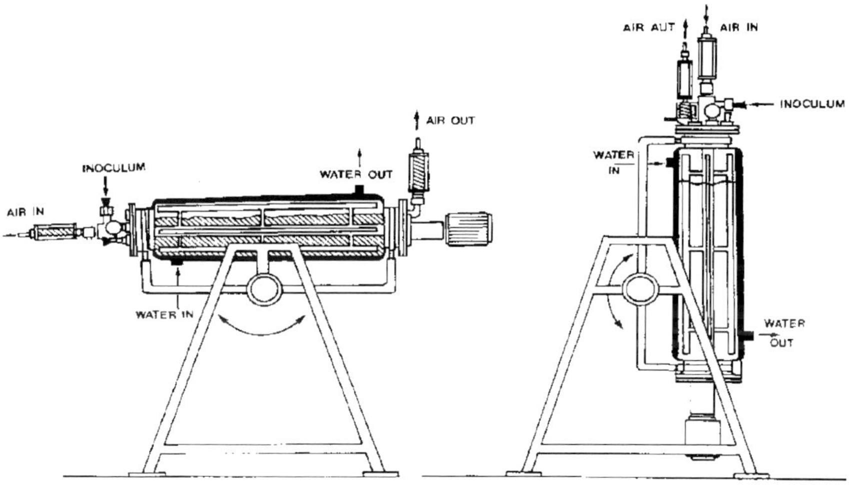

Figure 9. Stirred tank bioreactor [18].

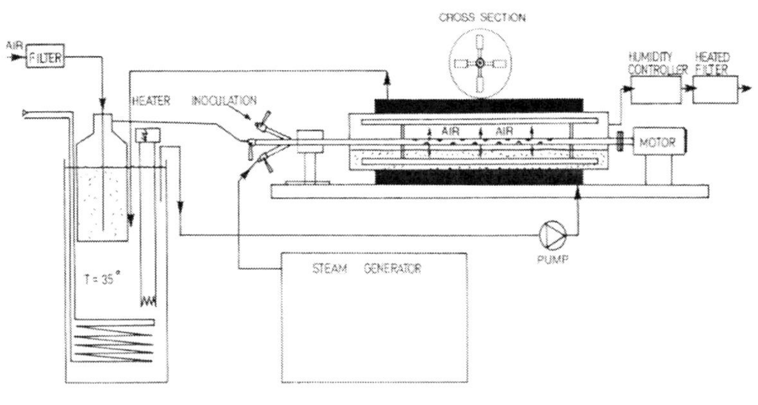

Figure 10. Horizontal stirred tank reactor and equipment, used for solid state cultivations of medicinal mushrooms *Ganoderma* and *Grifola* [19].

Vertical stirred bioreactors often have similarities with packed beds, since they are usually forcefully aerated. They differ from packet bed bioreactors by the fact that they are agitated, which may be continuous or intermittent. The advantage of being well mixed helps in avoiding the temperature and moisture gradients, which often occur in packed beds. In the design described by Mitchell et al., three screw agitators are mounted across the breadth of the reactor on a trolley above the substrate bed, with the screws extending down into the substrate bed [16]. Differences between submerged and solid state cultivations have been discussed and summarised in [20].

2. CULTIVATION AND APPLICATIONS OF *GANODERMA LUCIDUM*

Ganoderma lucidum (W.Curt.:Fr.) Lloyd (Aphyllophoromycetideae) is a white rot wood degrading basidiomycete with hard fruit bodies. In Asian traditional medicine, the fruit body of *G. lucidum* (Figures 11, 12), named *Ling-Zhi* in Chinese and *Reishi* in Japanese language, has been used for treatment of several diseases for thousands of years, including arteriosclerosis, heart diseases, hepatitis, arthritis, nephritis, bronchitis, asthma, hypertension, cancer and gastric ulcer [21, 22, 23]. An increasing modern systematic research into the *Ganoderma* active compounds elucidated its numerous pharmacological effects. Early scientific works reported on liver protection, detoxification, and effects on nervous system [24], anticancer and antitumor activity [25], antiallergenic effects [26], cardiovascular effects [27], anti viral activity [28, 29], and immunomodulation [30, 31]. During the last decade, scientific understanding of pharmacological effects, biochemical and pharmacological mechanisms has deepened and expanded with many in depth studies and trials.

Figure 11. *Ganodema lucidum* fruit bodies in artificial cultivation on sawdust in bags.

Figure 12. Production of *Ganodema lucidum* fruit bodies.

2.1. *G. Lucidum* Cultivation

G. lucidum is very scarce in nature. Artificial cultivation (Figure 13) has been introduced to meet the increasing demands for both, fruit bodies and mycelial biomass. Farming of *G. lucidum* fruit bodies on wood logs and in bags filled with wood or straw substrates has been known for decades, especially in China [32, 33]. Modern biotechnological cultivation in bioreactors, especially with submerged liquid substrate cultivation has been developed [5, 6, 9, 34, 35, 36, 37, 38, 39, 40, 41, 42, 43, 44, 45]. The quality and content of physiologically active substances vary from strain to strain, and also depend on location, culture conditions [46], the growth stage of the fungus [47], the processing procedures, and formulation preparation [48].

2.1.1. Cultivation of G. Lucidum Fruit Bodies

Production of *G. lucidum* fruit bodies can be performed outdoors as cultivation in hardwood stumps, buried logs or bed cultures, or indoors as cultivation in bags or bottles filled with sawdust. Detailed standard cultivation procedures have been published elsewhere, e.g. by Stamets [33]. Some authors report on *G. lucidum* cultivation on unusual agro-waste substrates, such as residue of citrus fruit mixed with rice bran [50], sunflower seed hulls [51], stillage grain from a rice-spirit distillery [52], and soy residue from the waste of tofu manufacturing [53].

2.1.2. Cultivation of G. Lucidum Biomass in Submerged Culture

Recently, submerged cultivation technology for mycelial cultures of *G. lucidum* has been developed as a promising cultivation alternative. In commercial applications, efficient cultivation processes combine the production of cellular biomaterials, and released ECP. Optimisation of substrate composition and cultivation conditions is vital for enhancing the production efficiency. Cultivation parameters, including inoculation density [54], pH [37], temperature, rate of rotation [5] and medium composition [38, 55] have been reported to affect the efficiency of cultivation of *G. lucidum*.

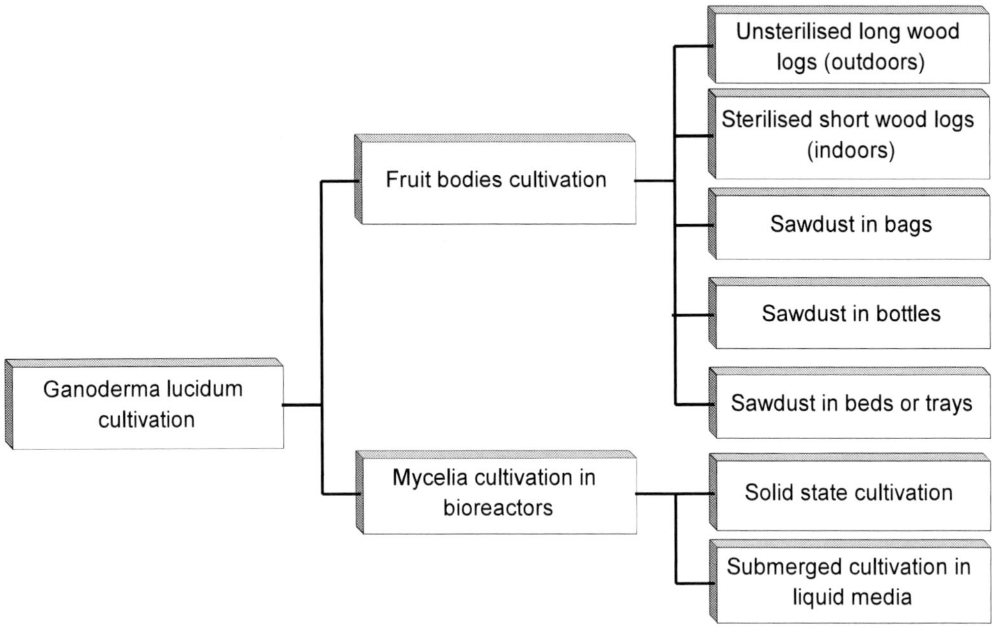

Figure 13. Main cultivation methods for the production of *G. Lucidum* fruit bodies and mycelia [49].

2.1.2.1. Innoculum Preparation and Density

There are two main strategies for the preparation of mycelial inocula for submerged cultivation, either preparation of the seed culture on agar, or preparation by submerged cultivation [13]. Zore et al. described a procedure for *G. lucidum* inoculum preparation in a shaked culture, and the production of mycelia by submerged fermentation in a bioreactor [35].

Simonic et al. [56] inoculated the bioreactor with 5.0% (v/v) of the homogenized seed culture, and cultivated at 30°C and an initial pH of 4. The optimum inoculum density for biomass and polysaccharides production in case of *G. lucidum* was 40%. These results were different from the results of Berovic et al. [6], which showed that 17% inoculum was the most suitable.

Fang et al. obtained the maximal cell concentration at an inoculation density of 330 mg dry weight/l [38, 54]. A larger inoculation density led to a small pellet size and high production of extracellular polysaccharides (ECP) and intracellular polysaccharides (ICP), while a relatively big pellet size and high accumulation of ganoderic acids were observed at a low inoculation density. Similar observation was reported by Hsieh et al., who noted that large inoculum density led to a small pellet size and a high ECP and ICP production [57].

2.1.2.2. Effect of Initial Medium pH

The initial medium pH can greatly affect cell membrane function, cell growth, morphology and structure, the salts solubility, the ionic state of substrates, the uptake of various nutrients, and product biosynthesis. Yang and Liau found out that pH values of glucose-NH_4Cl medium in a range from 4.0 to 4.5 were optimal for *G. lucidum* ECP production, while with a pH decrease, the amount of ECP declined [8]. Lee at al. reported that pH control affected *G. lucidum* mycelial growth and ECP production [58]. The bistage pH control technique, that is, shifting the pH from 3 to 6 at the initial phase of the exponential growth, retained the desirable morphologies of the mycelia during cultivation, and resulted in low viscosity of the culture broth. Fang and coworkers investigated the effects of initial pH on simultaneous production of *G. lucidum* polysaccharides and ganoderic acids. At pH of 6.5, a maximum in biomass was obtained. Lowering the initial pH from 6.5 to 3.5 gradually led to a higher production of ECP and ICP [37, 59].

By further studies, other authors confirmed the results of Fang and Zhong that ECP and specific ICP production increased with the decrease of the initial medium pH. This can be explained by the fact that cells continuously metabolised the carbon source and secreted ganoderic and lucidenic acids in the medium during the first few days of cultivation [6, 59, 60, 61].

2.1.2.3. Influence of Aeration and Agitation

Regarding the fact that oxygen affects cell growth, cellular morphology, nutrients uptake, and metabolite biosynthesis, agitation and aeration have a great influence on the biomass and polysaccharides production [5, 20]. Cultivation under constant aeration seems to be the best option for biomass and ECP production. Yang and Liau [5] studied the influence of cultivation parameters on polysaccharide formation by *G. lucidum* in submerged cultures, and found that agitation and aeration influenced the formation and secretion of polysaccharides. Although higher speeds enhanced mixing efficiency and polysaccharide release, higher shear stress had a detrimental effect on mycelial growth and polysaccharides formation.

2.1.2.4. Influence of Substrate Composition

Numerous studies showed that biomass and polysaccharides production by different *G. lucidum* strains depended on the type and concentration of a carbon source, such as glucose, maltose, or saccharose. Glucose and maltose were confirmed as a good carbon source for mycelial biomass production, contrary to the saccharose [62]. Using complex substrates, Fang and Zhong [37] noted the highest biomass and ICP production in the glucose-peptone-yeast extract medium, at initial glucose concentration of 50 g/l. The maximum value of ECP was observed at glucose concentration of 65 g/l. Berovič et al. observed a significant ECP and ICP production by *G. lucidum* in a liquid potato dextrose-olive oil medium [6]. Hsieh et al. reported that molasses was a good carbon source for biomass production, which increased significantly with the increase of molasses concentration, but the contrary was observed with ECP and ICP production [60]. With an increase of thin stillage concentration from 20 to 60%, the biomass and polysaccharides production grew, while with a further concentration increase, biomass production remained at the same level, and polysaccharides production decreased.

Some authors reported on submerged cultivation of *G. lucidum* mycelia on unconventional substrates, including liquid waste materials, such as thin stillage and deproteinated cheese whey. Lee et al. utilized deproteinated cheese whey for cultivating mycelia of the *G. lucidum* in a bioreactor by submerged cultivation, and concluded that cultivation of *G. lucidum* mycelia could offer a potential cost-effective solution for an alternative utilization of deproteinated cheese whey [63, 64]. The feasibility of reusing soy residue for the production of *G. lucidum* was investigated by Hsieh and Yang [53]. Soy residue was successfully reused as a substrate for the solid-state cultivation. However, proper mixing with other solid medium was necessary for adjusting the C/N ratio.

2.1.2.5. Influences of Carbon and Nitrogen Sources, and C/N Ratio

Tang and Zhong studied the effects of carbon sources and initial sugar concentration on the production of ganoderic acids and polysaccharides in a *G. lucidum* cultivation process [39]. Sucrose as a carbon source was suitable for the ECP production, although the cells could not grow well. Lactose was beneficial for the cell growth and production of ganoderic acid and ICP. The ganoderic acid production was remarkably improved by pulse feeding of lactose, when its residual concentration was between 10 and 5 g/l. Fang and Zhong studied the effects of nitrogen source and initial glucose concentration in submerged fermentation of *G. lucidum* for simultaneous production of bioactive ganoderic acids and polysaccharides [37, 38]. Hsieh et al. studied the production of polysaccharides by *G. lucidum in* shake flasks under various limitations of nutrients, including a carbon-source, nitrogen-source, phosphate-source, magnesium-source, and dissolved oxygen [57]. Different responses of polysaccharide production were observed under different limitations of nutrients. ECP production by *G. lucidum* in response to different culture conditions was studied also by Papinutti [45]. Cultures containing glucose and malt extract exhibited a marked increase (up to 29-fold) compared to the defined media.

2.1.2.6. Effect of Nitrogen Sources and Concentrations

Several studies showed that organic nitrogen sources were better than inorganic ones for biomass and polysaccharides production. This observation was reconfirmed in a cultivation of *G. lucidum,* where peptone was the optimum nitrogen source for biomass and for the ECP production [56]. According to Fang and Zhong [59], biomass and ICP production was the maximal at a combined usage of yeast extract and peptone in the ratio of 1:2, and ECP production in the ratio of 1:1.

2.1.2.7. Influence of Macro- and Microelements

Macroelements, especially P and Mg, have a significant effect on the biomass and polysaccharide production, as they are cofactors for some enzymes, catalyzing the polysaccharide synthesis [57]. However, Chang et al. reported that the addition of Ca caused a decrease of biomass and an increase of polysaccharide production [65]. The results of Simonic et al. on Fe influence on biomass and polysaccharides production by *G. lucidum* were in accordance with previous results of Park et al., who demonstrated that 0.01% $FeSO_4 \cdot 7H_2O$ was the best additive for ECP production by *Cordiceps militaris* [56, 66]. In a study by Song et al., when the cultivation medium was enriched with K, Na, and Mg, the mycelial biomass and ECP production was the highest. ICP production has increased with an increase of K and Na concentration, while in the case of Mg it was at the same level at concentrations of 1.0 and 2.0 mM, and than it rapidly decreased [62].

2.1.2.8. Effect of Plant Oils

Several plant oils, which can be used as antifoam agents, may be beneficial to the growth of *G. lucidum*. Lee *et al.* used a commercial antifoaming agent at volume fraction of 0.5 %, while Habijanic and Berovic used olive oil at a volume fraction of 2 % [58, 67]. The advantages of plant oils are their low cost and the possibility of higher growth yields due to the possible stimulatory properties of the oil. The effects of soy, peanut, safflower, corn, sunflower and olive oils were investigated by Yang et al. [55]. All of the oils tested stimulated growth of *G. lucidum*, with the highest biomass density was obtained by the olive oil.

2.1.2.9. Effect of Polymer Additives

Through the addition of a polymer into the medium, very uniform and tiny pellets could be formed during the inoculum preparation, which lead to the increase of mycelia growth rate. The addition of polymers such as Carbopol-934 (carboxypolymethylene), Junlon (polyacrylic acid) and Hostacerin (sodium polyacrylate) reduced agglomeration and decreased pellet formation in many filamentous fungi [68, 69, 70]. Junlon was especially effective in preventing aggregation of the mycelium of some basidiomycetes in both shake-flask and small-scale fermentor cultures [71].

In the *G. lucidum* cultivation, polymer additives such as sodium alginate, polyacrylmide, and agar were studied by Yang et al. [72] to vary mycelia morphology for higher mycelium production rates. The appearance of *G. lucidum* mycelia varied considerably, ranging from large globes to tiny pellets. Agar, even at a lower concentration, was more effective than sodium alginate and polyacrylamide in reducing the size of pellets.

2.1.2.10. G. Lucidum Cultivation in a Stirred Tank Reactor

The optimum agitation rate represents a balance between the oxygen transfer into the medium, and the shear stress, both of which increase with an increasing agitation rate. Zore et al. [35] cultivated *G. lucidum* in a 10 *l* laboratory stirred tank reactor, using a liquid substrate based on potato dextrose and olive oil. Tang and Zhong [39] studied a fed-batch *G. lucidum* cultivation process in a stirred tank bioreactor. Sucrose as a carbon source was suitable for the ECP production; lactose was beneficial for the cell growth, and the production of ganoderic acids and ICP. The ganoderic acid production was remarkably improved by pulse feeding of lactose. In a work of Berovic et al. [6], *G. lucidum* was cultivated in a stirred tank bioreactor in a liquid substrate based on potato dextrose and olive oil. The cultivation conditions were: temperature of cultivation T=30 °C; mixing N=300/min; aeration Qg=10 l /min; average values of pH 5.8-4.2; oxygen partial pressure 70-80%, and redox potential Eh=300-400 mV. Torres López et al. [73] used a stirred tank bioreactor with the aeration rate maintained at 0.5 vvm. Different agitation strategies were tested.

2.1.2.11. G. Lucidum Cultivation in Air-lift Reactor

Lee et al. [58] cultivated of *G. Lucidum* in an air-lift reactor. The fermentor was a concentric draught-tube internal loop type, which can greatly reduce the shear stress as compared to an impeller-type fermentor. 5% of the culture was inoculated into the bioreactor and cultivated at 25 °C with aeration rate of 2.5 vvm under batch conditions. With the bistage pH control technique, in which pH was shifted from 3 to 6 at the initial phase of the exponential growth, 20.1 g/l exopolysaccharldes were produced; the desirable mycelia morphologies were retained during the cultivation, and a low viscosity and yield stress of the culture broth were maintained.

2.1.2.12. Modes of Operation: Batch and Fed-bach Cultivation

Batch Cultivation

The studies with *G. lucidum* cultivation have been frequently done in Erlenmeyer flasks, although some were performed in bioreactors. Batch cultivation has been the simplest and most often used and reported technique.

Fed-batch Cultivation

A fed-batch process for simultaneous production of ganoderic acids and polysaccharides by *G. lucidum* was described by Tang and Zhong [39]. Ganoderic acids production was improved by the pulse feeding of lactose, when its residual concentration was between 10 and 5 g/l. Effects of oxygen supply and lactose feeding on fed-batch submerged culture of *G. lucidum* were also investigated [40]. A higher volumetric oxygen transfer coefficient (K_La) value led to a higher biomass density and a higher productivity of both ICP and ganoderic acids. Berovic et al. [9] reported on fed-batch cultivation in a 10 l laboratory stirred tank reactor. In the first phase *G. lucidum* was grown in a batch culture for 95 hours, until a lower value of dissolved oxygen was reached. In the second phase, nitrogen source (500 ml substrate containing 12 g/l yeast extract) and carbon source (500 ml substrate containing 80 g/l glucose) were added into the reactor. 15.2 g/l of dry *G. lucidum* biomass was produced in a fed-batch cultivation, compared to 9.6 g/l in the one-phase batch process. Tang et al. [41, 42] achieved a significant synergistic enhancement of ganoderic acids accumulation, up to 754.6 mg/l, which was the highest reported in the submerged fermentation of *G. lucidum*.

2.1.3. Cultivation of G. Lucidum Biomass by Solid State Cultivation

G. lucidum cultivation in a solid state culture has several advantages, such as low investment costs, easier operations, minimum waste production, as well as some disadvantages, including the longer cultivation time, and the difficulty of separating/isolating the biomass and pharmaceutically active compounds from the solid substrate. Data on *G. lucidum* solid state cultivation in bioreactors are scarce. Habijanic et al. [67] patented a process of growing *G. lucidum* on a solid cultivation substrate using beech sawdust as a solid cultivation substrate the solid-state cultivation in a horizontal stirred bioreactor. Large quantities of biomass were produced, including the pharmaceutically applicable products.

2.1.3.1. Inoculum Size

Control of inoculation density was significant for cell growth, morphology, production of polysaccharides and ganoderic acids [13, 46, 58].

2.1.3.2. Influence of Substrate Composition

An advantage of solid state cultivation over other techniques is that a concentrated product can be obtained from a cheap substrate, such as an agricultural residue, with little pretreatment or enrichment. However, the use of an undefined medium, such as sawdust, might make the product purification more difficult. For this reason, solid state cultivation seems to be most appropriate for the production of pharmaceutically active animal feed supplements, for which the whole fermented substrate can be used as the product.

An example of a solid substrate compositon, used in the production of *G. lucidum* mycelia in a bioreactor, was given by Habijanic et al. [67], consisting of beech sawdust, olive oil, $(NH_4)_2SO_4$, KH_2PO_4, $CaCl_2 \cdot 2H_2O$, $MgSO_4 \cdot 7H_2O$, $FeSO_4 \cdot 7H_2O$ and distilled water. During cultivation, the content of ECP in the solids increased rapidly during the first 7 days, remained relatively constant until 21 days, and then decreased, suggesting that the polysaccharide was actually degraded in the latter stages of the process. The period during which the polysaccharide content decreased, corresponded with the time in which the water mass fraction was falling rapidly. However, it was not clear whether there was a direct cause-and-effect relationship between these two observations [13, 66]. Hsieh and Yang [53] experimented with reusing soy residue for the production of *G. lucidum* in the solid-state cultivation. Proper mixing with other solid medium was necessary for adjusting the C/N ratio. Song et al. [62] used a whey permeate as an alternative aditive to a growth medium for the cultivation *G. lucidum* mycelia in a solid-state cultivation The optimum condition was found at pH 4.4 and at 29 °C.

2.1.3.3. Bioreactor

Very little information is available on the production of *G. lucidum* mycelium in a bioreactor by a solid state cultivation. In a rare report available by Habijanic et al., [67], the cultivation was carried out in a horizontal stirred tank reactor with a total working volume of 30 l. The conditions were controlled as follows: temperature 30 °C, airflow 2 l/min, agitation rate 80 rpm for 2 min every second day during the first 7 days, and every day during the latter stages of the cultivation. At least a_w 0.85 was necessary to give satisfactory rates of cell growth and polysaccharide production.

2.2. Main Pharmacological Effects and Active Compounds in *G. Lucidum*

Most important active constituents of *Ganoderma* mushrooms are triterpenoids and polysaccharides. Their pharmacological effects are summarised in Figure 14.

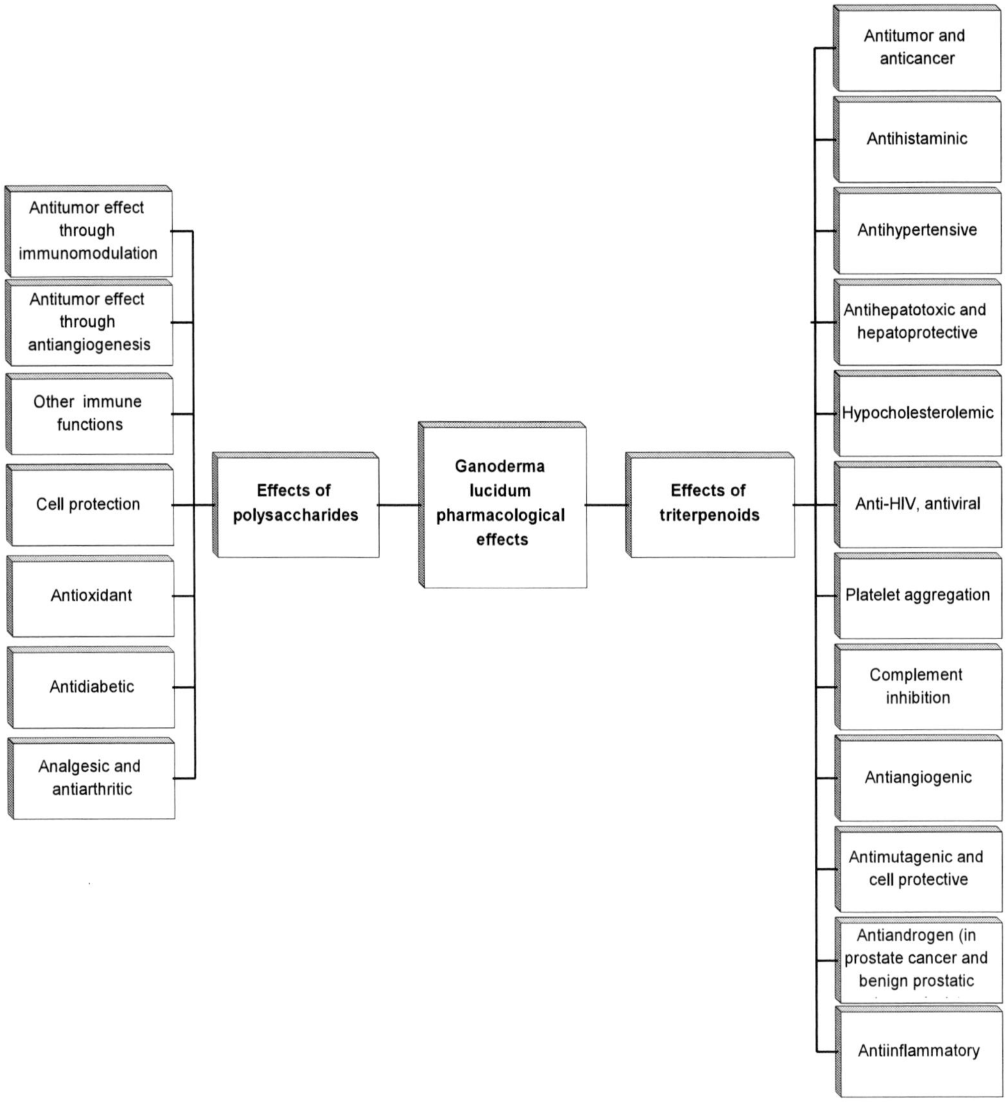

Figure 14. Main pharmacological effects of *G. lucidum.*

2.2.1. Triterpenoids

2.2.1.1. Triterpenoid Active Compounds

Over 150 triterpenoids were found in *Ganoderma* spp., such as ganoderic (highly oxygenated C_{30} lanostane-type triterpenoids), lucidenic, ganodermic, ganoderenic, ganolucidic and applanoxidic acids, lucidones, ganoderals and ganoderols [74, 75, 76, 77, 78, 79, 80, 81, 82, 83, 84, 85, 86, 87]. Representative examples are shown in Figures 15 - 23.

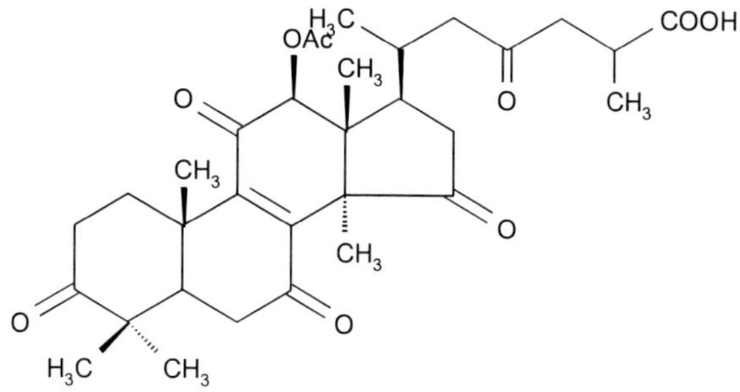

Figure 15. Ganoderic acid F 12β-acetoxy-3,7,11,15,23-pentaoxo-5α-lanost-8-en-26-oic acid).

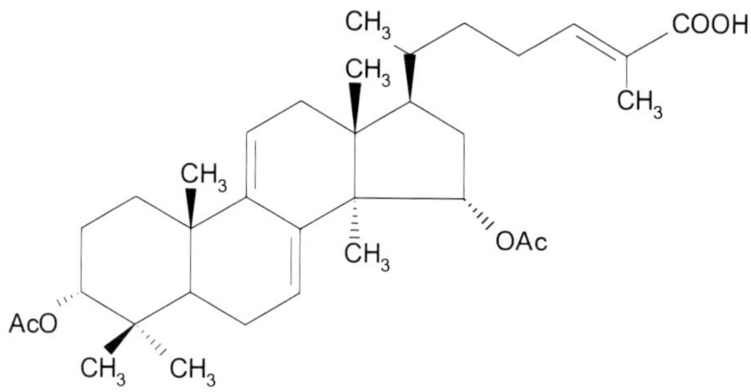

Figure 16. Ganodermic acid R (24E)-3a,15a-diacetoxy-5a-lanosta-7,9(11),24-triene-26- oic acid.

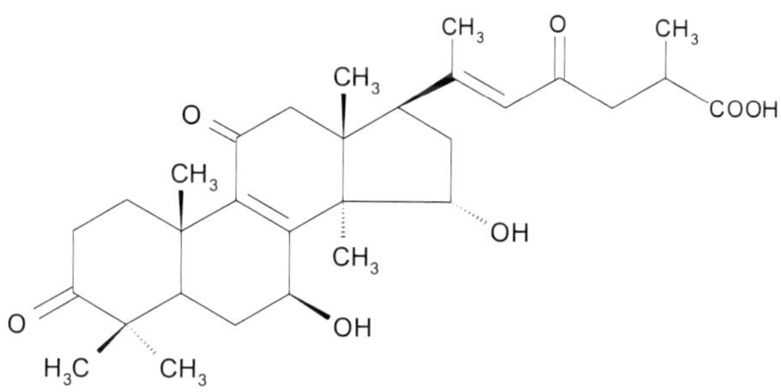

Figure 17. Ganoderenic acid A (20E)-7β,15α-dihydroxy-3,11,23-trioxo-5α-lanosta-8,20-dien-26-oic acid.

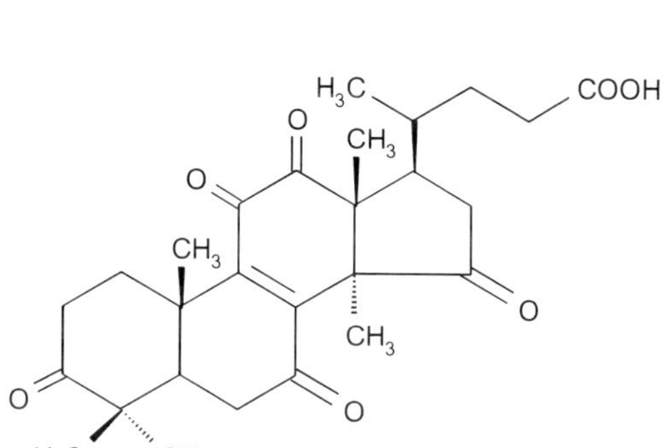

Figure 18. Lucidenic acid D1 4,4,14α-trimethyl-3,7,11,12,15-pentaoxo-5α-chol-8-en-24-oic acid.

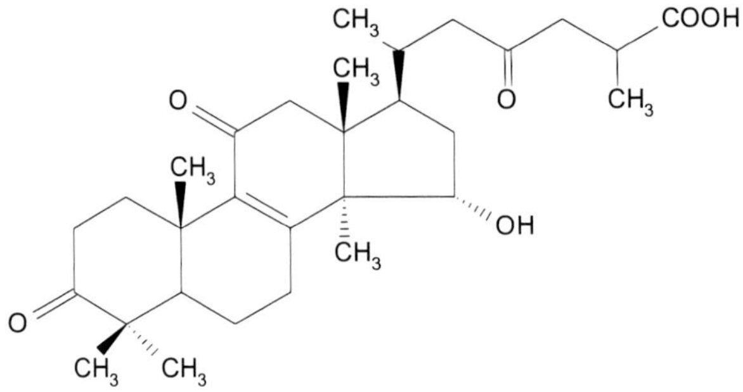

Figure 19. Ganolucidic acid A 15a-hydroxy-3,11,23-trioxo-5a-lanost-8-en-26-oic acid.

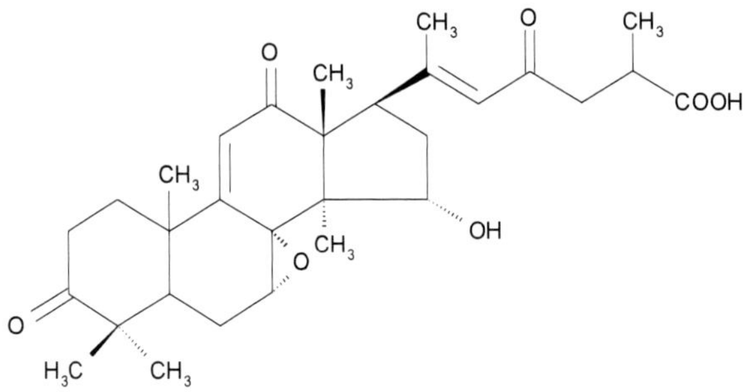

Figure 20. Applanoxidic acid A (20E)-15α-hydroxy-7α,8α-epoxy-3,11,23-trioxo-5α-lanosta-9(11),20-dien-26-oic acid.

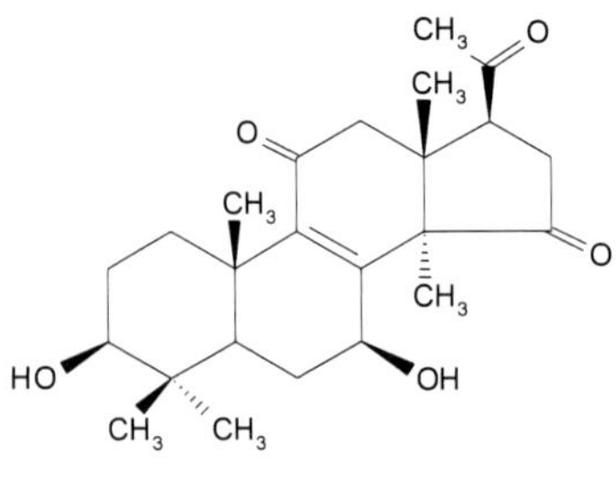

Figure 21. Lucidone 3b,7b-dihydroxy-4,4,14a-trimethyl-11,15,20-trioxo-5a-pregn-8-ene.

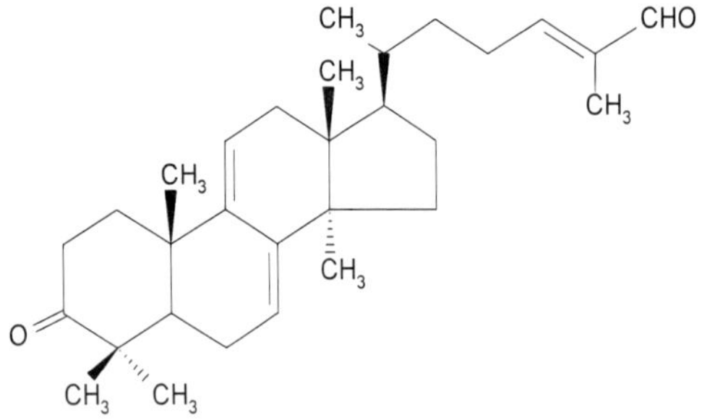

Figure 22. Ganoderal A(24E)-3-oxo-5a-lanosta-7,9(11),24-triene-26-al.

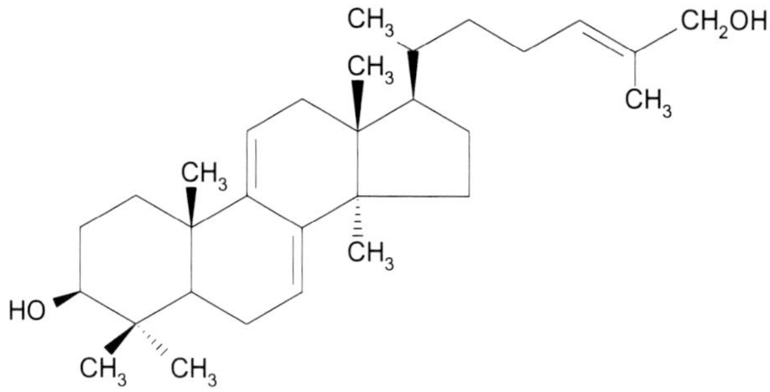

Figure 23. Ganoderol B ganodermadiol-5a-lanosta-7,9(11),24-triene-3b,26-diol.

2.2.1.2. Pharmacological Effects of Triterpenoids

Pharmacological effects of *G. lucidum* triterpenoids have been systematically studied for about 25 years. A literature surwey reveals the following effects of triterpenoid extracts or purified triterpenoid compounds:

- Antitumor and anticancer effects [36, 85, 86, 87, 88, 89, 90, 91, 92, 93, 94]
- Antihistaminic effects [95],
- Antihypertensive effects [76, 96],
- Antihepatotoxic and hepatoprotective effects [74, 97],
- Hypocholesterolemic effects [79, 80],
- AntiHIV activity [98, 99, 100],
- Platelet aggregation effects [101, 102, 103],
- Complement inhibition effects [104],
- Antiangiogenic effects [105],
- Antimutagenic and cell protective effect [106],
- Antiandrogen effect in prostate cancer and benign prostatic hyperplasia [107],
- Antiinflammatory effects [108, 109].

2.2.2. Polysaccharides

2.2.2.1. Polysaccharide Active Compounds

A lot of scientific attention has been focussed on *Ganoderma* polysaccharides, which represent a structurally diverse class of biological macromolecules with a wide-range of physicochemical properties. Studies have shown that the most active immunomodulatory polysaccharides are water-soluble β-1-3-D and β-1-6-D glucans, which can be precipitated by ethanol. Their prevailing structure is β-1-3 D-glucopyronan with 1–15 units of β-1-6 monoglucosyl side chains. Their 1,3-linked backbone, relatively small side chains, and an organized helical structure are beneficial for the immunostimulation [110].

Other immunomodulatory polysaccharides with different composition and structure have been reported, such as:

- galactose rich extracellular polysaccharide (GLP-2), which is composed of galactose, mannose, glucose, arabinose and rhamnose in the molar ratios 103:17:12:10:3 [111],

- heteropolysaccharide (LZ-C-1) with a molecular weight of 7 x10^3 Da, composed of L-Fuc, D-Glc and D-Gal [112],
- water-soluble branched glucan (GSG), extracted from the spores, with several different linkages [113],
- water insoluble anti-tumor polysaccharides [114]; such as a linear water-insoluble (1-3)-β-D-glucan (GL-IV-I), isolated from the fruit body of *G. lucidum* by NaOH solution extraction [115].

2.2.2. Pharmacological Effects of Polysaccharides

Reports on the pharmacological activity of *Ganoderma* polysaccharides mainly focus on antitumor effects, which are primarily linked to immunomodulation, although other effects, such as anti-angiogenic activity and protection of cells, have also been observed. Bioactive polysaccharides have been isolated from the fruit bodies, spores and from the mycelial biomass cultivated in liquid culture. Few ECP have been isolated from the culture medium. Scientific publications report on the following effects of *Ganoderma* polysaccharide extracts or purified compounds:

- Antitumor effects – observed as inhibition of tumor growth [116, 117, 118, 119, 120, 121], stimulation of immune functions of macrophages and T lymphocytes [122, 123, 124, 125, 126], cytokine release, such as tumor necrosis factor-α (TNF-α), interferon-γ (IF-γ) and interlukins (IL-1, IL-2, IL-3, IL-6) [92, 111, 122, 127, 128, 129, 130, 131, 132], and antitumor effects through anti-angiogenesis [137, 138, 139],
- Immunomodulation – reduction of the immunosuppresive response induced by anti-cancer drugs [133, 134]; enhancement of unspecific immune functions [135], induction of gene expression patterns that could reverse resistance to chemotherapeutic drugs [136],
- Protection of living cells – genoprotection, protecting cellular DNA from oxidative or radiation damage [94, 140, 141], protective effects on macrophages [142], protective effect on pancreatic islets [143], preservation of the synaptic density protein, synaptophysin, which is exterminated in Alzheimer's disease [144],
- Hepatoprotective effect [111],
- Analgesic and antiarthritic activity [145],
- Antidiabetic activity [145, 146, 147],
- Antioxidant activity [146, 148, 149].

2.2.3. Peptidoglycanes and Proteins

One of the oldest known proteins isolated from *G. lucidum* is LZ-8, with immunomodulatory and immunosuppressive activities [31]. Glycopeptides and proteoglycans were described also by other researchers [150, 151], such as a proteoglycan with a carbohydrate: protein ratio of 10.4:1 [152], a ribonuclease with a molecular mass of 42 kDa [153], a proteinase A inhibitor with a molecular mass of 38 kDa and with a remarkable heat stability [154], ganodermin, a 15-kDa protein with antifungal activity [155], a glycoprotein with 9.3% neutral sugar - 114 kDa hexameric lectin, which was stable in the pH range of 5–9 and temperature up to 50 °C, and expressed a hemagglutinating activity [156], a 1.2 x10^4 Da water-soluble glycopeptide (GLPCW-II), consisting of 90% carbohydrate and 8% protein

[157], low molecular weight peptides with hepatoprotective effects [158],and a water-soluble glycopeptide LZ-B-1 [159].

2.2.4. Other Compounds

Although triterpenes, polysaccharides and peptidoglycanes have been most thoroughly investigated, other active compounds from *G. lucidum* have been described, such as adenosine with antiplatelet aggregation effect, lectins with mitogenic effect, alkaloids, fatty acids, vitamins and essential minerals, hydrocarbons, monoterpenes and sesquiterpenes [114, 160].

2.3. Medicinal Applications of *G. Lucidum*

Methods for extraction or isolation of *G. lucidum* triterpenoids and polysaccharides have been described in several papers [27, 36, 47, 88, 89, 105, 106, 107, 115, 116, 119, 121, 129, 147, 166, 168, 170, 171]. Numerous formulations have been developed, patented and used as nutraceuticals, nutriceuticals and pharmaceuticals, mainly with *G. lucidum* fruit bodies, spores, and water or ethanol extracts, rarely with purified active compounds. Products became available commercially on the market, formulated as a capsule, tablet, extract, syrup, tea, bolus, or as food and feed additive [161, 162, 163]. A number of preparations have undergone clinical trials [164, 165, 166, 167, 168, 169, 170, 171].

Although preparations of *G. lucidum* have been a commonly used in China for millennia as self-medication for the prevention and treatment of various medical diseases, there were, until recently, no reports on their side effects. However, in 2004, Yuen et al. reported on a case of hepatotoxicity in a 78-year-old female, most likely due to the ingredients of the powder *G. lucidum* formulation [172]. Some internet resources [173] listed other possible side effects, such as dizziness, nausea, upset stomach and diarrhea, and warned that *G. lucidum* could interact with some medications, including high blood pressure medications, anticoagulants and hypoglycemic drugs.

Reports on beneficial pharmacological activity of extracts, partly purified preparations and isolated compounds from *G. lucidum* are convincing; there is abundant scientific evidence that triterpenoids, polysaccharides and proteoglycans have been effective *in vitro* and *in vivo* testing. Synergistic effects of mixtures of active components have been known, however, their biological activities need further assessment before they can be fully accepted not only by the traditional Asian medicine, but also by the Western science and medicine. In this respect, modern biotechnological cultivation methods in bioreactors enable fast, efficient and economical production of *G. lucidum* biomass in sufficient quantities for potential future pharmaceutical large-scale industrial production.

ACKNOWLEDGMENTS

Photographs in figures 1 - 5 and 11 - 12, previously unpublished, were donated by A. and M. Gregori.

REFERENCES

[1] Wasser, S. P. 2002. Medicinal mushrooms as a source of antitumor and immunomodulating polysaccharides. *Applied Microbiology And Biotechnology*, *60*, 258-274.

[2] Chen, J. Z. & Seviour, R. 2007. Medicinal importance of fungal ß-(1-3), (1-6)-glucans. *Mycological Research*, 111, 635-652.

[3] Moradali, M. F., Mostafavi, H., Ghods, S. & Hedjaroude, G. A. 2007. Immunomodulating and anticancer agents in the realm of macromycetes fungi (macrofungi), *International Immunopharmacology*, *7*, 701-724.

[4] FreePatentsOnline – Search patents, URL http://www.freepatentsonline.com/search.html. Accessed 25/1/2010.

[5] Yang, F. C. & Liau, C. B. 1998. The influence of environmental conditions on polysaccharide formation by *Ganoderma lucidum* in submerged cultures. *Process Biochemistry*, *33*, 547-553.

[6] Berovic, M., Habijanic, J., Zore, I., Wraber, B., Hodzar, D., Boh, B. & Pohleven, F. 2003. Submerged cultivation of *Ganoderma lucidum* biomass and immunostimulatory effects of fungal polysaccharides. *Journal of Biotechnology*, *103*, 77-86.

[7] Gibbs, P. A., Seviour R. J., & Schmid F. 2000. Growth of filamentous fungi in submerged culture: Problems and possible solutions. *Critical Reviews in Biotechnology*, *20*, 17-48.

[8] Papagianni, M. 2004. Fungal morphology and metabolite production in submerged mycelial processes. *Biotechnology Advances*, *22*, 189- 259.

[9] Habijanic, J. & Berovic, M. 2000. The relevance of solid-state substrate moisturing on *Ganoderma lucidum* biomass cultivation. *Food Technology and Biotechnology*, *38*, 225-228.

[10] Berovic, M. & Popovic M., 2001. Characterisation of gas mixed bioreactors in submerged citric acid fermentation. *Chemical* and *Biochemical Engineering Quarterly, 15*, 65-69.

[11] Schugerl, K., Wittler, R. & Lorenz, T. 1983. The use of mold in pellet form. *Trends in Biotechnology*, *1*, 120-123.

[12] Berovic, M., 2007. Bioreactor Types. In: *Encyclopedia of life support systems* [online] URL http://www.eolss.net/outlinecomponents/Biotechnology.aspx.. 31 pp. Oxford, UK: Eolss Publishers.

[13] Wagner R., Mitchell D. A., Lanzi Sassaki G., Lopes de Almeida Amazonas M. A. & Berovic, M. 2003. Submerged Cultivation of *Ganoderma lucidum*, *Food Technology and Biotechnology*, *41*, 371-382.

[14] Hu, C., Meguro, S. & Kawachi, S. 2004. Effects of physical properties of wood on the water activity of wood meal media for the cultivation of edible mushrooms. *Journal of Wood Science*, *50*, 365-370.

[15] Abdullah, A. L., Tengerdy, R. P. & Murphy, V. G. 1984. Optimization of solid substrate fermentation of wheat straw. *Biotechnology and Bioengineering*, *27*, 20-27.

[16] Mitchell, D. A., Berovic M. & Krieger N. 2006. Introduction to solid-state fermentation bioreactors. In: *Solid-State Fermentation Bioreactors: Fundamentals of Design and Operation* (edited by D. A., Mitchell, N. Krieger, M. Berovic). Pp. 33-44. Berlin, Germany: Springer.

[17] Lonsane, B. K., Saucedo-Castaneda, G., Raimbault, M., Roussos, S., Viniegra-Gonzalez, G., Ghildyal, N. P., Ramakrishna, M. & Krishnaiah, M. M. 1992. Scale-up strategies for solid state fermentation systems. *Process Biochemistry*, *27*, 259-273.

[18] Berovic, M. & Ostroversnik, H. 1997. Production of *Aspergillus niger* pectolytic enzymes by solid state bioprocessing of apple pomace. *Journal of Biotechnology*, *53*, 47-53.

[19] Svagelj, M., Berovic, M., Boh, B., Menard, A., Simcic, S. & Wraber, B. 2008. Solid-state cultivation of *Grifola frondosa* (Dicks: Fr) S.F. Gray biomass and immunostimulatory effects of fungal intra- and extracellular ß-polysaccharides. *New Biotechnology*, *25*, 150-155.

[20] Mitchell, D.A. & Berovic, M. 2005. Solid state bioprocessing. In: *Biochemical engineering principles* (edited by M. Berovic & A.W. Nienow). Pp. 204-236. Ljubljana, Slovenia: University of Ljubljana, FKKT Press.

[21] Jong, S. C. & Birmingham, J. M. 1992. Medicinal benefits of the mushrooms Ganoderma. *Advances in Applied Microbiology*, *37*, 101-134.

[22] Leung, S. W. S. 2002. Lingzhi (*Ganoderma*) research - the past, present and future perspectives. In *Ganoderma*: *Genetics, chemistry, pharmacology and therapeutics, Proceedings of International symposium on Ganoderma research, October 21-23, 2002* (edited by Z. B. Lin). Pp. 1-9. Shanghai, China: Beijing, Medical University Press.

[23] Kim, H.W. & Kim, B.K. 2002. Recent advances on the biologically active triterpenoids of *Ganoderma lucidum*. In Ganoderma: *Genetics, chemistry, pharmacology and therapeutics, Proceedings of International symposium on Ganoderma research, October 21-23, 2002* (edited by Z.B. Lin). Pp. 10-19. Shanghai, China: Beijing, Medical University Press.

[24] Liu, G. T., Bao, X., Niu, S., Li, Z. & Sung, Z. 1979. Some pharmacological actions of the spores of *Ganoderma lucidum* and the mycelium of *Ganoderma capense* (Lloyd) Teng cultivated by submerged fermentation. *Chinese Medical Journal*, *92*, 469-500.

[25] Maruyama, H., Yamazaki, K., Murofushi, S., Konda, C. & Ikekawa, T. 1989. Antitumor activity of *Sarcodon aspratus* (Berk.) S. Ito and *Ganoderma lucidum* (Fr.) Karst. *Journal of Pharmacobio-Dynamics*, *12*, 118-123.

[26] Tasaka, K., Akag, M., Miyoshi, K., Mio, M. & Makino, T. 1988. Anti-allergic constituents in the culture medium of *Ganoderma lucidum*. 1. Inhibitory effect of oleic acid on histamine release. *Agents and Action*, *23*, 153-156.

[27] Lee, S.Y. & Rhee, H. M. 1990. Cardiovascular effects of mycelium extract of *Ganoderma lucidum* - inhibition of sympathetic outflow as a mechanism of its hypotensive action. *Chemical & Pharmaceutical Bulletin*, *38*, 1359-1364.

[28] Hirose, K., Muto, S., Niimura, K., Ohara, M., Oguchi, Y., Matsunaga, K., Kadochi, J., Sugita, N., Furushu, T., Yoshikumi, C. & Takahashi, M. 1988. Antiviral agent. Japanese patent JP 63316734. 1988 December 26.

[29] Mizumoto, K., Yamashita, A., Kii, M. & Sumio, H. 1990. Antiretrovirus agent. Japanese patent JP 2032026. 1990 February 1.

[30] Kino, K., Sone, T., Watanabe, J., Yamashita, A., Tsuboi, H., Miyajima, H. & Tsunoo, H. 1991. Immunomodulator, LZ-8, prevents antibody-production in mice, *International Journal of Immunopharmacology,13*, 1109-1115.

[31] Van Der Hem, L. G., Van Der Vliet, J. A., Bocken, C. F. M., Kino, K., Hoitsma, A. J & Tax W.J.M. 1995. Ling zhi-8-studies of a new immunomodulating agent, *Transplantation (Baltimore)*, *60*, 438-443.

[32] Triratana, S., Thaithatgoon, S. & Gawgla, M. 1991. Cultivation of *Ganoderma lucidum* in sawdust bags. In: *Science and Cultivation of Edible Fungi, Proceedings of the 13th International Congress on the Science Cultivation of Edible Fungi, Dublin, 1-6 September 1991* (edited by M.J. Maher). Pp. 567-572, Rotterdam: A. A. Balkema.

[33] Stamets P. 2000. *Ganoderma lucidum* (Wm. Curtis: Fries) Karsten. In: *Growing Gourmet and Medicinal* Mushrooms. 3rd ed., Pp. 352-366. Berkeley, USA: Ten Speed Press.

[34] Kohlmünzer, S., Wegiel, J. & Sitarz, J. 1989. Polysaccharides in mycelial culture of *Ganoderma applanatum* (Pers.) Pat. *Herba Hungarica*, 28, 87-94.

[35] Zore, I., Berovic, M., Boh, B., Hodzar, D. & Pohleven, F. 1998. Procedure for preparation of inoculum for growing of fungus *Ganoderma lucidum* by submersion fermentation, Slovenian patent SI 9700014, 1998 August 31.

[36] Lin, S. B., Li, C. H., Chen, Y. R., Kan, L. S. & Lee, S. S. 2002. Triterpene extract from *Ganoderma lucidum* inhibits growth of hepatoma Huh7 cells: involvement of oxidative stress induction. In Ganoderma: *Genetics, chemistry, pharmacology and therapeutics, Proceedings of International symposium on Ganoderma research, October 21-23, 2002* (edited by Z.B. Lin). Pp. 176-182. Shanghai, China: Beijing, Medical University Press.

[37] Fang, Q.H. & Zhong, J.J. 2002. Effect of initial pH on production of ganoderic acid and polysaccharide by submerged fermentation of *Ganoderma lucidum. Process Biochemistry, 37*, 769-774.

[38] Fang, Q.H. & Zhong, J.J. 2002. Submerged fermentation of higher fungus *Ganoderma lucidum* for production of valuable bioactive metabolites - ganoderic acid and polysaccharide, *Biochemical Engineering Journal*, *10*, 61-65.

[39] Tang, Y. J. & Zhong, J. J. 2002. Fed-batch fermentation of *Ganoderma lucidum* for hyperproduction of polysaccharide and ganoderic acid, *Enzyme and Microbial Technology*, *31*, 20-28.

[40] Tang, Y. J. & Zhong, J. J. 2003. Role of oxygen supply in submerged fermentation of *Ganoderma lucidum* for production of Ganoderma polysaccharide and ganoderic acid, *Enzyme and Microbial Technology*, *32*, 478-484.

[41] Tang, Y. J., Zhang, W. & Zhong, J. J. 2009. Performance analyses of a pH-shift and DOT-shift integrated fed-batch fermentation process for the production of ganoderic acid and Ganoderma polysaccharides by medicinal mushroom *Ganoderma lucidum. Bioresource Technology, 100*,1852-1859.

[42] Tang, Y. J., Zhang, W. & Zhu, L. W. 2009. Hyperproduction of Ganoderma polysaccharides and ganoderic acid during fermentation of medicinal mushroom *Ganoderma lucidum* by multi-stage control and induced culture. *Journal of Bioscience and Bioengineering*, *108*, S114-S134.

[43] Xu, P., Ding, Z. Y., Qian, Z., Zhao, C. X. & Zhang, K. C. 2008. Improved production of mycelial biomass and ganoderic acid bysubmerged culture of *Ganoderma lucidum* SB97 using complex media, *Enzyme and Microbial Technology*, *42*, 325-331.

[44] Zhang, W. X. & Zhong, J. J. 2010. Effect of oxygen concentration in gas phase on sporulation and individual ganoderic acids accumulation in liquid static culture of *Ganoderma lucidum*. *Journal of Bioscience and Bioengineering*, *109*, 37-40.

[45] Papinutti L. 2010. Effects of nutrients, pH and water potential on exopolysaccharides production by a fungal strain belonging to *Ganoderma lucidum* complex. *Bioresource Technology*, *101*, 1941-1946.

[46] Mizuno, T., Wang, G., Zhang, J., Kawagishi, H., Nishitoba, T. & Li, J. 1995. Reishi, *Ganoderma lucidum* and *Ganoderma tsugae*: bioactive substance and medicinal effects, *Food Reviews International*, *11*,151-166.

[47] Miyahara, R., Yoshimoto, T. & Asawa, K. 1987. Chemical structures and changes of extracts during growth of reishi (*Ganoderma lucidum*), *Mokuzai Gakkaishi*, *33*, 416-422.

[48] Chan, K. 2002. Linking chemical and biological characteristics in assuring the quality of Chinese medicinal materials and OTC products. In Ganoderma: *Genetics, chemistry, pharmacology and therapeutics, Proceedings of International symposium on Ganoderma research, October 21-23, 2002* (edited by Z.B. Lin). Pp. 24-34. Shanghai, China: Beijing, Medical University Press.

[49] Boh, B., Berovic, M., Zhang, J. S. & Lin Z. B. 2007. *Ganoderma lucidum* and its pharmaceutically active compounds. In: *Biotechnology annual review, Vol. 13* (edited by M. R. El-Gewely) pp. 265-301. Amsterdam, Netherlands: Elsevier

[50] Shigeru, Y. 1991. Cultivation of Reishi, Japanese patent JP3083521. 1991 April 9.

[51] Gonzalez-Matute, R., Figlas, D., Devalis R., Delmastro, S. & Curvetto, N. 2002. Sunflower seed hulls as a main nutrient source for cultivating *Ganoderma lucidum*. *Micologia Aplicada International*, *14*, 19-24.

[52] Yang, F. C., Hsieh, C. & Chen, H. M. 2003. Use of stillage grain from a rice-spirit distillery in the solid state fermentation of *Ganoderma lucidum*. *Process Biochemistry*, *39*, 21-26.

[53] Hsieh, C. & Yang, F.C. 2004. Reusing soy residue for the solid-state fermentation of *Ganoderma lucidum*. *Bioresource Technology*, *91*, 105-109.

[54] Fang, Q.H., Tang, Y.J. & Zhong, J.J. 2002d. Significance of inoculation density control in production of polysaccharide and ganoderic acid by submerged culture of *Ganoderma lucidum*, *Process Biochemistry*, *37*, 1375-1379.

[55] Yang, F. C., Ke, Y. F. & Kuo, S. S. 2000. Effect of fatty acids on the mycelial growth and polysaccharide formation by *Ganoderma lucidum* in shake-flask cultures. *Enzyme and Microbial Technology*, *27*, 295-301.

[56] Simonic, J., Stajic, M., Glamoclija, J. Vukojevic, J., Duletic-Lausevic, S. & Brceski, I. 2008. Optimization of submerged cultivation conditions for extra- and intracellular polysaccharide production by medicinal Ling Zhi or Reishi mushroom *Ganoderma lucidum* (W. Curt.: Fr.) P. Karst. (Aphyllophoromycetideae). *International Journal of Medicinal Mushrooms*, *10*, 351-360.

[57] Hsieh, C., Tseng, M.H. & Liu, C.J. 2006b. Production of polysaccharides from *Ganoderma lucidum* (CCRC 36041) under limitations of nutrients. *Enzyme and Microbial Technology*, *38*, 109-117.

[58] Lee, K. M., Lee, S. Y. & Lee, H. Y. 1999. Bistage control of pH for improving exopolysaccharide production from mycelia of Ganoderma lucidum in an air-lift fermentor. *Journal of Bioscience and Bioengineering*, *88*, 646-650.

[59] Fang, J. N., Bao, X.F. & Yuen, W.H. 2002a. Studies on the polysaccharides from spores of *Ganoderma lucidum*. In Ganoderma: *Genetics, chemistry, pharmacology and therapeutics, Proceedings of International symposium on Ganoderma research, October 21-23, 2002* (edited by Z.B. Lin). Pp. 98-103. Shanghai, China: Beijing, Medical University Press.

[60] Hsieh, C., Hsu, T.H. & Yang, F.C. 2005. Production of polysaccharides of *Ganoderma lucidum* (CCRC36021) by reusing thin stillage, *Process Biochemistry*, *40*, 909-916.

[61] Hsieh, C., Liu, C.J., Tseng, M. H., Lo, C. T. & Yang, Y. C. 2006. Effect of olive oil on the production of mycelial biomass and polysaccharides of *Grifola frondosa* under high oxygen concentration aeration, *Enzyme and Microbial Technology*, *39*, 434-439.

[62] Song, M., Kim, N., Lee, S. & Hwang, S. 2007. Use of whey permeate for cultivating *Ganoderma lucidum* mycelia, *Journal of Dairy Science*, *90*, 2141-2146.

[63] Lee, H., Song, M., Yu, Y. & Hwang, S. 2003b. Optimizing bioconversion of deproteinated cheese whey to mycelia of *Ganoderma lucidum*, *Process Biochemistry*, *38*, 1685-1693.

[64] Lee, H., Song, M., Yu, Y. & Hwang, S. 2003c. Production of *Ganoderma lucidum* mycelium using cheese whey as an alternative substrate: response surface analysis and biokinetics, *Biochemical Engineering Journal*, *15*, 93-99.

[65] Chang, M. Y., Tsai, G. J. & Houng, J. Y. 2006. Optimization of the medium composition for the submerged culture of *Ganoderma lucidum* by Taguchi array design and steepest ascent method. *Enzyme and Microbial Technology*, *38*, 407–414.

[66] Park, J. P., Kim, S. W., Hwang, H. J. & Yun, J. W. 2001. Optimization of submerged culture conditions for the mycelial growth and exo-biopolymer production by *Cordyceps militaris*, *Letters in Applied Microbiology*, *33*, 76-81.

[67] Habjanic, J. & Berovic, M. 2002. Process of cultivation of fungus *Ganoderma lucidum* on a solid cultivation substrate. Patent SI 20923. 2002 December 31.

[68] Trinci, A. P. J, Wiebe, M. G. & Robson, G. D. 1994. Physiology of filamentous fungi. In: Modelling for improved bioreactor performance, 2. Modelling of filamentous fungi. International Conference, 15-16 September 1994, Otocec, Slovenia (edited by M. Berovič). Pp. 9. Ljubljana, Slovenia: National Institute of Chemistry Press.

[69] Trinci, A. P. J. 1983. Effect of Julon on morphology of *Aspergillus niger* and its use in making turbidity measurements of fungal growth, *Transactions of the British Mycological Society*, *81*, 408-412.

[70] Byrne, G. S. & Ward O. P. 1987. Effect of polymers on pelleting of *Rhizopus arrhizus*, *Transactions of the British Mycological Society*, *89*, 367- 371.

[71] Jones, P., Shabab B. A., Trinci, A. P. J. & Moore D. 1988. Effect of polymeric additives, especially Junlon and Hostacerin, on growth of some basidiomycetes in submerged culture, *Transactions of the British Mycological Society*, *90*, 577-583.

[72] Yang, F. C., Yang, M-J. & Cheng, S. H. 2009. A novel method to enhance the mycelia production of *Ganoderma lucidum* in submerged cultures by polymer additives and agitation strategies, *Journal of the Taiwan Institute of Chemical Engineers*, *40*,148-154.

[73] Torres Lopez, A. M., Gaviria Soto, J. A., Quintero Diaz J. C., Atehortua Garces L., Rojas Vahos D. F. 2009. Characterization of the medicinal mushroom *Ganoderma lucidum* submerged culture in stirred tank, *New Biotechnology*, *25*, S195-196.

[74] Hirotani, M., In,o C., Furuya, T. & Shiro, M. 1986. Ganoderic acids T, S, and R, new triterpenoids from the cultured media of *Ganoderma lucidum*, *Chemical & Pharmaceutical Bulletin*, *34*, 2282-2285.

[75] Kikuchi, T., Kanomi, S., Murai, Y., Kadota, S., Tsubono, K. and Ogita, Z. 1986. Constituents of the fungus Ganoderma lucidum (Fr.) Karst. I. Structures of ganoderic acids C2, E, I, and K, lucidenic acid F and related compounds, *Chemical & Pharmaceutical Bulletin*, *34*, 3695-3712.

[76] Morigiwa, A., Kitabatake, K., Fujimoto, Y. & Ikekawa, N. 1986. Angiotensin converting enzyme inhibitory triterpenes from *Ganoderma lucidum*, *Chemical & Pharmaceutical Bulletin*, *34*, 3025-3028.

[77] Nishitoba, T., Sato, H. & Sakamura, S. 1987. Triterpenoids from the fungus *Ganoderma lucidum*, *Phytochemistry*, *26*, 1777-1784.

[78] Nishitoba, T., Goto, S., Sato, H. & Sakamura, S. 1989. Bitter triterpenoids from the fungus *Ganoderma applanatum*, *Phytochemistry*, *28*, 193-197.

[79] Lin, L. J. & Shiao, M. S. 1988a. Seven new triterpenes from *Ganoderma lucidum*, *Journal of Natural Products*, *51*, 918-924.

[80] Lin, L.J., Shiao, M.S. & Yeh, S.F. 1988b. Triterpenes from *Ganoderma lucidum*, *Phytochemistry*, *27*, 2269-2271.

[81] Shiao, M. S., Lin, L. J., Yeh, S. F. & Chou, C. S. 1987. Two new triterpenes of the fungus *Ganoderma lucidum*, *Journal of Natural Products*, *50*, 886-890.

[82] Chairul, Tokuyama, T., Hayashi, Y., Nishizawa, M., Tokuda, H., Chairul, S. M. & Hayashi, Y. 1991. Applanoxidic acids A, B, C and D, biologically active tetracyclic triterpenes from *Ganoderma applanatum*, *Phytochemistry*, 30, 4105-4109.

[83] Chairul, Hayashi, Y., & Chairul, S. M. 1994. Lanostanoid triterpenes from *Ganoderma applanatum*, *Phytochemistry*, *35*, 1305-1308.

[84] Yang, M., Wang, X., Guan, S., Xia, J., Sun, J., Guo, H. & Guo, D. 2007. Analysis of triterpenoids in *Ganoderma lucidum* using liquid chromatography coupled with electrospray ionization mass spectrometry, *Journal of the Americal Society for Mass Spectrometry*, *18*, 927-939.

[85] Toth, J. O., Luu, B. & Ourisson, G. 1983. Les acides ganoderiques T a Z: triterpenes cytotoxiques de *Ganoderma lucidum* (Polyporaceae), *Tetrahedron Letters*, *24*, 1081-1084.

[86] Gao, J. J., Min, B. S., Ahn, E. M., Nakamura, N., Lee, H. K. & Hattori, M. 2002. New triterpene aldehydes, lucialdehydes A-C, from *Ganoderma lucidum* and their cytotoxicity against murine and human tumor cells, *Chemical & Pharmaceutical Bulletin (Tokyo)*, *50*, 837-840.

[87] Min, B. S., Gao, J. J., Nakamura, N. & Hattori, M. 2000. Triterpenes from the spores *of Ganoderma lucidum* and their cytotoxicity against meth-A and LLC tumor cells, *Chemical & Pharmaceutical Bulletin (Tokyo)*, *48*, 1026-1033.

[88] Lin, S. B., Li, C. H., Lee, S. S. & Kan, L. S. 2003. Triterpene-enriched extracts from *Ganoderma lucidum* inhibit growth of hepatoma cells via suppressing protein kinase C, activating mitogen-activatedprotein kinases and G2-phase cell cycle arrest, *Life Sciences*, *72*, 2381-2390.

[89] Zhu, H. S., Yang, X. L., Wang, L. B., Zhao, D. X. and Chen, L. 2000. Effects of extracts from sporoderm-broken spores of *Ganoderma lucidum* on HeLa cells, *Cell Biology and Toxicology*, *16*, 201-206.

[90] Yu, S. P., Zhang, J. S., Tang, Q. J., Shi, X. M., Liu, Y. F., Yan, Y. & Pan, Y. L. 2004. Correlation between intracellular triterpenes from mycelia of *Ganoderma lucidum* in different growth stages and inhibition effect on tumor cells, *Mycosystema*, *23*, 548-554.

[91] Liu, J., Shimizu, K., Konishi, F., Noda, K., Kumamoto, S., Kurashiki, K. & Kondo, R. 2007. Anti-androgenic activities of the triterpenoids fraction of *Ganoderma lucidum*, *Food Chemistry*, 100, 1691-1696.

[92] Mueller, C.I., Kumagai, T., Kelly, J.O., Seeram, N.P., Heber, D. & Koeffler, H.P. 2006. *Ganoderma lucidum* causes apoptosis in leukemia, lymphoma and multiple myeloma cells, *Leukemia Research*, 30, 841-848.

[93] Tang, W., Liu, J. W., Zhao. W. M., Wei, D. Z. & Zhong. J. J. 2006. Ganoderic acid T from *Ganoderma lucidum* mycelia induces mitochondria mediated apoptosis in lung cancer cells, *Life Sciences*, 80, 205-211.

[94] Yuen J.W.M., Gohel M.D.J. 2008. The dual roles of Ganoderma antioxidants on urothelial cell DNA under carcinogenic attack. *Journal of Ethnopharmacology*, *118*, 324-330.

[95] Kohda, H., Tokumoto, W., Sakamoto, K., Fujii, M., Hirai, Y., Yamasaki, K., Komoda, Y., Nakamura, H., Ishihara, S. & Uchida, M. 1985. The biologically active constituents of *Ganoderma lucidum* (Fr.) Karst histamine release-inhibitory triterpenes, *Chemical & Pharmaceutical Bulletin*, *33*, 1367-1374.

[96] Kabir, Y., Kimura, S. & Tamura, T. 1988. Dietary effect of *Ganoderma lucidum* mushroom on blood pressure and lipid levels in spontaneously hipertensive rats, *Journal of Nutritional Science and Vitaminology*, *34*, 433-438.

[97] Kim, D.H., Shim, S.B., Kim, N.J. & Jang, I.S. 1999. Beta-glucuronidase-inhibitory and hepatoprotective effect of *Ganoderma lucidum*, *Biological and Pharmaceutical Bulletin*, *22*, 162-164.

[98] Hattori, M., El-Mekkawy, S. & Meselhy, R. 1997. Inhibitory effects of components from *Ganoderma lucidum* on the growth of Human Immunodeficiency Virus (HIV) and the protease activity. In: *Proceedings of the 1st International Symposium on Ganoderma lucidum in Japan, Nov 17-18 1997* (edited by T. Mizuno, N. Ide, & Y. Hasegawa). Pp.128-135.

[99] El-Mekkawy, S., Meselhy, M. R., Nakamura, N., Tezuka, Y., Hattori, M., Kakiuchi, N., Shimotohno, K., Kawahata, T. & Otake, T. 1998.Anti-HIV-1 and anti-HIV-1 protease substances from *Ganoderma lucidum*, *Phytochemistry*, *49*, 1651-1657.

[100] Min, B. S., Nakamura, N., Miyashiro, H., Bae, K. W. & Hattori, M. 1998. Triterpenes from the spores of *Ganoderma lucidum* and their inhibitory activity against HIV-1 protease, *Chemical & Pharmaceutical Bulletin*, *46*, 1607-1612.

[101] Wang, C. N., Chen, J. C., Shiao, M. S. & Wang, C. T. 1989. The aggregation of human platelet induced by ganodermic acid, *S. Biochimica et Biophysica Acta*, *986*, 151-160.

[102] Su, C. Y., Shiao, M. S. & Wang, C. T. 1999. Predominant inhibition of ganodermic acid S on the thromboxane A2 -dependent pathway in human platelets response to collagen, *Biochimica et Biophysica Acta*, *1437*, 223-234.

[103] Su, C. Y, Shiao, M. S. & Wang, C. T. 2000. Potentiation of ganoderemic acid S on prostaglandin E(1)-induced cyclic AMP elevation in human platelets, *Thrombosis Research*, *99*, 135-145.

[104] Min, B. S., Gao, J. J., Hattori, M., Lee, H. K. & Kim, Y. H. 2001. Anticomplement activity of triterpenoids from the spores of *Ganoderma lucidum*, *Planta Medica*, *67*, 811-814.

[105] Kimura, Y., Taniguchi, M. & Baba, K. 2002. Antitumor and antimetastatic effects on liver of triterpenoid fractions of *Ganoderma lucidum*: Mechanism of action and isolation of an active substance, *Anticancer Research*, *22(6A)*, 3309-3318.

[106] Lakshmi, B., Ajith, T. A., Jose, N. & Janardhanan, K.K. 2006. Antimutagenic activity of methanolic extract of *Ganoderma lucidum* and its effect on hepatic damage caused by benzo[a]pyrene, *Journal of Ethnopharmacology*, *107*, 297–303.

[107] Liu, J., Shimizu, K., Konishi, F., Kumamoto, S. & Kondo, R. 2007. The anti-androgen effect of ganoderol B isolated from the fruiting body of *Ganoderma lucidum*, *Bioorganic & Medicinal Chemistry*, *15*, 4966-4972.

[108] Ko, H. H., Hung, C. F.,Wang, J. P. & Lin, C. N. 2008. Antiinflammatory triterpenoids and steroids from *Ganoderma lucidum* and *G. tsugae*, *Phytochemistry*, *69*, 234–239.

[109] Dudhgaonkar, S., Thyagarajan, A. & Sliva D. 2009. Suppression of the inflammatory response by triterpenes isolated from the mushroom *Ganoderma lucidum*, *International Immunopharmacology*, *9*, 1272-1280.

[110] Bao, X. F., Wang, X. S., Dong, Q., Fang, J. N. & Li, X. Y. 2002. Structural features of immunologically active polysaccharides from *Ganoderma lucidum*, *Phytochemistry*, *59*, 175-181.

[111] Li., Y.Q., Fang, L. & Zhang, K.C. 2007. Structure and bioactivities of a galactose rich, *Carbohydrate Polymers*, 68(2), 323-328.

[112] Ye, L.B., Zhang, J.S., Yang, Y., Zhou, S., Liu, Y.F., Tang, Q.J., Du, X.J., Chen, H. & Pan, Y.J. 2009. Structural characterisation of a heteropolysaccharide by NMR spectra, *Food Chemistry*, *112*, 962-966.

[113] Guo, L., Xie, J., Ruan, Y., Zhou, L., Zhu, H., Yun, X., Jian, Y., Lu, L., Chen, K., Min, Z., Wen, Y. & Gu, J. 2009. Characterization and immunostimulatory activity of a polysaccharide from the spores of *Ganoderma lucidum*, *International Immunopharmacology*, *9*, 1175-1182.

[114] Russell, R. & Paterson, M. 2006. Ganoderma - A therapeutic fungal biofactory. *Phytochemistry*, *67*,1985-2001.

[115] Wang, J. & Zhang, L., 2009. Structure and chain conformation of five water-soluble derivatives of a β-D-glucan isolated from *Ganoderma lucidum*, *Carbohydrate Research*, *344*, 105-112

[116] Lee, S. S., Chen, F. D., Chang, S. C., Wei, Y. H., Liu, I. & Chen, C. 1984. In vivo antitumor effect of crude extracts from the mycelium of *Ganoderma lucidum*, *Journal of Chinese Oncology Society*, *5*, 22-27.

[117] Hwang, S. F., Liu, K. J., Kuan, Y. H., Tung, K.S., Su, C. H. & Tung, T. C. 1989. The inhibitory effect on artificial pulmonary metastasis of murine S-180 Sarcoma cells by orally administered *Ganoderma lucidum*, *Journal of Chinese Oncology Society*, *5*, 10-15.

[118] Furusawa, E., Chou, S.C., Furasawa, S., Hirazum, A. & Dang, Y. 1992. Antitumor activity of *Ganoderma lucidum*, an edible mushroom, on intraperitoneally implanted Lewis lung carcinoma in synergeneic mice, *Phytotherapy Reearch*, 6, 300-304.

[119] Lu, H., Kyo, E., Uesaka, T., Katoh, O. & Watanabe, H. 2002. Prevention of development of N,N'-dimethylhydrazine-induced colon tumors by a water-soluble extract from cultured medium of *Ganoderma lucidum* (Rei-shi) mycelia in male ICR mice, *International Journal of Molecular Medicine*, *9*, 113-117.

[120] Liu, X., Yuan, J. P., Chuang, C. K. & Chen, X.J. 2002. Antitumor activity of the sporoderm-broken germinating spores of *Ganoderma lucidum*, *Cancer Letters*, *182*, 155-161.

[121] Lu, H., Kyo, E., Uesaka, T., Katoh, O. & Watanabe, H. 2003. A water-soluble extract from cultured medium of *Ganoderma lucidum* (Rei-shi) mycelia suppresses azoxymethane-induction of colon cancers in male F344 rats, *Oncology Reports*, *10*, 375-379.

[122] Wang, S. Y., Hsu, M. L., Hsu, H. C., Tzeng, C. H., Lee, S. S., Shiao, M. S., & Ho, C. K. 1997. The anti-tumor effect of *Ganoderma lucidum* is mediated by cytokines released from activated macrophages and T lymphocytes, *International Journal of Cancer*, *70*, 699-705.

[123] Hu, Y. H. & Lin, Z. B. 1999. Effects of polysaccharides isolated from mycelia of *Ganoderma lucidum* on HL-60 cell apoptosis, *Acta Pharmacologica Sinica*, *34*, 268-271.

[124] Hu, Y. H. & Lin, Z. B. 1999. Polysaccharides isolated from mycelia of *Ganoderma lucidum* induced HL-60 cell apoptosis by enhancing macrophage activity, *Chinese Pharmacology Bulletin*, *5*, 27-30.

[125] Lei, L. S., Li, M. C. & Sun, L. S. 2002. *Ganoderma lucidum* and its components on the function of macrophages. In Ganoderma: *Genetics, chemistry, pharmacology and therapeutics, Proceedings of International symposium on Ganoderma research, October 21-23, 2002* (edited by Z.B. Lin). Pp. 20-23. Shanghai, China: Beijing, Medical University Press.

[126] Cao, L. Z. & Lin, Z. B. 2003. Comparison of the effects of polysaccharides from wood-cultured and bag-cultured *Ganoderma lucidum* on murine spleen lymphocyte proliferation in vitro, *Acta Pharmacologica Sinica*, *38*, 92-97.

[127] Lee, S. S., Lee, P. L., Chen, C. F., Wang, S. Y. & Chen, K. Y. 2002. Antitumor effects of *Ganoderma lucidum*. In Ganoderma: *Genetics, chemistry, pharmacology and therapeutics, Proceedings of International symposium on Ganoderma research, October 21-23, 2002* (edited by Z.B. Lin). Pp. 46-57. Shanghai, China: Beijing, Medical University Press.

[128] Cao, L. Z. & Lin, Z. B. 2002. Regulation of *Ganoderma lucidum* polysaccharides on cytotoxic T lymphocytes induced by dendric cells in vitro. In Ganoderma: *Genetics, chemistry, pharmacology and therapeutics, Proceedings of International symposium on Ganoderma research, October 21-23, 2002* (edited by Z.B. Lin). Pp. 122-129. Shanghai, China: Beijing, Medical University Press.

[129] Boh, B., Berovic, M., Wraber, B., Hodzar, D., Habijanic, J., Pohleven, F. and Zore, I. 2004. *Ganoderma lucidum* (W.Curt.:Fr.) Lloyd and G. applanatum (Pers.) Pat. (Aphyllophoromycetideae) from Slovenian habitats: cultivation, isolation, and testing of active compounds, *International Journal of Medicinal Mushrooms*, *6*, 15-32.

[130] Chen, H. S., Tsai, Y. F., Lin, S., Lin, C. C., Khoo, K. H., Lin, C. H. & Wong, C. H. 2004. Studies on the immuno-modulating and anti-tumor activities of *Ganoderma lucidum* (Reishi) polysaccharides, *Bioorganic & Medicinal Chemistry*, *12*, 5595-5601.

[131] Zhu, X. L. & Lin, Z. B. 2005. Effects of *Ganoderma lucidum* polysaccharides on proliferation and cytotoxicity of cytokine-induced killer cells, *Acta Pharmacologica Sinica*, 26, 1130–1137.

[132] Zhu, X. L. & Lin, Z. B., 2006. Modulation of cytokines production, granzyme B and perforin in murine CIK cells by *G. lucidum* polysaccharides, *Carbohydrate Polymers*, 63, 188-197.

[133] Lei, L., Lin, Z., Chen, Q., Li, R. & He, Y. 1993. Antagonistic effect of *Ganoderma lucidum* polysaccharide on the immunosuppressive response induced by cyclosporin A, hydrocortisone and antitumor agents, *Chinese Journal of Pharmacology and Toxicology*, *7*, 183-185.

[134] Zhu, X. L., Chen, A. F. & Lin, Z. B. 2007. *Ganoderma lucidum* polysaccharides enhance the function of immunological effector cells in immunosuppressed mice, *Journal of Ethnopharmacology*, *111*, 219-226.

[135] Li, M., Lei, L. & Liang, D. 1999. Effect of Ganderma polysaccharide on intracellular free calcium in murine peritoneal macrophages, *Chinese Pharmaceutical Journal*, *34*, 805-807.

[136] Sadava, D., Still, D. W., Mudry, R. R. & Kane, S. E. 2009. Effect of Ganoderma on drug-sensitive and multidrug-resistant small-cell lung carcinoma cells, *Cancer Letters*, *277*,182-189.

[137] Cao, Q. Z. & Lin, Z. B. 2006. *Ganoderma lucidum* polysaccharides peptide inhibits the growth of vascular endothelial cell and the induction of VEGF in human lung cancer cell, *Life Sciences*, *78*, 1457-1463.

[138] Song, Y. S., Kim, S. H., Sa, J. H., Jin, C., Lim, C. L. & Park, E. H. 2004. Anti-angiogenic and inhibitory activity on inducible nitric oxide production of the mushroom *Ganoderma lucidum*, *Journal of Ethnopharmacology*, 90, 17-20.

[139] Stanley, G., Harvey, K., Slivova, V., Jiang, J. & Sliva, D., 2005. *Ganoderma lucidum* suppresses angiogenesis through the inhibition of secretion of VEGF and TGF-β1 from prostate cancer cells, *Biochemical and Biophyssical Research Communications*, *330*, 46-52.

[140] Shi, Y. L., James, A. E., Benzie, I. F. F. & Bushwell, J. A. 2002. Mushroom derived preparations in the prevention of H_2O_2-induced oxidative damage to cellular DNA, *Teratogenesis, Carcinogenesis, and Mutagenesis*, *22*,103-111.

[141] Pillai, T. G., Nair, C. K. K. & Janardhanan, K. K. 2010. Enhancement of repair of radiation induced DNA strand breaks in human cells by Ganoderma mushroom polysaccharides, *Food Chemistry*, *119*,1040-1043.

[142] You, Y. H. & Lin, Z. B. 2002. Protective effects of *Ganoderma lucidum* polysaccharides peptide on injury of macrophages induced by reactive oxygen species, *Acta Pharmacologica Sinica*, *23*, 878-791.

[143] Zhang, H. N., He, J. H., Yuan, L. & Lin, Z. B. 2003. In vitro and in vivo protective effect of *Ganoderma lucidum* polysaccharides on alloxan-induced pancreatic islets damage, *Life Sciences*, *73*, 2307-2319.

[144] Lai, C. S. A., Yu, M. S., Yuen, W. H., So, K. F., Zee, S. Y. & Chang, R. C. C. 2008. Antagonizing β-amyloid peptide neurotoxicity of the anti-aging fungus *Ganoderma lucidum*, *Brain Research*, *1190*, 215-224.

[145] Lam, F. F. Y., Ko, I. W. M., Ng, E. S. K., Tam, L. S., Leung, P. C. & Li, E. K. M. 2008. Analgesic and anti-arthritic effects of Lingzhi and San Miao San supplementation in a rat model of arthritis induced by Freund's complete adjuvant, *Journal of Ethnopharmacology*, *120*, 44-50.

[146] Jia, J., Zhang, J. J., Hu, Y. S., Wu, Y., Wang, Q. Z., Li, N. N., Guo, Q. C. & Dong, X. C. 2009. Evaluation of in vivo antioxidant activities of *Ganoderma lucidum* polysaccharides in STZ-diabetic rats, *Food Chemistry*, *115*, 32-36.

[147] Seto, S. W., Lam, T. Y., Tam, H. L., Au, A. L. S., Chan, S. W., Wu, J. H., Yu, P. H. F., Leung, G.P.H., Ngai, S.M., Yeung, J.H.K., Leung, P.S., Lee, S. M. Y. & Kwan, Y. W. 2009. Novel hypoglycemic effects of *Ganoderma lucidum* water-extract in obese/diabetic (+db/+db) mice, *Phytomedicine*, *16*, 426-436.

[148] Ping, C. X., Yan, C., Bing, L. S., Guo,C. Y., Yun, L. J. & Ping, L. L. 2009. Free radical scavenging of *Ganoderma lucidum* polysaccharides and its effect on antioxidant enzymes and immunity activities in cervical carcinoma rats, *Carbohydrate Polymers*, *77*, 389-393.

[149] Guo, C. Y., Ji, S. Z. & Ping, C. X. 2009. Modulatory effect of *Ganoderma lucidum* polysaccharides on serum antioxidant enzymes activities in ovarian cancer rats, *Carbohydrate Polymers*, *78*, 258-262.

[150] Zhang, J. S, Tang, Q. J., Zimmerman-Kordmann, M., Reuter, W. & Fan, H. 2002. Activation of B lymphocytes by GLIS, a bioactive proteoglycan from *Ganoderma lucidum*, *Life Sciences*, *71*, 623-638.

[151] Wang, Y. Y., Khoo, K. H., Chen, S. T., Lin, C. C., Wong, C. H. & Lin, C. H. 2002. Studies on the immuno-modulating and antitumor activities of *Ganoderma lucidum* (Reishi) polysaccharides: functional and proteomic analyses of a fucose-containing glycoprotein fraction responsible for the activities *Bioorganic & Medicinal Chemistry*, *10*, 1057-1062.

[152] Liu, J., Yang, F., Ye, L. B., Yang, X.J., Timani, K. A., Zheng, Y. & Wang, Y. H. 2004. Possible mode of action of antiherpetic activities of a proteoglycan isolated from the mycelia of *Ganoderma lucidum* in vitro, *Journal of Ethnopharmacology*, 95, 256-272.

[153] Wang, H. X., Ng, T. B. & Chiu, S. W. 2004. A distinctive ribonuclease from fresh fruiting bodies of the medicinal mushroom *Ganoderma lucidum*, *Biochemical and Biophysical Ressearch Communications*, 314, 519-522.

[154] Tian, Y. P. & Zhang, K. C. 2005. Purification and characterization of a novel proteinase A inhibitor from *Ganoderma lucidum* by submerged fermentation, *Enzyme and Microbial Technologynology*, *36*, 357-361.

[155] Wang, H. X. & Ng, T. B. 2006. Ganodermin, an antifungal protein from fruiting bodies of the medicinal mushroom *Ganoderma lucidum*, *Peptides*, 27, 27-30.

[156] Thakur, A., Rana, M., Lakhanpal, T. N., Ahmad, A. & Khan, M. I., 2007. Purification and characterization of lectin from fruiting body of Ganoderma lucidum Lectin from *Ganoderma lucidum, Biochimica et Biophysica Acta*, *1770*, 1404-1412.

[157] Ye, L. B., Zhang, J. S., Ye, X. Y., Tang, Q. J., Liu, Y. F.,Gong, C. Y., Du, X. J. & Pan Y. J. 2008. Structural elucidation of the polysaccharide moiety of a glycopeptide (GLPCW-II) from *Ganoderma lucidum* fruiting bodies, *Carbohydrate Research*, *343*, 746-752.

[158] Shi, Y., Sun, J., He, H, Guo, H. & Zhang, S. 2008. Hepatoprotective effects of *Ganoderma lucidum* peptides against D-galactosamine-induced liver injury in mice, *Journal of Ethnopharmacology*, *117*, 415-419.

[159] Ye, L., Li, J., Zhang, J. & Pan, Y. 2010. NMR characterization for polysaccharide moiety of a glycopeptide, *Fitoterapia*, *81*, 93-96.

[160] Ziegenbein, F. C., Hanssen, H. P. & Konig, W. A. 2006. Secondary metabolites from *Ganoderma lucidum* and *Spongiporus leucomallellus*, *Phytochemistry*, *67*, 202-211.

[161] Wasser, S. P. & Weiss, A. L. 1997. *Medicinal mushrooms - Ganoderma lucidum, Reishi mushroom*. Haifa, Israel: Pedeifus Publishing House. (no page number)

[162] Chang, S. T. & Buswell, J. A. 1996. Mushroom nutriceuticals, *World Journal of Microbiology and Biotechnology*, *12*, 473-476.

[163] Sanodiya, B. S., Thakur, G. S., Baghel, R. K., Prasad, G. B. & Bisen, P. S. 2009. *Ganoderma lucidum*: a potent pharmacological macrofungus, *Current Pharmaceutical Biotechnology*, 10, 717-42.

[164] Zhang, C. Y. & Li, Y. M. 2002. Clinical investigation of Green Valley Lingzhi capsule on type 2 diabetes mellitus. In Ganoderma: *Genetics, chemistry, pharmacology and therapeutics, Proceedings of International symposium on Ganoderma research, October 21-23, 2002* (edited by Z.B. Lin). Pp. 194-198. Shanghai, China: Beijing, Medical University Press.

[165] Shi, K. G. & Quing, L. H. 2002. The follow-up observation assessment of medium and late phases cancer treated by Chinese *Ganoderma lucidum* essence (CGLE). In Ganoderma: *Genetics, chemistry, pharmacology and therapeutics, Proceedings of International symposium on Ganoderma research, October 21-23, 2002* (edited by Z.B. Lin). Pp. 188-193. Shanghai, China: Beijing, Medical University Press.

[166] Gao, Y., Zhou, S., Jiang, W., Huang, M. & Dai, X. 2003. Effects of ganopoly (a *Ganoderma lucidum* polysaccharide extract) on the immune functions in advanced-stage cancer patients, *Immunological Investigations*, 32, 201-15.

[167] Gao, Y., Tang, W., Dai, X., Chen, G., Ye, J., Chen, E., Koh, H. L., Li, X. & Zhou, S. 2005. Effects of water-soluble *Ganoderma lucidum* polysaccharides on the immune functions of patients with advanced lung cancer, *Journal of Medicinal Food*, 8, 159-68.

[168] Noguchi, M., Kakuma, T., Tomiyasu, K., Konishi, F., Kumamoto, S., Kondo, R., Matsuoka, K. 2005. Phase I study of a methanol extract of *Ganoderma lucidum*, edible and medicinal mushroom, in men with mild symptoms of bladder outlet obstruction, *Urology*, *66*, Supplement3A, 21.

[169] Chen, X., Hu, Z. P., Yang, X. X., Huang, M., Gao, Y., Tang, W., Chan, S. Y., Dai, X., Ye, J., Ho, P.C., Duan, W., Yang, H. Y., Zhu, Y. Z. & Zhou, S. F. 2006. Monitoring of immune responses to a herbal immuno-modulator in patients with advanced colorectal cancer, *International Immunopharmacology*, *6*, 499-508.

[170] Chan, W. K., Lam, D. T., Law, H. K., Wong, W. T., Leung, K. M. W., Lau, Y. L. & Chan, G. C. 2005. *Ganoderma lucidum* mycelium and spore extracts as natural adjuvants for immunotherapy, *Journal of Alternative and Complementary Medicine*, 11, 1047-57.

[171] Noguchi, M., Kakuma, T., Tomiyasu, K., Itoh, K., Konishi, F., Kumamoto, S., Shimizu, K., Kondo, R. & Matsuoka, K. 2007. A randomized clinical trial of an ethanol extract of *Ganoderma lucidum* (edible and medicinal mushroom) in men with lower urinary tract symptoms, *Urology*, *70*, Supplement 3A, 254.

[172] Yuen, M. F., Ip, P., Ng, W. K. & Lai, C. L. 2004. Hepatotoxicity due to a formulation of *Ganoderma lucidum* (lingzhi). Letters to the editor, *Journal of Hepatology*, *41*, 685-690.

[173] eHow, 2010. Side Effects of *Ganoderma lucidum*. URL http://www.ehow.com/facts_5021871_side-effects-ganoderma-lucidum.html. Accessed 19/2/2010.

PART 4: PLANT CELLS AS SOURCES OF RAW MATERIAL

In: Bioprocess Sciences and Technology
Editor: Min-Tze Liong

ISBN 978-1-61122-950-9

Chapter 11

PLANT COMPOUNDS WITH *AEDES AEGYPTI* LARVICIDAL ACTIVITY AND OTHER BIOLOGICAL PROPERTIES

P. M. G. Paiva[1,*], T. H. Napoleão[1], N. D. L. Santos[1], M. T. S. Correia[1], D. M. A. F. Navarro[2] and L. C. B. B. Coelho[1]

[1]Departamento de Bioquímica, Centro de Ciências Biológicas, Universidade Federal de Pernambuco, Cidade Universitária, 50670-420, Recife, Pernambuco, Brazil.

[2]Departamento de Química Fundamental, Centro de Ciências Exatas e da Natureza, Universidade Federal de Pernambuco, Cidade Universitária, 50670-901, Recife, Pernambuco, Brazil

ABSTRACT

Chemical insecticides to control *Aedes aegypti* adversely affect the environment and also promote emergence of resistant mosquitoes. Plant compounds with damage effect on immature stage forms or adult insects can be an alternative to synthetic insecticides. This chapter shows essential oils, organic solvent extracts and lectins from plants as enviro-friendly agents against *A. aegypti* and other arthropods. Extraction, concentration and isolation of plant compounds are discussed; effect of essential oils, organic solvent extracts or lectins (hemagglutinating proteins) on *A. aegypti* larval development and survival as well as other biological properties are reported. Larvicidal mechanisms proposed for essential oils, organic solvent extr[1]acts and lectins are also reported. The chapter contributes to the study of alternative strategies to control arthropods. Additionally, it reports the medicinal and industrial uses of plant compounds with biotechnological applications.

[*] Corresponding author: *P.M.G. Paiva*. Email: ppaivaufpe@yahoo.com.br

1. *Aedes Aegypti*: Vector Mosquito of Etiologic Agents from Yellow Fever, Dengue Fever and Hemorrhagic Dengue Fever

A. aegypti Linn. belongs to the Phylum Arthropoda, Class Hexapoda, Order Diptera, Culicidae Family and its common name is dengue mosquito. *A. aegypti* is a mosquito with diurnal habit, with a preference to oviposit in clean water.

The control of *A. aegypti* can be directed to larvae (four instars: L1, L2, L3 and L4), pupa and adult insects. Prevention of infestation by larvae elimination is easier than control of adult insects since the larvae are confined in containers. The effect of plant compounds on *A. aegypti* larvae can be investigated by evaluation of larval development [1] and detection of larval mortality [2].

2. Essential Oils: Volatile Compounds from Plants

Essential oils are complex compounds synthesized as secondary metabolites by buds, flowers, leaves, stems, twigs, seeds, fruits, roots, wood or bark; they are characterized by two or three major volatile compounds (20–70%), lipid soluble and soluble in organic solvents [3]. The main chemical constituents found in essential oils are phenols, aldehydes, ethers, esters, alcohols, ketones, benzenoids, phenylpropanoids and terpenes. Isoprene is the building block of terpene and a number of authors use the designation "terpene" to include all terpenoids. Limonene, alpha-pinene, menthol, citronellal and citral (Figure 1) are good examples of monoterpenes (two isoprene units, C10).

Menthol is the main constituent in the essential oil of *Mentha piperita* and camphor was isolated from the essential oil in the wood of *Cinnamomum camphora* [4, 5]. The phenols carvacrol, 2-ethyl-4,5-dimethylphenol and thymol were isolated from shoots of *Thymus vulgaris* and eugenol is present in flower buds of *Eugenia caryophyllata* [6, 7]. Other chemical constituents of *T. vulgaris* essential oil are 1-octen-2-ol (alcohol), alpha-trimethyldodecane (hydrocarbon) and (Z)-3-hexenyl butyrate (ester). Benzyl benzoate (ester benzenoid) and octanal, nonanal and decanal (aldehydes) are found in essential oils of *Licaria canella* and oranges, respectively [8, 9].

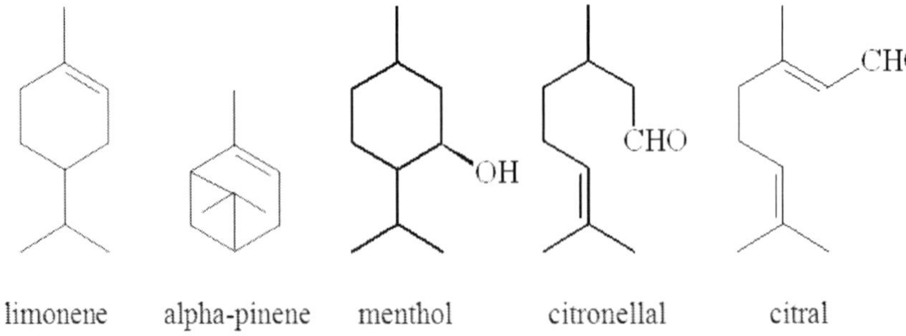

Figure 1. Monoterpene constituents of essential oils.

2.1. Methods for the Extraction, Concentration and Isolation of Compounds

The most useful extraction methods for obtaining essential oils from tissue plant are based on distillation. In distillation with cohobation, a solvent such as water is recycled in the process, returning to the still [10]. In hydro-distillation, the plant material is fully dipped in the water and placed into a still that has a heat source at the bottom and the top is connected to a condenser (Figure 2A); the disadvantage of this technique is that the thermally unstable compounds may decompose if remain into direct contact with the heat source. In steam distillation, the essential oil is isolated using a water boiler tank to produce steam; the steam passes through a flask containing the plant material and the oil is collected and separated to the oil receiver (Figure 2B). Hydro-steam distillation occurs when the vapor is generated below the plant material, which is suspended in the still on a support.

Enfleurage uses odorless animal fat to capture volatile compounds from the aromatic parts of plant; essential oil is obtained after washing the fat with alcohol or heating. Modern enfleurage uses vegetal fat instead of animal fat. Hot maceration extraction is very similar to this method, but is faster and uses hot fat or oil in which the plant material is immersed [10].

Microwave-assisted extraction has been used for the extraction of chemical constituents in plant material [11]; a small matrix particle size and high dielectric solvent can enhance the microwave-assisted extraction. The improved extraction yield, reduced extraction time and minimal amount (or even absence) of solvent are the greatest advantages of method. Solvent-free microwave extraction (Figure 3A) was used for isolation of essential oils from three aromatic herbs [12]; the advantage of this extraction method is the short extraction time and substantial energy savings. Microwave-assisted hydro-distillation (MAHD, Figure 3B) was used for extraction of essential oil from *Cinnamomum iners* leaves [13] and MAHD method using an adapted domestic microwave oven with the connection of a round glass flask in the oven to a condenser outside the oven was used for extraction of essential oils from *Satureja hortensis* and *Satureja montana* [14].

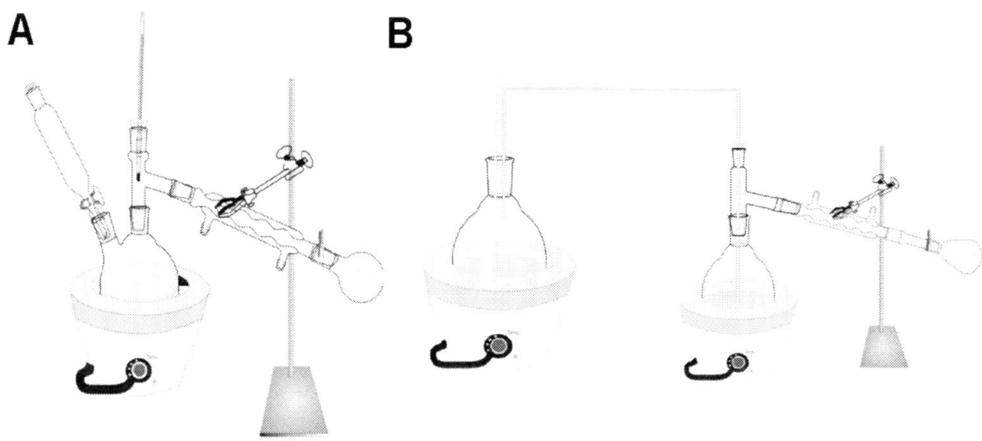

Figure 2. Laboratorial apparatus to extract essential oils by methods based on distillation. Hydro-distillation (A) and steam distillation (B).

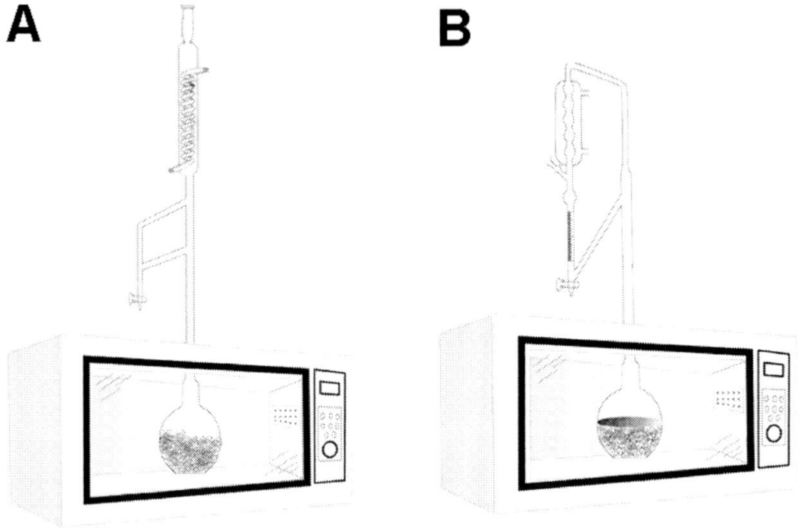

Figure 3. Laboratorial apparatus for solvent-free microwave-assisted extraction (A) and microwave assisted with Clevenger extraction (B).

Figure 4 shows a simple and open-system ultrasound-assisted extraction apparatus for extracting essential oils constituted by ultrasound bath, round glass flask and Liebig condenser [15]. Ultrasound-assisted extraction is more advantageous for the extraction of thermally unstable molecules.The advantage of ultrasound is that it provides a greater penetration of the solvent into the cell components and improves the transfer of mass from plant cells to the solvent. The ultrasound extraction applied to a large-scale is limited due to the higher costs; another disadvantage is that ultrasound can produce free radicals that can react with active molecules present in the plant, during ultrasound extraction.

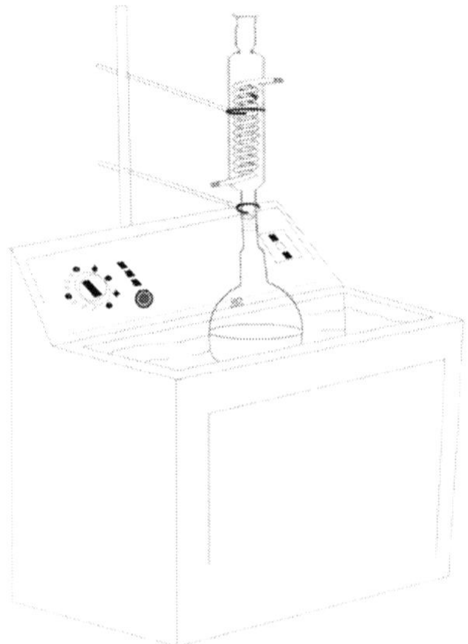

Figure 4. Simple ultrasound-assisted extraction apparatus with ultrasound bath, round glass flask and Liebig condenser.

Supercritical fluid extraction (Figure 5) has become attractive due to the lack of an organic solvent and extraction occurs near room temperature, thereby avoiding the thermal degradation of bioactive compounds. The most often employed supercritical fluid is carbon dioxide (CO_2) due to its moderate supercritical fluid conditions (*T*C=31.06 °C and *P*C=73.81 bar), high capacity to solubilize organic compounds under supercritical conditions and the fact that it is non-toxic, non-inflammable and inexpensive.

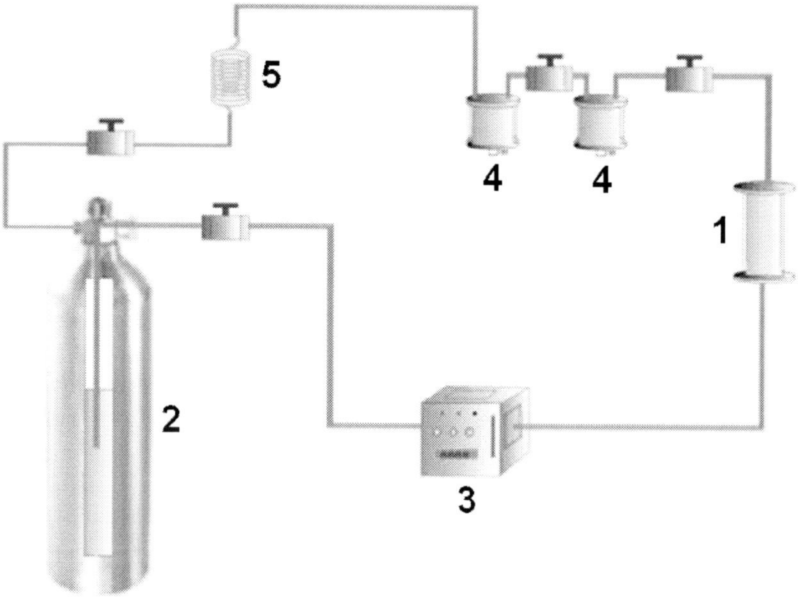

Figure 5. Supercritical fluid extraction apparatus. The raw plant material is placed in an extractor recipient (1) and the supercritical fluid is produced from a reservoir (2) connected to a powerful pump (3); the system is pressurized and controlled by valves. After passing through the extractor, the fluid and dissolved compounds are transported to separators (4). The dissolution of the organic matter is precipitated by a decrease inpressure and the product is collected via a valve located at the bottom of the separators. A condenser (5) regenerates the fluid, which is returned to the fluid reservoir.

2.2. Larvicidal Activity against *Aedes aegypti*

Essential oils are permeable to cellular membrane due to their lipophilic nature; cell alterations in plasma membrane, cytoplasm and nucleus are suggested to be responsible by their cytotoxic effects [3]. Essential oils with larvicidal activity on *A. aegypti* are a mixture of active compounds and, in general, the oil is more efficient in promoting larvae mortality than isolated constituents.

Larvae treated with essential oils from leaves of *Chloroxylon swietenia* (LC_{50} of 16.5 $\mu g.mL^{-1}$), *Lippia gracilis* (LC_{50} of 98 ppm), *Hyptis fruticosa* (LC_{50} of 502 ppm) and *Hyptis pectinata* (LC_{50} pf 366 ppm) showed abnormal wagging and later died. The sesquiterpenes pregeijerene, geijerene and germacrene D isolated from *C. swietenia* were the active principles with LC_{50} of 28.3, 43.4, and 63.6 $\mu g.mL^{-1}$, respectively; the major constituent of *L.*

gracilis oil was carvacrol (LC_{50} of 70 ppm) although the most active (LC_{50} of 37 ppm) constituent was *R*-limonene [16, 17].

Larvicidal activity on L3 was reported for essential oils from leaves of *Piper permucronatum*, *Piper hostmanianum*, *Piper humaytanum*, and *Piper gaudichaudianum* with LC_{50} values of 36, 54, 121 and 156 $\mu g.mL^{-1}$, respectively. The oils from *P. permucronatum* and *P. hostmanianum* showed high content of arylpropanoids [18]. Essential oil from *Lippia sidoides* and its hydrolate were strongly active on *A. aegypti* L3 promoting 100% mortality in 1-5 min (pure oil and hydrolate), 20 min (1:5 dilution), and 24 h (1:10 and 1:20 dilutions). In addition, the hydrolate showed considerable persistence remaining active even after 60 days and did not cause adverse effects on mice. Thymol was identified as the active principle promoting 100% mortality at 0.017% (w/v) in 1 ½ h [19]. *Mentha piperita* essential oil (peppermint oil) was also larvicidal for L3. The bioassay was performed applying the oil in an enamel containing water; the oil remained in the surface as a layer. Peppermint oil caused 100% mortality in 24 h at 4 $mL.m^{-2}$ and reduction in adult emergence [20].

Larvicidal activity on *A. aegypti* L4 was detected to essential oils from *Apium graveolens* (LC_{50} of 42.07 ppm), *Carum carvi* (54.62 ppm), *Curcuma zedoaria* (31.87 ppm), *Foeniculum vulgare* (49.32 ppm) and *Zanthoxylum limonella* (24.61 ppm). The authors observed that, in addition to mortality, the exposure to the oils provoked restlessness, sluggishness, tremors, convulsions and larvae paralysis [21]. Essential oil from foliage of *Tagetes patula* was also able to kill *A. aegypti* L4 (LC_{50} of 13.57 ppm); larvae showed restlessness, abnormal wagging and did not develop into pupae [22]. Essential oils from *Pimpinella anisum* seeds (LC_{95} of 115.7 $\mu g.mL^{-1}$) and *Zingiber officinale* rhizome (193.9 $\mu g.mL^{-1}$) also were toxic on L4 [23].

Larvicidal activity against L4 has been found in multiple compounds isolated from essential oils. The most abundant constituents (*E*)- or (*Z*)-asarone and patchouli alcohol from *Piper marginatum* leaf, stem and inflorescence oils were toxic on L4 with LC_{50} of 23.8, 19.9, and 19.9 ppm, respectively [24]. Essential oil from *Dendropanax morbifera* flowers contain the larvicidal agents (LC_{50} of 62.32 ppm) γ-elemene, tetramethyltricyclohydrocarbon, β-selinene, α-zingibirene, 2-isopropyl-5-methylbicylodecen and β-cubebene, while α-pinene and α-terpinene were suggested as the main larvicidal compounds of *Eucalyptus camaldulensis* and *Eucalyptus grandis* oils which LC_{50} were 14.7 $\mu g.mL^{-1}$ and 32.4 ppm, respectively [25, 26, 27]. Fourth instar larvae were also susceptible to *Ocimum sanctum* essential oil (LC_{50} of 85.11 ppm), which main components are oxygenated monoterpenes and sesquiterpene hydrocarbons [28].

2.3. Other Biological Properties

Essential oils with larvicidal activity on *A. aegypti* are also harmful to other mosquito's species (Table 1). Essential oil from *Cinnamomum osmophloeum* leaf was larvicidal on L4 of *Aedes albopictus* (LC_{50} of 40.8 $\mu g.mL^{-1}$) and the active constituents were benzaldehyde and trans-cinnamaldehyde; the last was also toxic to fourth-instar larvae of *Culex quinquefasciatus* (lymphatic filariasis vector) and *Armigeres subalbtus* [29]. Essential oils from *Apium graveolens*, *Carum carvi*, *Curcuma zedoaria*, *Foeniculum vulgare*, and *Zanthoxylum limonella* are larvicidal agents on both fourth instar larvae of *A. aegypti* and *Anopheles dirus* although larvae showed different susceptibility to them; *A. graveolens, C. carvi, Z. limonella* oils were mainly effective on *A. aegypti* while the others were on *A. dirus*

[21]. Similarly, larvae of *Anopheles stephensi* (malaria vector) were more susceptible than that of *A. aegypti* to oils from *Chloroxylon swietenia* leaf and stem [16].

Table 1. Plant essential oils able to kill arthropod species

Arthropod	Plant
Acari	
Tetranychus urticae	*Micromeria fruticosa, Nepeta racemosa, Origanum vulgare*
Diptera	
Aedes aegypti	*Apium graveolens, Carum carvi, Chloroxylon swietenia, Curcuma zedoaria, Dendropanax morbifera, Eucalyptus camaldulensis, Eucalyptus grandis, Foeniculum vulgare, Hyptis fruticosa, Hyptis pectinata, Lippia gracilis, Lippia sidoides, Mentha piperita, Ocimum sanctum, Pimpinella ansium, Piper gaudichaudianum, Piper hostmanianum, Piper humaytanum, Piper marginatum, Piper permucronatum, Tagetes patula, Zanthoxylum limonella, Zingiber officinale*
Aedes albopictus	*Cinnamomum osmophloeum*
Anopheles dirus	*Apium graveolens, Carum carvi, Curcuma zedoaria, Foeniculum vulgare, Zanthoxylum limonella*
Anopheles stephensi	*Chloroxylon swietenia, Eucalyptus tereticornis, Mentha piperita, Tagetes patula*
Armigeres subalbtus	*Cinnamomum osmophloeum, Mentha piperita*
Culex quinquefasciatus	*Cinnamomum osmophloeum, Tagetes patula*
Isoptera	
Coptotermes formosanus	*Calocedrus macrolepis, Chamaecyparis obtusa, Cryptomeria japonica*
Homoptera	
Bemisia tabaci	*Micromeria fruticosa, Nepeta racemosa, Origanum vulgare*

References: [16, 17, 18, 19, 20, 21, 22, 23, 24, 25, 26, 27, 28, 29, 31, 32].

Essential oils, beyond the larvicidal activity, may interfere with other stages of the insect life cycle; ovicidal, oviposition-deterrent activity, pupicidal and adulticidal activities have also been reported. Essential oils showed repellent activity on Coleoptera, Diptera, Isoptera, Lepidoptera, Phthiraptera and Thysanoptera; mixed oils are more efficient than pure oils probably due to synergistic effect among particular compounds [30]. *Mentha piperita* oil has insecticidal potential against *A. aegypti*, *Anopheles stephensi* and *Culex quinquefasciatus* by its strong repellent action against adult mosquitoes; loss of fertility and fecundity of adults emerged from larvae exposed to oil [20].

Essential oils can promote termite and acari mortality (Table 1) as well as termite and tick repellency. Essential oils of *Calocedrus macrolepis* var. *formosana* (heartwood), *Chamaecyparis obtusa* var. *formosana* (leaf) and *Cryptomeria japonica* (sapwood) showed anti-termitic activity on termite *Coptotermes formosanus* due to the toxicity and repellent action [31]. Essential oil vapors from leaves of aromatic plants *Micromeria fruticosa*, *Nepeta racemosa* and *Origanum vulgare* showed acaricidal activity (95% mortality at 120 h) on *Tetranychus urticae* (adult and nymphal stage) and insecticidal action (100% mortality at 120 h) against *Bemisia tabaci* (Homoptera), pests of greenhouse vegetables and ornamentals [32]. Essential oils of *Dianthus caryophyllum* and *Cymbopogon nardus* (citronella) have repelling properties on *Aedes* species and tick *Ixodes ricinus*; phenylethanol (*D. caryophyllum*) and citronellol (oil of citronella) were the most active compounds [33, 34].

3. Organic Solvent Extracts

Larvicidal plant extracts have been obtained from bark, fruits, leaves, seeds, rhizomes and woods. These organic solvent extracts can constitute an economically feasible strategy for the control of insects; some of them can be associated with other control methods to obtain a relatively lasting effect.

3.1. Preparation of Extracts

Organic solvent extracts from plant tissues, in general, are prepared using air-dried material in order to obtain reasonable extraction yields. Cutting, breaking, powdering and cryogenic milling are convenient methods to increase tissue surface area; cryogenic milling, in addition, minimizes the losses of volatile organic compounds [35]. Vegetal tissues with high lipid, pigment and polyphenol contents can be treated before extraction procedure to eliminate these components; defatting can be made using non-polar organic solvents such as hexane; pigment and polyphenol can be removed by extracting them with acetone, methanol, petroleum ether and detergents.

Extraction with organic solvents is the separation process of a constituent from plant material; the component is removed from a phase and brought to another immiscible liquid phase. Solute will be more concentrated in the phase in which it has higher solubility; solvent is chosen according to the characteristics of constituents being extracted. Polar compounds like sugars and alkalis will dissolve in polar solvents (such as organic acids), intermediate or low polar compounds (tannins, phenolics, and terpenes) will dissolve in solvents of intermediate polarity (such as methanol, ethanol, acetone, and dichloromethane) and non-polar compounds (oils, resins, and waxes) are extracted using solvents of low polarity (such as hexane and chloroform). Plant tissue can be sequentially extracted with different solvents in increasing order of polarity or a unique extract obtained using a polar solvent is later partitioned in different fractions using solvents in increasing order of polarity.

The time and temperature of extraction is variable; thermal instability of some compounds may limit the temperature of extraction. Steam distillation is a technique used for extraction of volatile plant compounds and the compounds obtained are separated in the oil and water phases; vegetal tissue is continuously flushed with steam and volatile compounds are taken up by the vapor phase [35].

Maceration and percolation are useful methods to obtain organic solvent extracts when the compounds of interest show high solubility in the solvent. Maceration consists of soaking the plant material in the solvent, using agitation or heat to increase the solubility of the compounds and the rate of mass transfer. A single maceration step does not result in the total extraction of the compounds and for this reason residues are generally re-extracted in two or three more steps.

Soxhlet extractor is used when the compound of interest shows low solubility in organic solvent or the compound is more soluble in water than in the organic solvent. In a conventional Soxhlet extractor, the plant tissue is placed in a thimble-holder which is gradually filled with the solvent present in a distillation flask and there is a reflux of the solvent in an intermittent process. The liquid that overflows returns to the distillation flask through a siphon and carries the substances that have been extracted. This operation is repeated until the extraction is complete. Soxhlet extraction can be coupled to high-pressure, automated, ultrasound-assisted, and microwave-assisted systems intended to automating the extraction and shortening percolation time by using different forms of energy [36].

3.2. Larvicidal Activity against *Aedes aegypti*

The damage effect of organic solvent plant extracts on *A. aegypti* larvae can be detected by evaluation of larval development in presence of extracts; ethanolic extract from *Melia azedarach* leaf promoted delay in *A. aegypti* larval development and larvae needed twice than usual the number of days to complete the third instar [37].

Bioassays for larvicidal activity of organic extracts are performed using defined larval instars. Larvicidal mechanisms proposed are antichitinogenic activity, toxic effect on larvae nervous system and changes in digestive tract including total or partial destruction of cells, high cytoplasmic vacuolation, increased subperitrophic space, cell hypertrophy, and apparent epithelial stratification [38, 39, 40].

Larvicidal methanolic extracts of *Albizzia amara* and *Ocimum basilucum* leaves [41] were more harmful to L1 (LC_{50} of 5.243% and 3.734%, respectively) than L2 (6.480% and 4.154%), L3 (7.106% and 4.664%) and L4 (7.515% and 5.124%) instars. Dichloromethane extracts of *Aristolochia triangularis*, *Baccharis cordifolia*, *Eupatorium hecatanthum*, *Pterocaulon purpurascens*, *Xanthium spinosum*, and *Abuta grandifolia* were active on L2 [42]; the efficiency (LC_{50}) of extracts to promote larval mortality varied from 2.6 (*A. grandifolia*) to 373.3 $\mu g.mL^{-1}$ (*B. cordifolia*). Acetone fraction obtained with petroleum ether extract of *Argemone mexicana* seeds was also larvicidal for *A. aegypti* L2 in laboratory and under field conditions at concentrations of 25, 50, 100 and 200 ppm; larvae were more sensitive under field conditions [39].

Ethanol was able to extract compounds with larvicidal activity on L3 and L4 [38, 43, 44] from *Magonia pubescens* (stem bark), *Spondias mombin* (seed), *Annona muricata* (leaf), *Bauhinia cheilantha* (wood), *Caesalpinia echinata* (leaf and wood), *Operculina macrocarpa* (tuber), *Eugenia uniflora* (leaf), *Annona crassiflora* (root, LC_{50} of 0.71 $\mu g.mL^{-1}$) and *Annona glabra* (seeds, LC_{50} of 0.06 $\mu g.mL^{-1}$). Methanolic extract from *Atlantia monophylla* leaves harmful to L4 (LC_{50} of 0.09 $mg.L^{-1}$) was safe to aquatic mosquito predators [45]. Petroleum ether extracts of *Jatropha curcas*, *Pedilanthus tithymaloides*, *Phyllanthus amarus*, *Euphorbia hirta* and *Euphorbia tirucalli* showed larvicidal activity on L4 with LC_{50} of 11.34, 76.61, 113.40, 424.94, and 5.52 ppm, respectively. Based on these results, the authors indicated *E. tirucalli* extract as an ideal potential larvicide against *A. aegypti* [46].

Several plant compounds have been identified as larvicidal agent on *A. aegypti* larvae including tectoquinone (LC_{50} of 3.3 $\mu g.mL^{-1}$) from methanolic extract of *Cryptomeria japonica* sapwood; cubebol (LC_{50} of 60.1 $\mu g.mL^{-1}$) from ethanolic extract of *C. japonica* wood; cordiaquinones (LC_{100} ranging of 12.5 to 25 $\mu g.mL^{-1}$) from dichloromethane extract of *Cordia curassavica* root; limonoids from ethanolic extracts of *Melia azedarach* and *Azadirachta indica* fruit endocarps; dicentrine from ethanolic extract of *Ocotea valloziana valloziana* trunk bark (LC_{50} of 213.70 $\mu g.mL^{-1}$) and saponins from methanolic extract of *Balanites aegyptiaca* callus [40, 47, 48, 49, 50, 51]. Larvicidal organic solvent extracts also can be harmful to larvae before pupation and *A. aegypti* adult [37, 39, 41, 45]. Pupicidal activity was detected in leaf extracts from *Melia azedarach, Albizzia amara, Ocimum basilicum* and *Atlantia monophylla.* Acetone fraction from petroleum ether extract of *Argemone mexicana* induced reduction in blood meal utilization and fecundity as well as sterility of first generation eggs [39].

3.3. Other Biological Properties

Organic solvent extracts are effective in killing mosquitoes other than *A. aegypti* (Table 2). Tectoquinone from methanolic extract of *Cryptomeria japonica* sapwood is a larvicidal agent (LC_{50} of 5.4 $\mu g.mL^{-1}$) against L4 larvae of *Aedes albopictus*; petroleum ether extracts from *Jatropha curcas*, *Pedilanthus tithymaloides*, *Phyllanthus amarus*, *Euphorbia hirta*, and *Euphorbia tirucalli* were lethal for L4 larvae of *Culex quinquefasciatus* and methanolic extract from *Atlantia monophylla* leaves was pupicidal for *Anopheles stephensi* and *C. quinquefasciatus* [45, 46, 49].

Insecticidal activity has also been reported for methanolic extracts of *Ocimum basilicum*, *Origanum majorana* and *Salvia officinalis* on third instar larvae of the moth *Spodoptera littoralis* (LC_{50} of 0.17, 0.36, and 0.47 %, respectively); petroleum ether extract from *Curcuma aromatica* rhizome was a larvicidal agent against *Culex quinquefasciatus* (LC_{50} of 11.42 ppm) and active principles identified were 9-oxoneoprocurcumenol and neoprocurcumenol [52, 53, 54]. Table 2 shows other insects that were affected by organic solvent extracts including flies (*Bactrocera oleae* and *Ceratitis capitata*), aphids (*Pterochloroides persicae*), bugs (*Nilaparvata lugens*), and beetles [55, 56, 57, 58]. The flavanone pinocembrin from *Flourensia oolepis* ethanolic extract showed antifeedant activity (ED_{50} 7.98 $\mu g.cm^{-2}$) on *Epilachna paenulata* [59].

Plant extracts insecticidal on Diptera insects are harmful to other arthropods (Table 2). Methanolic extracts of *Centella asiatica* leaves showed larvicidal effect against *Anopheles subpictus* and *Culex tritaeniorhynchus* as well as adulticidal acitivity on ticks *Haemaphysalis bispinosa* and *Paramphistomum cervi* [60]. Toluene extract from *Artemisia abrotanum* leaves showed repelling property on *A. aegypti* and nymphs of tick (*Ixodes ricinus*); eugenol was identified as the constituent of highest activity [34].

Saponin and thujone were identified as active molluscicidal compounds in ethanolic extracts from *Saraca asoca* bark and *Thuya orientalis* leaf [61]. The alkaloid dicentrine is known as an antitumor agent on human hepatoma cell line HuH-7 [62].

Table 2. Plant organic solvent extracts able to kill arthropod species

Arthropod	Plant
Acari	
Panonychus citri	*Boenninghausenia sessilicarpa*
Coleoptera	
Tribolium confusum	*Mantisalca duriaei*, *Rhaponticum acaule*
Diptera	
Aedes aegypti	*Abuta grandifolia, Albizzia amara*, *Annona crassiflora*, *Annona glabra*, *Annona muricata*, *Argemone mexicana*, *Aristolochia triangularis*, *Atlantia monophylla*, *Azadirachta indica*, *Baccharis cordifolia*, *Balanites aegyptiaca*, *Bauhinia cheilantha*, *Caesalpinia echinata*, *Cryptomeria japonica*, *Eugenia uniflora*, *Eupatorium hecatanthum*, *Euphorbia hirta*, *Euphorbia tirucalli*, *Jatropha curcas*, *Magonia pubescens*, *Melia azedarach*, *Pedilanthus tithymaloides*, *Phyllanthus amarus*, *Pterocaulon purpurascens*, *Ocimum basilucum*, *Ocotea valloziana*, *Operculina macrocarpa*, *Spondias mombin*, *Xanthium spinosum*
Aedes albopictus	*Cryptomeria japonica*
Anopheles subpictus	*Centella asiatica*
Anopheles labranchiae	*Calotropis procera*, *Cotula cinerea*
Anopheles stephensi	*Atlantia monophylla*
Bactrocera oleae	*Citrus aurantium*
Ceratitis capitata	*Citrus aurantium*
Culex quinquefasciatus	*Atlantia monophylla*, *Curcuma aromatica*, *Euphorbia hirta*, *Euphorbia tirucalli*, *Jatropha curcas*, *Pedilanthus tithymaloides*, *Phyllanthus amarus*
Culex tritaeniorhynchus	*Centella asiatica*
Hemiptera	
Nilaparvata lugens	*Euphorbia kansui*

Table 2. Plant organic solvent extracts able to kill arthropod species (Continued)

Homoptera *Pterochloroides persicae*	*Rhamnus dispermus*
Isoptera *Nasutitermes corniger*	*Myracrodruon urundeuva*
Ixodida *Haemaphysalis bispinosa*	*Centella asiatica*
Lepidoptera *Spodoptera littoralis*	*Ocimum basilicum*, *Origanum majorana*, *Salva officinalis*

4. Lectins: Carbohydrate-binding Proteins

The term lectin (originated from the Latin *lectus* that means selected) was introduced by Boyd and Shapleigh in 1954, because of the ability of these proteins to bind carbohydrates selectively. Lectins interact with glycoconjugates present on cell surface inducing agglutination and the term agglutinin is also used to designate lectin; hemagglutinating activity occurs when the lectin binds to carbohydrate from erythrocyte surface promoting a network among them (Figure 6A). Plant lectins have been isolated from bark, cladodes, flowers, leaves, rhizomes, roots and seeds.

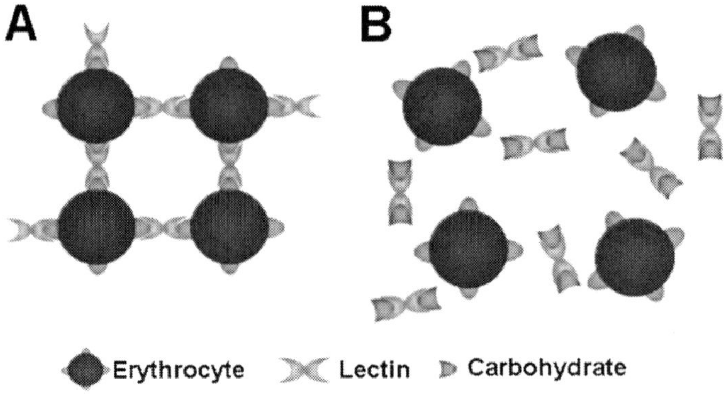

Figure. 6. Schematic representation of erythrocyte network promoted by lectin binding to surface carbohydrate (A) and inhibition of hemagglutinating activity by free carbohydrate (B).

4.1. Methodologies for Extraction, Concentration and Purification

The fragmentation of plant tissue is performed before the extraction step and aims to promote greater solubilization by increasing contact surface between the lectin and extraction solution. Treatment of crude extract with ammonium sulfate [63] promotes differential protein precipitation since proteins rich in hydrophilic amino acids are more soluble than those rich in hydrophobic amino acids and require highest concentration of ammonium sulfate to precipitate from the solution. Fractionation of lectin can also be obtained by heating the crude extract at elevated temperatures (thermal precipitation); denatured proteins precipitate and thermo-stable lectins remain in solution [64].

Lectin present in a mixture of proteins can be isolated by chromatographic methods (Figure 7). Ion exchange chromatography is able to separate proteins that differ in isoelectric points. Basic and acidic lectins bind, respectively, to matrices negatively (cation exchange column) and positively (anion exchange column) charged and lectin desorption can be made using stepwise or linear gradients of pH or NaCl. Size exclusion chromatography is based on percolation of a protein sample in the pores of inert matrices of controlled porosity with subsequent separation of the components by differential elution, according to the molecular size. The separation criterion in affinity chromatography is the property of lectins to bind specifically to carbohydrates. The lectin adsorbed on affinity matrix is recovered by elution with solution containing the hapten saccharide, or changing pH and ionic strength of chromatographic medium.

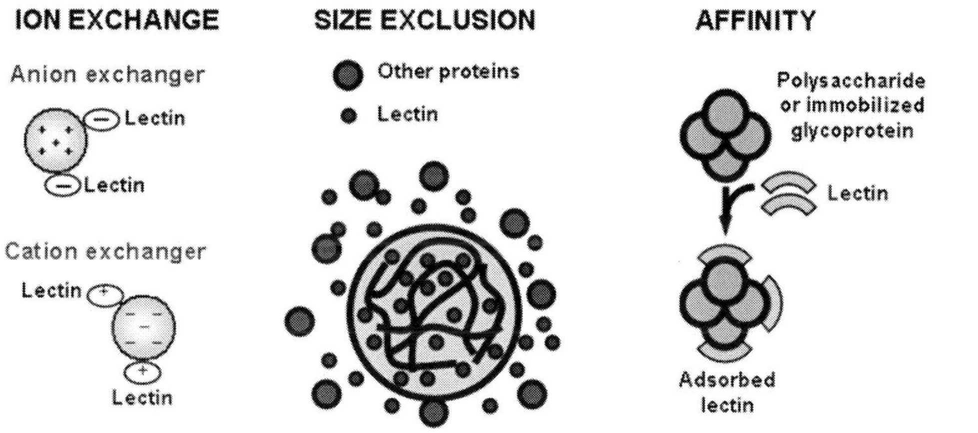

Figure 7. Schematic representation of chromatographic methods used for lectin isolation.

Lectin carbohydrate specificity can be easily defined by hemagglutinating activity assay when it is done in the presence of different carbohydrates. The lectin binding to carbohydrate free in solution is not able to interact with glycoconjugates from surface of erythrocytes and the hemagglutinating activity is inhibited (Figure 6B).

Electrophoretic methods are efficient tools for structural characterization and purity evaluation of lectins. Polyacrylamide gel electrophoresis (PAGE) for native proteins reveals the protein charge nature; in the presence of sodium dodecyl sulfate and reducing agent the protein is unfolded in its constituent subunits and the molecular mass is defined (Figure 8). Protein is usually detected on gel by Coomassie Brilliant Blue, amido black or silver staining. Electrophoresis is also efficient for identification and characterization of carbohydrate moiety from glycosylated lectins.

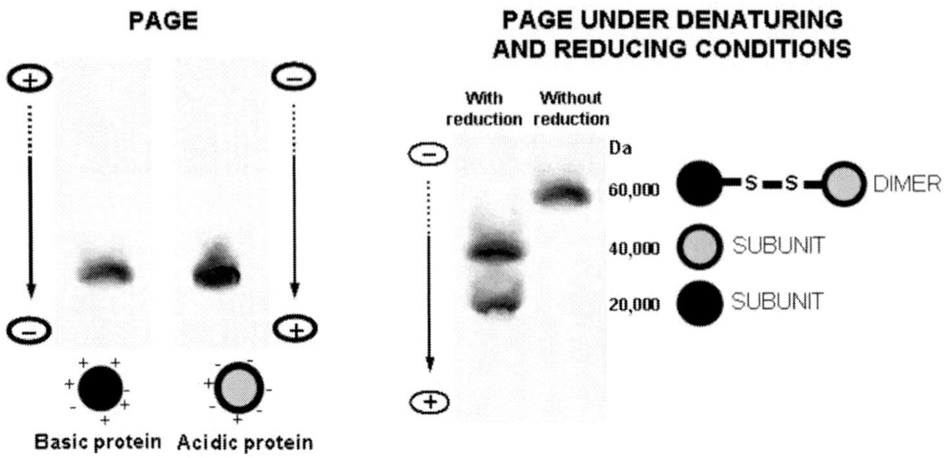

Figure 8. Schematic representation of polyacrylamide gel electrophoresis (PAGE) for native proteins and PAGE in presence of sodium dodecyl sulfate and reducing agent.

4.2. Larvicidal Activity against *Aedes aegypti*

Insect gut contains the peritrophic matrix which is constituted by proteins, glycoproteins, proteoglycans, and chitin. The matrix separates gut lumen contents from digestive epithelial cells; matrix integrity is essential for insect survival by important roles in digestive processes as well as insect protection from invasion by microorganisms and parasites. It has been suggested that chitin at peritrophic membrane, glycosylated receptors at the stomach epithelial cell surface as well as sugar moiety of glycosylated digestive enzymes are also targets for lectin binding [65, 66, 67 68, 69].

Recently, three lectins with detrimental action on *A. aegypti* larvae were reported. One lectin was isolated from *Moringa oleifera* seeds, the so-called water-soluble *M. oleifera* lectin (WSMoL); two were obtained from *Myracrodruon urundeuva* bark (MuBL) and heartwood (MuHL). *M. oleifera* belongs to the Moringaceae family and some of its vernacular names are ''moringa" in Portuguese, ''árbol del ben" in Spanish, and horseradish tree in English. Seeds of *M. oleifera* are widely used in developing countries as a natural coagulant to treat water for human consumption. *M. urundeuva* belongs to the Anacardiaceae family and its vernacular names are "aroeira do sertão" in Portuguese and "urundel" in Spanish. The plant is broadly distributed in Brazil and its heartwood is considered a hardwood very resistant to termites. WSMoL, MuBL and MuHL were purified from tissue extracts by ammonium sulphate precipitation followed by chromatography on chitin column; the hemagglutinating activity of these lectins was inhibited by *N*-acetylglucosamine.

Aqueous extracts of 1, 3, 6 and 15 seeds of *Moringa oleifera* (SE1, SE3, SE6 and SE15) delayed *A. aegypti* larval development (L1 to L4) and the effect was more pronounced in SE6 and SE15 [70]. Figure 9 shows that after 24 h a greater number of L1 was found in treatment with *M. oleifera* extracts than in distilled water (negative control); the life cycle at 72 h has reached the last larval stage (L4) only in control, SE1 and SE3. The extracts SE6 and SE15, with quantity of seeds higher than SE1 and SE3, prolonged mainly L1 and L2 instars revealing a correlation between larval stage and sensitivity to the active principle.

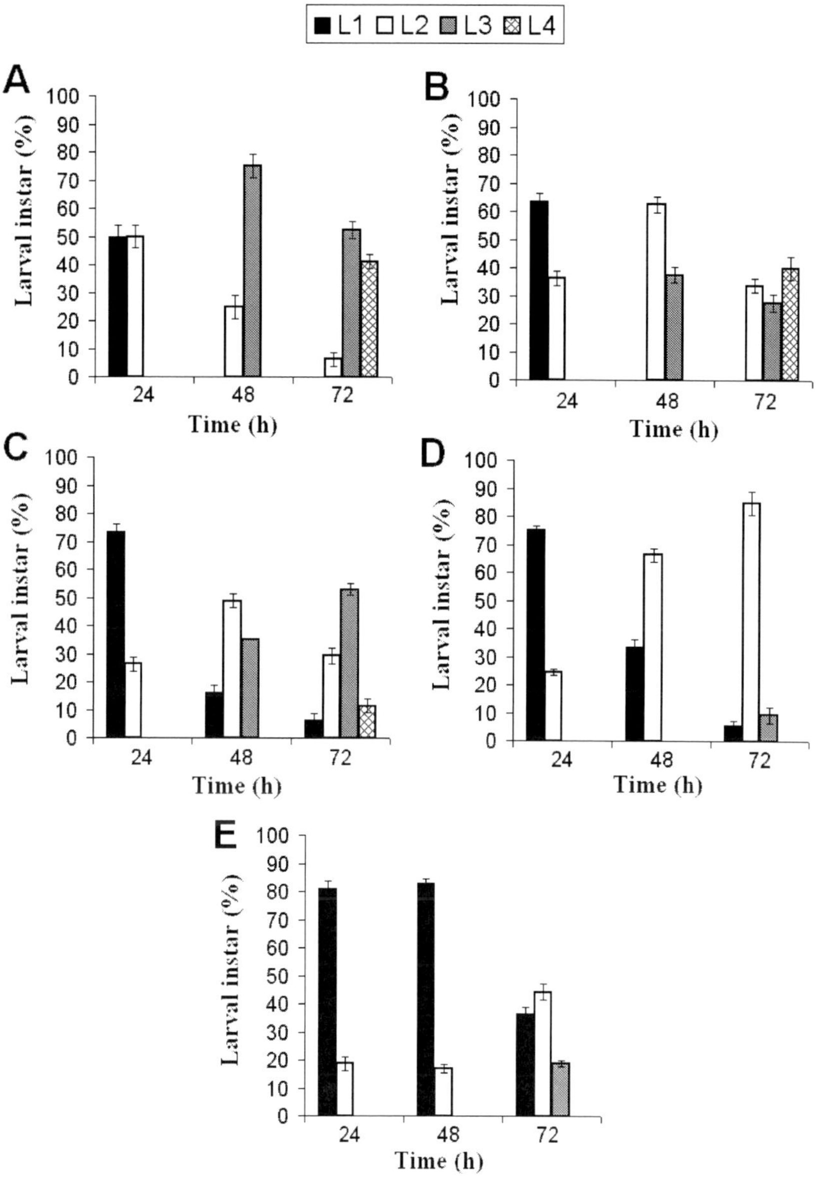

Figure 9. Larval instars (%) in incubation with control (A) and seed extracts SE1 (B), SE3 (C), SE6 (D) and SE15 (E). The bars are the mean ± S.D.

Moringa oleifera and *Myracrodruon urundeuva* lectins promoted mortality of L4 and the effect was proportional to the concentration of proteins (Figure 10); lethal concentrations of WSMoL, MuBL and MuHL required to kill 50% of the larvae (LC_{50}) in 24 h were 0.197, 0.125 and 0.04 $mg.mL^{-1}$, respectively [70, 71]. The lectins were thermo-stable proteins. Evaluation by optical microscopy of dead L4 from WSMoL treatment showed morphological changes with hypertrophy of the segments, gut volume increase and absence of epithelial layer that delimits the gut. *M. oleifera* seed extract was also a larvicidal agent on *A. aegypti*

L3 (LC_{50} of 1,260 μg.mL^{-1}) and evaluation of extract effect on non-target organisms revealed low toxicity on *Daphnia magna*, *Mus musculus* and *Rattus novergicus* [72].

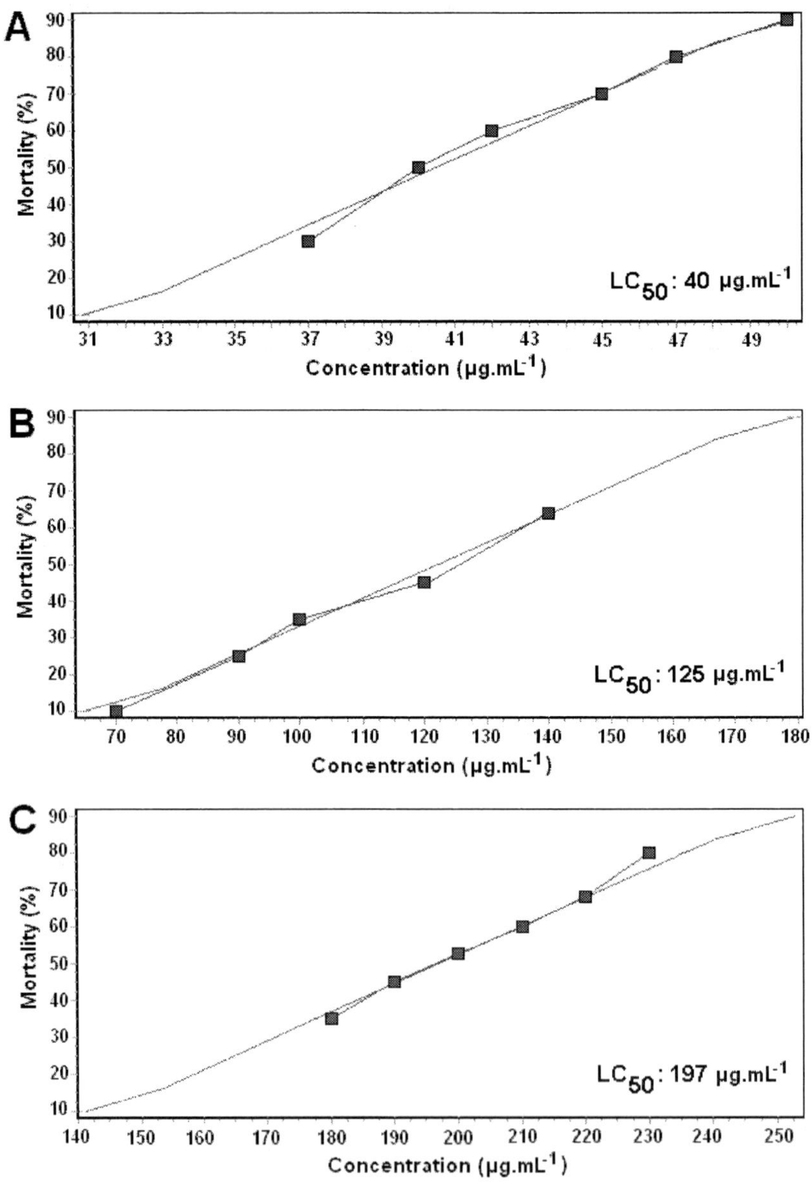

Figure 10. Mortality of *A. aegypti* L4 in incubation with lectins from *Myracrodruon urundeuva* heartwood (A) and bark (B) as well as *Moringa oleifera* seeds (C). Lethal lectin concentration required to kill 50% (LC_{50}) of larvae in 24 h was determined by probit analysis with a reliability interval of 95%.

4.3. Other Biological Properties

Plant lectins are toxic on several insect orders (Table 3). Lectin binding to glycosylated receptors at the surface of stomach epithelial cells of *Acyrthosiphon pisum* nymphs induced epithelial cell distention, enlargement, and shedding in the midgut region [73]. The involvement of different targets in the mechanism of insecticidal lectins justifies the fact that lectins of different specificities are harmful to the same insect species and also different insecticidal efficiency of lectins with similar carbohydrate-binding ability. Termites consume many types of food, for example humus and celullosic materials such as paper and wood; these insects are responsible for deterioration of wood, paintings, ancient books, monuments and buildings. *Cladonia verticillaris* lichen and *Myracrodruon urundeuva* heartwood lectins showed insecticidal activity on *Nasutitermes corniger* and are promising for population control of this insect pest [74, 75].

The availability of a great number of lectins with distinct carbohydrate specificities has resulted in the use of these proteins as tools in medical and biological research [76]. Lectins can be utilized to detection, isolation and characterization of glycoconjugates and to explore cellular surfaces binding to the carbohydrate moiety of glycoproteins or glycolipids projected from the cell. Lectin applications with therapeutic purposes have been suggested.

Lectin affinity adsorbents can be prepared by lectin immobilization to insoluble supports such as silica and agarose and these matrices are able to fractionate both nitrogen and oxygen-linked sugar chain glycoproteins [77, 78, 79, 80]. Lectin-Sepharose column can be used for isolation of glycosylated molecules; the method was able to isolate active trypsin inhibitor without structural modification and did not have the pitfalls (irreversible inhibitor-enzyme binding and/or partial inhibitor proteolysis) of inhibitor isolation by trypsin-Sepharose column [81, 82]. Lectin affinity matrices separate glycoproteins by natural pattern of glycosylation and therefore can be used for characterization of serum glycoproteins [83].

The high carbohydrate specificity of lectin does allow its use for diagnostic purposes. Lectin immobilized on gold nanoparticles with polyvinyl butyral, and adsorbed on the surface of gold electrodes can be used to detect lectin–sugar interaction. The modified electrode could be applicable to the construction of a biosensor to isolate glycoproteins present in human blood serum [84]. The system is also able to detect different glycoprotein patterns in the sera from patients infected by dengue fever and hemorrhagic dengue fever [85]. Peroxidase-labelled lectin and lectin-acridinium ester conjugate (chemiluminescent marker) did detect changes in glycosylation pattern in cell surface by differential recognition patterns (differential staining) of normal and neoplasic tissues [86, 87].

Lectins are potential therapeutic agents in cancer treatment and promote tumor growth reduction; since some antitumor lectins are toxic to normal cells they can be encapsulated into liposomes. *Cratylia mollis* isolectins (Cramoll 1,4) showed antitumor activity against Sarcoma 180; however lectin hepatotoxicity was revealed by lymphocyte infiltration in the liver of treated animals. The lectin was then loaded into liposomes and the preparation was an antitumor agent more efficient (71%) than free Cramoll 1,4 (30%); liver abnormalities and morphological alterations were not detected in spleen or kidneys after treatment with Cramoll 1,4-loaded liposomes [88].

Table 3. Insecticidal plant lectins

Insect	Plant (lectin abreviation)
Coleoptera	
Callosobruchus maculatus	*Bauhinia monandra* (BmoLL), *Griffonia simplicifolia* (GSII), *Solanum tuberosum* (STA), *Triticum aestivum* (WGA)
Zabrotes subfasciatus	BmoLL
Diptera	
Aedes aegypti	*Myracrodruon urundeuva* (MuBL, MuHL), *Moringa oleifera* (WSMoL)
Bactrocera curcubitae	*Arisaema helleborifolium* (AHL)
Lucilia cuprina	*Canavalia ensiformis* (ConA), *Lens culinaris* (lentil lectin), *Triticum aestivum* (WGA)
Hemiptera	
Acyrthosiphon pisum	ConA
Nilaparvata lugens	ConA, *Galanthus nivalis* GNA
Homoptera	
Dysdercus cingulatus	*Allium sativum* (ASAL)
Lypaphis erysimi	ASAL
Hymenoptera	
Atta opaciceps	*Canavalia brasiliensis* (ConBr)
Isoptera	
Nasutitermes corniger	MuHL
Lepidoptera	
Anagasta kuehniella	*Annona coriacea* (ACLEC), BmoLL, KpLec
Corcyra cephalonica	ACLEC
Neuroptera	
Chrysoperla carnea	GNA

References: [68, 69, 70, 71, 74, 90, 91, 92, 93, 94, 95, 96, 97].

5. Conclusion

The search for environmental safety and biodegradability stimulates the evaluation of plant versatility for several biological properties. In general, *A. aegypti* mortality promoted by essential oils, solvent organic extracts and lectins is directly proportional to compound concentration and the first two stages of larval development are more sensible to plant compounds than following larval stages. Larvicidal mechanisms proposed for essential oils and organic solvent extracts are cell alterations, antichitinogenic activity and toxic effect on larvae nervous system. Changes in digestive tract is a sign that chitinous structures at the peritrophic matrix and/or receptors at epithelial cells present in *A. aegypti* midgut are the lectin targets; common characteristics of insecticidal lectins include thermo-stability, resistance to enzyme hydrolysis and chitin-binding ability. Plant tissues are sources of promising essential oils, solvent organic extracts and lectins for pharmaceutical and cosmetic uses, as well as control of arthropods.

Acknowledgment

The authors express their gratitude to the Conselho Nacional de Desenvolvimento Científico e Tecnológico (CNPq) for research grants and fellowships (LCBBC, MTSC and PMGP), Coordenação de Aperfeiçoamento de Pessoal de Nível Superior (CAPES) and Fundação de Amparo à Ciência e Tecnologia do Estado de Pernambuco (FACEPE – PPSUS - PRONEX) for research grants. We also thank Dr. Lothar Wilhelm Bieber for scientific discussions and Emmanuel Dias da Silva for design of the figures 1, 2, 3, 4 and 5.

References

[1] World Health Organization. 1996. Report of the WHO informal consultation on the evaluation and testing of insecticides. CTD/WHOPES/IC/96.1, P. 69.

[2] Navarro, D. M. A. F., Oliveira, P. E. S., Potting, R. J. P., Brito, A. C., Fital, S. J. F. & Sant'Ana, A. E. G. 2003. The potential attractant or repellent effects of different water types on oviposition in *Aedes aegypti* L. (Dipt., Culicidae). *Journal of Applied Entomology*, 127, 46-50.

[3] Bakkali, F., Averbeck, S., Averbeck, D. & Idaomar, M. 2008. Biological effects of essential oils – A review. *Food and Chemical Toxicology*, 46, 446-475.

[4] Frizzo, C. D., Santos, A. C., Paroul, N., Serafini, L. A., Dellacassa, E., Lorenzo, D. & Moyna, P. 2000. Essential oils of camphor tree (*Cinnamomum camphora* Nees & Eberm) cultivated in southern Brazil. *Brazilian Archives of Biology and Technology*, 43, 313-316.

[5] Scavroni, J., Boaro, C. F. S., Marques, M. O. M. & Ferreira, L. C. 2005. Yield and composition of the essential oil of *Mentha piperita* L. (Lamiaceae) grown with biosolid. *Brazilian Journal of Plant Physiology*, 17, 345-352.

[6] Reddy, M. V. B., Angers, P., Gosselin, A. & Arul, J. 1998. Characterization and use of essential oil from *Thymus vulgaris* against *Botrytis cinerea* and *Rhizopus stolonifer* in strawberry fruits. *Phytochemistry*, 47, 1515-1520.

[7] Kelecom, A., Rocha, M. A., Majdalani, E. C., Gonzalez, M. S. & Mello, C. B. 2002. Novas atividades biológicas em antigos metabólitos: ácido oleanólico e eugenol de *Eugenia caryophyllata*. *Revista Brasileira de Farmacognosia*, 12, 70-71.

[8] Haypek, E., Silva, L. H. M., Batista, E., Marques, D. S., Meireles, M. A. A. & Meirelles, A. J. A. 2000. Recovery of aroma compounds from orange essential oil. *Brazilian Journal of Chemical Engineering*, 17, 705-712.

[9] Silva, J. R. A., Carmo, D. F. M., Reis, E. M., Machado, G. M. C., Leon, L. L., Silva, B. O., Ferreira, J. L. P. & Amaral, A. C. F. 2009. Chemical and biological evaluation of essential oils with economic value from Lauraceae species. *Journal of Brazilian Chemical Society*, 20, 1071-1076.

[10] Handa, S. S., Khanuja, S. P. S., Longo, G. & Rakesh, D. D. 2008. Extraction Technologies for Medicinal and Aromatic Plants. Trieste, Italy: ICS-UNIDO.

[11] Craveiro, A. A., Matos, F. J. A., Alencar, J. W. & Plumel, M. M. 1989. Microwave extraction of an essential oil. *Flavour and Fragance Journal*, 4, 43-44.

[12] Lucchesi, M. E., Chemat, F. & Smadja, J. 2004. An original solvent free microwave extraction of essential oils from spices. *Flavour and Fragance Journal*, 19, 134-138.

[13] Phutdhawong, W., Kawaree, R., Sanjaiya, S., Sengpracha, W. & Buddhasukh, D. 2007. Microwave-assisted isolation of essential oil of *Cinnamomum iners* Reinw. Ex Bl: comparison with conventional hydrodistillation. *Molecules*, 12, 868-877.

[14] Rezvanpanah, S., Rezaei, K., Razavi, S. H. & Moini, S. 2008. Use of microwave-assisted hydrodistillation to extract the essential oils from *Satureja hortensis* and *Satureja Montana*. *Food Science and Technology Research*, 14, 311-314.

[15] Asfaw, N., Licence, P., Novitskii, A. A. & Poliakoff, M. 2005. Green chemistry in Ethiopia: the cleaner extraction of essential oils from *Artemisia afra*: a comparison of clean technology with conventional methodology. *Green Chemistry*, 7, 352–356.

[16] Kiran, S. R., Bhavani, K., Devi, P. S., Rao, B. R. R. & Reddy, K. J. 2006. Composition and larvicidal activity of leaves and stem essential oils of *Chloroxylon swietenia* DC against *Aedes aegypti* and *Anopheles stephensi*. *Bioresource Technology*, 97, 2481-2484.

[17] Silva, W. J., Dória, G. A. A., Maia, R. T., Nunes, R. S., Carvalho, G. A., Blank, A. F., Alves, P. B., Marçal, R. M. & Cavalcanti, S. C. H. 2008. Effects of essential oils on *Aedes aegypti* larvae: Alternatives to environmentally safe insecticides. *Bioresource Technology*, 99, 3251-3255.

[18] Morais, S. M., Facundo, V. A., Bertini, L. M., Cavalcanti, E. S. B., Júnior, J. F. A., Ferreira, S. A., Brito, E. S. & Neto, M. A. S. 2007. Chemical composition and larvicidal activity of essential oils from *Piper* species. *Biochemical Systematics and Ecology*, 35, 670-675.

[19] Carvalho, A. F. U., Melo, V. M. M., Craveiro, A. A., Machado, M. I. L., Bantim, M. B. & Rabelo, E. F. 2003. Larvicidal activity of the essential oil from *Lippia sidoides* Cham. against *Aedes aegypti* Linn. *Memórias do Instituto Oswaldo Cruz*, 98, 569-571.

[20] Ansari, M. A., Vasudevan, P., Tandon, M. & Razdan, R. K. 2000. Larvicidal and mosquito repellent action of peppermint (*Mentha piperita*) oil. *Bioresource Technology*, 71, 267-271.

[21] Pitasawat, B., Champakaew, D., Choochote, W., Jitpakdi, A., Chaithong, U., Kanjanapothi, D., Rattanachanpichai, E., Tippawangkosol, P., Riyong, D., Tuetun, B. & Chaiyasit, D. 2007. Aromatic plant-derived essential oil: An alternative larvicide for mosquito control. *Fitoterapia*, 78, 205-210.

[22] Dharmagadda, V. S. S., Naik, S. N., Mittal, P. K. & Vasudevan, P. 2005. Larvicidal activity of *Tagetes patula* essential oil against three mosquito species. *Bioresource Technology*, 96, 1235-1240.

[23] Prajapati, V., Tripathi, A. K., Aggarwal, K. K. & Khanuja, S. P. S. 2005. Insecticidal, repellent and oviposition-deterrent activity of selected essential oils against *Anopheles stephensi*, *Aedes aegypti* and *Culex quinquefasciatus*. *Bioresource Technology*, 96, 1749-1757.

[24] Autran, E. S., Neves, I. A., Silva, C. S. B., Santos, G. K. N., Câmara, C. A. G. & Navarro, D. M. A. F. 2009. Chemical composition, oviposition deterrent and larvicidal activities against *Aedes aegypti* of essential oils from *Piper marginatum* Jacq. (Piperaceae). *Bioresource Technology*, 100, 2284-2288.

[25] Lucia, A., Audino, P. G., Seccacini, E., Licastro, S., Zerba, E. & Masuh, H. 2007. Larvicidal effect of *Eucalyptus grandis* essential oil and turpentine and their major components on *Aedes aegypti* larvae. *Journal of the American Mosquito Control Association*, 23, 299-303.

[26] Cheng, S., Huang, C., Chen, Y., Yu, J., Chen, W. & Chang, S. 2009. Chemical compositions and larvicidal activities of leaf essential oils from two eucalyptus species. *Bioresource Technology*, 100, 452-456.

[27] Chung, I., Seo, S., Kang, E., Park, S., Park, W. & Moon, H. 2009. Chemical composition and larvicidal effects of essential oil of *Dendropanax morbifera* against *Aedes aegypti* L. *Biochemical Systematics and Ecology*, 37, 470-473.

[28] Gbolade, A. A. & Lockwood, G. B. 2008. Toxicity of *Ocimum sanctum* L. essential oil to *Aedes aegypti* larvae and its chemical composition. *Journal of Essential Oil Bearing Plants*, 11, 148-153.

[29] Cheng, S., Liu, J., Huang, C., Hsui, Y., Chen, W. & Chang, S. 2009. Insecticidal activities of leaf essential oils from *Cinnamomum osmophloeum* against three mosquito species. *Bioresource Technology*, 100, 457-464.

[30] Nerio, L. S., Olivero-Verbel, J. & Stashenko, E. 2010. Repellent activity of essential oils: A review. *Bioresource Technology*,101, 372-378.

[31] Cheng, S., Chang, H., Wu, C. & Chang, S. 2007. Anti-termitic activities of essential oils from coniferous trees against *Coptotermes formosanus*. *Bioresource Technology*, 98, 456-459.

[32] Çalmaşur, Ö., Aslan, İ., & Şahin, F. 2006. Insecticidal and acaricidal effect of three Lamiaceae plant essential oils against *Tetranychus urticae* Koch and *Bemisia tabaci* Genn. *Industrial Crops and Products*, 23, 140-146.

[33] Thorsell, W., Mikiver, A. & Tunón, H. 2006. Repelling properties of some plant materials on the tick *Ixodes ricinus* L. *Phytomedicine*, 13, 132-134.

[34] Tunón, H., Thorsell, W., Mikiver, A. & Malander, I. 2006. Arthropod repellency, especially tick (*Ixodes ricinus*), exerted by extract from *Artemisia abronatum* and essential oil from flowers of *Dianthus caryophyllum*. *Fitoterapia*, 77, 257-261, 2006.

[35] Starmans, D. A. J. & Nijhuis, H. H. 1996. Extraction of secondary metabolites from plant material: a review. *Trends in Food Science & Technology*, 7, 191-197.

[36] Castro, M. D. L. & Priego-Capote, F. 2010. Soxhlet extraction: Past and present panacea. *Journal of Chromatography A*, 1217, 2383-2389.

[37] Coria, C., Almiron, W., Valladares, G., Carpinella, C., Ludueña, F., Defago, M. & Palacios, S. 2008. Larvicide and oviposition deterrent effects of fruit and leaf extracts from *Melia azedarach* L. on *Aedes aegypti* (L.) (Diptera: Culicidae). *Bioresource Technology*, 99, 3066-3070.

[38] Arruda, W., Oliveira, G. M. C. & Silva, I. G. 2003. Toxicidade do extrato etanólico de *Magonia pubescens* sobre larvas de *Aedes aegypti*. *Revista da Sociedade Brasileira de Medicina Tropical*, 36, 17-25.

[39] Sakthivadivel, M. & Thilagavathy, D. 2003. Larvicidal and chemosterilant activity of the acetone fraction of petroleum ether extract from *Argemone Mexicana* L. seed. *Bioresource Technology*, 89, 213-216.

[40] Wandscheer, C. B., Duque, J. E., Silva, M. A. N., Fukuyama, Y., Wohlke, J. L., Adelmann, J. & Fontana, J. D. 2004. Larvicidal action of ethanolic extracts from fruit endocarps of *Melia azedarach* and *Azadirachta indica* against the dengue mosquito *Aedes aegypti*. *Toxicon*, 44, 829-835.

[41] Murugan, K., Murugan, P. & Noortheen, A. 2007. Larvicidal and repellent potential of *Albizzia amara* Boivin and *Ocimum basilicum* Linn against dengue vector, *Aedes aegypti* (Insecta:Diptera:Culicidae). *Bioresource Technology*, 98, 198-201.

[42] Ciccia, G., Coussio, J. & Mongelli, E. 2000. Insecticidal activity against *Aedes aegypti* larvae of some medicinal South American plants. *Journal of Ethnopharmacology*, 72, 185-189.

[43] Luna, J. S., Santos, A. F., Lima, M. R. F., Omena, M. C., Mendonça, F. A. C., Bieber, L. W. & Sant'Ana, A. E. G. 2005. *Journal of Ethnopharmacology*, 97, 199-206.

[44] Omena, M. C., Navarro, D. M. A. F., Paula, J. E., Luna, J. S., Lima, M. R. F. & Sant'Ana, A. E. G. 2007. Larvicidal activities against *Aedes aegypti* of some Brazilian medicinal plants. *Bioresource Technology*, 98, 2549-2556.

[45] Sivagnaname, N. & Kalyanasundaram, M., 2004. Laboratory evaluation of methanolic extract of *Atlantia monophylla* (Family: Rutaceae) against immature stages of mosquitoes and non-target organisms. *Memórias do Instituto Oswaldo Cruz*, 99, 115-118.

[46] Rahuman, A. A., Gopalakrishnan, G., Venkatesan, P. & Geetha, K. 2008. Larvicidal activity of some Euphorbiaceae plant extracts against *Aedes aegypti* and *Culex quinquefasciatus* (Diptera: Culicidae). *Parasitology Research*, 102, 867-873.

[47] Ioset, J., Marston, A., Gupta, M. P. & Hostettmann, K. 2000. Antifungal and larvicidal cordiaquinones from the roots of *Cordia curassavica*. *Phytochemistry*, 53, 613-617.

[48] Chapagain, B. P., Saharan, V. & Wiesman, Z. 2008. Larvicidal activity of saponins from *Balanites aegyptica* callus against *Aedes aegypti* mosquito. *Bioresource Technology*, 99, 1165-1168.

[49] Cheng, S., Huang, C., Chen, W., Kuo, Y. & Chang, S. 2008. Larvicidal activity of tectoquinone isolated from red heartwood-type *Cryptomeria japonica* against two mosquito species. *Bioresource Technology*, 99, 3617-3622.

[50] Garcez, W. S., Garcez, F. R., Silva, L. M. G. E. & Hamerski, L. 2009. Larvicidal activity against *Aedes aegypti* of some plants native to the West-Central region of Brazil. *Bioresource Technology*, 100, 6647-6650.

[51] Gu, H., Cheng, S., Huang, C., Chen, W. & Chang, S. 2009. Mosquito larvicidal activities of extractives from black heartwood-type *Cryptomeria japonica*. *Parasitology Research*, 105, 1455-1458.

[52] Markouk, M., Bekkouche, K., Larhsini, M., Bousaid, M., Lazrek, H. B. & Jana, M. 2000. Evaluation of some Moroccan medicinal plant extracts for larvicidal activity. *Journal of Ethnopharmacology*, 73, 293-297.

[53] Pavela, R. 2004. Insecticidal activity of certain medicinal plants. *Fitoterapia*, 75, 745-749.

[54] Madhu, S. K., Shaukath, A. K. & Vijayan, V. A. 2010. Efficacy of bioactive compounds from *Curcuma aromatic* against mosquito larvae. *Acta Tropica*, 113, 7-11.

[55] Boussaada, O., Kamel, M. B. H., Ammar, S., Haouas, D., Mighri, Z. & Helal, A.N. 2008. Insecticidal activity of some Asteraceae plant extracts against *Tribolium confusum*. *Bulletin of Insectology*, 61, 283-289.

[56] Ateyyat, M. A. & Abu-Darwish, M. S. 2009. Insecticidal activity of different extracts of *Rhamnus dispermus* (Rhamnaceae) against peach trunk aphid, *Pterochloroides persicae* (Homoptera: Lachnidae). *Spanish Journal of Agricultural Research*, 7, 160-164.

[57] Dang, Q. L., Choi, Y. H., Choi, G. J., Jang, K. S., Park, M. S., Park, N., Lim, C. H., Kim, H., Ngoc, L. H. & Kim, J. 2010. Pesticidal activity of ingenane diterpenes isolated from *Euphorbia kansui* against *Nilaparvata lugens* and *Tetranychus urticae*. *Journal of Asia-Pacific Entomology*, 13, 51-54.

[58] Siskos, E. P., Konstantopoulou, M. A. & Mazomenos, B. E. 2009. Insecticidal activity of *Citrus aurantium* peel extract against *Bactrocera oleae* and *Ceratitis capitata* adults (Diptera: Tephritidae). *Journal of Applied Entomology*, 133, 108-116.

[59] Napal, G. N. D., Carpinella, M. C. & Palacios, S. M. 2009. Antifeedant activity of ethanolic extract from *Flourensia oolepis* and isolation of pinocembrin as its active principle compound. *Bioresource Technology*, 100, 3669-3673.

[60] Bagavan, A., Kamaraj, C., Elango, G., Zahir, A. A. & Rahuma, A. A. 2009. Adulticidal and larvicidal efficacy of some medicinal plant extracts against tick, fluke and mosquitoes. *Veterinary Parasitology*, 166, 286-292.

[61] Singh, A. & Singh, V. K. 2009. Molluscicidal activity of *Saraca asoca* and *Thuja orientalis* against the fresh water snail *Lymnaea acuminata*. *Veterinary Parasitology*, 164, 206-210.

[62] Huang, R. L., Chen, C. C., Huang, Y. L., Ou, J. C., Hu, C. P., Chen, C. F. & Chang, C. 1998. Anti-tumor effects of d-dicentrine from the root of *Lindera megaphylla*. *Planta Medica*, 64, 212-215.

[63] Green, A. A. & Hughes, L. 1955. Protein fractionation on the basis of solubility in aqueous solution of salts and organic solvents. In: Methods in Enzymology (edited by S. Colowick and N. Kaplan). Pp. 67-90. New York, USA: Academic Press.

[64] Santana, M. A., Santos, A. M. C., Oliveira, M. E., Oliveira, J. S., Baba, E. H., Santoro, M. M. & Andrade, M. H. G. 2008. A novel and efficient and low-cost methodology for purification of *Macrotyloma axillare* (Leguminosae) seed lectin. *International Journal of Biological Macromolecules*, 43, 352-358.

[65] Tellam, R. L., Wijffels, G. & Wiladsen, P. 1999. Peritrophic matrix proteins. *Insect Biochemistry and Molecular Biology*, 29, 87-101.

[66] Fitches, E. & Gatehouse, J. A. 1998. A comparison of the short and long term effects of insecticidal lectins on the activities of soluble and brush border enzymes of tomato moth larvae (*Lacanobia oleracea*). *Journal of Insect Physiology*, 44, 1213-1224.

[67] Harper, M. S., Hopkins, T. L. & Czapla, T. H. 1998. Effect of wheat germ agglutinin on formation and structure of the peritrophic membrane in European corn borer (*Ostrinia nubilalis*) larvae. *Tissue and Cell*, 30, 166-176.

[68] Coelho, M. B., Marangoni, S. & Macedo, M. L. R. 2007. Insecticidal action of *Annona coriacea* lectin against the flour moth *Anagasta kuehniella* and the rice moth *Corcyra cephalonica* (Lepidoptera: Pyralidae). *Comparative Biochemistry and Physiology – Part C: Toxicology & Pharmacology*, 146, 406-414.

[69] Macedo, M. L. R., Freire, M. G. M., Silva, M. B. R. & Coelho, L. C. B. B. 2007. Insecticidal action of *Bauhinia monandra* leaf lectin (BmoLL) against *Anagasta kuehniella* (Lepidoptera: Pyralidae), *Zabrotes subfasciatus* and *Callosobruchus maculatus* (Coleoptera: Bruchidae). *Comparative Biochemistry and Physiology – Part A: Molecular & Integrative Physiology*, 146, 486-498.

[70] Coelho, J. S., Santos, N. D. L., Napoleão, T. H., Gomes, F. S., Ferreira, R. S., Zingali, R. B., Coelho, L. C. B. B., Leite, S. P., Navarro, D. M. A. F. & Paiva, P. M. G. 2009. Effect of *Moringa oleifera* lectin on development and mortality of *Aedes aegypti* larvae. *Chemosphere*, 77, 934-938.

[71] Sá, R. A., Santos, N. D. L., Silva, C. S. B., Napoleão, T. H., Gomes, F. S., Cavada, B. S., Coelho, L. C. B. B., Navarro, D. M. A. F., Bieber, L. W. & Paiva, P. M. G. 2009. Larvicidal activity of lectins from *Myracrodruon urundeuva* on *Aedes aegypti*. *Comparative Biochemistry and Physiology – Part C: Toxicology & Pharmacology*, 149, 300-306.

[72] Ferreira, P. M. P., Carvalho, A. F. U., Farias, D. F., Cariolano, N. G., Melo, V. M. M., Queiroz, M. G. R., Martins, A. M. C. & Machado-Neto, J. G. 2009. Larvicidal activity of the water extract of *Moringa oleifera* seeds against *Aedes aegypti* and its toxicity upon laboratory animals. *Anais da Academia Brasileira de Ciências*, 81, 207-216.

[73] Sauvion, N., Nardonb, G., Febvay, G., Gatehouse, A. M. R. & Rahbé, Y. 2004, Binding of the insecticidal lectin Concanavalin A in pea aphid *Acyrthosiphon pisum* (Harris) and induced effects on the structure of midgut epithelial cells. Journal of Insect Physiology, 50, 1137–1150.

[74] Sá, R. A., Napoleão, T. H., Santos, N. D. L., Gomes, F. S., Albuquerque, A. C., Xavier, H. S., Coelho, L. C. B. B., Bieber, L. W. & Paiva, P. M. G. 2008. Induction of mortality on *Nasutitermes corniger* (Isoptera, Termitidae) by *Myracrodruon urundeuva* heartwood lectin. *International Biodeterioration & Biodegradation*, 62, 460-464.

[75] Silva, M. D. C., Sá, R. A., Napoleão, T. H., Gomes, F. S., Santos, N. D. L., Albuquerque, A. C., Xavier, H. S., Paiva, P. M. G., Correia, M. T. S. & Coelho, L. C. B. B. 2009. Purified *Cladonia verticillaris* lichen lectin: Insecticidal activity on *Nasutitermes corniger* (Isoptera: Termitidae). *International Biodeterioration & Biodegradation*, 63, 334-340.

[76] Correia, M. T. S., Coelho, L. C. B. B. & Paiva, P. M. G. 2008. Lectins, carbohydrate recognition molecules: Are they toxic? In: Recent Trends in Toxicology (edited by Y. H. Siddique). Pp. 47-59. Kerala, India: Transworld Research Network.

[77] Kobata, A. 1994. Affinity chromatography with use of immobilized lectin columns. *Biochemical Society Transactions*, 22, 360–364.

[78] Kennedy, J. F., Palva, P. M. G., Corrella, M. T. S., Cavalcanti, M. S. M., & Coelho, L. C. B. B. 1995. Lectins, versatile proteins of recognition: a review. *Carbohydrate Polymers*, 26, 219-230.

[79] Fraguas, L. F., Carlsson, J. & Lönnberg, M. 2008. Lectin affinity chromatography as a tool to differentiate endogenous and recombinant erythropoietins. *Journal of Chromatography A*, 1212, 82-88.

[80] Helmholz, H., Naatz, S., Lassen, S. & Prange, A. 2008. Isolation of a cytotoxic glycoprotein from the Scyphozoa *Cyanea lamarchii* by lectin-affinity chromatography and characterization of molecule interactions by surface Plasmon resonance. *Journal of Chromatography B*, 871, 60-66.

[81] De Martino, N. 1989. Purification of proteolytic enzymes. In: Proteolytic Enzymes (edited by R.J. Beynon and J.S. Bond). Pp. 15-23. England: I.R.L. Press.

[82] Paiva, P. M. G., Souza, A. F., Oliva, M. L. V., Kennedy, J. F., Cavalcanti, M. S. M., Coelho, L. C. B. B. & Sampaio, C. A. M. 2003. Isolation of a trypsin inhibitor from *Echinodorus paniculatus* seeds by affinity chromatography on immobilized *Cratylia mollis* isolectins. *Bioresource Technology*, 88, 75-79.

[83] Lima, V. L. M., Correia, M. T. S., Cechinel, Y. M. N., Sampaio, C. A. M., Owen, J. S. & Coelho, L. C. B. B. 1997. Immobilized *Cratylia mollis* lectin as a potential matrix to isolate plasma glycoproteins, including lecithin-cholesterol acyltransferase, *Carbohydrate Polymers*, 33, 27-32.

[84] Oliveira, M. D. L., Correia, M. T. S., Coelho, L. C. B. B. & Diniz, F. B. 2008. Eletrochemical evaluation of lectin-sugar interaction on gold electrode modified with colloidal gold and polyvinyl butyral. *Colloids and Surfaces B: Biointerfaces*, 66, 13-19.

[85] Oliveira, M. D. L., Correia, M. T. S. & Diniz, F. B. 2009. Concanavalin A and polyvinil butyral use as a potential dengue electrochemical biosensor. *Biosensors and Bioelectronics*, 25, 728-732.

[86] Beltrão, E. I. C., Correia, M. T. S., Figueredo-Silva, J. & Coelho, L. C. B. B. 1998. Binding evaluation of isoform 1 from *Cratylia mollis* lectin to human mammary tissues. *Applied Biochemistry and Biotechnology*, 74, 125-134.

[87] Campos, L. M., Cavalcanti, C. L. B., Lima-Filho, J. L., Carvalho Jr., L. B. & Beltrão, E. I. C. 2006. Acridinium ester conjugated to lectin as chemiluminescent histochemistry marker. *Biomarkers*, 11, 480-484.

[88] Andrade, C. A. S., Correia, M. T. S., Coelho, L. C. B. B., Nascimento, S. C. & Santos-Magalhães, N. S. 2004. Antitumor activity of *Cratylia mollis* lectin encapsulated into liposomes. *International Journal of Pharmaceutics*, 278, 435-445.

[89] Yang, H., Hu, J., Li, Q., Liu, H., Yao, T., Ran, C. & Lei, H. 2009. Acaricidal activity of *Boenninghausenia sessilicarpa* against *Panonychus citri*. *Agricultural Sciences in China*, 8, 1097-1102.

[90] Zhu-Salzman, K., Shade, R. E., Koiwa, H., Salzman, R. A., Narasimhan, M., Bressan, R. A., Hasegawa, P. M., Murdock, L. L. 1998. Carbohydrate binding and resistance to proteolysis control insecticidal activity of *Griffonia simplicifolia* lectin II. *Proceedings of the National Academy of Sciences of the United States of America*, 95, 15123–15128.

[91] Bandyopadhyay, S., Roy, A. & Das, S. 2001. Binding of garlic (*Allium sativum*) leaf lectin to the gut receptors of homopteran pests is correlated to its insecticidal activity. *Plant Science*, 161, 1025-1033.

[92] Eisemann, C. H., Donaldson, R. A., Pearson, R, D., Cadogan, L. C., Vuocolo, T. & Tellam, R. L. 1994. Larvicidal activity of lectins on *Lucilia cuprina*: mechanism of action. *Entomologia Experimentalis et Applicata*, 72, 1–10.

[93] Hogervorst, P. A. M., Ferry, N., Gatehouse, A. M. R., Wäckers, F. L. & Romeis, J. 2006. Direct effects of snowdrop lectin (GNA) on larvae of three aphid predators and fate of GNA after ingestion. *Journal of Insect Physiology*, 52, 614-624.

[94] Isidro, R., Sales, F. J. M., Cavada, B. S., Grangeiro, T. B. & Moreira, R. A. 2001. Ação de lectina de sementes de *Canavalia brasiliensis* Mart. sobre o comportamento da saúva do Nordeste (*Atta opaciceps* Borgmeier, 1939). *Revista de la Facultad de Agronomía*, 27, 77-89.

[95] Kaur, M., Singh, K., Rup, P. J., Kamboj, S. S., Saxena, A. K., Sharma, M., Bhagat, M., Sood, S. K. & Singh, J. 2006. A tuber lectin from *Arisaema helleborifolium* Schott with anti-insect activity against melon fruit fly, *Bactrocera cucurbitae* (Coquillett) and anti-cancer effect on human cancer cell. *Archives of Biochemistry and Biophysics*, 445, 156-165.

[96] Murdock, L. L., Huesing, J. E., Nielsen, S. S., Pratt, R. C. & Shade, R. E. 1990. Biological effects of plant lectins on the cowpea weevil. *Phytochemistry*, 29, 85-89.

[97] Powell, K. S., Gatehouse, A. M. R., Hilder, V. A. & Gatehouse, J. A. 1995. Antifeedant effects of plant-lectins and an enzyme on the adult stage of the rice brown planthopper, *Nilaparvata lugens*. *Entomologia Experimentalis et Applicata*, 75, 51-59.

In: Bioprocess Sciences and Technology
Editor: Min-Tze Liong
ISBN 978-1-61122-950-9

Chapter 12

PLANTS AS PLATFORM FOR PRODUCTION [1]OF PHARMACEUTICAL COMPOUNDS

Miguel Angel Gomez-Lim*
Departamento de Ingeniería Genética de Plantas, Libramiento Norte Carretera Irapuato-León, Cinvestav Campus Guanajuato. Km 9.6, Apdo. Postal 629, Irapuato, 36500, Guanajuato, Mexico

ABSTRACT

Plants are gaining widespread acceptance for the large-scale production of recombinant proteins as they provide a number of advantages over conventional systems including low cost, increased safety and scalable production, among others. Over the last few years, there have been significant successes in overcoming some of the limitations of plants as bioreactors, such as low yields and inconsistent product quality that have limited the approval of plant-derived pharmaceuticals. Indeed, one of the most important driving factors has been yield improvement, as product yield has a significant impact on economic feasibility. Strategies to improve the recombinant protein yield in plants include the development of novel promoters, the improvement of protein stability and accumulation and the improvement of downstream processing technologies. Attention is now shifting from basic research towards commercial exploitation, and molecular farming is reaching the stage at which it may challenge established production technologies based on bacteria, yeast and cultured mammalian cells. There are already several plant-produced proteins on the market including one at large scale. Several plant-derived recombinant pharmaceutical proteins are reaching the final stages of clinical evaluation, and more are in the development pipeline. In this review, recent progress in molecular farming will be highlighted, including the potential for commercial drug development and production.

* Corresponding author: *Miguel Angel Gomez-Lim* E-mail: mgomez@ira.cinvestav.mx

1. Expression of Recombinant Proteins in Transgenic Plants

There is an increasing demand for production of recombinant proteins on an industrial scale because of their utility as diagnostic reagents, vaccines and therapeutic agents. Recombinant DNA technology has allowed expression of heterologous proteins in different systems [1]. Although mammalian cell cultures and yeast have been the main expression systems employed, they have some downsides in terms of cost, scalability, risk of pathogenicity and authenticity [2]. Since cost is still one of the persistent barriers for development and distribution of safe and effective new vaccines and pharmaceutical compounds to populations in dire need of them, plants represent a simple and inexpensive alternative allowing a scalable production system for recombinant proteins. Plant bioreactors have been estimated to yield over 10 kg of therapeutic protein per acre in tobacco, maize, soybean and alfalfa [3]. In comparison with conventional bioreactors and mammalian cells or microorganisms, the cost of producing a protein under good manufacturing practice conditions is reduced to perhaps one-tenth [4, 5].

Genetically engineered plants have many advantages as sources of recombinant proteins such as low cost, rapid scale-up or down, no need for syringes and needles, antigens easy to store and transport without the need for a cold chain, polyvalent vaccines are quite feasible, increased efficacy against mucosal-borne pathogens and reduced concerns over contamination with human pathogens. Depending upon the promoters used, the recombinant proteins can accumulate throughout the plant, in specific organs (e.g. in seeds) or in specific organelles (e.g. chloroplasts) within a plant cell. However, seeds have been identified as an ideal target for recombinant protein accumulation because of the high content of protein, adequate storage properties and flexibility in processing management and batch production [6].

Furthermore, proteins produced in seed exhibit high stability; for example, enzymes and antibodies expressed in seed and stored for more than three years in the refrigerator retain full enzymatic or binding activity [7].

2. Methods for Gene Transfer into Plants

There are four main approaches for expression of recombinant proteins in plants: (a) nuclear transformation and regeneration of transgenic plant lines; (b) transfer and expression of transgenes into the chloroplast genome and (c) by the use of plant viral vectors (transient expression). Each will be examined in turn.

2.1. Nuclear Transformation

Successful expression of transgenes in plants is possible thanks to the unique capability of single plant cells to regenerate into whole plants while keeping all the genetic features of the parent plant. In the transgenic plant, foreign genes are stably incorporated into the plant genome, transcribed and inherited in a Mendelian fashion [8]. In addition, gene transfer into a plant could be mediated by a plant-infecting bacterium, *Agrobacterium tumefaciens*, which is able to transfer DNA into the plant genome [9].

Microprojectile bombardment or biolistics (direct DNA transfer) has been extensively employed [10]. The method is so versatile that multiple genes (>10] coding for complex recombinant macromolecules can be transferred simultaneously into the plant genome. Interestingly, even though the genes may be delivered on different plasmids, such multiple transgenes are frequently inherited in a linked fashion [11]. Nicholson et al. (2005) successfully introduced four genes coding for components of a secretory antibody into rice and approximately 20% of the resulting plants contained all four genes [12]. This represents an advantage over alternative gene transfer methods that involve the stepwise introduction of individual components, followed by successive rounds of crossing to generate plants containing the fully assembled molecule.

Direct DNA transfer also allows the introduction into plants of minimal expression cassettes containing only the promoter, open reading frame and terminator sequences. As no vector backbone sequences are transferred, this approach increases transgene stability and expression levels considerably by preventing the integration of potentially recombinogenic sequences [13]. The described methods are applicable to a wide range of species and this explains why the majority of plant-derived recombinant pharmaceutical proteins have been produced by nuclear transformation and regeneration of transgenic plant lines.

2.1.1. Chloroplast Transformation

An alternative approach to express recombinant proteins in a stable manner is by gene transfer into the plant chloroplasts. Chloroplasts are plant cellular organelles with their own genome and transcription-translation machinery. The chloroplast genome is a highly polyploid, circular double-stranded DNA of 120 kb to 180 kb in size, encoding approximately 120 genes [14]. Each chloroplast carries a number of identical genome copies, which are attached to membranes in clusters called nucleoids. A tobacco leaf cell may contain 100 chloroplasts, with 10-14 nucleoids each, and about 10,000 copies of the genome per cell [15]. A gene may be introduced into a spacer region between the functional genes of the chloroplast by homologous recombination, targeting the foreign gene to a precise location. Gene silencing has not been observed with chloroplast transformation, whereas it is a common phenomenon with nuclear transformation.

Chloroplast transformation has been achieved in several plant species such as carrot, tomato, *Brassica oleracea*, petunia, soybean, lettuce, the liverwort *Marchantia polymorpha* and the green algae *Chlamydomonas reinhardtii*, but transformation is routine only in tobacco [16]. Currently, the most employed method for chloroplast transformation is by microprojectile bombardment. The ability to transform the chromoplasts of fruit and vegetable crops represent an interesting possibility for the expression of subunit vaccines [17].

There are several advantages of chloroplast transformation: high-level expression of the recombinant proteins (10,000 copies of trans-gene per cell), the recombinant proteins will accumulate within the chloroplast thus limiting toxicity to the host plant, multiple genes can be expressed in operons [18] and the absence of functional chloroplast DNA in pollen of most crops may provide natural transgene containment [19]. To date, 23 antigens against 16 different bacterial, viral or protozoan pathogens have been expressed in chloroplasts [16].

Transgene expression in tobacco plastids reproducibly yields protein levels in the 5% to 20% range, however, levels up to 47% of the total soluble protein, can be achieved [16].

Even though there is no protein glycosylation in chroloplasts [20]. The functionality of chloroplast-derived antigens and therapeutic proteins has been demonstrated by in vitro assays and animal protection studies [21].

2.1.1.1. Expression Based on Plant Viruses

Another tobacco transient-expression technology is based on the use of plant viruses as expression vectors. Viruses can gain entry into a plant cell where they accumulate and then spread throughout the entire plant. They remain in the cytoplasm throughout infection and do not incorporate into the genome of the susceptible host and, thus, are not inherited by the next generation. Several features make plant viruses well suited as transient expression vectors: high-level expression of the introduced genes (up to 2 g/kg of plant tissue, in the case of human antibodies) within a short period (1-2 weeks after inoculation), rapid accumulation of the appropriate products and the fact that more than one vector can be used in the same plant, allowing multimeric proteins to be assembled [22] with the additional benefit of biological containment on the viral sequences. Post-transcriptional gene silencing, a natural mechanism of defense by the plant against viruses, can be avoided by expressing a replicase from some inducible promoters [23].

Plant-virus genomes can be composed of DNA or RNA but the main virus systems from which efficient expression systems are being developed mainly consist of positive-sense RNA genome, single-stranded DNA geminiviruses and double-stranded non-integrating DNA pararetroviruses [24] but RNA viruses can multiply to very high titres in infected plants, which makes them better suited for protein expression vectors. For genetic manipulation, viral RNA genomes are reverse transcribed *in vitro* and cloned as full-length cDNAs. There are at least three approaches for insertion of foreign genes into plant viral genomes: (a) gene replacement, when non-essential viral genes, like those coding for coat proteins, are replaced by the gene of interest. Since some viruses have limitations as to the size of the molecules than can be incorporated into their genomes this would be the best strategy. (b) gene insertion, where the gene of interest is placed under the control of an additional promoter; this approach would be advisable where large coding sequences have to be expressed and (c) gene fusion, when the gene of interest is translationally fused with a viral gene; the use of coat-protein genes has allowed an efficient method for presentation of foreign peptide sequences on the surfaces of viral particles [25].

Numerous reports have confirmed that plant viruses can be effective vectors for expression of antigens and can provide complete protection in challenge trials [26]. Based on the approach described above for gene replacement, Icon Genetics (Halle, Germany) has developed extremely efficient viral replicons that can be delivered through infection with *Agrobacterium*, a process termed 'magnifection' [27].

3. Production of Antigens in Plants

Vaccines, the most effective medical intervention to prevent disease, have been based on live, attenuated organisms, purified antigens (subunit vaccines) or DNA coding for specific antigens. Purified antigens are usually delivered at a set dose and have constituted a relatively simple and uniform material for administration, generally prepared from recombinant sources. Considerable progress has been made since Charles Arntzen first envisaged the idea of transgenic plant vaccines in the early 1990s. There are now a number of examples demonstrating the successful expression of subunit candidate vaccines both for humans and animals in transgenic plants (Table 2). These include antigens from bacterial and viral sources that infect humans, domestic or wild animals and representing secreted toxins and cell or viral surface antigens.

Table 2. Antigens expressed in transgenic plants

Human Antigen	Plant system	Expression levels*	References
Yersinia pestis F1, V, and F1-V antigens	*Nicotiana benthamiana*	1 mg/g	[28]
Cholera toxin B subunit protein	Tomato	0.02-0.04%	[29,30]
Cholera toxin B subunit-human glutamate decarboxylase fusion protein	Potato	0.4%	[31]
Human secreted alkaline phosphatase	Tobacco NT1 cell	27 mg/L	[32]
Measles virus hemagglutinin glycoprotein	Carrot	2%	[33]
Japanese encephalitis virus envelope protein	Rice	1.1-1.9 μg/mg of TSP	[34]
Heat labile enterotoxin from *E. coli*	Somatic embryos of siberian ginseng	0.36%	[38]
Rotavirus VP7 fused to the cholera toxin B subunit gene	Potato	0.01%	[39]
Human lactoferrin	Potato	0.1%	[41]
Human milk protein beta-casein	Potato	0.01	[43]
M. tuberculosis ESAT6 antigen	*Nicotiana benthamiana*	800 μg/g	[45]
Mannheimia haemolytica GS60 outer membrane lipoprotein	Alfalfa	0.02%	[48]
Hepatitis B virus surface antigen	*Nicotiana benthamiana*	7.14%	[49]
Human growth hormone	*Nicotiana benthamiana*	10%	[53]
ApxIIA, a bacterial exotoxin *Actinobacillus pleuropneumoniae*	Tobacco	0.1%	[54]
A1 leukotoxin from *Mannheimia haemolytica*	White clover	1%	[55]

Table 2. (Continued)

Major F4ac fimbrial subunit protein from *E. coli*	Tobacco	1%	[58]
Fimbrial adhesin FaeG from *E. coli*	Potato	1%	[59]
Major K99 fimbrial subunit, FanC from *E. coli*	Soybean	0.5%	[60]
Middle protein of HBV containing the surface S and preS2 antigen	Potato	<0.1%	[61]
Carboxy-terminal host cell-binding domain of *E. coli* O157:H7 intimin	Tobacco	3-13 μg/g of total plant material	[62]
House dust mite (HDM) allergen, Der p 1	Rice	58 μg/grain	[63]
HIV-Tat	Tomato	1 μg/mg dry weigth	[65]
HIV-ENV and GAG fused to the surface protein antigen of hepatitis B virus	Tomato	~0.3 μg/g	[66]
HIV-1 p24 capsid protein	Arabidopsis	0.2 μg - 0.5 μg/g fresh tissue	[67]
Fusion HIV-1 gp120 V3/cholera toxin B subunit	Potato	0.002-0.004%	[68]
HPV16 E7	*Nicotiana benthamiana*	0.4 μg/g	[69]
Rabies glycoprotein	Tomato	1-5%	[71]
	Corn	1%	[72]
Fusion of epitopes from rabies virus glycoprotein and nucleoprotein	Alfalfa	0.4±0.07 mg/g of fresh leaf tissue.	[73]
Cholera toxin B subunit-ubiquitin fusion	Tobacco	0.9%	[74]
Fusion (F) protein from respiratory syncytial virus	Tomato	0.003%	[76]
Porphyromonas gingivalis fimbrial antigen fused to the cholera toxin B subunit	Potato	0.33%	[77]
Sunflower seed albumin	Lupine	5%	[78]
Colorectal cancer antigen GA733-2	Tobacco	0.4%	[79]
Structural subunit A of the bundle-forming pilus from *E. coli*	Tobacco	7.7%	[80]
Immunocontraceptive epitope fused to the heat labile enterotoxin from *E. coli*	Tomato	37.8 μg/g	[81]
AB5 toxin from *E. coli*	Tobacco NT-1 cells	6.5- 8.2 μg/g	[82]
Fusion of the Cholera toxin B and A2 subunis with a rotavirus NSP4 enterotoxin and *E. coli* fimbrial antigen genes	Potato	1μg/3 g fresh weight	[85]
Heat-shock protein A from *H. pylori*	Tobacco	<1%	[86]
Ascaris suum As 14 protein fused to cholera toxin B subunit	Rice seeds	50 μg/g seed	[87]

Veterinary Antigen	Plant system	Expression levels*	References
Major structural protein VP60 of rabbit hemorrhagic disease virus	Potato	0.30%	[35,36]
VP1 protein, an epitope and the coat protein of *Enterovirus* 71	Tomato	N.R.	[37]
eBRV4 antigen of the VP4 protein from bovine rotavirus fused to the *uidA* gene	Alfalfa	0·4 and 0·9 mg/g	[40]
Chlamydophila psittaci antigen (MOMP) fused to the B subunit of Escherichia coli heat labile protein	Rice	N.R.	[42]
Fusion of a gut adhesion molecule and linear B- and T-cell epitopes from a fish virus	Potato	N.R.	[44]
Epitope from foot and mouth disease virus fused to the glucuronidase (*uidA*) reporter gene	Alfalfa	N.R.	[46]
VP1 antigen from foot-and-mouth disease virus	Alfalfa	2-3%	[47]
21-mer epitope from parvovirus	Arabidopsis	3%	[52]
Immunogenic proteins of Newcastle disease virus	Potato	0.3 to 0.6 μg/mg of total leaf protein	[56]
Fusion protein of Newcastle disease virus	Maize	0.95-3%	[57]
Fusion protein gene of Newcastle disease virus	Potato	0.25–0.55 g/100 g of fresh tuber tissue	[83]
Synthetic neutralizing epitope gene of Porcine epidemic diarrhea virus	Tobacco	2.1%	[64]
Hemagglutinin protein of rinderpest virus	Peanut	10μg/g of peanut leaves	[70]
Murine rotavirus VP6 protein	Potato	0.01%	[84]

* % from total soluble proteins, unless indicated otherwise. List is grouped according to antigen.

All along, the main goal has been to attain high levels of expression but, usually, levels of expression vary greatly depending on various factors. Different approaches have resulted in high levels of expression of several antigens but it is difficult to make comparisons since specific antigens have rarely been tested in multiple systems. A reduction in the amount of material to be consumed is anticipated by increasing the yield, because usually, oral immunization requires higher doses of antigen than parenteral immunizations [88].

3.1. The Importance of Virus-Like Particles

Antigen presentation to the mucosa associated lymphoid tissue is a key element in the response of the mucosal immune system [88,89]. One major obstacle in the development of mucosal vaccines is to be able to induce systemic as well as mucosal responses and one way to overcome it has been by the use of adjuvants, which can promote the generation of antibodies and T cells to an antigen following immunization [90,91]. However, not all the adjuvants enhance priming of cytotoxic T lymphocytes [92-94].

Recombinant virus-like particles (VLPs) represent an attractive alternative for antigen presentation since they are more immunogenic than recombinant proteins alone and are able to stimulate both the humoral and cellular arms of the immune system [95]. These self-assembling, non-replicating viral core structures consisting of one or more viral coat proteins, can act as an adjuvant by carrying peptide sequences inside the antigen presenting cells and feeding into the endogenous processing pathway [96], a phenomenon known as "cross-priming" [97]. The efficacy of immunization with VLPs is best illustrated by the success of the HBV-like particles produced in Saccharomyces cerevisiae, which was the first recombinant vaccine ever developed [98].

VLPs are especially interesting from a mucosal vaccine point of view as they offer the opportunity to deliver the virus employing the natural route of transmission [94]. Oral delivery of VLP has been shown to induce both systemic and mucosal IgA responses both to the virus particles and to foreign epitopes expressed as chimeric proteins on the VLP surface, without the need for external adjuvants [99]. VLPs have been produced from the capsid or envelope components of a wide variety of viruses to study virus assembly and for development of vaccines [95]. Vaccines from HBV and HPV VLPs have been successful, but VLPs from pathogens affecting immune cells and those that successfully evade the immune system, such as HIV-1 and hepatitis C virus have proven to be more challenging.

Production of properly folded VLPs in plants has been extensively reported. In most cases, they have been shown to mimic the immunological properties of native VLPs and stimulate antibody and T cell responses in mice [100]. Antigens having the ability to assemble into VLPs have been generally employed such as the surface [66,101] and core antigens of Hepatitis B virus [51], the capsid protein of Norwalk Virus [102], the L1 protein of human papillomavirus [103], the haemagglutinin/neuraminidase of paramyxoviruses [57], two capsid proteins from rotavirus [75] and chimeric HIV (ENV and GAG). In fact Hepatitis B VLPs have been successfully obtained in a variety of plants using the surface antigen [104]. Recently, VLPs from Hepatitis C virus, employing core, E1 and E2 proteins were obtained for the first time in plants (Lopez Castillo and Gómez Lim, manuscript in preparation). Many of these antigens as well as several capsid or core proteins from various plant viruses have also been employed as carriers to express a wide variety of antigenic peptides. However, not all antigens may form VLPs in plants. The hepatitis E virus capsid protein, which assemblies readily into VLPs in a baculovirus system, did not assembly adequately into VLPs in potato and this lead to a failure to elicit detectable antibodies in mice serum [105]. Clearly formation of VLPs is essential for immune recognition.

3.1.1. Human Clinical Trials with Oral Plant-Based Vaccines

The functionality of plant-based antigens has been tested in a number of experiments in mice and other animals. In the vast majority of the cases, plant-based antigens have been able to elicit a strong immune response and to confer protection against challenge with the pathogen [106]. These results have paved the way for several clinical trials aimed to assessing human immune responses to plant-produced recombinant proteins. The heat-labile enterotoxin from *E. coli* has been tested twice in human clinical trials, in the first trial, 14 adult volunteers ingested three doses of transgenic potatoes (containing 3.7-15.7 μg/g of antigen) or control wild-type potato. Serum antibody responses were detected in 10 out of the 11 volunteers and eight out of the 11 developed neutralization titers of more than 1:100 [107]. This trial was the proof-of-principle that humans could develop serum and mucosal immune response to antigen delivered in transgenic plants. Recently, the same antigen was delivered to volunteers in transgenic corn with seven out of the nine volunteers developing increases in serum IgG and four of the volunteers also developing stool IgA [108].

In another trial, 20 human volunteers ate two or three doses of transgenic potatoes expressing 215-751 μg of the Norwalk virus capsid protein [109]. Out of the 20 volunteers, 19 developed significant increases in the numbers of specific IgA antibody-secreting cells, four developed serum IgG and six developed specific stool IgA, although the levels of serum antibody were not high. This variation in immune response was probably due to the inconsistent assembly of the antigen into virus-like particles and to the possible presence of pre-existing antibodies to the antigen, which might have had an effect.

The surface antigen from the Hepatitis B virus was utilized in a randomized, placebo-controlled, double-blind trial conducted in volunteers previously immunized, 1-15 years earlier, with the licensed hepatitis B vaccine [110]. A total of 42 subjects were enrolled in the study and received either three doses of placebo potatoes or transgenic potatoes (two or three doses). None of the volunteers who ingested control potatoes had any change in their antibody titers during the study. On the other hand, 63% of volunteers (10 out of 16) who consumed three doses of HBsAg-containing transgenic potatoes showed marked increases in antibody titers compared with titers at day 0 [110]. Thus, an antigen from a nonenteric pathogen, was capable to elicit a immune response with no buffering of stomach pH and without the presence of a mucosal adjuvant. The same antigen had been expressed in lettuce and fed to three seronegative volunteers [50]. The volunteers received about 0.2-1 μg of antigen in a 200 g dose of lettuce. In comparison, the commercial Hepatitis B vaccine contains 10 μg of antigen per adult dose [108]. Two of the three volunteers produced short-lived anti-HBs antibody titers, which were detectable 2 weeks after the second immunization and were no longer detectable after the additional 2 weeks.

Mucosal immunologists recognize that the heat-labile enterotoxin from *E. coli* and its relative the B subunit of cholera toxin are particular antigens with highly immunogenic properties. The fact that two different antigens were also immunogenic in humans after oral administration was rather encouraging.

4. Production of Antibodies in Plants (Plantibodies)

Antibodies were the first bioactive molecules expressed in transgenic plants [111]. Since then, numerous antibodies have been produced in a variety of plants, and expressed in different tissues and different sub-cellular compartments (Table 3). This is not surprising, considering the wide range of uses for antibodies in the diagnostic, industrial, medical and research fields.

Table 3. Antibodies produced in transgenic plants

Type of antibody	Plant system	Expression levels*	References
Streptococcal antigen I or II	Tobacco	500 μg/g	[112]
Streptoccocus mutans adhesion protein antibody	Tobacco	80 μg/g	[113]
Herpes simplex-2 glycoprotein B antibody	Tobacco	N.R.	[114]
Herpes simplex-2 glycoprotein B antibody	Rice	N.R.	[115]
Recombinant T84.66/GS8 diabody	Tobacco	~ 0.5-1.5 μg/g	[116]
Single chain Fv antibodies	Tobaco	4-6.8%	[117]
Non-Hodgkins lymphoma idiotypes	Tobacco	30.2μg/g	[118]
Human IgG	Alfalfa	1.0%	[3]
Human carcinoembryonic antigen	*Nicotiana benthamiana*	<1%	[22]
Human carcinoembryonic antigen	Rice	29.0 μg/g leaves; 32.0 μg/g seed	[119]
Human carcinoembryonic antigen	Wheat	0.1 μg/g leaves; 1.5 μg/g seed	[119]
Several scFv antibodies	Tobaco	3-6.8%	[117]
ScFv antibodies against a phytoplasma	Tobacco	0.04%	[120]
Anti-zearalenone single-chain Fv (scFv)	*Arabidopsis*	11.2 ng/ml	[121]

* % from total soluble proteins, unless indicated otherwise. List is grouped according to antigen.NR, not reported.

Plants were first envisaged as systems to produce antibodies for one reason: cost. Transgenic plants represent the most productive and economical system for making recombinant antibodies. The cost of producing monoclonal antibodies in plants is significantly less than by traditional fermentation methods. Smith and Glick [122] have made an estimate that assuming a best-case yield of two grams of recoverable protein/L of bacteria, the minimum cost of production (assuming $2/L for the growth medium and purification costs) would be approximately $1000/kg, not including salaries. Since multimeric proteins, such as antibodies, are assembled with very low efficiency in bacteria [122], the yield could be much lower which would have a direct impact on costs. Plant-produced antibodies are predicted to cost as little as $10–100/g, approximately 10-100 times less expensive than antibodies produced in bacteria [123], mainly attributed to existing land, infrastructure, and expertise necessary for harvesting and processing large volumes of plant materials.

The costs of producing an IgG from alfalfa grown in a 250 m^2 greenhouse are estimated to be within US$500-600/g, compared with US$5000 per gram for the hybridoma produced antibody [3]. In another study, the cost of producing one gram of purified IgA in plants was estimated to be well below US$50/g, which compares favorably with the costs of cell culture (US$1000/g) or transgenic animal production systems (US$100/g) [4].

Corn has been the system of choice for production of many recombinant proteins including antibodies. The main reason is the high protein content of the corn kernel, which accounts for approximately 10% of the dry weight. Even if the antibodies accounted for only 0.5% of the total grain protein, it would cost less than US$200 to produce one kilogram of antibody in corn kernels. However, it has already been shown that corn can produce recombinant proteins up to 5.7% of the total protein [124].

In all likelihood, the biggest component of cost of antibodies will be purification. In theory, purification of plantibodies should be straightforward, using standard procedures. Protein degradation that may occur during extraction can be minimized by the addition of protein stabilizers and proteinase inhibitors, although proteolysis *in planta* represents a challenge [125]. Stevens et al. [126] had suggested that proteolytic degradation in leaves is, in part, linked to the natural process of senescence [126]. This indicates that the physiological state of the plant may have an impact on antibody integrity. On the other hand, it is conceivable that for some uses the plantibody would not even need to be purified. Delivery of the antibody could be achieved by direct consumption of the plant tissue containing the plantibody [127]. This is very important from the standpoint of safety as plants do not serve as hosts for human viruses or prions, unlike hybridomas.

There are no plantibodies yet in commercial production, however there are several candidates that are potentially useful as human therapeutics. The most advanced is a chimeric secretory IgG–IgA antibody called CaroRx*TM*, against a surface antigen of *Streptococcus mutans*, the bacteria that causes tooth decay, which prevents it from binding to teeth, thereby reducing cavities [82]. Unexpectedly, application protects from recolonization and this may last for up to 2 years, although the antibody was applied for only 3 weeks and functional antibody was detected on the teeth for only 3 days following the final application of the antibody. This antibody has reached a pilot Phase II trial. Phase I/II confirmatory clinical trials are underway.

Another antibody, a humanized anti-herpes-simplex virus (HSV) antibody prepared in soybean, was effective in the prevention of vaginal HSV-2 transmission in mouse [114]. A third antibody was developed in rice and wheat against a carcinoembryonic antigen [128]. This antigen, a cell-surface glycoprotein, is one of the best-characterized tumor-associated antigens and antibodies against it are usually employed for diagnostic and therapy. Levels of the antibody in seeds did not show a significant decline after storage at room temperature for six months [128].

Finally, a tumor-specific vaccine was prepared in tobacco for the treatment of lymphoma using a modified plant virus [129]. The antibody genes for expression of an scFv were derived from a mouse B-cell lymphoma. Mice were immunized with the plant-produced scFv and anti-idiotypic antibodies (antibodies against the binding portion of the antibody) were generated. The mice were protected against infection by the lymphoma that produced the original antibody [129].

Although plant antibodies are normally found to be properly folded and functional [127], differential glycosylation by the plant has raised concern over the potential immunogenicity of plant-specific complex N-glycans, which are present on the heavy chain of plant-derived antibodies (see section 8). One approach is the removal of peptide recognition sequences for N-glycosylation. Another is the addition of a ER retention C-terminal (KDEL in amino acid code) sequence which avoids Golgi-mediated modifications [119]. A final approach is the humanization of plant glycans, and to this end, human β-1,4-galactosyltransferase was stably expressed in tobacco plants which were crossed with plants expressing a murine antibody which resulted in a plantibody with partially galactosylated N-glycans [130]. Recently a novel strategy was reported where the HIV-p24 antigen was expressed as a genetic fusion with two constant region sequences from human Ig α-chain and targeted to the endomembrane system. This allowed to increase the expression approximately 13-fold higher than with HIV p24 expressed alone and to enhance the immunological properties. The fusion elicited T-cell and antibody responses in immunized mice [131].

5. Issues Regarding Protein Production in Plants

Even though plant-based compounds present a number of benefits, some potential issues of concern have been identified. These issues differ depending on whether the vaccine candidate is to be purified from plant tissues prior to formulation and delivery or whether it is to be administered orally as recombinant plant material.

5.1. Molecular Strategies to Improve Protein Yields

Production costs are the key issues for recombinant protein production and they include the generation, growing and harvesting of the recombinant material and downstream processing and purification, which tend to increase with tissue complexity. The achievement of high expression levels is a major goal in all systems. Several strategies have been developed to increase the levels of recombinant proteins and they have focused on transcription, transcript stability and translation.

Increasing the level of transcription of integrated sequences has been the main approach to boost expression. To this end, the choice of an adequate promoter is critical. Several promoters have been tested from pathogens and plants as well as some synthetic promoters. The promoters tested included those of plant viruses and from genes highly expressed in dicot leaves. The cauliflower mosaic virus 35S promoter [132] has been the main promoter employed to drive the expression of recombinant proteins in a number of plants. Using this promoter, fairly high levels of accumulation of recombinant proteins have been achieved such as interleukin 12 in tomato (7.3 μg/g of fresh weight) and scFv antibodies in tobacco (4–6.8% TSP) [117,133]. Other plant virus promoters employed include those of cassava vein mosaic virus, with expression comparable to that obtained with 35S promoter [134], the C1 promoter of cotton leaf curl Multan virus, which yielded an expression up to 10-fold higher than that obtained with 35S promoter [135] and the promoter of component 8 of milk vetch dwarf virus, which was not comparatively better than the 35S promoter [136]. Promoters from *Agrobacterium tumefaciens* have also been used to express recombinant proteins in plants, such as the nopaline synthase promoter [137].

Some of the plant promoters employed include that of the small subunit of ribulose-bisphosphate carboxylase [138] and the maize polyubiquitin-1 promoter. Since many pharmaceutical proteins may have possible toxic side effects in plant cells, expression of recombinant proteins should be restricted to the target tissue to prevent toxicity. In this sense, tissue-specific promoters have been employed such as the embryo-specific maize globulin-1 promoter [139], the endosperm-specific maize 27-kDa zein and barley D hordein promoters [140,141]. Expression levels in all these cases were comparable to that yielded by the 35S promoter with the additional advantage of being restricted to a particular tissue. Inducible promoters also allow for expression in a target tissue following a specific treatment, such as chemical spraying [142,143]. In this case, protein accumulation can be restricted to a define timeframe to limit the possible detrimental effects on plant health.

Another approach to boost transcript levels has been to stack transcription units. They combine the most active sequences of well-characterized promoters such as the cauliflower mosaic virus 35S promoter or the *Agrobacterium* Ti plasmid mannopine synthetase promoter. The synthetic promoter may be several-fold more active than the natural promoter [144,145].

The inclusion of regulatory sequences on expression cassettes such as scaffold attachment or matrix attachment regions, may also boost expression. These sequences have been reported to interact with plant nuclear scaffolds *in vitro*, and are thought to be involved in the recruitment of transcription factors to promoters [146]. Although these sequences sometimes can slightly boost the expression of plant genes, their presence does not necessarily result in an increase in gene expression but rather in a reduction in gene silencing [147].

As a recombinant message is transcribed, the stability of that message is important to ensure high-level expression. The non-translated region located downstream of the translation stop codon is critical for processing and should include signals targeting the message for polyadenylation. Message destabilizing sequences in the downstream region can greatly affect stability [148], and these sequences must be avoided when preparing gene constructs to express high levels of recombinant proteins. Several plant viral and plant 3′ untranslated regions have been utilized to process messages for recombinant proteins without destabilizing these messages. They include the cauliflower mosaic virus 35S terminator and the potato proteinase inhibitor II terminator. For plant-based expression systems, eukaryotic genes are generally re-synthesized or engineered to remove introns. Sequences such as consensus intron splice sites, message destabilizing sequences and transcript termination sequences should be avoided in constructing synthetic genes. However, some plant intron sequences contribute positively to the expression level observed for their native genes, and these sequences can boost expression if inserted as synthetic introns into genes [149].

To increase translational efficiency of recombinant sequences several approaches have been tested. The use of both plant and viral 5′ non-translated regions [150], optimization of sequences located around the translation start to fit the consensus initiation sequence of plants [151], codon-optimization to suit the expression host, removal of predicted mRNA secondary structures that might hinder translation, of internal ribosome entry sites that might compete for translation of the full message and of rare codons, together contribute to increasing the translatability of the message.

Expression of a foreign protein in plants as a fusion to a plant protein present at high levels or to a recombinant protein that has been shown to be stably expressed in plants, usually results in the stable accumulation of the foreign protein. In addition, it presents a number of advantages in terms of stability, subcellular targeting, purification and activity of product. In the case of plant-based vaccines, the fusion may enhance the immunogenicity of the incorporated antigens. Since the cholera toxin and its close relative the heat-labile toxin from *Escherichia coli*, have been shown to be effective in enhancing mucosal immune responses to vaccines, they have been employed extensively for fusing recombinant proteins. Fusion is usually done at the C-terminus of the B subunit.

Fusions of antigens to other proteins have been also produced in plants and include the HIV-Tat gene fused to a rotavirus enterotoxin [153], ricin-B subunit from *Ricinus communis* fused to rotavirus NSP4 [154] or to core protein p24 from HIV-1 and the E7 coding sequence from HPV16 fused to beta-1,3-1,4-glucanase of *Clostridium thermocellum* [69].

5.1.1. Downstream Processing

If purification is involved there are several approaches that could be implemented: One is to engineer the protein, to be secreted into the culture medium. Secretion systems are convenient because no disruption of plant cells is necessary during protein recovery, hence, release of phenolic compounds is avoided. Nevertheless, the recombinant proteins might be unstable in the culture medium. Another approach is the use of affinity tags to facilitate the recovery of proteins as long as the tag be removed after purification to restore the native structure of the protein. In either case, good manufacturing practices will be needed and possible lot-to-lot variability will need to be closely monitored. By contrast, if the recombinant antigen is to be delivered in a processed plant product as an oral vaccine, production would be based on food processing technology rather than protein purification schemes, but good manufacturing practices will still apply.

Consistency of product (homogeneity) is very important for plant-based vaccines as well as for purified antigens. Therefore, rather than administering whole plant organs (fruits, or grains) directly, as it was originally envisaged, it may be better to process the plant material into a uniform state and to be stable to the food processing technology. This has been assessed using recombinant corn expressing the B subunit of the heat labile toxin, and the antigen has been shown to be stable to milling and modified extrusion conditions and to be evenly distributed in the products [155]. This will be very important if the product is to be combined with non-transgenic material or with transgenic material expressing different antigens, or even protein adjuvants, for even dosing.

Stability of antigens over time in processed food products stored at different temperatures will also need to be assessed, especially for plant-based products needing cold chain storage and distribution. This feature is particularly important in poor countries with limited resources to provide a cold chain and the equipment and personnel needed for injections. The low cost of plant-based vaccines make them ideal for large-scale programs in poor countries.

Since the general eukaryotic protein synthesis pathway is conserved between plants and animals, folding and assembly, as well as transfer of an oligosaccharide precursor to *N*-glycosylation sites can be correctly accomplished in transgenic plant systems [156]. However, addition of plant-specific residues α-1,3-fucose and β-1,2-xylose, might pose problems for the production of fully functional therapeutic proteins, even though this may not seem not to occur with all plant-based recombinant proteins [157].

Both full-sized antibodies and various functional antibody derivatives have been produced successfully in plants, including Fab fragments, scFvs, bispecific scFvs, single domain antibodies and antibody fusion proteins. Some studies have shown that while there were some differences in the glycan groups present on plant-based recombinant antibodies, neither the antibody nor the glycans were immunogenic [2]. In an elegant study, Chargelegue et al. [158] tested the immunogenic effect in mice of plant glycans of transgenic murine monoclonal IgG antibodies and horseradish peroxidase [158]. Because the same mouse strain was used for generating the original mAb, the study specifically compared the immunogenicity of a self-protein and a plant protein displaying foreign plant glycans, with the self-protein displaying mammalian glycans. Encouragingly, the plant glycans of both the self and foreign (horseradish peroxidase) proteins were poorly immunogenic even when parenterally administered with alum adjuvant.

6. Future Prospects

Considering that a wide variety of recombinant proteins can be produced in a safe and inexpensive manner, it comes as no surprise the interest that the technology has generated from a commercial standpoint. Several companies have been founded based on this technology and others are actively engaged investigating the potential. There are already on the market several plant-produced proteins employed in the area of diagnostics such as beta-glucuronidase [159], avidin [160] and aprotinin [161]. Several plant-derived recombinant pharmaceutical proteins are reaching the final stages of clinical evaluation, and more are in the development pipeline. However, there is some regulatory uncertainty, particularly for proteins requiring approval for human use [162]. It is important that these regulatory issues be resolved before pursuing product licensure. So far, only one plant-derived vaccine targeting a viral disease of poultry (Newcastle disease) has reached licensure (Dow Agrosciences, USA). Although vaccines intended for use in humans will probably require a more laborious path to licensure, this milestone establishes a credible foothold for plant-derived vaccines [163].

Even though only a few products have reached the market so far, there are more approaching commercialization, after meeting the technological challenges and clearing the regulatory hurdles. In addition to antigens and antibodies, plants have also been employed for production of a wide range of proteins with pharmaceutical or industrial applications and high added value (Table 4). In the short term, enzymes for large-scale industrial processes and antigens for oral animal vaccines are the most likely plant-expressed products to be commercially viable as the first ever licensure of a plant-derived vaccine targeting a viral disease of poultry described before has confirmed. Attention will be required to ensure correct post-translational processing and protein stability in plant tissues. Advances are required to boost expression further and stacking of many of the available tools discussed above will probably be required to produce commercial products.

As with all biotechnological developments, most of the technology for plant-derived vaccines has been patented in industrialized countries. Poor countries, which usually have a high disease burden, often have poor or inexistent IP protection rules and lack of adequate knowledge and infrastructure to protect and commercialize a biotechnological product. It has been postulated that plant-derived vaccines may be approved in an industrialized country and then be more broadly used in poor countries [163]. There is a need to develop plant-based low-cost medicines and vaccines to poor countries. It is hoped that this technology will eventually help those who needed it the most and that the issue of IP does not represent an insurmountable obstacle. Putting the collective benefit ahead of the personal gain, will be the key for the full realization of this technology.

Table 4. Pharmaceutical and industrial proteins produced in transgenic plants

Protein	Host	Expression levels	References
Human enkephalins	*Arabidopsis*	0.10% seed protein	[164]
Human lactoferrin	Potato	0.10% TSP	[41]
Human aprotinin	Maize	N.R.	[165]
Human epidermal growth-factor	Tobacco	0.001%	[166]
Human acid ß-glucosidase	Tobacco	3%	[167]
Human hemoglobin α, β	Tobacco	0.05%	[168]
Human serum albumin	Tobacco	0.02%	[164]
Human hirudin	Canola (*Brassica napus*)	0.30%	[168]
Human interferon β	Tobacco	<0.01%	[164]
Human erythropoietin	Tobacco	<0.01	[164]
Human α-1-antitrypsin	Rice	N.R.	[165]
Human protein C	Tobacco	<0.01%	[168]
Human homotrimeric collagen I	Tobacco	10 mg/100	[169]
Acetylcholinesterase	Tomato	100 mU/g leaf (specific acivity)	[170]
Angiotensin-converting enzyme	Tobacco, tomato	N.R.	[165]
α-Tricosanthin from TMV-U1 Subgenomic coat protein	Nicotiana bethamiana	2.00%	[165]
Glucocerebrosidase	Tobacco	1.00–10.00%	[168]
Transglutaminase	Rice	N.R.	[171]
Human interleukin 12	Tobacco	17.5-40 ng/g of fresh weight	[172]
Mouse interleukin-12	Tomato	3.4-7.3 μg/g fresh weight	[133]
E. coli ß-glucuronidase and chicken egg-white avidin	Maize	0.7%and 5.7% respectively	[124]
Human cytotoxic T-lymphocyte antigen 4-immunoglobulin	Rice cell suspension culture	31.4 mg/L	[173]
Human growth hormone	*Nicotuana benthamiana*	10%	[53]

* % from total soluble proteins, unless indicated otherwise. List is grouped according to antigen.NR, not reported.

ACKNOWLEDGMENTS

Research support from CONACYT (grant 86965), Cinvestav, and The Welcome Foundation in the author's laboratory is gratefully acknowledged.

REFERENCES

[1] Frank, B. H. 1998. The development of recombinant human insulin. *Developments in Biological Standardization*, 96, 91-95.

[2] Balen, B. & Krsnik-Rasol, M. 2007. N-Glycosylation of Recombinant Therapeutic Glycoproteins in Plant Systems Plant Specific N-Glycosylation. *Food Technology and Biotechnology*, 45, 1-10.

[3] Khoudi, H., Laberge, S., Ferullo, J. M., Bazin, R., Darveau, A., Castonguay, Y., Allard, G., Lemieux, R. & Vézina L. P. 1999. Production of a diagnostic monoclonal antibody in perennial alfalfa plants. *Biotechnology and Bioengineering*, 64, 135–143.

[4] Larrick, J. W. & Thomas, D. W. 2001. Producing proteins in transgenic plants and animals. *Current Opinion in Biotechnology*, 12, 411-418.

[5] Daniell, H., Streatfield, S. J. & Wycoff, K. 2001. Medical molecular farming: production of antibodies, biopharmaceuticals and edible vaccines in plants. *Trends in Plant Science*, 6, 219-226.

[6] Takaiwa, F., Takagi, H., Hirose, S. & Wakasa, Y. 2007. Endosperm tissue is good production platform for artificial recombinant proteins in transgenic rice. *Plant Biotechnology Journal*, 5, 84-92

[7] Larrick, J. W., Yu, L., Naftzger, C., Jaiswal, S. & Wycoff, K. 2001. Production of secretory IgA antibodies in plants. *Biomolecular Engineering*, 18, 87-94.

[8] Vain, P. 2007. Thirty years of plant transformation technology development. *Plant Biotechnology Journal*, 5, 221-229.

[9] Gelvin, S. B. 2003. Agrobacterium-mediated plant transformation: the biology behind the 'gene-jockeying' tool. *Microbiology and Molecular Biology Reviews*, 67, 16-37.

[10] Altpeter, F., Baisakh, N., Beachy, R., Bock, R., Capell, T., Christou, P., Daniell, H., Datta, K., Datta, S., Dix, P. J., Fauquet, C., Huang, N., Kohli, A., Mooibroek, H., Nicholson, L., Nguyen, T. T., Nugent, G., Raemakers, K., Romano, A., Somers, D. A., Stoger, E., Taylor, N. & Visser R. 2005. Particle bombardment and the genetic enhancement of crops: myths and realities. *Molecular Breeding*, 15, 305-327.

[11] Chen, L. L., Marmey, P., Taylor, N. J., Brizard, J. P., Espinoza, C., D'Cruz, P., Huet, H., Zhang, S. P., de Kochko, A., Beachy, R. N. & Fauquet, C. M. 1998. Expression and inheritance of multiple transgenes in rice plants. *Nature Biotechnology*, 16, 1060-1064.

[12] Nicholson, L., Gonzalez-Melendi, P., van Dolleweerd, C., Tuck, H., Perrin, Y., Ma, J. K. C., Fischer, R., Christou, P. & Stoger, E. 2005. A recombinant multimeric immunoglobulin expressed in rice shows assembly-dependent subcellular localization in endosperm cells. *Plant Biotechnology Journal*, 3, 115-127.

[13] Loc, N. T., Tinjuangjun, P., Gatehouse, A. M. R., Christou, P. & Gatehouse, J. A. 2002. Linear transgene constructs lacking vector backbone sequences generate transgenic rice plants which accumulate higher levels of proteins conferring insect resistance. *Molecular Breeding*, 9, 231-2

[14] Maliga, P. 2004. Plastid transformation in higher plants. *Annual Review of Plant Biology*, 55, 289-313.

[15] Bock, R. & Khan, M. S. 2004. Taming plastids for a green future. *Trends in Biotechnology*, 22, 311-318.

[16] Davoodi-Semiromi, A., Samson, N. & Daniell, H. 2009. The green vaccine: A global strategy to combat infectious and autoimmune diseases. *Human Vaccines*, 5, 488-493.

[17] Ruf, S., Hermann, M., Berger, I. J., Carrer, H. & Bock, R. 2001. Stable genetic transformation of tomato plastids and expression of a foreign protein in fruit. *Nature Biotechnology*, 19, 870-875.

[18] Quesada-Vargas, T, Ruiz, O. N. & Daniell, H. 2005. Characterization of heterologous multigene operons in transgenic chloroplasts. Transcription, processing, and translation. *Plant Physiology*, 138, 1746-1762.

[19] Daniell, H. 2007. Transgene containment by maternal inheritance: Effective or elusive? *Proceedings of the National Academy of Sciences of the United States of America*, 104, 6879-6880.

[20] Tregoning, J. S., Nixon, P., Kuroda, H., Svab, Z., Clare, S., Bowe, F., Fairweather, N., Ytterberg, J., van Wijk, K. J., Dougan, G. & Maliga, P. 2003. Expression of tetanus toxin Fragment C in tobacco chloroplasts. *Nucleic Acids Research*, 31, 1174-1179

[21] Koya, V., Moayeri, M., Leppla, S. H. & Daniell, H. 2005. Plant-based vaccine: mice immunized with chloroplast-derived anthrax protective antigen survive anthrax lethal toxin challenge. *Infection and Immunity*, 73, 8266-8274.

[22] Verch, T., Yusibov, V. & Koprowski, H. 1998. Expression and assembly of a full-length monoclonal antibody in plants using a plant virus vector. *Journal of Immunological Methods*, 220, 69-75.

[23] Mori, M., Fujihara, N., Mise, K. & Furusawa, I. 2001. Inducible high-level mRNA amplification system by viral replicase in transgenic plants. *Plant Journal*, 27, 79-86.

[24] Porta, C. & Lomonossoff, G. P. 1996. Use of viral replicons for the expression of genes in plants. *Molecular Biotechnology*, 5, 209-221.

[25] Johnson, J., Lin, T. & Lomonossoff, G. 1997. Presentation of heterologous peptides on plant viruses: Genetics, structure, and function. *Annual Review of Phytopathology*, 35, 67-86.

[26] Dalsgaard, K., Uttenthal, A., Jones, T. D., Xu, F., Merryweather, A., Hamilton, W. D. O., Langeveld, J. P. M., Boshuizen, R. S., Kamstrup, S., Lomonossoff, G. P., Porta, C., Vela, C., Casal, J. I., Meloen, R. H. & Rodgers, P. B. 1997. Plant-derived vaccine protects target animals against a viral disease. *Nature Biotechnology*, 15, 248-252.

[27] Gleba, Y., Klimyuk, V. & Marillonnet, S. 2005. Magnifection - a new platform for expressing recombinant vaccines in plants. *Vaccine*, 23, 2042-2048.

[28] Santi, L., Huang, Z. & Mason, H. 2006b. Virus-like particles production in green plants. *Methods*, 40, 66-76.

[29] Jani, D., Meena, L. S., Rizwan-ul-Haq, Q. M., Singh, Y., Sharma, A. K. & Tyagi, A. K. 2002. Expression of cholera toxin B subunit in transgenic tomato plants. *Transgenic Research*, 11, 447-454.

[30] Jani, D., Singh, N. K., Bhattacharya, S., Meena, L. S., Singh, Y., Upadhyay, S. N., Sharma, A. K. & Tyagi, A. K. 2004. Studies on the immunogenic potential of plant-expressed cholera toxin B subunit. *Plant Cell Reports*, 22, 471-477.

[31] Arakawa, T. Chong, D. K. X., Yu, J., Hough, J., Engen, P. C., Elliot, J. F. & Langridge, W. H. 1999. Suppression of autoimmune diabetes by a plant-delivered cholera toxin B subunit-human glutamate decarboxylase fusion protein. *Transgenics*, 3, 51-60.

[32] Becerra-Arteaga, A., Mason, H. S. & Shuler, M. L. 2006. Production, secretion, and stability of human secreted alkaline phosphatase in tobacco NT1 cell suspension cultures. *Biotechnology progress*, 22, 1643-1649.

[33] Marquet-Blouin, E., Bouche, F. B., Steinmetz, A. & Muller, C. P. 2003. Neutralizing immunogenicity of transgenic carrot (Daucus carota L.)-derived measles virus hemagglutinin. *Plant Molecular Biology*, 51, 459-469.

[34] Wang, Y., Deng, H., Zhang, X., Xiao, H., Jiang, Y., Song, Y., Fang, L., Xiao, S., Zhen, Y. & Chen, H. 2009. Generation and immunogenicity of Japanese encephalitis virus envelope protein expressed in transgenic rice. *Biochemistry and Biophysical Research Communications*, 380, 292-297.

[35] Gil, F., Titarenko, E., Terrada, E., Arcalís, E. & Escribano, J. M. 2006. Successful oral prime-immunization with VP60 from rabbit haemorrhagic disease virus produced in transgenic plants using different fusion strategies. *Plant Biotechnology Journal*, 4, 135-143.

[36] Martin-Alonso, J. M. Castanon, S., Alonso, P., Parra, F. & Ordas, R. 2003. Oral immunization using tuber extracts from transgenic potato plants expressing rabbit hemorrhagic disease virus capsid protein. *Transgenic Research*, 12, 127-130.

[37] Chen, H. F., Chang, M. H., Chiang, B. L, & Jeng, S. T. 2006. Oral immunization of mice using transgenic tomato fruit expressing VP1 protein from enterovirus 71. *Vaccine*, 24, 2944-2951.

[38] Kang, T. J., Lee, W. S., Choi, E. G., Kim, J. W., Kim, B. G. & Yang, M. S. 2006. Mass production of somatic embryos expressing Escherichia coli heat-labile enterotoxin B subunit in Siberian ginseng. *Journal of Biotechnology*, 121, 124-33.

[39] Choi, N. W., Estes, M. K. & Langridge, W. H. 2005. Synthesis and assembly of a cholera toxin B subunit-rotavirus VP7 fusion protein in transgenic potato. *Molecular Biotechnology*, 31, 193-202.

[40] Wigdorovitz, A.,Mozgovoj, M., Santos, M. J., Parreno, V., Gomez, C. Perez-Filgueira, D. M., Trono, K. G., Rios, R. D., Franzone, P. M., Fernandez, F., Carrillo, C., Babiuk, L. A., Escribano, J. M. & Borca, M. V. 2004. Protective lactogenic immunity conferred by an edible peptide vaccine to bovine rotavirus produced in transgenic plants. *Journal of General Virology*, 85, 1825-1832.

[41] Chong, D. K. X. & Langridge, W. H. R. 2000. Expression of full-length bioactive antimicrobial human lactoferrin in potato plants. *Transgenic Research*, 9, 71-78.

[42] Zhang, X., Yuan, Z., Duan, Q., Zhu, H., Yu, H. & Wang, Q. 2009. Mucosal immunity in mice induced by orally administered transgenic rice. *Vaccine*, 27, 1596-1600.

[43] Philip, R., Darnowski, D. W., Maughan, P. J., Vodkin, L. O. 2001. Processing and localization of bovine beta-casein expressed in transgenic soybean seeds under control of a soybean lectin expression cassette. *Plant Science*, 161, 323-335.

[44] Companjen, A. R., Florack, D. E., Bastiaans, J. H., Matos, C. I., Bosch, D. & Rombout, J. H. 2005. Development of a cost-effective oral vaccination method against viral disease in fish. *Developmental Biology* (Basel), 121, 143-150.

[45] Dorokhov, Y. L., Sheveleva, A. A., Frolova, O. Y., Komarova, T. V., Zvereva, A. S., Ivanov, P. A. & Atabekov, J.G. 2007. Superexpression of tuberculosis antigens in plant leaves. *Tuberculosis*, 87, 218-224.

[46] Dus Santos, M. J. & Wigdorovitz, A. 2005. Transgenic plants for the production of veterinary vaccines. *Immunology and Cell Biology*, 83, 229-238.

[47] Wigdorovitz, A., Carrillo, C., Dus Santos, M. J., Trono, K., Peralta, A., Gomez, M. C., Rios, R. D., Franzone, P. M., Sadir, A. M., Escribano, J. M. & Borca, M. V. 1999. Induction of a protective antibody response to foot and mouth disease virus in mice following oral or parenteral immunization with alfalfa transgenic plants expressing the viral structural protein VP1. *Virology*, 255, 347-353.

[48] Lee, R. W., Cornelisse, M., Ziauddin, A., Slack, P. J., Hodgins, D. C., Strommer, J. N., Shewen, P. E. & Lo, R. Y. 2008. Expression of a modified Mannheimia haemolytica GS60 outer membrane lipoprotein in transgenic alfalfa for the development of an edible vaccine against bovine pneumonic pastereullosis. *Journal of Biotechnology*, 135, 224-231.

[49] Huang, Z., Santi, L., LePore, K., Kilbourne, J., Arntzen, C. J. & Mason, H. S. 2006. Rapid high-level production of hepatitis B core antigen in plant leaf and its immunogenicity in mice. *Vaccine*, 24, 2506-2513.

[50] Kapusta, J., Modelska, A., Figlerowicz, M., Pniewski, T., Letellier, M., Lisowa, O., Yusibov, V., Koprowski, H., Plucienniczak, A. & Legocki, A. B. 1999. A plant-derived edible vaccine against hepatitis B virus. *The FASEB journal*, 13, 1796-1799.

[51] Huang, Z., Santi, L., LePore, K., Kilbourne, J., Arntzen, C. J. & Mason, H. S. 2006. Rapid high-level production of hepatitis B core antigen in plant leaf and its immunogenicity in mice. *Vaccine*, 24, 2506-2513.

[52] Gil, F., Brun, A., Wigdorovitz, A., Catalá, R., Martínez-Torrecuadrada, J., Casal, I., Salinas, J., Borca, M. & Escribano, J. M. 2001. High-yield expression of a viral peptide vaccine in transgenic plants. *FEBS Letters*, 488, 13-17.

[53] Gils, M., Kandzia, R., Marillonnet, S., Klimyuk, V. & Gleba, Y. 2005. High-yield production of authentic human growth hormone using a plant virus-based expression system. *Plant Biotechnology Journal*, 3, 613-620.

[54] Lee, K. Y., Kim, D. H., Kang, T. J., Kim, J., Chung, G. H., Yoo, H. S., Arntzen, C. J., Yang, M. S. & Jang, Y. S. 2006. Induction of protective immune responses against the challenge of *Actinobacillus pleuropneumoniae* by the oral administration of transgenic tobacco plant expressing ApxIIA toxin from the bacteria. *FEMS Immunology & Medical Microbiology*, 48, 381-389.

[55] Lee, R. W., Strommer, J., Hodgins, D., Shewen, P. E., Niu, Y. & Lo, R. Y. 2001. Towards development of an edible vaccine against bovine pneumonic pasteurellosis using transgenic white clover expressing a *Mannheimia haemolytica* A1 leukotoxin 50 fusion protein. *Infection and Immunity*, 69, 5786.

[56] Berinstein, A., Vazquez-Rovere, C., Asurmendi, S., Gómez, E., Zanetti, F., Zabal, O., Tozzinia, A., Conte Granda, D., Taboga, O., Calamante, G., Barrios, H., Hopp E. & Carrillo, E. 2005. Mucosal and systemic immunization elicited by Newcastle disease virus (NDV) transgenic plants as antigens. *Vaccine*, 23, 5583-5589.

[57] Guerrero-Andrade, O., Loza-Rubio, E., Olivera-Flores, T., Fehervari-Bone, T. & Gomez-Lim, M. A. 2006. Expression of the Newcastle disease virus fusion protein in transgenic maize and immunological studies. *Transgenic Research*, 15, 455-463.

[58] Joensuu, J. J., Kotiaho, M., Riipi, T., Snoeck, V., Palva, E. T., Teeri, T. H., Lang, H., Cox, E., Goddeeris, B. M. & Niklander-Teeri, V. 2004. Fimbrial subunit protein FaeG expressed in transgenic tobacco inhibits the binding of F4ac enterotoxigenic *Escherichia coli* to porcine enterocytes. *Transgenic Research*, 13, 295-298.

[59] Liang, W., Huang, Y., Yang, X., Zhou, Z., Pan, A., Qian, B., Huang, C., Chen, J. & Zhang, D. 2006. Oral immunization of mice with plant-derived fimbrial adhesin FaeG induces systemic and mucosal K88ad enterotoxigenic *Escherichia coli*-specific immune responses. *FEMS Immunology and Medical Microbiology*, 46, 393-399.

[60] Piller, K. J., Clemente, T. E., Jun, S. M., Petty, C. C., Sato, S., Pascual, D. W. & Bost, K. L. 2005. Expression and immunogenicity of an *Escherichia coli* K99 fimbriae subunit antigen in soybean. *Planta*, 222, 6-18.

[61] Joung, Y. H., Youm, J. W., Jeon, J. H., Lee, B. C., Ryu, C. J., Hong, H. J., Kim, H. C., Joung, H. & Kim, H. S. 2004. Expression of the hepatitis B surface S and preS2 antigens in tubers of Solanum tuberosum. *Plant Cell Reports*, 22, 925-930.

[62] Judge, N. A., Mason, H. S. & O'Brien, A. D. 2004. Plant cell-based intimin vaccine given orally to mice primed with intimin reduces time of *Escherichia coli* O157:H7 shedding in feces. *Infection and Immunity*, 72, 168-175.

[63] Suzuki, K., Kaminuma, O., Yang, L., Motoi, Y., Takai, T., Ichikawa, S., Okumura, K., Ogawa, H., Mori, A., Takaiwa, F. & Hiroi, T. 2009. Development of transgenic rice expressing mite antigen for a new concept of immunotherapy. *International Archives of Allergy and Immunology*, 149, 21-24.

[64] Kang, T. J. Kim, Y. S., Jang, Y. S. & Yang, M. S. 2005. Expression of the synthetic neutralizing epitope gene of porcine epidemic diarrhea virus in tobacco plants without nicotine. *Vaccine*, 23, 2294-2297.

[65] Pena Ramirez, Y. J., Tasciotti, E., Gutierrez-Ortega, A., Donayre Torres, A. J., Olivera Flores, M. T., Giacca, M. & Gomez Lim, M. A. 2007. Fruit-specific expression of human immunodeficiency virus 1-Tat gene in tomato plants and its immunogenic potential in mice. *Clinical and Vaccine Immunol*ogy, 14, 685-692.

[66] Shchelkunov, S. N., Salyaev, R. K., Pozdnyakov, S. G., Rekoslavskaya, N. I., Nesterov, A. E., Ryzhova, T. S., Sumtsova, V. M., Pakova, N. V., Mishutina, U. O., Kopytina, T. V. & Hammond, R. W. 2006. Immunogenicity of a novel, bivalent, plant-based oral vaccine against hepatitis B and human immunodeficiency viruses. *Biotechnology Letters*, 28, 959-967.

[67] Lindh, I., Kalbina, I., Thulin, S., Scherbak, N., Sävenstrand, H., Bråve, A., Hinkula, J., Strid, A. & Andersson, S. 2008. Feeding of mice with Arabidopsis thaliana expressing the HIV-1 subtype C p24 antigen gives rise to systemic immune responses. *Acta Pathologica, Microbiologica et Immunologica Scandinavica*, 116, 985-994.

[68] Kim, T. G., Gruber, A. & Langridge, W. H. 2004. HIV-1 gp120 V3 cholera toxin B subunit fusion gene expression in transgenic potato. *Protein Expression and Purification*, 37, 196-202.

[69] Massa, S., Franconi, R., Brandi, R., Muller, A., Mett, V., Yusibov, V. & Venuti, A. 2007. Anti-cancer activity of plant-produced HPV16 E7 vaccine. *Vaccine*, 25, 3018-3021.

[70] Khandelwal, A., Renukaradhya, G. J., Rajasekhar, M., Sita, G. L. & Shaila, M. S. 2004. Systemic and oral immunogenicity of hemagglutinin protein of rinderpest virus expressed by transgenic peanut plants in a mouse model. *Virology*, 323, 284-291.

[71] Perea-Arango, I., Loza, E., Rojas, E., Olivera, T., Gonzalez, L. & Gomez Lim, M.A. 2007. Expression of the rabies virus nucleoprotein in plants at high-levels and evaluation of immune responses in mice. *Plant Cell Reports*, 27, 677–685.

[72] Loza-Rubio, E., Rojas, E., Gómez, L., Olivera, M. T. & Gómez-Lim, M. A. 2008. Development of an edible rabies vaccine in maize using the Vnukobo strain. *Developmental Biology* (Basel), 131, 477-482.

[73] Yusibov, V., Hooper, D. C., Spitsin, S. V., Fleysh, N., Kean, R. B., Mikheeva, T., Deka, D., Karasev, A., Cox, S., Randall, J. & Koprowski, H. 2002. Expression in plants and immunogenicity of plant virus-based experimental rabies vaccine. *Vaccine*, 20, 3155-3164.

[74] Mishra, S., Yadav, D. K. & Tuli, R. 2006. Ubiquitin fusion enhances cholera toxin B subunit expression in transgenic plants and the plant-expressed protein binds GM1 receptors more efficiently. *Journal of Biotechnology*, 127, 95-108.

[75] Saldaña-Trinidad, S. Esquivel-Guadarrama, F., Olivera-Flores, M. T., Arias, N., López-Charretón, S., Arias, C., Ruiz-Medrano, R. Mason, H. S., Mor, T., Richter, L., Arntzen, C. J. & Gómez-Lim, M. A. 2006. Production of rotavirus-like particles in tomato (*Lycopersicon esculentum* L.) fruit by expression of capsid proteins VP2 and VP6 and immunological studies. *Viral Immunology*, 19, 42-53.

[76] Sandhu, J. S., Krasnyanski, S. F., Domier, L. L., Korban, S. S., Osadjan, M. D. & Buetow, D. E. 2000. Oral immunization of mice with transgenic tomato fruit expressing respiratory syncytial virus-F protein induces a systemic immune response. *Transgenic Research*, 9, 127-135.

[77] Shin, E. A., Lee, J. Y., Kim, T. G., Park, Y. K. & Langridge, W. H. 2006. Synthesis and assembly of an adjuvanted Porphyromonas gingivalis fimbrial antigen fusion protein in plants. *Protein Expression and Purification*, 47, 99-109.

[78] Smart, V., Foster, P. S., Rothenberg, M. E., Higgins, T. J. & Hogan, S. P. 2003. A plant-based allergy vaccine suppresses experimental asthma via an IFN-gamma and CD4+CD45RBlow T cell-dependent mechanism. *Journal of Immunology*, 171, 2116-2126.

[79] Verch, T., Hooper, D. C., Kiyatkin, A., Steplewski, Z. & Koprowski, H. 2004. Immunization with a plant-produced colorectal cancer antigen. *Cancer Immunology Immunotherapy*, 53, 92-99.

[80] Vieira da Silva, J., Garcia, A. B., Vieira Flores, V. M., Sousa de Macedo, Z. & Medina-Acosta, E. 2002. Phytosecretion of enteropathogenic Escherichia coli pilin subunit A in transgenic tobacco and its suitability for early life vaccinology. *Vaccine*, 20, 2091-2101.

[81] Walmsley, A. M., Alvarez, M. L., Jin, Y., Kirk, D. D., Lee, S. M., Pinkhasov, J., Rigano, M. M., Arntzen, C. J. & Mason, H. S. 2003. Expression of the B subunit of Escherichia coli heat-labile enterotoxin as a fusion protein in transgenic tomato. *Plant Cell Reports*, 21, 1020-1026.

[82] Wen, S. X., Teel, L. D., Judge, N. A. & O'Brien, A. D. 2006. A plant-based oral vaccine to protect against systemic intoxication by Shiga toxin type 2. *Proceedings of the National Academy of Sciences of the United States of America*, 103, 7082-7087.

[83] Yang, Z. Q., Liu, Q. Q., Pan, Z. M., Yu, H. X. & Jiao, X. A. 2007. Expression of the fusion glycoprotein of Newcastle disease virus in transgenic rice and its immunogenicity in mice. *Vaccine*, 25, 591-598.

[84] Yu, J. & Langridge, W. H. 2003. Expression of rotavirus capsid protein VP6 in transgenic potato and its oral immunogenicity in mice. *Transgenic Research*, 12, 163-.169.

[85] Yu, J. & Langridge, W. H. 2001. A plant-based multicomponent vaccine protects mice from enteric diseases. *Nature Medicine*, 19, 548-552.
[86] Zhang, H., Zhang, X., Liu, M., Zhang, J., Li, Y. & Zheng, C. C. 2006. Expression and characterization of Helicobacter pylori heat-shock protein A (HspA) protein in transgenic tobacco (*Nicotiana tabacum*) plants. *Biotechnology and Applied Biochemistry*, 43, 33-38.
[87] Nozoye, T., Takaiwa, F., Tsuji, N., Yamakawa, T., Arakawa, T., Hayashi, Y. & Matsumoto, Y. 2009. Production of Ascaris suum AS14 protein and its fusion protein with cholera toxin B subunit in rice seeds. *Journal of Veterinary Medical Science*, 71, 995-1000.
[88] Mestecky, J., Nguyen, H., Czerkinsky, C. & Kiyono, H. 2008. Oral immunization: an update. *Current Opinion in Gastroenterology*, 24, 713-719.
[89] Dietrich, G., Griot-Wenk, M., Metcalfe, I. C., Lang, A. B. & Viret, J. F. 2003. Experience with registered mucosal vaccines. *Vaccine*, 21, 678-683.
[90] Cheroutre, H. 2004. Starting at the beginning: New perspectives on the biology of mucosal T cells. *Annual Review Immunology*, 22, 217-246.
[91] Neutra, M. R. & Kozlowski, P. A. 2006. Mucosal vaccines: the promise and the challenge. *Nature Immunology*, 6, 148-158.
[92] Stevceva, L. & Ferrary, M. G. 2005. Mucosal adjuvants. *Current pharmaceutical design*, 11, 801-811.
[93] Lavelle, E. C. & O'Hagan, D. T. 2006. Delivery systems and adjuvants for oral vaccines. *Expert Opinion on Drug Delivery*, 3, 747-762.
[94] Gilbert, S. C. 2001. Virus-like particles as vaccine adjuvants. *Molecular Biotechnology*, 19, 169-177.
[95] Grgacic, E. V. L. & Anderson, D. A. 2006. Virus-like particles: Passport to immune recognition. *Methods*, 40, 60-65.
[96] Schirmbeck, R., Dem, L., Melber, K., Wolf, H., Wagner, R. & Reimann, J. 1995. Priming of class I-restricted cytotoxic T lymphocytes by vaccination with recombinant protein antigens. *Vaccine*, 13, 857-86.
[97] Schwarz, K., Meijerink, E., Speiser, D. E., Tissot, A. L. C., Cielens, I., Renhof, R., Dishlers, A., Pumpens, P. & Bachmann, M. F. 2005. Efficient homologous prime-boost strategies for T cell vaccination based on virus-like particles. *European Journal of Immunology*, 35, 816-821.
[98] Mcaleer, W. J., Buynak, E. B., Maigetter, R. Z., Wampler, D. E., Miller, W. J. & Hilleman, M. R. 1984. Human Hepatitis-B Vaccine from Recombinant Yeast. *Nature*, 307, 178-180.
[99] Niikura, M., Takamura, S., Kim, G., Kawai, S., Saijo, M., Morikawa, S., Kurane, I., Li, T. C., Takeda, N. & Yasutomi, Y. 2002. Chimeric recombinant hepatitis E virus-like particles as an oral vaccine vehicle presenting foreign epitopes. *Virology*, 293, 273-280.
[100] Santi, L., Giritch, A., Roy, C. J., Marillonnet, S., Klimyuk, V., Gleba, Y., Webb, R., Arntzen, C. J. & Mason, H. S. 2006. Protection conferred by recombinant Yersinia pestis antigens produced by a rapid and highly scalable plant expression system. *Proceedings of the National Academy of Sciences of the United States of America*, 103, 861-866.

[101] Sunil Kumar, G. B., Ganapathi, T. R., Revathi, C. J., Prasad, K. S. N. & Bapat, V. A. 2003. Expression of hepatitis B surface antigen in tobacco cell suspension cultures. *Protein Expression and Purification*, 32, 10-17.

[102] Huang, Z., Elkin, G., Maloney, B. J., Beuhner, N., Arntzen, C. J., Thanavala, Y. & Mason, H. S. 2005. Virus-like particle expression and assembly in plants: hepatitis B and Norwalk viruses. *Vaccine*, 23, 1851-1858.

[103] Maclean, J., Koekemoer, M., Olivier, A. J., Stewart, D., Hitzeroth, II., Rademacher, T., Fischer, R., Williamson, A. L. & Rybicki, E. P. 2007. Optimization of human papillomavirus type 16 (HPV-16) L1 expression in plants: comparison of the suitability of different HPV-16 L1 gene variants and different cell-compartment localization. *Journal of General Virology*, 88, 1460-1469.

[104] Kumar, G. B., Ganapathi, T. R. & Bapat, V. A. 2007. Production of hepatitis B surface antigen in recombinant plant systems: an update. *Biotechnology Progress*, 23, 532-539.

[105] Maloney, B. J., Takeda, N., Suzaki, Y., Ami, Y., Li, T. C., Miyamura, T., Arntzen, C. J. & Mason, H. S. 2005. Challenges in creating a vaccine to prevent hepatitis E. *Vaccine*, 23, 1870-1874.

[106] Tacket, C. O. 2009. Plant-based oral vaccines: results of human trials. *Current Topics in Microbiology Immunology*, 332, 103-117.

[107] Tacket, C. O., Mason, H. S., Losonsky, G., Estes, M. K., Levine, M. M. & Arntzen, C. J. 1998. Human immune responses to a novel Norwalk virus vaccine delivered in transgenic potatoes. *Journal of Infectious Diseases*, 182, 302-305.

[108] Tacket, C. O., Pasetti, M. F., Edelman, R., Howard, J. A. & Streatfield, S. 2004. Immunogenicity of recombinant LT-B delivered orally to humans in transgenic corn. *Vaccine*, 22, 4385-4389.

[109] Tacket, C. O., Mason, H. S., Losonsky, G., Estes, M. K., Levine, M. M. & Arntzen, C. J. 2000. Human immune responses to a novel norwalk virus vaccine delivered in transgenic potatoes. *Journal of Infectious Diseases*, 182, 302-305.

[110] Thanavala Y., Mahoney M., Pal S., Scott A., Richter L., Natarajan N., Goodwin P., Arntzen C. J. & Mason H. S. 2005. Immunogenicity in humans of an edible vaccine for hepatitis B. *Proceedings of the National Academy of Sciences of the United States of America*, 102, 3378-3382.

[111] Hiatt, A., Cafferkey, R. & Bowdish, K. 1989. Production of antibodies in transgenic plants. *Nature*, 342, 76-78.

[112] Ma, J. K. C., Hiatt, A., Hein, M., Vine, N. D., Wang, F., Stabila, P., Van Dolleweerd, C., Mostov, K. & Lehner, T. (1995) Generation and assembly of secretory antibodies in plants. *Science* 268,716-719.

[113] Ma, J. K. C., Hikmat, B. Y., Wycoff, K., Vine, N. D., Chargelegue, D. Y., Hein, M. B. & Lehner, T. 1998. Characterization of a recombinant plant monoclonal secretory antibody and preventive immunotherapy in humans. *Nature Medicine*, 4, 601-606.

[114] Zeitlin, L., Olmsted, S. S., Moench, T. R., Co, M. S., Martinell, B. J., Paradkar, V. M., Russell, D. R., Queen, C., Cone, R. A. & Whaley, K. J. 1998. A humanized monoclonal antibody produced in transgenic plants for immunoprotection of the vagina against genital herpes. *Nature Biotechnology*, 16, 1361-1364.

[115] De Muynck, B., Navarre, C. & Boutry, M. 2010. Production of antibodies in plants: status after twenty years. *Plant Biotechnology Journal*, 8, 1–35.

[116] Vaquero, C., Sack, M., Schuster, F., Finnern, R., Drossard, J., Schumann, D., Reimann, A. & Fischer, R. 2002. A carcinoembryonic antigen-specific diabody produced in tobacco. *FASEB* Journal, 16, 408-410.

[117] Fiedler, U., Phillips, J., Artsaenkoc, O. & Conrad, U. 1997. Optimization of scFv antibody production in transgenic plants. *Immunotechnology*, 3, 205-216.

[118] McCormick, A. A., Kumagai, M. H., Hanley, K., Turpen, T. H., Hakim, I., Grill, L. K., Tuse, D., Levy, S. & Levy, R. 1999. Rapid production of specific vaccines for lymphoma by expression of the tumor-derived single-chain Fv epitopes in tobacco plants. *Proceedings of the National Academy of Sciences of the United States of America*, 96, 703-708.

[119] Stoger, E., Sack, M., Perrin, Y., Vaquero, C., Torres, E., Twyman, R. M., Christou, P. & Fischer, R. 2002. Practical considerations for pharmaceutical antibody production in different crop system. *Molecular Breeding*, 9, 149-158.

[120] Malembic-Maher, S., Le Gall, F., Danet, J. L., de Borne, F. D., Bove, J. M. & Garnier-Semancik, M. 2005. Transformation of tobacco plants for single-chain antibody expression via apoplastic and symplasmic routes, and analysis of their susceptibility to stolbur phytoplasma infection. *Plant Science*, 168, 349-358.

[121] Yuan, Q. P., Hu, W. Q., Pestka, J. J., He, S. Y. & Hart, L. P. 2000. Expression of a functional antizearalenone single-chain Fv antibody in transgenic Arabidopsis plants. *Applied Environmental Microbiology*, 66, 3499-3505.

[122] Smith, M. D. & Glick, B. R. 2000. The production of antibodies in plants: An idea whose time has come? *Biotechnology Advances*, 18, 85–89.

[123] Potera, C. 1999. EPIcyte produces antibodies in plants - Plantibodies step in to fulfill the promise of mabs. *Genetic Engineering News*, 19, 22-23.

[124] Kusnadi, A. R., Hood, E. E., Witcher, D. R., Howard, J. A. & Nikolov, Z. L. 1998. Production and purification of two recombinant proteins from transgenic corn. *Biotechnology Progress*, 14, 149-155.

[125] Sharp, J. M. & Doran, P. M. 2001. Characterization of monoclonal antibody fragments produced by plant cells. *Biotechnology and Bioengineering*, 73, 338-346.

[126] Stevens, L. H., Stoopen, G. M., Elbers, I. J. W., Molthoff, J. W., Bakker, H. A. C., Lommen, A., Bosch, D. & Jordi, W. 2000. Effect of climate conditions and plant developmental stage on the stability of antibodies expressed in transgenic tobacco. *Plant Physiology*, 124, 173-182.

[127] Ma, J. K. & Hein, M. B. 1995. Immunotherapeutic potential of antibodies produced in plants. *Trends in Biotechnology*, 13, 522-527.

[128] Stoger, E., Vaquero, C., Torres, E., Sack, M., Nicholson, L., Drossard, J., Williams, S., Keen, D., Perrin, Y., Christou, P. & Fischer, R. 2000. Cereal crops as viable production and storage systems for pharmaceutical scFv antibodies. *Plant Molecular Biology*, 42, 583-590.

[129] Savelyeva, N., Munday, R., Spellerberg, M. B., Lomonossoff, G. P. & Stevenson, F. K. 2001. Plant viral genes in DNA idiotypic vaccines activate linked CD4+ T-cell mediated immunity against B-cell malignancies. *Nature Biotechnology*, 19, 760-764.

[130] Bakker, H., Bardor, M., Molthoff, J. W., Gomord, V., Elbers, I., Stevens, L. H., Jordi, W., Lommen, A., Faye, L., Lerouge, P. & Bosch, D. 2001. Galactose-extended glycans of antibodies produced by transgenic plants. *Proceedings of the National Academy of Sciences of the United States of America*, 98, 2899-2904.

[131] Obregon, P., Chargelegue, D., Drake, P. M., Prada, A., Nuttall, J., Frigerio, L. & Ma, J. K. 2006. HIV-1 p24-immunoglobulin fusion molecule: a new strategy for plant-based protein production. *Plant Biotechnology Journal*, 4, 195-207.

[132] Odell, J. T., Nagy, F. & Chua, N. H. 1985. Identification of DNA-sequences required for activity of the cauliflower mosaic virus-35S promoter. *Nature*, 313, 810-812.

[133] Gutierrez-Ortega, A., Sandoval-Montes, C., Olivera-Flores, T. J., Santos Argumedo, L. & Gomez Lim, M. A. 2005. Expression of functional interleukin-12 from mouse in transgenic tomato plants. *Transgenic Research*, 14, 877-885.

[134] Verdaguer, B., deKochko, A., Beachy, R. N. & Fauquet, C. 1996. Isolation and expression in transgenic tobacco and rice plants, of the cassava vein mosaic virus (CVMV) promoter. *Plant Molecular Biology*, 31, 1129-1139.

[135] Xie, Y., Liu, Y., Meng, M., Chen, L. & Zhu, Z. 2003. Isolation and identification of a super strong plant promoter from cotton leaf curl Multan virus. *Plant Molecular Biology*, 53, 1-14.

[136] Shirasawa-Seo, N., Sano, Y., Nakamura, S., Murakami, T., Gotoh, Y., Naito, Y., Hsia, C. N., Seo, S., Mitsuhara, I., Kosugi, S. & Ohashi, Y. 2005. The promoter of Milk vetch dwarf virus component 8 confers effective gene expression in both dicot and monocot plants. *Plant Cell Reports*, 24, 155-163.

[137] Shaw, C. H., Carter, G. H., Watson, M. D. & Shaw, C. H. 1984. A Functional Map of the Nopaline Synthase Promoter. *Nucleic Acids Research*, 12, 7831-7846.

[138] Dai, Z., Hooker, B. S., Anderson, D. B. & Thomas, S. R. 2000. Expression of Acidothermus cellulolyticus endoglucanase E1 in transgenic tobacco: biochemical characteristics and physiological effects. *Transgenic Research*, 9, 43-54.

[139] Belanger, F. C. & Kriz, A. L. 1991. Molecular basis for allelic polymorphism of the maize globulin-1 gene. *Genetics*, 129, 863.

[140] Horvath, H., Huang, J. T., Wong, O., Kohl, E., Okita, T., Kannangara, C. G. & von Wettstein, D. 2000. The production of recombinant proteins in transgenic barley grains. *Proceedings of the National Academy of Sciences of the United States of America*, 97, 1914-1919.

[141] Russell, D. A. & Fromm, M. E. 1997. Tissue-specific expression in transgenic maize of four endosperm promoters from maize and rice. *Transgenic Research*, 6, 157-168.

[142] Aoyama, T. & Chua, N. H. 1997. A glucocorticoid-mediated transcriptional induction system in transgenic plants. *Plant Journal*, 11, 605-612.

[143] Zuo, J., Niu, Q. W. & Chua, N. H. 2000. Technical advance: An estrogen receptor-based transactivator XVE mediates highly inducible gene expression in transgenic plants. *Plant Journal*, 24, 265-273

[144] Comai, L., Moran, P. & Maslyar, D. 1990. Novel and Useful Properties of a Chimeric Plant Promoter Combining Camv-35S and Mas Elements. *Plant Molecular Biology*, 15, 373-381.

[145] Rance, I., Norre, F., Gruber, V. & Theisen, M. 2002. Combination of viral promoter sequences to generate highly active promoters for heterologous therapeutic protein over-expression in plants. *Plant Science*, 162, 833-842.

[146] Spiker, S. & Thompson, W. F. 1996. Nuclear matrix attachment regions and transgene expression in plants. *Plant Physiology*, 110, 15-21.

[147] Allen, G. C., Spiker, S. & Thompson, W. F. 2000. Use of matrix attachment regions (MARs) to minimize transgene silencing. *Plant Molecular Biology*, 43, 361-376.

[148] Green, P. J. 1993 Control of mRNA stability in higher plants. *Plant Physiology*, 102, 1065-1070.
[149] Fiume, E., Christou, P., Gianì, S. & Breviario, D. 2004. Introns are key regulatory elements of rice tubulin expression. *Planta*, 218, 693-670.
[150] Pooggin, M. M. & Skryabin, K. G. 1992. The 5'-Untranslated Leader Sequence of Potato Virus-X Rna Enhances the Expression of a Heterologous Gene Invivo. *Molecular & General Genetics*, 234, 329-331.
[151] Joshi, C. P., Zhou, H., Huang, X. & Chiang, V. L. 1997. Context sequences of translation initiation codon in plants. *Plant Molecular Biology*, 35, 993-1001.
[152] Kim, T. G. & Langridge, W. H. 2003. Assembly of cholera toxin B subunit full-length rotavirus NSP4 fusion protein oligomers in transgenic potato. *Plant Cell Reports*, 21, 884-890.
[153] Kim, T. G. & Langridge, W. H. 2004. Synthesis of an HIV-1 Tat transduction domain-rotavirus enterotoxin fusion protein in transgenic potato. *Plant Cell Reports*, 2, 382-387.
[154] Choi, N. W., Estes, M. K. & Langridge, W. H. R. 2006. Ricin Toxin B Subunit Enhancement of Rotavirus NSP4 Immunogenicity in Mice. *Viral Immunology*, 19, 54-63.
[155] Tiwari, S., Verma, P. C., Singh, P. K. & Tuli, R. 2009. Plants as bioreactors for the production of vaccine antigens. *Biotechnol Advances*, 27, 449-467.
[156] Rayon, C., Lerouge, P. & Faye, L. 1998. The protein N-glycosylation in plants. *Journal of Experimental Botany*, 49, 1463-1472.
[157] Sriraman, R., Bardor, M., Sack, M., Vaquero, C., Faye, L., Fischer, R., Finnern, R. & Lerouge, P. 2004. Recombinant anti-hCG antibodies retained in the endoplasmic reticulum of transformed plants lack core-xylose and core-alpha(1,3)-fucose residues. *Plant Biotechnology Journal*, 2, 279-287.
[158] Chargelegue, D., Drake, P. M., Obregon, P., Prada, A., Fairweather, N. & Ma, J. K. 2005. Highly immunogenic and protective recombinant vaccine candidate expressed in transgenic plants. *Infection and Immunity*, 73, 5915-5922.
[159] Witcher, D. R., Hood, E. E., Peterson, D., Bailey, M., Bond, D., Kusnadi, A., Evangelista, R., Nikolov, Z., Wooge, C., Mehigh, R., Kappe, W., Register, J. & Howard, J. A. 1998. Commercial production of beta-glucuronidase (GUS): a model system for the production of proteins in plants. *Molecular Breeding*, 4, 301-312.
[160] Zhong, G. Y., Peterson, D., Delaney, D. E., Bailey, M., Witcher, D. R., Register III, J. C., Bond, D., Li, C. P., Marshall, L., Kulisek, E., Ritland, D., Meyer, T., Hood, E. E. & Howard, J. A. 1999. Commercial production of aprotinin in transgenic maize seeds. *Molecular Breeding*, 5, 345-356.
[161] Hood, E. E., Witcher, D. R., Maddock, S., Meyer, T., Baszczynski, C., Bailey, M., Flynn, P., Register, J., Marshall, L., Bond, D., Kulisek, E., Kusnadi, A., Evangelista, R., Nikolov, Z., Wooge, C., Mehigh, R. J., Hernan, R., Kappel, W. K., Ritland, D., Li, C. P. & Howard, J. A. 1997. Commercial production of avidin from transgenic maize: characterization of transformant, production, processing, extraction and purification. *Molecular Breeding*, 3, 291-306.
[162] Kirk, D. D. & Webb, S. R. 2005. The next 15 years: taking plant-made vaccines beyond proof of concept. *Immunology and Cell Biology*, 83, 248-256.

[163] Thanavala, Y., Huang, Z. & Mason, H. S. 2006. Plant-derived vaccines: a look back at the highlights and a view to the challenges on the road ahead. *Expert Review of Vaccines*, 5, 249-260.

[164] Kusnadi, A. R., Nikolov, Z. L. & Howard, J. A. 1997. Production of recombinant proteins in transgenic plants: Practical considerations. *Biotechnology and Bioengineering*, 56, 473-484.

[165] Giddings, G., Allison, G., Brooks, D. & Carter, A. 2000. Transgenic plants as factories for biopharmaceuticals. *Nature Biotechnology*, 18, 1151-1155.

[166] Higo, K., Saito, Y. & Higo, H. 1993. Expression of a Chemically Synthesized Gene for Human Epidermal Growth-Factor under the Control of Cauliflower Mosaic Virus-35S Promoter in Transgenic Tobacco. *Bioscience Biotechnology and Biochemistry*, 57, 1477-1481.

[167] Reggi, S., Marchetti, S., Patti, T., De Amicis, F., Cariati, R., Bembi, B. & Fogher, C. 2005. Recombinant human acid beta-glucosidase stored in tobacco seed is stable, active and taken up by human fibroblasts. *Plant Molecular Biology*, 57, 101-113.

[168] Cramer, C. L., Weissenborn, D. L., Oishi, K. K., Grabau, E. A., Bennett, S., Ponce, E., Grabowski, G.A. & Radin, D. N. 1996. Bioproduction of human enzymes in transgenic tobacco. *Engineering Plants for Commercial Products and Applications*, 792, 62-71.

[169] Ruggiero, F., Exposito, J. Y., Bournat, P., Gruber, V., Perret, S., Comte, J., Olagnier, B., Garrone, R. & Theisen, M. 2000. Triple helix assembly and processing of human collagen produced in transgenic tobacco plants. *FEBS Letters*, 469, 132-136.

[170] Mor, T. S., Sternfeld, M., Soreq, H., Arntzen, C. J. & Mason, H. S. 2001. Expression of recombinant human acetylcholinesterase in transgenic tomato plants. *Biotechnology and Bioengineering*, 75, 259-266.

[171] Claparols, M. I., Bassie, L., Miro, B., Del Duca, S., Rodriguez-Montesinos, J., Christou, P., Serafini-Fracassini, D. & Capell, T. 2004. Transgenic rice as a vehicle for the production of the industrial enzyme transglutaminase. *Transgenic Research*, 13, 195-199.

[172] Gutierrez-Ortega, A., Avila Moreno, F., Saucedo-Arias, L. J., Sánchez Torres, C. & Gómez-Lim, M. A. 2004. Expression of a single-chain human interleukin 12 gene in transgenic tobacco plants and functional studies. *Biotechnology and Bioengineering*, 85, 734-740.

[173] Lee, S. J., Park, C. I., Park, M. Y., Jung, H. S., Ryu, W. S., Lim, S. M., Tan, H. K., Kwon, T. H., Yang, M. S. & Kim, D. I. 2007. Production and characterization of human CTLA4Ig expressed in transgenic rice cell suspension cultures. *Protein Expression and Purification*, 51, 293-302.

In: Bioprocess Sciences and Technology
Editor: Min-Tze Liong

ISBN 978-1-61122-950-9

Chapter 13

TOWARDS A PRACTICAL AND AFFORDABLE ORAL PAPAYA-BASED VACCINE: A CRUCIAL TOOL FOR TAENIASIS CYSTICERCOSIS CONTROL PROGRAMS

E. Sciutto[1,*], M. Hernández[1], A. Díaz-Orea[2], J. Cervantes[1], G. Rosas-Salgado[3], J. Morales[1], N. Villalobos[4], A. Aluja[4], A. Blancas[1], R. Aquino[1], O. Fuentes[5], H. Jiménez[6], C. Trejo-Moreno[3], M. Méndez-Martínez[3], K. Manoutcharian[1], G. Gevorkian[1], L. Herrera-Estrella[6, 7], J. L. Cabrera Ponce[6], C. Larralde [1], G. Fragoso[1], M. Huerta[5]

[1]Instituto de Investigaciones Biomédicas, 4Facultad de Medicina Veterinaria y Zootecnia Universidad Nacional Autónoma de México, México, [1]D. F., 04510, México

[2]Centro de Investigación Biomédica de Oriente, Instituto Mexicano del Seguro Social, 4.5 Km, Carretera Atlixco-Metepec, C.P. 74360, Puebla, México

[3]Facultad de Medicina, Universidad Autónoma del Estado de Morelos. Cuernavaca, Morelos, México

[4]Facultad de Medicina, [5]Escuela de Medicina Veterinaria y Zootecnia, Benemérita Universidad Autónoma de Puebla, 13 Sur No. 2702, Col. Volcanes, Puebla, Pue. México.

[6]Departamento de Ingeniería Genética. Centro de Investigación y de Estudios Avanzados del IPN, Unidad Irapuato, Irapuato, Guanajuato, CP 36821 México

[7]Laboratorio Nacional de Genómica para la Biodiversidad, CINVESTAV Irapuato, Km 9.6 Libramiento Norte Carretera Irapuato-León, 36500 Irapuato, Mexico

ABSTRACT

Cysticercosis is a parasitic disease caused by *Taenia solium* which affects rurally bred pigs and humans in developing countries. The adult tapeworm parasite lives in the

---

[*]Correspondent author: *Edda Sciutto Conde, Instituto de Investigaciones Biomédicas, Universidad Nacional Autónoma de México, México, D.F., 04510, México. Tel: +52-5556-223-153 Fax: +52-5556-223-369. E-mail: edda@servidor.unam.mx*

humans′ intestines, where it sheds thousands of eggs. Cysticercosis occurs after eggs are ingested by pigs or humans. In humans, cysticerci may cause neurocysticercosis, the most severe and frequent form of the human disease. In pigs, cysticerci can lodge in different tissues. Deficient hygiene, inadequate feces disposal, outdoor defecation, freely roaming pigs, and inadequate meat inspection promote transmission.

Since pigs are indispensable intermediate host, it is conceivable to curb transmission by reducing pig cysticercosis through vaccination. A vaccine against porcine cysticercosis to be extensively used in non-developed countries must not only be immunologically effective, but must also be inexpensive and simple to administrate to millions of rustically bred pigs, which are renewed every year and are dispersed in thousands of villages across large and variable geographic territories.

We developed an anti-cysticercosis vaccine (S3Pvac) composed by three peptides originally identified in *Taenia crassiceps*. The first version of S3Pvac synthetically produced reduced by 50% the percent of infected pigs, and by 97% the number of cysticerci established, as assessed in a field trial. In search of reducing vaccine production costs, a version of S3Pvac recombinantly expressed in filamentous phages was developed. An improved procedure was standardized for its scale-up production and inactivation. S3Pvac-phage reduced by 54% the number of cysticercotic pigs and by 87% the total number of cysticerci, based on necropsy studies, in a field trial that included 1,047 rural pigs of central Mexico. However, the parenteral administration of the vaccine was found to be a major logistic and economic difficulty, limiting its use in yearly nation's wide programs. Such vaccine requires a risky capture of wild roaming pigs by not less than three or four trained persons.

On the other hand, an orally administered vaccine would elude these difficulties, since it could be delivered by the pigs' owners when fed. An oral vaccine has additional attractive features, especially for the prevention of mucosally transmitted parasites as *Taenia solium*. Transgenic plants are ideal systems to produce oral vaccines, since the walls of the plant cells may protect antigenic proteins from digestion in the intestinal tract, enabling them for effective presentation to mucosal-associated lymphoid tissues. Moreover, the expression of foreign proteins in transgenic plants or in cultured plant cells offers an inexpensive system for the massive production of a vaccine. Thus, the S3Pvac peptides were expressed in transgenic embryogenic papaya cells. The vaccine peptides were stably expressed and incorporated into the papaya nuclear genome by particle bombardment. Orally administered S3Pvac-papaya, composed by transgenic embryogenic cell lines pKETc12_6, pKETc1_9, and pKETc723, effectively protect against experimental murine cysticercosis and are immunogenic in pigs. A safe, easy and low-cost system based on the culture of the S3Pvac-papaya cell suspension is being optimizing for massively producing the vaccine. Altogether, results point to the feasibility of having a cost-effective oral vaccine field trial conducted in the near future.

Abbreviations

Transgenic embryogenic cell lines: (ETgpC)
Embryogenic papaya cell lines: (EmpCL)

Papaya clones constitued by the ETgpC pKETc12$_6$, pKETc1$_9$, pKETc7$_{23}$ (S3Pvac-papaya)

1. DEVELOPMENT OF ORAL VACCINES

Many pathogens, among them helminths, enter the human body through the mucosal surfaces; thus, mucosal immunization could be a promising strategy to prevent infections. Moreover, the needle-free oral vaccines are an attractive method of vaccine delivery, which also avoid costly logistics problems in their administration, making them suitable for massive immunization programs.

However, just few oral vaccines have been commercialized to date, possibly due to the lack of effective oral delivery systems capable of preventing its digestion and hydrolysis in the gastrointestinal tract [1]. This limitation could be overridden by antigen expression within plant cells, naturally protected by their walls [2, 3]. In addition, antigen expression by plants could reduce production costs, avoiding the expensive step of antigen purification and costly artificial antigen encapsulation technologies.

Thus, the use plant derived vaccines provide a means to deliver proteins for oral vaccination, although concerns about immune tolerance and allergy to plant-based vaccines has been expressed and needs to be addressed suitably [4). However, not all ingested antigens are seen as food proteins, which are well known to induce clonal deletion or anergy. In fact, there are some oral antigens that are able to bind to epithelial cells or can replicate in the gut, exhibiting high immunogenic properties [5]. In addition, several plant-based vaccine have been successfully tested orally administered i.e. Hepatitis B virus [6], gastroenteritis swine virus [7] and *E. coli* derived toxin [8] among others. Thus, it seems that plant modification in some way may circumvent the tolerogenic response. It has also to mention that oral immunity is quite sensible to the dose of antigen employed and the number of doses, both parameters that can be modified for the particular immunogenicity of the employed antigen. In general, it has been observed that after multiple doses of oral antigen, spaced at appropriate intervals, the responses to the antigen are absent when the subject is latter exposed to the same antigen parenterally administered [9]. However, using plant vaccines, it is expected that with only two or three doses of an edible vegetable vaccine in an adequate dose, such tolerance would not occur [4].

Herein, we revise our advances towards an oral vaccine against *Taenia solium* cysticercosis, expressed in transgenic embryogenic papaya cell lines (ETgpC) and its protective capacity oral and parenteral administered in mice against Taenia crassiceps experimental murine cysticercosis. Additional data of the immunogenic properties of S3Pvac-papaya vaccine orally administered to pigs are also shown.

2. *TAENIA SOLIUM*

Taenia solium is a parasite whose larvae (cysticerci) may locate in the central nervous system of humans, causing neurocysticercosis (NC). This is the most frequent, severe and even fatal form of the infection in humans [10]. Extra-central nervous system infections were also frequently reported in Asia and Africa [11], but much less in America [12]. The great majority of cases occur in developing countries [13], but it is increasingly considered as an emerging disease in the developed world because of human migration [14]. Cysticerci may also establish in the skeletal muscles of pigs (intermediary hosts) as an obligatory step of the tapeworms life cycle. Porcine cysticercosis results in substantial economic loss by banning cysticercotic swine at slaughters.

Prevalence rates vary in the endemic countries, but are estimated at 1-10% for human cysticercosis and up to 20-40% for pig cysticercosis in rustically bred pigs [15]. Variations depend on sampling and survey designs (i.e. rural vs. urban), the techniques used for detection (i.e. image, serologic) and local sanitary conditions in the area and inhabitants. In other studies, human NC was found in 1-5% of large autopsy series conducted in Mexico [16, 17]; it accounts for about 10% in rural areas, according to two epidemiological studies based on CTscan [18, 19], and it is the most frequent cause of epilepsy in neurological patients [20].

Humans are also the only definitive host for the tapeworm stage, and become infected by ingesting raw or inadequately cooked cysticercotic meat. The intestinal tapeworm causes minor gastrointestinal disturbances to its host, but it is of major importance for cysticercosis transmission, as it daily produces tens of thousands of eggs, shed to the environment by defecation [13].

3. Effective Vaccine Candidates against Cysticercosis

3.1. Strategies Employed to Identify Protective Antigens against Cysticercosis

Two main strategies have been used to identify antigens of interest to be employed as a source of vaccine against *T. solium*: one considering their expression in oncospheres and/or early developed cysticerci (based on the argument that these are the most vulnerable stages of the parasite to the immune attack), and the other identifying protective antigens, disregarding their location, by experimental murine models. These two strategies have been used with encouraging results [21-26]. Different parasite antigens originally derived from related cestodes (*T. solium*, *T. crassiceps*, *T. saginata*, *T. ovis*) purified, recombinant or synthetic *produced* have been proven as vaccines and shown to reduce the rate of infection and the parasite load in vaccinated pigs challenged with *T. solium* eggs under experimental conditions [21-26]. Moreover, it has been recently reported an amazing 100% high level of protection against the homologous infection using the oncospheral antigen TSOL18 antigen [26]. Also, it is conceivable that the use of multiple protective epitopes may improve their efficiency, considering their protective capacity, their presence in the different parasite stages and structures, as demonstrated by using different vaccine candidates together against malaria and cysticercosis [27, 28]. In addition, it would be possible to kill oncospheres as well as early cysticerci, which may evade the initial immune attack. Taken together, these results let encourage us to express TSOL18 in embryogenic papaya calli to include in the S3Pvac-papaya formulation in order to improve the vaccine efficacy.

3.2. Identification of the S3Pvac Vaccine Epitopes

Three protective peptides against cysticercosis were identified by the Mexican research group in a cDNA *Taenia crassiceps* library [29]. The three peptides which constitute the S3Pvac anti-cysticercosis vaccine are 8 (KETc12), 12 (KETc1) and 110 amino acids (KETc7) long [30-32]. Two new protective peptides have been identified, GK1, 18 aa [31], and PT1, 10 aa [33].

All three peptides belong to native antigens present in the different developmental stages of *Taenia crassiceps*. The homologous antigens are also found in *Taenia solium* [30-32] in different anatomical structures of the cysticerci, the eggs and the adult tapeworm. More recently, highly homologous sequences were found in *Taenia saginata, Echinoccocus granulosus* and *Echinoccocus multilocularis* [34].

3.3. Field Trial Vaccination

S3Pvac synthetically expressed was tested on the field in communities of the State of Puebla [22]. Vaccination reduced by 50% the number of infected pigs, and by 98% the amount of *Taenia solium* cysticerci potentially able to transform into tapeworms. Upon re-evaluation in a different community of the State of Morelos, similar results were obtained [35]. Thus, the effectiveness of S3Pvac as a preventing agent against naturally acquired porcine cysticercosis has been demonstrated.

More recently, a second version of the vaccine based on the S3Pvac peptides but expressed in filamentous phages (S3Pvac-phage) was produced and tested as well on the field [25]. S3Pvac-phage reduced by 54% the number of cysticercotic pigs and by 87% the total number of cysticerci, based on necropsy studies, in a field trial including 16 communities and 1,047 rural pigs in central Mexico. This new version may now be produced at very low costs (aprox 1 dl per pig), but it still demands the work on the field of many trained persons to implement a yearly vaccination program in millions of free roaming pigs. It would be desirable to circumvent this major limitation by the design of an oral vaccine to be administered by the pig owners when feeding the pigs.

3.4. Expression of the S3Pvac Vaccine Epitopes in Embryogenic Transgenic Papaya Clones (ETgpC)

In order to deliver the S3Pvac anti-cysticercosis vaccine in a system adequate for oral immunization we decided to use plant cells. Papaya was selected among others considering the availability and experience to produce transgenic papaya in Mexico [36, 37], the intrinsic papaya anthelminthic properties due to its high content of proteolitic enzymes [38-40] and due to the advantage of having sequenced the papaya genome [41]. In addition, papaya has been employed to produce an anti-tuberculosis vaccine, even though its immunogenicity remains to be tested [42].

3.4.1. Design of the Vaccine Genetic Constructions and Transformation Efficiency

The three S3Pvac peptides (KETc1, KETc12 or KETc7) were cloned in the pUI235-5.1 plant expression vector, which drives the strong and constitutive expression of transgenes in plant tissues and contains the kanamycin resistance gene (Km^R). KETc1 and KETc12 were modified by adding additional six histidine residues in 3'-COOH to increase their size and aid in their identification (Figure 1). In order to facilitate the detection and selection of transgenic tissues, pWRG1515 plasmid, which contains the β-glucuronidase reporter gene (GUS gene) besides hygromycin resistance gene (Hyg^R) as a selectable marker, were used for co-transformation with each construction [43].

Seeds of papaya 'Maradol' Tabasco (*Carica papaya* L.) were isolated from young green fruits and used as a source of zygotic embryos for the induction of embryogenic papaya cells line (EmpCL) according to Cabrera-Ponce *et al.,* [36].

EmpCL at the globular stage was co-transformed using the particle bombardment system PDS-1000/He (BioRAD) following the Sanford *et al.*[44] and Tomes *et al.,* [45] protocols with some modifications [36, 37]. Briefly, EmpCL were bombarded once at P = 800 psi. Selection of the ETgpC was done according to Cabrera-Ponce *et al.* [36], with modifications [36]. Medium with hygromicin (50 mg/L) and kanamycin (100 mg/L) were used as selectable markers. Four subcultures (every 45-60 days) on this medium were applied in order to obtain ETgpC.

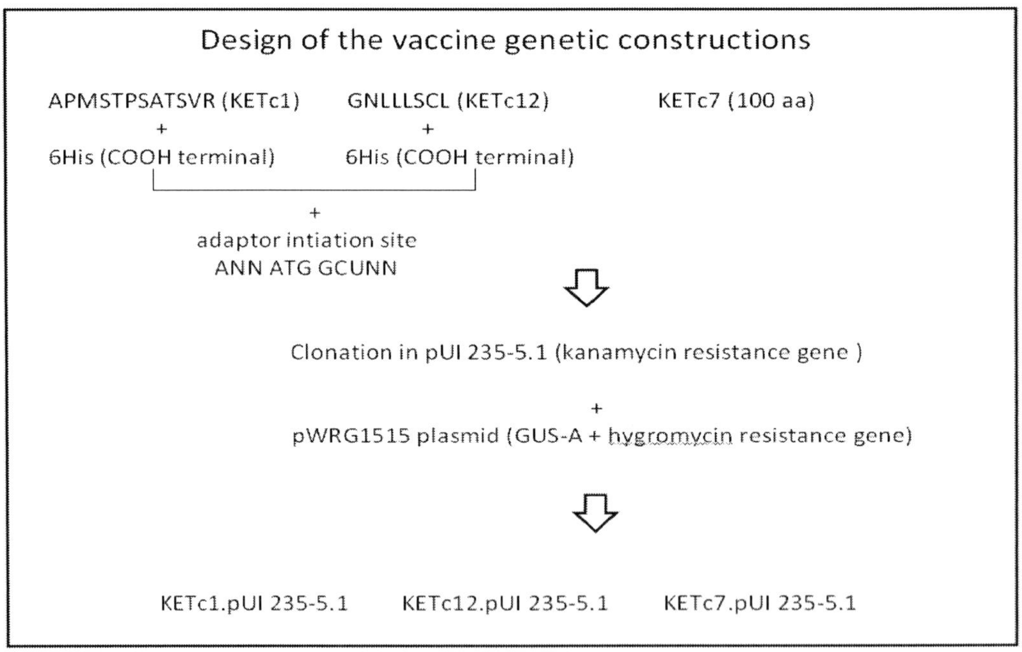

Figure 1. Design of genetic constructions in pUI235-5.1expression vector.

The ETgpC, resistant to Hyg^R and Km^R were progressively numbered, adding an initial p for papaya, e.g. $pKETc7_{21}$ [46]. Some ETgpC embryos were incubated with X-GLUC (5-bromo-4-chloro-3-indolyl D-glucuronic acid) for periods ranging from a few minutes through 24 hours at room temperature, and forty-one ETgpC clones exhibited stable expression of the β-glucuronidase reporter gene (Figure 2).

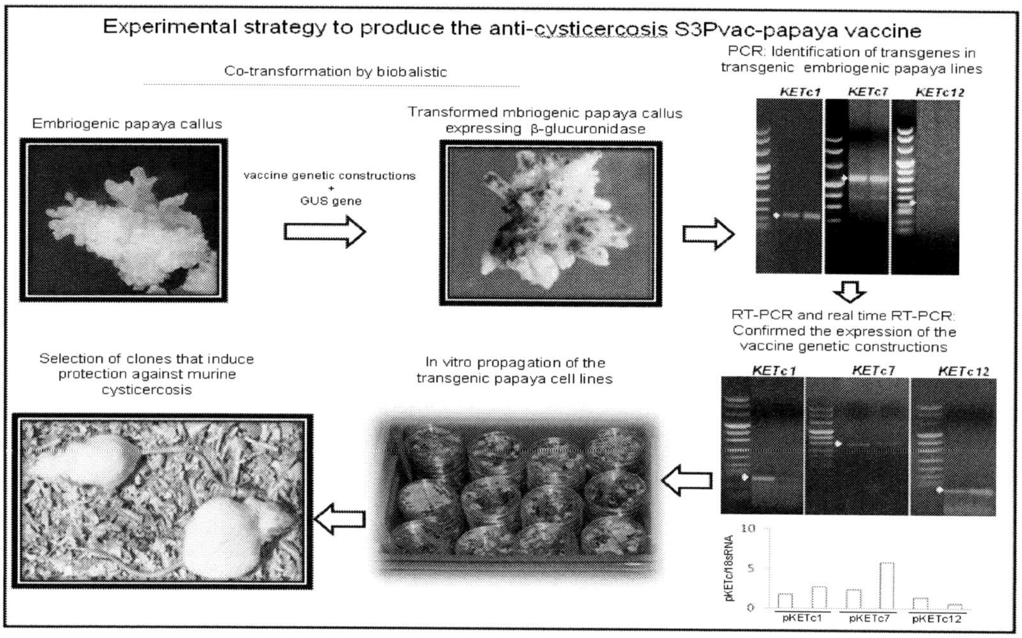

Figure 2. General strategy for the expression of the S3Pvac-papaya vaccine.

3.4.2. Molecular Characterization of ETgpC

Table 1 shows the number of transformed callus with each genetic construction, and the number of ETgpC obtained. As it is shown, the transformation efficiency varied from 5.2 to 13.5%.

The presence and expression of the each transgen was determined by PCR and RT-PCR (Figure 2), and the transgenic peptides content in the different clones were further estimated using real time PCR (Figure 2).

Table 1. Efficiency of transformation of papaya callus by co-bombardment with genetic constructions in pUI235-5.1 and pWGR1515

pUI235-5.1 to express the peptide	Number of transformed callus	Number of clones obtained	Transformation efficiency (%)
KETc7	210	11	5.2
KETc1.6His	280	38	13.5
KETc12.6His	175	12	6.8

3.4.3. In Vitro Propagation of ETgpC

Two clones expressing KETc7, two expressing KETc12.6His and 15 clones expressing KETc1.6His were propagated by *in vitro* subculture for vaccine experiments. Although there is a risk of losing the transgene with this form of maintenance, the inserts remained in ETgpC after five years of transformation.

3.4.4. Experiences in Vaccination with S3Pvac-Papaya against Taenia Crassiceps Murine Cysticercosis

Immunization experiments were performed using soluble extracts or suspensions from EmpCL and ETgpC.

To obtain soluble extracts of ETgpC (5 g) and EmpCL were powdered in liquid nitrogen and homogenized in extraction buffer (PBS, 50 mM sodium ascorbate, 1mM EDTA pH 8.0, 0.2% Triton X-100 and 10μl/ml of protease inhibitor to His-tag protein (Sigma, St. Louis, MI). Thereafter, the material was centrifuged at 21,320 × *g* at 4 °C for 20 min. to remove insoluble debris [46]. Supernatant was collected and protein concentration was determined by Lowry method [47].

Cell suspensions were prepared by grinding EmpCL and ETgpC in liquid nitrogen and freeze-dried. The obtained tissue was stored at –80 °C until used [48].

Subcutaneous immunization with the soluble extracts from ETgpC pKETc7_{23} (200 to 1000 μg) provided complete protection in 50 and 63% of immunized mice, respectively [49]. Subcutaneous immunization with ETgpC pKETc1_9 soluble extract (200 μg) or the equivalent amount of suspension (40 mg per mouse) also provided total protection in 70 to 90% of immunized mice without the use of any adjuvant and with only one booster [46]. Moreover, as Figure 3 shows, oral immunization with S3Pvac-papaya composed by the ETgpC, pKETc1_9, pKETc12_6 and pKETc7_{23} induced total protection in 75% of immunized mice at a dose of 0.1 mg/mouse of cell suspension. Higher doses of S3Pvac-papaya cell suspension progressively lowered the percent of protection and eventually (at 5 and 10 mg per mouse) facilitated the infection. This is especially important, since lower S3Pvac-papaya amount would be required for oral effective immunization, thus reducing the cost of the vaccine. It is important to remark that the finding of a lower dose required to induce protection by oral immunization is in agreement with other experiences in inducing mucosal immunity [50, 51].

It is interesting to remark the non-specific protection induced by subcutaneous immunization of mice with only a suspension or a soluble extract of EmpCL, as it is shown in Table 2. It is conceivable that this effect could be due by the anti-parasitic components included in papaya. In this respect, it has been reported anti-parasite effects in papaya mainly attributed to its content of carpasemina [39]. The possibility that papaya contained components that improve the non-specific immunity, resulting in an increase in resistance against cysticercosis, remains to be explored.

Individual number of cysticerci recovered from BALB/cAnN mice after 30 days of intraperitoenal infection with 10 cysticerci per mouse. P values were obtained comparing the parasite load between immunized and non-immunized control mice by the Mann Whitney test.

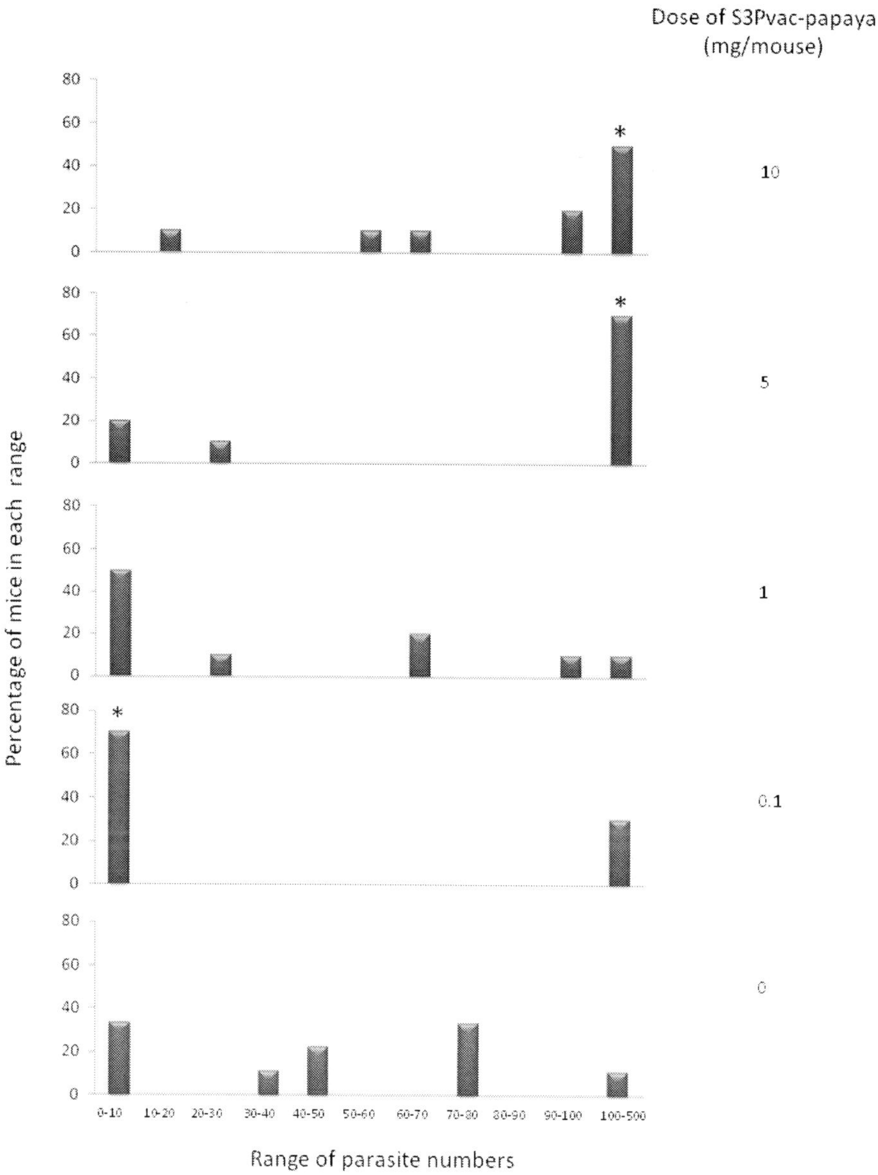

Figure 3. Distribution of parasite load recovered after 30 days of intraperitoneal infection with 10 cysts per female BALB/cAnN mouse. *Indicates significant differences with the respective percent of mice exhibiting zero ($P = 0.07$) or high number (100-500) of parasites in the control group ($P < 0.05$) using the Fisher test.

3.4.5. Oral Vaccination with Cell Suspension Elicited a Specific Proliferative Peripheral Response in Pigs

In order to determine the immunogenicity of the S3Pvac-papaya vaccine, a total of 27 rural pigs were orally immunized twice with a suspension at a dose of 750 mg (250 mg of ETgpC pKETc1_9, pKETc12_6 and pKETc7_{23}) each, every two weeks. Blood samples from pigs were obtained before immunization and peripheral cells purified in a Ficoll gradient as previously reported [24]. Purified mononuclear cells were suspended at 5 x 10^5 cells per well in RPMI 1640 media supplemented with 10% heat-inactivated fetal calf serum, 25 m M HEPES (Sigma), antibiotics (100 U/mL penicillin, 100 U/mL streptomycin), L-glutamine (2 mM) and essential amino acids (10%). Cells were then incubated with S3Pvac-papaya at 1 µg/well at 37 °C in 5% CO_2. The colorimetric assay previously reported by Mossmann was employed to determine the proliferative capacity of the mononuclear cells [52]. Briefly, forty-eight hours later 20 µL of a solution of methyl-thiazole-(5 mg/mL) was added to 200 µl of cultured cells, followed by incubation by four hours. Sodium dodecyl sulfate (20%) and 50% of N'N-dimethylformamide was added to the cultures, and they were incubated for twelve hours. Finally, the optical density (OD) was read at 570 nm. Figure 4 shows the higher level of specific proliferative response observed in pigs receiving S3Pvac-papaya with respect to those immunized with saline only. A significant increased proliferation was observed two weeks after the first immunization. These results indicate that the S3Pvac-papaya oral vaccine is immunogenic in the pigs, which are the natural intermediate host of *T. solium.* This is particularly relevant considering that the increased proliferative cellular response has been associated with the protection induced by vaccination [24]. However, the effectiveness of the S3Pvac-papaya oral vaccine remains to be tested on the field, under natural conditions of *T. solium* transmission.

Table 2. Effect of immunization with soluble extract from EmpCL against experimental murine cysticercosis

Treatment	Individual number of cysticerci per mouse (Mean ± SD)	P
Non-Immunized	493, 166, 291, 130, 69, 173, 480, 503, 293 (287 ± 168)	
Immunization route:		
Subcutaneous	0, 0, 0, 0, 0, 0, 2, 118 (15 ± 41)	0.001
Oral	0, 0, 0, 0, 122, 0, 0, 0 (15± 43.1)	0.001

3.4.6. Massive Production of S3Pvac-Papaya: Safety to the Environment

To avoid any non-desirable impact of transgenic plants on the environment, ETgpC has been maintained by propagation by *in vitro* subcultures under controlled conditions in the laboratory for five years.

Cell suspension cultures of the most promising ETgpC ($pKETc12_6$, $pKETc1_9$, $pKETc7_{23}$) expressing KETc12.6His, KETc1.6His and KETc7 peptides are being initiated for the massive production of the S3Pvac-papaya.

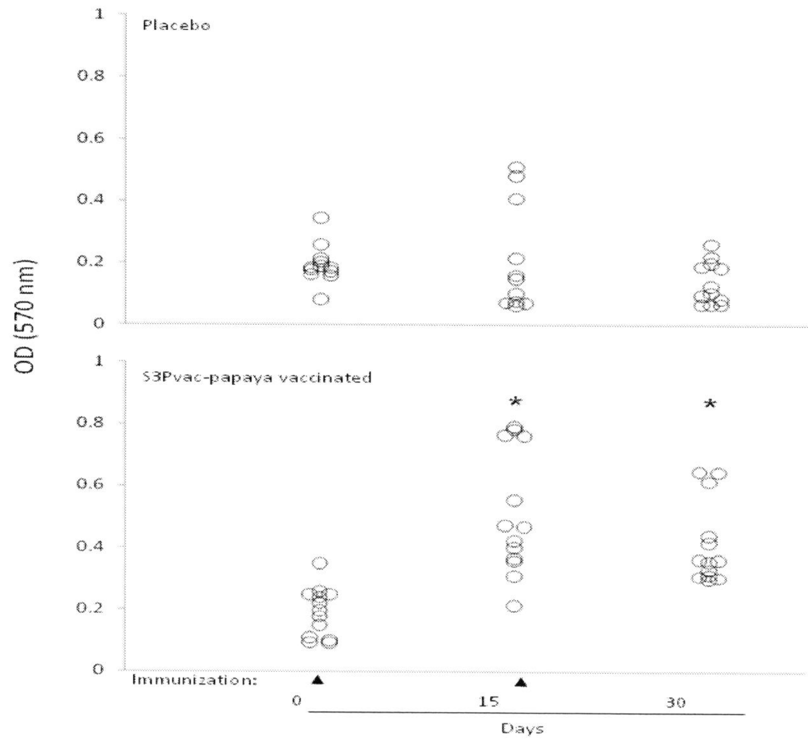

Figure 4. Increased S3Pvac-papaya specific proliferative response of peripheral mononuclear cells (PMNs) in S3Pvac-papaya immunized pigs than in non-immunized pigs. Each value corresponds to the OD readings at 570 nm of each non-immunized or immunized pigs. *Indicate significant differences of the mean of the OD with the respective controls ($P < 0.05$, Tukey-Kramer Multiple Comparisons Test).

4. Concluding Remarks and Perspectives

ETgpC proved to be a stable form to maintain the transgenes. Evidences summarized in this article indicate that S3Pvac-papaya is capable to elicit systemic immunity through oral vaccination and protection against murine cysticercosis under experimental conditions. Results also support the use of ETgpC for antigen production and delivery.

Further studies testing the immunogenic and protective properties of S3Pvac-papaya against cysticercosis in pigs rustically bred still have to be performed to validate the usefulness of this new oral vaccine.

The expression efficiency of foreign antigens ETgpC as well as papayas's intrinsic anti-parasitic properties point to its usefulness to express other parasite antigens for vaccine development against intestinal pathogens.

Acknowledgments

The authors thank Gerardo Arrellin and Georgina Díaz for technical support. Juan Francisco Rodriguez proofread the English version of this manuscript. This study was supported by the Dirección General de Asuntos de Personal Académico (IN 217908) and CONACYT (2008, 70072 and 62471-M).

REFERENCES

[1] Cripps, A. W., Kyd, J. M. & Foxwell, A. R. 2001. Vaccines and mucosal immunization. *Vaccine*, 19, 2513-2515.

[2] Walmsley, A. M. & Arntzen, C. J. 2003. Plant cell factories and mucosal vaccines. *Current Opinion in Biotechnology*, 14, 145-150.

[3] Hooper, D. C. 2009. Plant vaccines: an immunological perspective. *Current Topics in Microbiology and Immunology*, 332, 1-11.

[4] Tacket, C.O., & Mason, H. S. 1999. A review or oral vaccination with transgenic vegetables. *Microbes and Infection,* 1, 777-783.

[5] Strober, W., Kelsall, B. & March, T. 1998. Oral Tolerance. *Journal of Clinical Immunology*, 18, 1-30

[6] Kong, Q., Richter, L., Yang, Y. F., Arntzen, C.J., Mason, H. S. & Thanavala, Y. 2001. Oral immunization with hepatitis B surface antigen expressed in transgenic plants. *Proceedings of the National Academy of Sciences of the United States of America*, 98, 11539-11544.

[7] Streatfield, S. J. 2006. Mucosal immunization using recombinant plant-based oral vaccines. *Methods*, 38, 150-157.

[8] Wen, S. X., Teel, L. D., Judge, N. A., & O'Brien, A.D. 2006. Plant-based oral vaccine to protect against systemic intoxication by Shiga toxin type 2. *Proceedings of the National Academy of Sciences of the United States of America,* 103, 7082-7087.

[9] Weiner, H. L. 1994. Oral Tolerance. *Proceedings of the National Academy of Sciences of the United States of America*, 91, 10762-10765.

[10] Fleury, A., Escobar, A., Fragoso, G., Sciutto, E. & Larralde, C. 2010. Clinical heterogeneity of human neurocysticercosis results from complex interactions among parasite, host and environmental factors. *Transactions of the Royal Society of Tropical Medicine and Hygiene*, 104, 243-250.

[11] Willingham, A. L., De, N. V., Doanh, N. Q., Cong le, D., Dung, T. V., Dorny, P., Cam, P. D. & Dalsgaard, A. 2003. Current status of cysticercosis in Vietnam. *The Southeast Asian Journal of Tropical Medicine and Public Health*, 34, 35-50.

[12] Cruz, I., Cruz, M. E., Teran, W., Schantz, P. M., Tsang, V. & Barry, M. 1994. Human subcutaneous *Taenia solium* cysticercosis in an Andean population with neurocysticercosis. *American Journal of Tropical Medicine and Hygiene*, 51, 405-407.

[13] Sciutto, E., Fragoso, G., Fleury, A., Laclette, J. P., Sotelo, J., Aluja, A., Vargas, L. & Larralde, C. 2000. *Taenia solium* disease in humans and pigs: an ancient parasitosis disease rooted in developing countries and emerging as a major health problem of global dimensions. *Microbes and Infection*, 2, 1875-1890.

[14] Schantz, P. M., Wilkins, P. P. & Tsang, V. C. W. 1998. Immigrants, imaging and immunoblots: the emergence of neurocysticercosis as a major public health problem. In: *Emerging Infections 2* (edited by Scheld, Craig & Hughes). Pp. 213-242. ASM Press, US: Washington, DC.

[15] Villarán, M. V., Montano, S. M., Gonzalvez, G., Moyano, L. M., Chero, J. C., Rodriguez, S., Gonzalez, A. E., Pan, W., Tsang, V. C., Gilman, R. H., Garcia, H. H. & Cysticercosis Working Group in Peru. 2009. Epilepsy and neurocysticercosis: an incidence study in a Peruvian rural population. *Neuroepidemiology*, 33, 25-31.

[16] Villagran, U. J. & Olvera, J. E. 1988. Cisticercosis humana. Estudio clínico y patológico de 481 casos de autopsia. *Patología*, 26, 149-156.

[17] Rabiela, M. T., Lombardo Rivera, L. & Flores Barroeta, F. 1972. Cisticercosis cerebral. Análisis de 68 casos de autopsia. *Patología*, 10, 27-37

[18] Fleury, A., Gomez, T., Alvarez, I., Meza, D., Huerta, M., Chavarria, A., Carrillo Mezo, R. A., Lloyd, C., Dessein, A., Preux, P. M., Dumas, M., Larralde, C., Sciutto, E. & Fragoso, G. 2003. High prevalence of calcified silent neurocysticercosis in a rural village of Mexico. *Neuroepidemiology*, 22, 139-145.

[19] Fleury, A., Morales, J., Bobes, R. J., Dumas, M., Yánez, O., Piña, J., Carrillo-Mezo, R., Martínez, J. J., Fragoso, G., Dessein, A., Larralde, C. & Sciutto, E. 2006. An epidemiological study of familial neurocysticercosis in an endemic Mexican community. *Transactions Royal Society of Tropical Medicine and Hygiene*, 100, 551-558.

[20] Dumas, M. & Preux, P. M. 2008. [Epilepsy in tropical areas]. *Bulletin de l'Academie Nationale de Médecine*, 192, 949-960.

[21] Plancarte, A., Flisser, A., Gauci, C. G. & Lightowlers, M. W. 1999. Vaccination against *Taenia solium* cysticercosis in pigs using native and recombinant oncosphere antigens. *International Journal of Parasitology*, 29, 643-647.

[22] Huerta, M., de Aluja, A. S., Fragoso, G., Toledo, A., Villalobos, N., Hernandez, M., Gevorkian, G., Acero, G., Diaz, A., Alvarez, I., Avila, R., Beltran, C., Garcia, G., Martinez, J. J., Larralde, C. & Sciutto, E. 2001. Synthetic peptide vaccine against *Taenia solium* pig cysticercosis: successful vaccination in a controlled field trial in rural Mexico. *Vaccine*, 20, 262-266.

[23] Flisser, A., Gauci, C. G., Zoli, A., Martinez-Ocaña, J., Garza-Rodriguez, A., Dominguez-Alpizar, J. L., Maravilla, P., Rodriguez-Canul, R., Avila, G., Aguilar-Vega, L., Kyngdon, C., Geerts, S. & Lightowlers, M. W. 2004. Induction of protection against porcine cysticercosis by vaccination with recombinant oncosphere antigens. *Infection and Immunity*, 72, 5292-5297.

[24] Manoutcharian, K., Diaz-Orea, A., Gevorkian, G., Fragoso, G., Acero, G., Gonzalez, E., De Aluja, A., Villalobos, N., Gomez-Conde, E. & Sciutto, E. 2004. Recombinant bacteriophage-based multiepitope vaccine against *Taenia solium* pig cysticercosis. *Veterinary Immunology and Immunopathology*, 99, 11-24.

[25] Morales, J., Martínez, J. J., Manoutcharian, K., Hernández, M., Fleury, A., Gevorkian, G., Acero, G., Blancas, A., Toledo, A., Cervantes, J., Maza, V., Quet, F., Bonnabau, H., de Aluja, A. S., Fragoso, G., Larralde, C., Sciutto, E. 2008. Inexpensive anti-cysticercosis vaccine: S3Pvac expressed in heat inactivated M13 filamentous phage proves effective against naturally acquired *Taenia solium* porcine cysticercosis. *Vaccine*, 26, 2899-2905.

[26] Assana, E., Kyngdon, C.T., Gauci, C.G., Geerts, S., Dorny, P., De Deken, R., Anderson, G.A., Zoli, A.P. & Lightowlers, M. W. 2010. Elimination of *Taenia solium* transmission to pigs in a field trial of the TSOL18 vaccine in Cameroon. *International Journal of Parasitology*, 40, 515-519.

[27] Patarroyo, M. E., Cifuentes, G., Bermúdez, A. & Patarroyo, M. A. 2008. Strategies for developing multi-epitope, subunit-based, chemically synthesized anti-malarial vaccines. *Journal of Cellular and Molecular Medicine*, 12, 1915-1935.

[28] Cervantes-Torres, J. 2010. Vacunación contra la cisticercosis murina: caracterización inmunogénicas inducida por el péptido KETc1 sintético y recombinante utilizando diferentes sistemas de expresión Tesis. Facultad de Química,UNAM.

[29] Manoutcharian, K., Rosas, G., Hernandez, M., Fragoso, G., Aluja, A., Villalobos, N., Rodarte, L. F. & Sciutto, E. 1996. Cysticercosis: identification and cloning of protective recombinant antigens. *Journal of Parasitology*, 82, 250-254.

[30] Rosas, G., Cruz-Revilla, C., Fragoso, G., López-Casillas, F., Pérez, A., Bonilla, M. A., Rosales, R. & Sciutto, E. 1998. *Taenia crassiceps* cysticercosis: humoral immune response and protection elicited by DNA immunization. *Journal of Parasitology*, 84, 516-523.

[31] Toledo, A., Larralde, C., Fragoso, G., Gevorkian, G., Manoutcharian, K., Hernandez, M., Acero, G., Rosas, G., Lopez-Casillas, F., Garfias, C. K., Vazquez, R., Terrazas, I. & Sciutto, E. 1999. Towards a *Taenia solium* cysticercosis vaccine: an epitope shared by *Taenia crassiceps* and *Taenia solium* protects mice against experimental cysticercosis. *Infection and Immunity*, 67, 2522-2530.

[32] Toledo, A., Fragoso, G., Rosas, G., Hernandez, M., Gevorkian, G., Lopez-Casillas, F., Hernandez, B., Acero, G., Huerta, M., Larralde, C. & Sciutto, E. 2001. Two epitopes shared by *Taenia crassiceps* and *Taenia solium* confer protection against murine *T. crassiceps* cysticercosis along with a prominent T1 response. *Infection and Immunity*, 69, 1766-1773.

[33] Manoutcharian, K., Terrazas, L. I., Gevorkian, G., Acero, G., Petrossian, P., Rodriguez, M. & Govezensky, T. 1999. Phage-displayed T-cell epitope grafted into immunoglobulin heavy-chain complementarity-determining regions: an effective vaccine design tested in murine cysticercosis. *Infection and Immunity*, 67, 4764-4770.

[34] Rassy, D., Bobes, R. J., Rosas, G., Anaya, V. H., Brehm, K., Hernández, B., Cervantes, J., Pedraza, S., Morales, Villalobos, N., de Aluja, A., Laclette, J. P., Nunes, C., Biondi, G., Fragoso, G., Hernández, M. & Sciutto, E. 2010. Characterization of S3Pvac anti-cysticercosis vaccine components: implications for the development of an anti-cestodiasis vaccine. *PLoS One*, (submitted).

[35] Sciutto, E., Morales, J., Martínez, J. J., Toledo, A., Villalobos, M. N., Cruz-Revilla, C., Meneses, G., Hernández, M., Díaz, A., Rodarte, L. F., Acero, G., Gevorkian, G., Manoutcharian, K., Paniagua, J., Fragoso, G., Fleury, A., Larralde, R., De Aluja, A. S. & Larralde, C. 2007. Further evaluation of the synthetic peptide vaccine S3Pvac against *Taenia solium* cysticercosis in pigs in an endemic town of Mexico. *Parasitology*, 134, 129-133.

[36] Cabrera-Ponce, J. L., Vegas, J. A. & Herrera-Estrella, L. 1995. Herbicide resistant transgenic papaya plants produced by and efficient particle bombardments transformation methods. *Plant Cell Reports*, 15, 1-7.

[37] Cabrera-Ponce, J. L., López, L., Assad-García, C., Medina-Arevalo, N., Bailey, A. & Herrera-Estrella, L. 1997. An efficient particle bombardment system for the genetic transformation of asparagus (*Asparagus officinalis* L.). *Plant Cell Reports*, 16, 255-260.

[38] Berger, J. & Asenjom, C. F. 1940. Anthelmintic activity of crystalline papain. *Science*, 91, 387-388.

[39] Okeniyi, J. A., Ogunlesi, T. A., Oyelami, O. A. & Adeyemi, L. A. 2007. Effectiveness of dried *Carica papaya* seeds against human intestinal parasitosis: a pilot study. *Journal of Medicinal Food*, 10, 194-196.

[40] Stepek, G,. Behnke, J. M., Buttle., D. J. & Duce, L. R. 2004. Natural plant cisteine proteinases as anthelmintics? *Trends in Parasitoogyl*, 20, 322-327.

[41] Ming, R., Hou, S., Feng, Y., Yu, Q., Dionne-Laporte, A., Saw, H. J., Senin, P., Wang, W. et al., 2008. The draft genome of the transgenic tropical fruit tree papaya (*Carica papaya* Linnaeus). *Nature,* 452, 991-996.

[42] Zhang, G. L., Zhou, Z., Guo, A. P., Shen, W. T. & Li, X. Y. 2003. An initial study of transgenic *Carica papaya* used as a kind of vaccine for anti-tuberculosis. *Acta Botanica Yunnanica*, 2, 223-229.

[43] Christou, P., Ford, T. L. & Kofron, M. 1992. The development of a variety independent gene-transfer method for rice. *Trends in Biotechnology*, 10, 239-246.

[44] Sanford, J. C., De Vit, M. J., Rusell, J. A., Smith, F. D., Harpending, P. R., Roy, M. K. & Jonhson, S. A. 1991. An improvement helium-drive bioballistic device. *Techniques*, 3, 3-16.

[45] Tomes, D. T., Ross, M. C. & Songstad, D. D. 1995. Direct DNA transfer into intact plant cells via microprojectile bombardment. In: *Plant Cell Tissue and Organ Culture: Fundamental Methods* (edited by O. L. Gamborg & C. G. Phillips). Pp. 197-213. Springer-Verlag, Germany: Berlin, Heidelberg.

[46] Hernández, M., Cabrera-Ponce, J. L., Fragoso, G., López-Casillas, F., Guevara-García, A., Rosas, G., León-Ramírez, C., Juárez, P., Sánchez-García, G., Cervantes, J., Acero, G., Toledo, A., Cruz, C., Bojalil, R., Herrera-Estrella, L. & Sciutto, E. 2007. A new highly effective anticysticercosis vaccine expressed in transgenic papaya. *Vaccine,* 25, 4252-4260.

[47] Lowry, O. H., Rosebrough, N. J., Farr, S. L. & Randall, R. J. 1951. The folin phenol protein quantification method of protein, *Journal Biological Chemistry*. 193, 265-275.

[48] Molina, A. & Veramendi, J. 2005. Induction of neutralizing antibodies by a tobacco chloroplast-derived vaccine based on a B cell epitope from canine parvovirus. *Virology*, 342, 266-275.

[49] Hernández, M., Cervantes, J., Cabrera Ponce, J. L., Herrera Estrella, L., Rosas, G., Villarreal, M. L., Fragoso, G. & Sciutto, E. 2009. Development of an oral anti-cysticercosis vaccine delivered in genetically modified embryogenic callus of papaya. *Transgenic Plant Journal*, 3, In-Press.

[50] Kapusta, J., Pniewski, T., Wojciechowicz, J., Bociag, P. & Płucienniczak, A. 2010. Nanogram doses of alum-adjuvanted HBs antigen induce humoral immune response in mice when orally administered. *Archivum Immunologiae et Therapice Experimentalis*, 58, 143-151.

[51] Kostrzak, A., Cervantes Gonzalez, M., Guetard, D., Nagaraju, D. B., Wain-Hobson, S., Tepfer, D., Pniewski, T. & Sala, M. 2009. Oral administration of low doses of plant-based HBsAg induced antigen-specific IgAs and IgGs in mice, without increasing levels of regulatory T cells. *Vaccine*, 27, 4798-4807.

[52] Mossmann, T. 1983. Rapid colorimetric assay for cellular growth and survival: application to proliferation and cytotoxicity assays. *Journal of Immunological Methods*, 65, 55-63.

In: Bioprocess Sciences and Technology
Editor: Min-Tze Liong

ISBN 978-1-61122-950-9

Chapter 14

PRODUCTION OF RECOMBINANT PHARMACEUTICALS USING PLANT BIOTECHNOLOGY

M. Golovkin[*]
Biotechnology Foundation Laboratories,
Thomas Jefferson University, Philadelphia, PA 19107, USA

ABSTRACT

Progress in plant biology and genetic engineering identified plants as an excellent resource for efficient production of recombinant proteins. This new area of plant biotechnology, plant molecular farming or plant bio-factories, as it is called today, has grown over the past 20 years and become a sound technology that is most often used in serious commercialization processes today. It is finally getting recognition for its advantages and much praised potential for large scale up and low maintenance, as well as validating the alleged lower costs of production processes. The public health's view on plants as naturally safe and pathogen-free is also being embraced. The current growing interest in plant-derived vaccines and other pharmaceuticals is hopefully gaining attention to technology from the industry. The key here is to get a competitive edge with better yields and reduced costs in the downstream processes and deliver superior results with feasible products. Out of the academic settings, pharmaceutical plant biotechnology is now getting a thorough tune up. It is all about the conventional techniques and assessments, the selection of fit-host species, implementation of expedient transformation applications, and improvement of genetic engineering and targeting gears. The robust and cost-effective technology is getting much needed regulatory support which helps to resolve impending issues with clinical trials and improving the public-ill perception of presumed environmental risks. Highlights of this research are summarized here with references to the most recent reviews and pivotal literature sources hopefully making it easier for the reader to get particulars on the prevailing strategies and trends in this field.

[*] Corresponding author. E-mail: Maxim.Golovkin@jefferson.edu

1. Plants as a Novel Production System

Plants are ubiquitous and pivotal organisms cultivated by humans for centuries. They serve almost the entire specter of human needs and are the main source of food. For this plants basically require only a bare minimum: light, water and soil. Utilization of plants in recombinant biotechnology, often called plant molecular farming or plant biofactories/bioreactors - names frequently seen in the literature, is the new research area that has evolved based on recent advances in molecular biology, genetic engineering and plant physiology breeding, somatic and cell propagation techniques. A schematic representation of the concept is shown in Figure 1 that omits some steps and procedures of tedious and empirical nature.

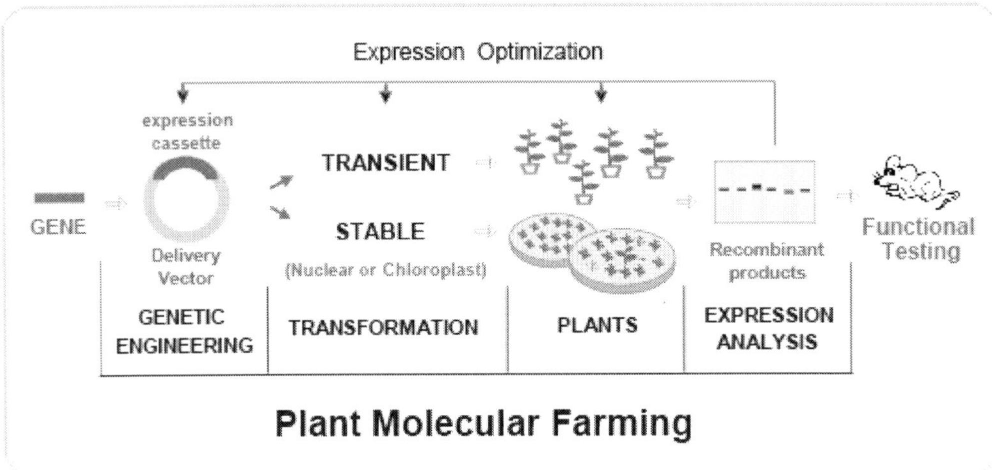

Figure 1. Schematic diagram depicting the key stages of "Plant Molecular Farming" in production of recombinant pharmaceutical proteins *in planta,* starting with gene selection, genetic engineering and expression cassette/contract arrangement, plant transformation and expression analysis/ production of recombinant protein, followed with functional testing.

Plant were identified as a promising alternative system for production of a wide range of potentially valuable recombinant proteins – pharmaceuticals: subunit vaccines, antibodies and various other therapeutic proteins. In the late 1980s, it all started with the first publications on expression of Human Growth Hormone [1], Hepatitis B surface antigen [2] and functional antibodies [3]. These studies [4, 5, 6, 7, 8] came right after the first reports on successful stable genetic transformation of plants (tobacco) were published in 1983 [9, 10].

Experimental work is primarily aimed at achieving higher yields of important therapeutic agents that display functional activity. The conventional procedure employs plant-specific expression cassettes harboring the corresponding genes and tissue specific and/or intracellular targeting signals that are assembled and tested in planta for detectable expression, followed by the functional analysis in vitro (molecular analysis) and in vivo (animals).

Until then, the recombinant production of pharmaceuticals was a prerogative of prokaryotic bacterial organisms, like E.coli; or eukaryotic yeast, insect, animal or human cell systems used as bioreactors [11, 12]. Even today, these commercial technologies remain the key production source of all recombinant pharmaceuticals and the industry behind them remains at the core of developing the new generation of recombinant products. Recombinant subunit vaccines are a new generation of vaccines that are generally comprised of immunoprotective target proteins or peptides, called antigens, which are capable of inducing strong and protective immune responses sufficient to neutralize the pathogens [13]. Research in this direction is currently getting a boost to replace the existing old vaccines (usually live/attenuated or inactivated form of vaccines developed many decades ago) and is rushing toward the expeditious development of new ones, especially ones against the emerging and reemerging infectious diseases, subunit vaccine against cancer, autoimmune diseases and more, hence the proportional sizes of the pieces from the pharmaceutical pie of 2006, see Figure 2 [14, 15].

Any contemplation for new production platforms of recombinant proteins to be taken seriously should first and specifically show the potential to reduce the cost of pharmaceutical production including, if possible, the assumed processes like storage, packaging, transport, trained personnel for injections, additional equipment and more. These are often overlooked by researchers conceiving the project, though could amass for a lot of hidden expenses. In addition, a brave new technology or production concept (platform) is also expected to help resolve the safety issues and, preferably, overcome the hurdles of scale-up and other conventional limitations, like the need for correct post translational modifications of many recombinant proteins to be functional, as well as, folding requirements and more [16, 17, 18].

This novel area of plant biotechnology has all that it is necessary to address the above-mentioned queries of low cost, better safety, ease of scaling up (Table 1) and is probably capable of addressing some other concerns, such as product instability, ease of storage, possible needle-free administration and many more.

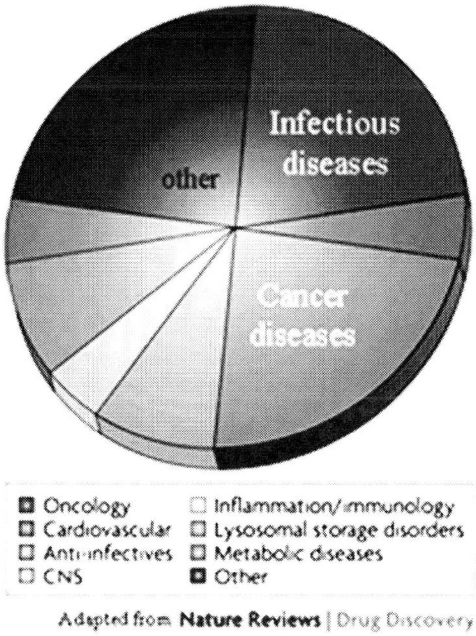

Figure 2. Important pharmaceuticals launched by the industry in 2006 [15]. The emphasis is on Infectious and Cancer diseases since both are the areas of primary interest for developers of new drugs and vaccines.

Table 1. Brief comparison of the current recombinant production systems

System	Cost	Production time	Scale up potential	Glycosylation	Contamination risk
Bacteria	Low	Short	High	None	Endotoxins
Yeast	Medium	Medium	High	Yes /partially	Low risk
Baculovirus	High	Short	Low	Yes / minor differences	Low risk
Mammalian cells	High	Long	Low	Yes	Viruses, prions, oncogenic DNA
Transgenic animals	High	Very long	Low	Yes	Viruses, prions oncogenic DNA
Transgenic plants	**Very low**	**Long**	**Very high**	**Yes / minor differences**	**Low risk**
Transient plants	**Medium**	**Medium**	**Medium**	**Yes / minor differences**	**Low risk**

This area became an essential part of plant biotechnology (Table 2), though remains a vigilant integrative research area that has brought together molecular biology, genetics, biochemistry, plant physiology, immunology, pharmacology, industrial biotechnology, agriculturale etc [6, 19 - 27].

Table 2. Production of pharmaceuticals *in planta* as a part of plant biotechnology

Modern plant biotechnology		
Agriculture	**Molecular Farming**	**Other**
Herbicide/Parasite Resistance	**Pharmaceutical proteins**	Phyto-remediation
Pathogen Resistance	Medicinal compounds	Energy sources & biofuel
Development control	Industrial Proteins	Forestry & cellulose industry
Yields & product quality	Polymers	Cosmetics
Senescence & Stress control		
Seed & Germination		
Industrial products		

2. The "Edible" Plant Vaccine Concept

In any process of recombinant production the downstream processes, firstly purification, are the major factors contributing to high cost of final products. And it is exactly where the plants, as natural bioreactors, have offered an interesting solution. Plants are usually perceived as being generally "friendly" and safe and particularly the crop plants have a great potential to become eukaryotic recombinant pharmaceutical production system that is praised for its unbiased ease of growth, harvest, storage, distribution and, in prospective, direct consumption and other easy (mucosal, ectopic etc.) applications [28, 29].

Plants have become the focus of extensive studies for vaccine development with the "edible", "oral" or "green vaccine" breakthrough approach as it is often referred to in the literature [30 - 37]. The main idea was to get a protective antigen into edible plants, like vegetables or fruits (potato tuber, banana) to be grown and consumed by animals or humans as an immune "shot". That concept was pioneered by the work of scientist, Dr. Charles Arntzen and his colleague Dr. Mason, and that has given rise to a very enthusiastic wave of studies that have brought the world's attention to this field and helped establish it as a novel branch of plant biotechnology [23, 26, 38, 39]. The concept has found its firm proof with a few successful oral feeding experiments when using some strong antigenic proteins: *E. coli* enterotoxin B – LT-B; *Cholera* toxin B – CT-B and Hepatitis B surface antigen – HBsAg [18, 40] that do not require immune stimulators – adjuvants [41, 42], the empirical agents usually necessary to get a pronounced mucosal responses. Together with another type of recombinant product - tplant-derived antibodies [43, 44, 45, 46, 47, 48], also called "plantibodies", these breakthrough studies have reached the initial stages of testing into animals and humans, as clinical trials (Table 3). Just for the record, the first officially approved trial was conducted in 1997 with LTB in potato tubers [49].

These ground-breaking studies and many other experiments that followed have shown that edible transgenic plants can produce immunologically active recombinant antigens and induce immune responses in humans and animals. Hence, plants were attested as safe and probably cheaper contenders for quality vaccine production that can be used as processed material and/or even directly in needle-free immunization [18, 27, 40, 50, 51].

Table 3. Plant-derived pharmaceuticals in clinical trials

Plant-derived pharmaceuticals in clinical testing		
Product	**Subject of Treatment**	**Source**
Vaccines		
E. coli enterotoxin B (LTB)	Diarrhea	Potato & Maize
Hepatitis B surface antigen (HBsAg)	Hepatitis B	Potato & Lettuce
Norwalk virus	Sickness / diarrhea	Potato
Rabies glycoprotein G	Rabies	Viral / spinach
Antibodies		
LSBC scFVs	Non-Hodgkin's lymphoma	Viral / tobacco
Avicidin	Colorectal cancer	Maize
CaroRX	Dental caries	Tobacco
Other products		
Gastric lipase	Cystic fibrosis, pancreatitis	Maize
Human intrinsic factor	Vitamin B12 deficiency	Arabidopsis
Lactoferrin	Gastrointestinal infections	Maize
Human Glucocerebrosidase enzyme	Gaucher disease	Carrot

By the early 2000's many field researchers came to understand that, despite its obvious appeal, the "edible" concept in its original form is rather simplistic [52]. One of the obvious obstacles for direct consumption is the overlooked issue of dosage control in the uneven quantities of recombinant protein in plant tissues (fruits, vegetables, grains etc.) given for consumption to humans or animals. Needless to say, variability of expression/accumulation levels is dependent on many empirical factors including the smart design of the expression cassettes suiting the task and plant specifics, variability of growth conditions, harvesting conditions and many other factors. In addition, if dealing with transgenic plants, foreign gene stable integration is a random, non-homologous, recombination event (see the next chapter) that does affects the foreign gene expression by a so called "gene-position" effect [53].

Other major concerns to mention here are the plausible environmental risks and public concerns [54, 55, 56, 57, 58]. That is called environmental release where edible crops, like vegetables, fruits and grains used for production of recombinant pharmaceuticals could possibly threaten the entire food chain of humans and animals by genetic cross-contamination with interrelated plant sources [59, 60]. One could only imagine the consequences of such an outcome, therefore a special regulatory approval for such pharmaceuticals was put in place [61, 62]. Presently it is all implemented in the form of "current good manufacturing practice" (cGMP) aimed to guide the entire research/production process (see www.fda.gov/downloads/drugs/.../guidances/ucm070273.pdf). That was another important outcome, the development of much needed regulatory measures that are ordained specifically to the plant-based pharmaceutical products. It was done at the regional and national levels, US, Canada, European Union and some other countries, as well as the international levels, like World Health Organization and others.

3. Plant Transformation and Expression Methods

Transformation and expression systems contribute significantly to the project feasibility with the yield, stability, functional quality and many other economical parameters. The right choice of plant is important and could be crucial presently for commercial project success. The major transformation strategies presented here are defined by the means of the delivery of the genetic material into plant cells that is done either by stable integration of trans-gene into the nuclear genome [53, 63, 64], or into the plant-specific organelle - plastid (chloroplast) genome [65 - 69]. The other process is transient, non-integrative, transformation methods of delivery and expression of genetic information that utilizes plant viruses as vectors [7, 70 - 72]. These plant transformation systems and their key features are presented for comparison in Table 4. Please note, that each method has its benefits and drawbacks and all methods are currently widely used in plant biotechnology, as well as for development of new recombinant pharmaceuticals.

The most popular transformation method today is the Agrobacterium-mediated approach. The method employs the natural plant pathogen *Agrobacterium tumefaciens* that efficiently attaches to plant cells (mostly dicots and some monocots) followed by transfer of a segment of its coding region, T DNA, by random non-homologous recombination integration events into the nuclear genome. For some recalcitrant (hard to transform) plants it is possible to use other methods of direct transformation, DNA uptake delivery, like "biolistic" transformation, that essentially is a microprojectile bombardment of plant cells/tissues with the DNA-bound metal micro-particles. Please note that this method is also frequently used in the labs for a fast (24-48 hours) transient expression-evaluation of expression cassette variants. One should not confuse it with the transient plant-virus system.

Table 4. Plant transformation and expression systems

	Major System	Expression levels	Advantages	Disadvantages
STABLE	**Nuclear**		inheritable gene integration broad host range	low expression & accumulation
	constitutive expression	Modest	possible direct consumption plant-specific PTM (glycosylation)	lengthy procedure positional effect & gene silencing
	transgenic plants & cell cultures	< **1 %** TSP (Seeds >1%)	high yield and scale-up	environmental risks
	Chloroplast		high expression & accumulation	prokaryotic machinery
	constitutive expression	High	inheritable gene integration no gene transfer though pollen	no glycosylation narrow host range somatic propagation
	transplastomic plants	> **5%** TSP	position effect & gene silencing scale-up capacity	lengthy procedure
TRANSIENT	**Plant virus**		high expression & accumulation	not inheritable some hosts restrictions
	transient expression	High	easy molecular manipulations simple and quick transformation	need of purification (endotoxins)
	/modern systems/ wild-type plants	>> **10%** TSP	short production times high scale-up capacity less environmental risks	some environmental risks

Once transgenic lines of plants or cells are selected, they are called "genetically stable" and used further for expression analysis and, if positive, for mass propagation. Thus, the best-producing line(s) are becoming an infinite source (seeds or somatic propagation) of the low-cost biomass production. However, some problem in genetic propagation could occur that could lead sometimes to a complete cessation of recombinant production. The main reasons for that are known as gene silencing or complex genetic rearrangements (due to position effect, multiple integration events etc.) that also usually cause the loss of function, or dramatic variability of gene expression in offspring's [23, 53]. That is why the empirical analysis of large numbers of independent transgenic lines after each stable transformation has become a necessary routine. Many other important issues also should be taken into consideration that could be helped by the modern tools of molecular biology to improve the expression, prevent the degradation of the product and help to target peptides into various plant tissues and cells [73, 74]. The current molecular biology and genetic engineering strategies that help to build the potent expression cassettes are shown in Table 5.

Table 5. Considerations for design of expression-cassettes

Genetic Engineering of Plant Expression Cassettes
Coding region: full size, parts, epitopes
Codon optimization for empirical testing
Promoters, enhancers, 3'- polyadenylation signals, introns etc.
Intra- and inter-cellular targeting signals
Extra hinges (fusions), affinity tags (purification) etc.
Identification/selection markers and reporters

Together with codon optimization of target genes and their parts researchers can use the synthetic genes and fusions, as well as, a plethora of specific regulatory elements, such as constitutive and inducible promoters, enhancers that are aiding in tissue/organ compartmentalization of recombinant products into plant leaves, stems; roots, tubers, fruits, seeds [18, 75, 76] and cell cultures [23, 77, 78]. Some researchers tends to believe that plant cell-wall compartmentalization could also assist with the protection of vaccine components from degradation in digestive tract and other mucosal surfaces helping with anticipated therapeutic response. The use of specific targeting signals has also allowed scientists to achieve direct expression/accumulation of recombinant products in different plant organelles or cell-compartments, such as plant-specific chloroplasts, mitochondria, storage vacuoles, cytoplasm, endoplasmic reticulum, apoplast. It could be used to help to secrete the products out from the cell as root exudates in hairy root cultures [18, 76]. A wide range of therapeutic proteins have been already expressed in transgenic plants using these systems [79]. Many active constitutively expressed recombinant proteins have been reported, though most of them are associated together by the relatively low yields and accumulation of the products upon recovery, usually way below the economically feasible value of 1% of total soluble protein (TSP), with exception for seeds accumulating far more then 1% of TSP. Hence, the additional enrichment, concentration and/or purification of the hereby products are usually necessary, adding to the cost.

This biolistic transformation has become the method of choice mostly for the integrative transformation of plant organelles, like chloroplasts. Expression of the first case of recombinant antigenic product in "transplastomic" plants was reported from Dr. Daniell's group in 2001 [32]. The chloroplast-based systems offer significantly higher yields that could be ten times more than nuclear 1% TSP. That is what makes this technology well suited for possible direct mucosal delivery, including the oral route [65, 66, 68, 80, 81]. Nowdays the "transplastomic" technology is claiming the main plant-production-platform advantages, adding also to the post research and development production cost-efficiency and the overall safety. This method allows for any foreign genes to integrate into the chloroplast genome by site-specific homologous recombination unlike the nuclear transformation. Hence, there is no obstructive genetic and expression deviation, like gene-silencing, in transplastomic lines [69, 82]. To date, more then twenty vaccine candidates were reported to be successfully expressed in chloroplasts. Just a few technological drawbacks to mention here. One is the prokaryotic type of expression environment it provides. The chloroplast organelles are essentially the archae-bacteria and the proteins there are not glycosylated at all. Another problem is the current narrow range of plant hosts that were developed as production platforms, the major ones are tobacco, carrot and recently lettuce [81].

Expression of foreign proteins using plants as transient vectors started with a short epitopes expressed as part of viral coat proteins usually in tobacco hosts [83]. The development of modern plant viral transformation systems has significantly improved since then. Nowdays the product is not the 25 amino acid epitopes as they used to be, but full size functional proteins. The entire process also is much faster due to the use of Agro-based delivery and production yields are much higher. This method of transformation specifically relies on efficient self-replication of recombinant virus templates that can be delivered to plant cells either directly, initially as RNA inoculums, or as in the recent technological developments, via the Agro-infiltration treatments. The first commercial modern Agro-transient system "MagnIcon" is comprised of multiple plant-virus-based modules. It was first reported on the highest yields of production of recombinant protein achieved by any transient system [84]. With that the long-standing system-specific drawback of acute size limitation was removed. The other hurdle to be considered these days is the removal of the endotoxins, the leftovers from the Agrobacteria inoculums, hat is currently the downstream purification almost a requirement. Nevertheless, in some particular cases it could be avoided and recombinant vaccines can be produced for direct mucosal applications in the form of virus-like particles (VLPs). They are easy to harvest and concentrate. Several commercially-developed transient systems are currently reported in use [51, 70, 71, 85, 86].

4. The Choice of Plants and Recombinant Targets

It is that obvious that there is no ideal plant for the job in terms of recombinant pharmaceuticals production. If transformable, the plant for direct consumption or for veterinary use, for example, should be an edible crop; though for purification purposes it should be a plant without all the starches or other obstructing compounds. The other reflection should be about the tissues and plant organs. The green leaf tissues utilized frequently by many may not be the best choice after all the better possible large biomass alternatives are probably seeds/grains plants that are currently praised by many [75]. In addition, plant individual cell as bioreactors - in the form of suspensions or hairy root cultures, for example, could be a better choice for contained industrial production [87].

It is well known that plants and plant cells are all offering the eukaryotic-type of protein production machinery and only the efficiency of expression and accumulation is different. In the case of chloroplast-transformed (transplastomic) plants, the higher production yields are imminent and cost efficient, though it might not be enough and would not compensate for the possible loss of function (immunological activity) due to the lack of glycosylation modifications (Table 6). Therefore, it is always important to beforehand attest as many parameters as possible to help the decision on expression system for a particular protein target [23]. The proteins usually originate from different (heterologous) types of organisms, other than plants, hence, they are often named the "foreign" genes. The recombinant proteins might be in monomeric or multimeric forms and therefore could have a complex folding assembly that is especially important when it is essential for its proper activity and efficacy. With all that said it is for the plant of selected specie to do the work and help prove that the researcher's choice was right. Many plants were already explored for the purpose of pharmaceutical production [79]. A selection of some plants that were used successfully and reported in peer-review publications are presented here, starting with the model plant Arabidopsis and followed by the conventional crop plants (Table 7).

Table 6. Effects of glycosylation on recombinant product

The Role of Glycosylation on Recombinant Protein Production	
Role/effect	**Comment**
Folding	Affects protein secondary structure
Targeting/trafficking	Sorting/directing of protein to its final destination
Ligand recognition/binding	Affects antibody function
Biological activity	Essential to the activation of some proteins
Stability	Help enhancing protein solubility, protecting from proteolysis etc.
Half-life	Help to increase/decrease a glycoprotein' clearing from the system
Immunogenicity	Plant glycans (fucose and xylose residues) could affect immune responses

Table 7. Plant hosts for biopharmaceutical production

Plant Species	Characteristics
Model / research	strong scientific background
	available molecular & genetic material
Arabidopsis thaliana	available mutants & databases
	easy transformation and lab manipulations
	low biomass
Leafy plants	easy to harvest high biomass
	established transformation
Tobacco (model / crop)	scale-up potential
Alfalfa, Clover, Spinach,	somatic propagation
Lettuce, Collard	mucosal vaccine potential (exc. Tobacco)
	low-alkaloid tobacco available
Vegetables	large vegetable biomass
	storage & processing potential
Tomato, Carrot, Potato,	scale-up potential
Cauliflower, Squash	established transformations
	human and veterinary vaccine potential
	long product wait
	protein accumulation instability
Cereals (grains)	high expression/accumulation yields

	protein stability in seeds
Maize/corn, Rice, Wheat,	long-term storage
Barley	established transformations
	mucosal vaccine potential
Legumes	high protein content
	economic growth
Soybean, Pea, Peanut	coated seed accumulation
	established transformations
	human and veterinary vaccine potential
Oil crops	oleosin-fusion platform
Canola, *Camelina sativa*	sprouting system
Simple plants	easy containment
Moss/*Physcomitrella*	clonal propagation
Unicellular/*Chlamydomonas*	possible secretion
	regulatory compliance
Aquatic plant	bioreactor-type
Duckweed/*Lemna*	

Most of the pharmaceutical proteins are expressed *in planta* in their original or genetically modified (optimized, chimeric, as fusions or tagged) form. The choices of the target are entirely up to the scientist(s), who earnestly try to do a proper preliminary evaluation of the targets from different angles and, once chosen, consider the best way to produce it *in planta.* The molecular design of expression cassettes is really dependent on the choice of expression system reviewed [23, 63]. It is still necessary to do many experimental expression tests in plants for each individual cassette to get the answers which is best usually at a mini-scale level and transiently (expedite). There is a great assortment of modern molecular tools to help with genetic arrangements (Table 5). They could be useful in the view on the type of the target protein; its size, solubility, folding parameters; some special considerations - glycosylation and/or other post-translational modifications; plus one can make the fusions or chimeras to help enhance the performance, and/or add the special tags to ease detection and purification processes (Table 6) [17, 88, 89].

A vast variety of target proteins (~150) were already tested for production *in planta* and even more products are getting there through the current research pipeline. The target selection of some proteins was done simply because of their obvious and immediate benefit for humans or in veterinary applications. Some other targets were identified by the anticipated promise they hold to deliver the benefits in the future. Recently many targets have been chosen because of the enormous commercial or, near potential, interest they hold. Please regard that with any choices it is still a real hardship to get it right.

As mentioned before, the subunit vaccine components have two largest areas of interest: cancer and infectious disease (Figure 1). And the corresponding target proteins for recombinant production are usually originating from viruses or bacteria that are causing diseases like Hepatitis B, HPV, Rabies, Herpes, HIV, Norwalk virus, Rotavirus, RSV, SARS, Measles, Cholera, Tetanus, Diphtheria, Pertussis, staphylococcus etc. [50, 79]. More recently some other agents against the bioterrorism have come on the agenda such as: anthrax, smallpox, plague, haemorrhagic fever, Denge and more.

From all the pharmaceuticals, the antibodies have become the real favorites of plant biofarming. The current production of Abs in mammalian systems has proven to be costly and is posing a major limitation for its wide practical use. It was rapidly discovered that any type of antibodies (therapeutic, prophylactic or diagnostic) and in various forms (full sized, single-chain and chimeras etc.) can be readily expressed in plants at relatively higher yields holding its functional properties uncompromised. With much pioneering work in the field, antibodies are thought to become the first plant-based recombinant products approved for human use and most probably for cancer treatment. So far, plant-derived antibodies expressed *in planta* were against the following diseases: dental caries, rheumatoid arthritis, cholera, E.coli diarrhea, malaria, certain cancers, Norwalk virus, HIV, rhinovirus, influenza, Hepatitis B virus, and herpes simplex virus, and many others [35, 47, 90, 91, 79].

Plants were also efficiently employed to produce a wide range of other types of recombinant proteins of general medicinal or commercial interest, the so called therapeutic agents, like epidermal growth factor, trichosanthin, serum albumin, interferons, interleukins, lysozyme, lactoferrin etc. (Tables 1, 2, 3) [50, 79].

5. Future Perspectives

A recent inspiring notice has come in the year 2006 and brought us long anticipated news about the approval of the first US plant-based (tobacco) recombinant vaccine against Newcastle disease in chickens [58]. This was mostly a regulatory breakthrough, but also a huge victory for the field, since it officially confirmed that plant vaccines are safe and efficacious at least for veterinary use. All the advancements that followed have shifted the research priorities toward the key issues of crucial high yields and design of downstream processing/ purification stages. They have to be now executed preferably according to the cGMP rules and regulations ought to secure commercial product acceptance [55, 92, 60].

With all that said, the main paradigm of the plant biotechnological advantages is well afloat, thanks to the compartmentalization in transgenic seeds/grains (ideal accumulation/storage system) and the chloroplast research, both having the highest accumulation yields and excellent safety record [18, 76, 79]. More researchers now are in a quest for some kind of mucosal vaccine endeavor *in planta*, hence, they are extending their research to find the adjuvants that would help them make their formulations evoke formidable immune responses. Moreover, some are exploring the possibility of producing adjuvants also *in planta* as a separate product, or linked together with the vaccine component [42]. The "green vaccine" is now mostly recognized as a private case of mucosal vaccine development, linking it together with the newly emerging interest for ectopic and transdermal drug biopharmaceutical products [41, 93, 94]. On that note, the other simple adjustment to the old concept could be the enrichment of plant-based products like VLP's, or a simple dry or dry/freezing of the biomass helping to normalize the active recombinant components (e.g., needed at low concentration microbicides) before mixing them with the adjuvants or other agents. The final products could then be seen as extracts, powders, capsules, tablets, ointments, or any other feasible forms that may also be rising from the yet to be explored medicinal plants, see Table 8 [18, 51, 52, 79, 81, 91, 95, 96].

Table 8. Outlines of the plant-based vaccine future perspectives

Future of Plant-based Vaccines
Cost-competitive & environmentally friendly
Products made from fresh & processed biomass
Normalized vaccine dosages & obligatory cGMP quality
Liquids, powders, tablets, capsules, ointments for direct delivery
Ease of storage & downstream purification
Composite: mixes with adjuvants & other agents, pharmaceuticals
Transient approach as alternative to GMOs
Transplastomic technology potential for direct oral vaccines
The use of specialized, genetically enhanced & medicinal plants

In conclusion, the last two decades have seen the plant pharmaceutical biotechnology rising to commercial and public success. Scientists remain cautiously optimistic that the latest developments in the field will bring interest from the pharma industry together with much needed support to establish plant biofarming worldwide. The growing need for modern drugs and vaccines should also help plant biofarming to move from "holding a huge promise" to actual production of high-quality and much-needed medicinal proteins that are safer and cheaper. And that would be a chance to finally direct long anticipated help toward the resource-limited regions and, hopefully, to the entire world as well.

REFERENCES

[1] Barta, A., Sommergruber, K., Thompson, D., Hartmuth, K., Matzke, M. A. & Matzke, A. J. M. 1986. The expression of a nopaline synthase-human growth hormone chimaeric gene in transformed tobacco and sunflower callus tissue. *Plant Molecular Biology,* 6, 347-357.

[2] Mason, H. S., Lam, D. M. & Arntzen, C. J. 1992. Expression of hepatitis B surface antigen in transgenic plants. *Proceedings of the National Academy of Sciences U.S.A,* 89, 11745-11749.

[3] Hiatt, A., Caffertey, R. & Bowdish, K. 1989. Production of antibodies in transgenic plants. *Nature,* 342, 76-78.

[4] Giddings, G., Allison, G., Brooks, D. & Carter, A. 2000. Transgenic plants as factories for biopharmaceuticals. *Nature Biotechnology,* 18, 1151–1155.

[5] Doran, P. M. 2000. Foreign protein production in plant tissue cultures. *Current Opinion in Biotechnology,* 11, 199–204.

[6] Fisher, R. & Emans, N. 2000. Molecular farming of pharmaceutical proteins. *Transgenic Research,* 9, 279-299.

[7] Koprowski, H. & Yusibov, V. 2001. The green revolution: plants as heterologous expression vectors. *Vaccine,* 19, 2735-2741.

[8] Streatfield, S. J., Jilka, J. M., Hood, E. E., Turner, D. D., Bailey, M. R., Mayor, J. M., Woodard, S. L., Beifuss, K. K., Horn, M. E., Delaney, D. E., Tizard, I. R. & Howard, J. A. 2001. Plant-based vaccines: unique advantages. *Vaccine,* 19, 2742-2748.

[9] Zambryski, P., Joos, H., Gentello, C., Leemans J., Van Montagu, M. & Schell, J. 1983. Ti-plasmid vector for the introduction of DNA into plant cells without alteration of their normal regeneration capacity. *EMBO Journal,* 2, 2143-2150.

[10] Fraley, R. T., Rogers, S. G., Horsch, R. B., Sanders, P. R., Flick, J. S., Adams, S., Bittner, M., Brand, L., Fink, C. L., Fry, J., Galluppi, G., Goldberg, S., Hoffman, N.L. & Woo, S. 1983. Expression of bacterial genes in plant cells. *Proceedings of the National Academy of Sciences* of *United States of America,* 80, 4803-4807.

[11] Dertzbaugh, M. 1998. Genetically Engineered Vaccines: An Overview. *Plasmid,* 39, 100-113.

[12] Hansson, M., Nygren, P-A & Stahl, S. 2000. Design and production of recombinant subunit vaccines. *Biotechnology and Applied Biochemistry,* 32, 95-107.

[13] Bae, K., Choi, J., Jang, Y., Ahn, S. & Hur, B. 2009. Innovative vaccine production technologies : the evolution and value of vaccine production technologies. *Archives of Pharmacal Research,* 32, 465-480.

[14] Ulmer, J. B., Valley, U. & Rappuoli, R. 2006. Vaccine manufacturing: challenges and solutions. *Nature Biotechnology,* 24, 1377-1383.

[15] Owens, J. 2007. 2006 drug approvals: finding the niche. *Nature Reviews Drug Discovery,* 6, 99-101.

[16] Twyman, R. M., Schillberg, S. & Fischer, R. 2005. The transgenic plant market in the pharmaceutical industry. *Expert Opinion on Emerging Drugs,* 10, 185–218.

[17] Walsh, G. & Jefferis, R. 2006. Post-translation modifications in the context of therapeutic proteins. *Nature biotechnology,* 24, 1241-1252.

[18] Tiwari, S., Verma, P. C., Singh, P. K. & Tuli, R. 2009. Plants as bioreactors for the production of vaccine antigens. *Biotechnology Advances,* 27, 447-467.

[19] Giddings, G. 2001.Transgenic plants as protein factories. *Current Opinion in Biotechnology,*12, 450-454.

[20] Twyman, R. M., Stoger, E., Schillberg. S., Christou. P. & Fischer, R. 2003. Molecular farming in plants: host systems and expression technology. *Trends in Biotechnology,* 21, 570–577.

[21] Horn, M. E., Woodward, S. L. & Howard, J. A. 2004. Plant molecular farming: systems and products. *Plant Cell Reports,* 22, 711–720.

[22] Fischer, R., Stoger, E., Schillberg, S., Christou, P. & Twyman, R. M. 2004. Plant-based production of biopharmaceuticals. *Current Opinion in Plant Biology,* 7, 152-158.

[23] Schiermeyer, A., Dorfmuller, S. & Schinkel, H. 2004. Production of pharmaceutical proteins in plants and and plant cell suspension cultures. In: Fischer, R. & Schillberg, S. (eds) Molecular Farming: Plant-made Pharmaceuticals and Technical Proteins. Wiley, Weinheim, pp.91-109.

[24] Giddings, G., Allison, G., Brooks, D. & Carter, A. 2005. Transgenic plants as factories for biopharmaceuticals. *Nature Biotechnology,* 18, 1151-1155.

[25] Dus Santos, M. J. & Wigdorovitz, A. 2005. Transgenic plants for the production of veterinary vaccines. *Immunology and Cell Biology,* 83, 229–238.

[26] Ma, J. K., Chikwamba, R., Sparrow, P., Fischer, R., Mahoney, R., Twyman, R. M. 2005 Plant-derived pharmaceuticals-the road forward. *Trends in Plant Science,* 20, 580-585.

[27] Ma, J. K., Barros, E., Bock, R., Christou, P., Dale, P. J., Dix, P. J., Fischer, R., Irwin, J., Mahoney, R., Pezzotti, M., Schillberg, S., Sparrow, P., Stoger, E. & Twyman, R. M. 2005. Molecular farming for new drugs and vaccines. Current perspectives on the production of pharmaceuticals in transgenic plants. *EMBO Reports,* 6, 593–599.

[28] Graff, G. & Moschini, G. 2004. Pharmaceuticals and industrial products in crops: economic prospects and impact on agriculture. *Iowa Ag Review,* 10, 4-11.

[29] Boehm, R. 2007. Bioproduction of Therapeutic proteins in the 21st century and the role of plants and plant cells as production platforms. *Annals of the New York Academy of Sciences,* 1101, 121-134.

[30] Mor, T. S., Gomez-Lim, M. A. & Palmer, K. E. 1998. Perspective: edible vaccines – a concept coming of age. *Trends in Microbiology,* 6, 449–453.

[31] Tacket, C. O. & Mason, H. 1999. A review of oral vaccination with transgenic vegetables. *Microbes and Infection,* 1, 777-783.

[32] Daniell, H., Streatfield, S. J. & Wycoff, K. 2001. Medical molecular farming: production of antibodies, biopharmaceuticals and edible vaccines in plants. *Trends in Plant Science,* 6, 219-226.

[33] Mason, H. S., Warzecha, H., Mor, T. & Arntzen C. J. 2002. Edible plant vaccine: applications for prophylactic and therepeutic molecular medidcine. *Trends in Molecular.Medicine,* 8, 324-329.

[34] Walmsley, A. M. & Arntzen, C. J. 2003. Plant cell factories and mucosal vaccines. *Current Opinion in Biotechnology,* 14, 145–150.

[35] Ma, J. K. C., Drake, P. M. W. & Christou, P. 2003. The production of recombinant pharmaceutical proteins in plants. *Nature Genetics,* 4, 794-805.

[36] Korban, S. S., Krasnyanski, S. F. & Buetow, D. E.. 2002. Foods as production and delivery vehicles for human vaccines. *Journal of the American College of Nutrition,* 21, 212-217.

[37] Streatfield, S. J. & Howard, J. A. 2003. Plant-based vaccines. *International Journal of Parasitology,* 33, 479–493.

[38] Goldstein, D. A. & Thomas, J. A. 2004. Biopharmaceuticals derived from genetically modified plants. *QJM: An International Journal of Medicine*, 97, 705-716.

[39] Teli, N. P. & Timko, M. P. 2004. Recent developments in the use of transgenic plants for the production of human therapeutics and biopharmaceuticals. *Plant Cell, tissue and Organ Culture,* 79, 123-145.

[40] Streatfield, S. J. 2006. Mucosal immunization using recombinant plant-based oral vaccines. *Methods,* 38, 150-157.

[41] Holmgren, J. & Czerkinsky, C. 2005. Mucosal immunity and vaccines. *Nature Medicine,* 11, 45-53.

[42] Reed, S. G., Bertholet, S., Coler, R. N. & Friede, M. 2008. New horizons in adjuvants for vaccine development. *Trends in Immunology,* 30, 23-32.

[43] Stoger, E., Sack, M., Fischer, R. & Christou, O. 2002. Plantibodies: application, advantages and bottnecks. *Current Opinion in Biotechnology,* 13, 161-166.

[44] Stoger, E., Sack, M., Perrin, Y. & Vaquero, C. 2002. Practical considerations for pharmaceutical antibody production in different crop systems. *Molecular Breeding,* 9, 149–158.

[45] Hood, E. E., Woodard, S. L. & Horn, M. E. 2002. Monoclonal antibody manufacturing in transgenic plants myths and realities. *Current Opinion in Biotechnology,* 13, 630–635.

[46] Fischer, R., Twyman, R. M. & Schillberg, S. 2003. Production of antibodies in plants and their use for global health. *Vaccine,* 21, 820-825.

[47] Ko, K & Koprowski, H. 2005. Plant biopharming of monoclonal antibodies. *Virus Research,* 111, 93-100.

[48] Wycoff, K. L. 2005. Secretory IgA antibodies from plants. *Current Pharmaceutical Design,* 11, 2429-2437.

[49] Tacket, C. O., Mason, H. S., Losonsky, G., Clemens, J. D., Levine, M. M. & Arnthen, C. J. 1998. Immunogenicity in humans of a recombinant bacterial antigen delivered in a transgenic potato. *Nature Medicine,* 4, 607-609.

[50] Floss, D. M., Falkenburg, D. & Conrad, U. 2007. Production of vaccines and therapeutic antibodies for veterinary applications in transgenic plants. *Transgenic Research,* 16, 315-332.

[51] Yusibov, V. & Rabindran S. 2008. Recent progress in the development of plan-derived vaccines. *Expert Reviewsof Vaccines,* 7, 1173-1183.

[52] Rybicki, E., 2009. Plant-produced vaccine: promise and reality. *Drug Discovery Today,* 14, 1 16-22.

[53] Slater, A., Scott, N. & Fowler, M. 2003. Plant Biotechnology. The Genetic Manipulation of Plants. Oxford, UK: Oxford University Press. 346 p.

[54] Commandeur, U., Twyman, R. M. & Fischer, R. 2003. The biosafety of molecular farming in plants. *AgBiotechNet,* 5, 1–9.

[55] Mascia, P. N. & Flavell, R. B. 2004. Safe and acceptable strategies for producing foreign molecules in plants. *Current Opinion in Plant Biology,* 7, 189–195.

[56] Peterson, R. K. D. & Arntzen, C. J. 2004. On risk and plant-based biopharmaceuticals. *Trends in Biotechnology,* 22, 64–66.

[57] Kirk, D. D., McIntosh, K., Walmsley, A. M. & Peterson, R. K. 2005. Risk analysis for plan-made vaccines.*Transgenic Research,* 14, 449-462.

[58] Vermij, P. & Waltz, E. 2006. USDA approves the first plant-based vaccine. *Nature,* 24, 233–234.

[59] Sparrow, P. A. & Twyman, R. M. 2009. Biosafety, risk assessment and regulation of plant-made pharmaceuticals. *Methods molecular Biology,* 483, 341-353.

[60] Breyer, D., Goossens, M., Herman, P. & Sneyers, M. 2009. Biosafety considerations associated with molecular farming in genetically modified plants. *Journal of Medicinal Plant Research,* 3, 825-838.

[61] FDA. 2002. Draft guidance for industry. Drugs, biologics, and medical devices. Derived from bioengineered plants for use in humans and animals. URL

[62] http://www.fda.gov/downloads/Drug/GuidanceComplianceRegulatoryInformation/Gui dances 124811.pdf

[63] USDA. 2008. Guidance for APHIS Permits for field testing or movement of organisms intended for pharmaceutical or industrial use. URL http://www.aphis. usda.gov/brs/pdf/Pharma_Guidance.pdf

[64] Sala, F, Rigano, M. M., Barbante, A., Basso, B., Walmsely, A. M. & Castiglioni, S. 2003. Vaccine antigen production in transgenic plants: strategies, gene constructs and perspectives. *Vaccine,* 21, 803–808.

[65] Dunwell, J. M. 2005. Transgenic crops: the current and next generations, *Methods in Molecular Biology,* 286, 377-398.

[66] Maliga, P. 2002. Engineering the plastid genome of higher plants. *Current Opinion in Plant Biology,* 5, 164–172.

[67] Maliga, P. 2003. Progress towards commercialization of plastid transformation technology. *Trends in Biotechnology,* 21, 20–28.

[68] Daniell, H., Khan, M. S. & Alisson, L. 2002. Milestones in chloroplast genetic engineering: an environmentally friendly era in biotechnology. *Trends in Plant Science,* 7, 84-91.

[69] Daniell, H. 2006. Production of biopharmaceuticals and vaccines in plants via the chloroplast genome. *Biotechnology Journal,* 1, 1071-1079.

[70] Verma, D. & Daniell, H. 2007. Chloroplast vector systems for biotechnology applications. *Plant Physiology,* 145, 1129-1143.

[71] Gleba, Y., Klimyuk, V. & Marillonnet, S. 2005. Magnifection-a new platform for expressing recombinant vaccines in plants. *Vaccine,* 23, 2042-2048.

[72] Marillonet, S., Giritch, A., Gils, M., Kandzia, R., Klimyuk, V. & Gleba, Y. 2005. *In planta* engineering of viral RNA replicons: efficient assembly by recombination of DNA modules delivered by *Agrobacterium. Proceedings of the National Academy of Sciences* of *United States of America,* 101, 6852-6857.

[73] Jones, H. D., Doherty, A. & Sparks, C. 2009. Transient transformation of plants. Methods in Molecular Biology, In: *Plant Genomics: Methods and Protocols* (edited by D. J. Somers, P. Langridge & J. P. Gustafson). Pp. 131-152. New Jersey, USA: Humana Press.

[74] Doran, M. D. 2006. Foreign protein degradation and instability in plant tissues cultures. *Trends in Biotechnology,* 24, 426-432.

[75] Lienard, D., Sourrouille, C., Gomord, V. & Faye, L.. 2007. Pharming and transgenic plants. *Biotechnology Annual Review,*13, 115-47.

[76] Stoger, E., Ma, J. K-C., Fischer, R. & Christou, P. 2005. Sowing the seeds of success: pharmaceutical proteins form plants. *Current Opinion in Biotechnology,* 16, 167-173

[77] Miao, Y., Ding, Yu., Sun, Q-Y., Xu, Z.. F. & Jiang, L. 2008. Plant bioreactors for pharmaceuticals. *Biotechnology and Genetic Engineering Reviews,* 25, 363-380.

[78] Hellwig, S, Drossard, J., Twyman, R. M. & Fischer, R. 2004. Plant cell cultures for the production of recombinant proteins. *Nature Biotechnology,* 22, 1415-1422.

[79] Plasson, C., Michel, R., Lienard, D., Saint-Jore-Dupas, C., Sourrouille, C., de March, G. G. & Gomord, V. 2009. Production of recombinant proteins in suspension-cultured plant cells. *Methods in Molecular Biology,* 483, 145-161.

[80] Daniell, H., Singh, N., Mason, H. & Streatfield, S. 2009. Plant-made vaccine antigens and biopharmaceuticals. *Trends in Plant Science,* 14, 669-79.

[81] Daniell, H., Chebolu, S., Kumar, S., Sigleton, M. & Falconer, R. 2005. Chloroplast-derived vaccine antigens and other therapeutic proteins. *Vaccine,* 23, 1779-1783.

[82] Davoodi-Semiromi, A., Samson, N. & Daniell, H. 2009. The green vaccine: A global strategy to combat infectious and autoimmune diseases. *Human vaccines,* 5, 488-93.

[83] Grevich, J. J. & Daniell, H. 2005. Chloroplast genetic engineering: Recent advances and future perspectives. *Plant Sciences,* 24, 83-107.

[84] Haynes, J. R., Cunningham, J., von Seefried, A., Lennick, M., Garvin, R. T. & Shen, S. H. 1986. Development of genetically engineered, candidate polio vaccine employing the self-assembling properties of the tobacco mosaic virus coat protein. *Nature Biotechnology,* 4, 637-641.

[85] Santi, L., Girich, A., Chad, R., Marillonnet, S., Klimyuk, V., Glebs, Y., Webb, R., Arntzen C. & Mason, H. S. 2006. Protection conferred by recombinant *Yersinia pestis* antigens produced by a rapid and highly scalable plant expression system. *Proceedings of the National Academy of Sciences of United States of America,* 103, 861-866.

[86] Porta, C. & Lomonosoff, G. P. 2002. Use of viral replicons for the expression of genes in plant. *Biotechnology Genetic Engineering Reviews.* 19, 245-291.

[87] Yusibov, V., Rabindran, S., Commandeur, U., Twyman, R. M. & Fischer, R. 2006. The potential of plant virus vectors for vaccine production. *Drugs in R&D,* 7, 203-217.

[88] Benchabane, M., Goulet, C., Rivard, D., Faye, D., Gomord, V. & Michaud, D. 2008. Preventing unintended proteolysis in plant protein boifactories. *Plant Biotechnology Journal,* 6, 633-648

[89] Gomord, V, Sourrouille, C, Fitchette, A. C., Bardor, M., Pagny, S., Lerouge, P. & Faye, L. 2004. Production and glycosylation of plant-made pharmaceuticals: the antibodies as a challenge. *Plant Biotechnology Journal,* 2, 83–100.

[90] Fitchette, A. C., Dinh, O. T., Faye, L. & Bandor, M. 2007. Plant proteomics and glycosylation. *Methods in Molecular Biology,* 335, 317-342.

[91] Scott, C. T. 2005. The problem with potency. *Nature Biotechnology,* 23, 1037-1039.

[92] Pujol, M., Gavilondo, A., Ayala, M., Rodriques, M., Gonzales, E. & Perer L. 2007. Fighting cancer with plant expressed pharmaceuticals. *Trends in Biotechnology,* 25, 454-459.

[93] Sparrow, P. A., Irwin, J. A., Dale, P. J., Twyman, R. M. & Ma, J. K. M. 2007. Pharma-Planta: Road testing the developing regulatory guidelines for plant-made pharmaceuticals. *Transgenic Research,* 16, 147-161

[94] Mitragotri, S. 2005. Immunization without needles. *Nature Reviews Immunology,* 5, 905-916.

[95] Prausnitz, M. R. & Langer, R. 2008. Transdermal drug delivery. *Nature Biotechnology,* 26, 1261-1268.

[96] Koprowski, H. 2005. Vaccines and sera through plant biotechnology. *Vaccine,* 23, 1757-1763.

[97] Arntzen C., Plotkin S., Dodet B. 2005. Plant-derived vaccines and antibodies: potential and limitatins. Vaccine, 23, 1753-1756.

To Read: Recent Books on the Subject

Hefferon, K.L. 2009. *Biopharmaceuticals in Plants:* Toward the Next Century of Medicine. CRC Press. Taylor& Fransis, Inc. 204 p.

Rehbinder, E., Engelhard, M., Hagen, K., Jorgensen, R.B., Pardo-Avellaneda R., Schnieke, A. & Thiele, F. 2009. *Promises and Risks of Biopharmaceuticals Derived from Genetically Modified Plants and Animals.* Springer-Verlag Berlin Heidelberg. 334 p.

Kontemann, R. & Dubel, S. 2010. *Expression of Antibody Fragments in Transgenic Plants.* Antibody Engineering. Springer Berlin Heidelberg. 386 p.

In: Bioprocess Sciences and Technology
Editor: Min-Tze Liong
ISBN 978-1-61122-950-9

Chapter 15

BOTANICALS IN BIOCONTROL AND APPLICABILITY OF BIOPROCESSING FOR THEIR PRODUCTION

S. Hettiarachi[1,*] ***and P. L. Hettiarachchi***[2]
[1]Department of Botany, University of Ruhuna, Matara, Sri Lanka
[2]Department of Botany, University of Sri Jayawardenapura, Nugegoda, Sri Lanka

ABSTRACT

Plants have coevolved with their pests and pathogens. As a consequence they display resistance against harmful organisms manifested through various mechanisms including production of toxic chemicals. Some of these chemicals are produced constitutively whereas others are elicited as a result of injury or infection. As they are derived from plants and kill, inhibit, reduce fertility or repel other organisms, such chemicals are botanicals useful as biocontrol agents.

A considerable number of trees, shrubs and herbs are known as sources of botanicals that can be used in biocontrol of a variety of pests and pathogens ranging from insects, nematodes, fungi and bacteria. Further botanicals also have a potential in weed control and under certain cropping systems and management of parks and orchards this is being successfully practiced.

Plants and their extracts have long been used in biological control in traditional agricultural practices and in traditional medicine based on herbal treatment. With the realization of the harmful impacts of synthetic chemicals used in agriculture and medicine, chemicals extracted from plants have been viewed as the best alternatives.

Certain botanicals are secreted or released into the environment which may have biocontrol capability. Further such plants can attract and retain other organisms such as predators and parasitoids which may in turn reduce the populations of pests and pathogens. Therefore increasing the crop diversity within an agricultural field may also help reduce damage caused by harmful organisms.

Although botanicals have often been suggested as an alternative for chemical control, it is unlikely that botanicals alone would be able to produce required level of

* Corresponding author. Email: sanath@bot.ruh.ac.lk

control. Botanicals may be effectively used in integrated management systems where a combination of traditional practices, cultural practices, use of chemicals and biocontrol brought about by other agents.

Herbal medicine has been used in control of diseases caused by infectious agents. Traditional medicine uses a holistic approach including pathogen control with simultaneous strengthening of bodily functions and hence use of purified chemicals may not be as effective as extracts. Organisms develop resistance to chemicals particularly when used singly and continuously. Therefore sustainable use of botanicals against pests and diseases of plants, animals and humans and against weeds will depend on suitable plant extracts, rather than use of purified or synthesized active ingredients.

On the other hand identification of active compounds may lead to their chemical synthesis and also to use biotechnological approaches to enhance productivity and to minimize cost of production.

As diverse plants produce diverse chemicals, conservation of biodiversity becomes extremely important. Once the plants producing useful chemicals are identified, it may be necessary to cultivate the plant in commercial scale in order to reduce the pressure on natural populations.

Further modern biotechnology has been employed in certain cases where the product maybe extracted from cell, callus and hairy root cultures. Certain useful botanicals are tissue specific and produced through complex biochemical pathways. Genetic manipulation can also be useful in generating new variants of plants producing chemicals that are originally produced by other organisms and also to generate overproducers of existing botanicals.

1. Introduction

Prior to the period of modern medicine, the people depended almost entirely on the plants and various preparations made of plant extracts for treating ailments and also for improving general health. According to the estimates by WHO (2008), about 80% of the modern population in some Asian and African countries still uses the herbal treatment for their primary health care. With the recognition of manifold environmental [1] and health hazards [2] of synthetic chemicals, use of chemicals with a plant origin in control of pests and diseases and even weeds in the agricultural fields and forestry set up is of much interest. Due to long coexistence of all these organisms over the long history of evolution, plants have developed various defenses that largely depend on chemicals. Therefore it is not surprising that plants have become chemical factories that produce many different chemicals against pests and diseases, herbivory and competition by other plants. Traditionally the term "botanicals" has been used to encompass drugs, medicinal preparations, or similar substances obtained from a plant or plants due to their significant role in traditional or folk medicine. However it easily covers other chemicals or products in purified or crude forms or released from live plants that can be used in agriculture to control pests, diseases and weeds and to increase general health of plants.

In its basic sense, biological control or biocontrol is applied to manage the populations of harmful organisms below a level that does not pose a threat rather than their complete eradication. For this purpose natural enemies of those harmful organisms such as, parasitoids, predators, pathogens or weed feeders are employed. As the populations of such organisms depend on host/ prey populations, a balance of the two kinds will establish. If the populations of the harmful organisms were maintained below the economic threshold level, a successful biological control has been achieved. In its broadest sense, biological control may be achieved by the use of chemicals, products or preparations of biological origin. As mentioned above, plants have developed a myriad of chemical weapons against a variety of their enemies, and thus are extremely useful in biological control. In addition to the direct effect on the harmful organisms, certain plants can be used to attract natural enemies of such organisms.

Chemical control is very effective within a short period of time and death of insects can be seen within seconds after application and in the case of weeds, the killing occurs within hours. This and massive propaganda by chemical companies promote the use of chemicals resulting in abuse and overuse leading to rapid development of resistance in the pests, pathogens and weeds rendering some chemicals ineffective. The environmental problems include build up of toxic chemicals in the environment and killing of nontarget, often beneficial organisms. Application of chemicals poses health hazards to users as well as consumers.

2. Biological Control

Biological control is broadly understood today as the use of natural or modified organisms, their genes or gene products to control the effect of undesirable organisms while promoting the desirable organisms. This means the application of action of parasite, predators or pathogens in order to maintain the populations of undesirable organisms below a certain threshold level which does not cause injury to a harmful scale to the useful organisms. At the first glance of above statements, it may appear to suggest exclusion of plants as biocontrol agents. However plants, through their chemical weapons, do have an important position in biocontrol. Further, use of harmless strains of an organism with devastating effects for preemptive colonization and competition should also be considered within biological control. The modern approaches such as engineering coat protein of a plant virus in to the crop plant is a biocontrol strategy against the infection by that particular virus. Induced resistance, managing the environment (eg. soil), transgenic plants and biopesticides are the other areas which come under biocontrol.

2.1. Biocontrol of Insect Pests

Ladybird beetle has received highest attention over a period of at least a century as a control agent of insect pests in agriculture. The first intentional biocontrol was achieved by introducing ladybird beetles from Australia to orange groves in California in 1989. A comprehensive review of different aspects including taxonomy, biology and its use in biocontrol has been compiled by Koch (2003) [3]. It feeds on eggs, larvae and adults of insect pests. Many other insects that predate or parasitize pests are listed and several of these are encouraged by manipulation of the environment and by artificial introductions. As plants are important in supporting predators and parasitoids, this will be discussed elsewhere in this chapter.

Exploiting entomophagous fungi has also been under consideration. At present several commercial formulations of entomopathogenic fungi are available. *Metarhizium anisopliae* is perhaps the most extensively studied fungal species. After analyzing all pros and cons, Zimmermann (2007) suggested this species to be safe with minimal risks to vertebrates, humans and the environment [4]. Other available options are bacteria and viruses [5].

Although nematodes are also capable of keeping insects and mites under check, the emphasis on exploitation of nematodes is meager. Pathogens and nematodes have been a 100% success in some occasions, for example control of saw flies and wood wasps [6].

2.2. Biocontrol of Mollusk Pests

Mollusks are more commonly known as destructive pests in agriculture, but they also are important as hosts of parasites of humans and animals. They are also troublesome in cooling towers and other such installation where water is circulated. On the other hand mollusks are a delicacy and an expensive item in some parts of the world. They should also be viewed as an important component of ecosystem. Therefore the measures for control of mollusk pests should be specific as far as possible. As they are found in both aquatic and terrestrial habitats and belong to different taxonomic categories with a variety of behavioral and morphological features, one common practice would not be possible. Furthermore chemical control can be hazardous especially in aquatic habitat.

Biological control of mollusks has been attempted by using fish [7], bacteria [8], nematodes [9], leeches [10, 11], and fire ants [12]. Although a formulation of nematodes as biological control agent has been commercialized in 1994, further investigations are needed to strengthen and justify the need of predators as an effective and ecofriendly measure of biocontrol.

2.3. Biocontrol of Nematode Pests

Most commonly known biocontrol of nematodes is attained by nematophagous Hypomycetes fungi. Some of these fungi can effectively catch nematodes by setting traps. Several others can live endoparasitically through germination of ingested spores or spores that stick on the nematode [13]. Kerry (1997) provided a good account on exploitation of natural enemies of nematodes for their biocontrol [14]. Except for the use of rhizosphere bacteria, all other examples are different fungi acting against nematodes as predators or parasites and root associated fungi competing with nematodes and producing toxins. However Tian *et al.* (2007) pointed out *Pasteuria* as a good bacterial candidate for nematode control [15]. Several opportunistic bacterial pathogens have been identified, but present knowledge does not indicate a successful utilization of such bacteria. The possibility of using parasitic nematodes for controlling plant parasitic nematodes has also been considered, and some preliminary research works have generated positive results [16]. For example, among the different types of predators, diplogasterids are appraised as the most suited for biocontrol of nematodes, because of their short life cycles, easy culture, prey-specificity, chemotaxis sense and resistance to adverse conditions. However more research is necessary to conclusively decide whether this approach would be a success.

2.4. Biocontrol of Diseases

Bacteria, viruses and fungi are the major causes of diseases in plants and animals. Other category of organisms receiving recognition as important pathogens is the Mycoplasmas. As many pathogens have a growth phase before entering into the host, they have to compete with other microorganisms that are well adapted to live on the surfaces of hosts. Therefore much attention is given to find microbial agents that are aggressive colonizers and therefore exclude the pathogens development. Other microorganisms that can fight pathogens are the antagonistic flora which compete for nutrients and space and also which produce antimicrobial substances such as antibiotics and toxins. They can inhibit germination of spores and elongation of germ tubes and also cause lysis. There are also hyperparasites which parasitize the causal agents of disease

The first biocontrol agents for postharvest diseases entered the market in 1995 [17]. As biocontrol agents are also microorganisms they must be used in an integrated system where the environmental conditions should be manipulated to favour those agents while creating unfavourable conditions for pathogens. Over the last decade, at least 80 products reached the market for biocontrol of plant diseases [18]. *Coniothyrium minitans* and nonpathogenic *Fusarium* are exploited against major soilborne pathogens such as *Sclerotinia*, *Pythium*, *Rhizoctonia* and *Fusarium*. Products containing *Trichoderma*, *Ampelomyces quisqualis*, *Bacillus*, and *Ulocladium* are aimed at controlling the primary foliar diseases, *Botrytis* and powdery mildew. *Pseudomonas* for the control of *Pythium* diseases in hydroponics and *Pseudozyma flocculosa* for the control of powdery mildew have been developed.

In other occasions, the environment is manipulated in order to encourage the flora having negative effects on the pathogen. For example coating seeds of sugar beet with nonsterilized powered straw of different crops exploited to control *Pythium* [19]. The negation of the effect by sterilizing the straw powder is a good indication of the biological nature of the control. Wheat straw powder increased the incidence of *Pythium* infections, but co-inoculation of the straw powder with the biocontrol agent *Pseudomonas fluorescens* 708 was found effective.

2.5. Biocontrol of Weeds

In order to achieve permanent control of weeds, release of exotic insects, mites, or pathogens had been the method of choice for safe, economic and environmentally sustainable solution [20]. However this is not a totally foolproof option. The effect on nontarget plants is a serious concern and thus assessing biosecurity risk is essential [21]. In an earlier occasion Fowler *et al.* (2001) argued that if host range was identified, the nontargt effects could be predicted [22]. They found information on only two instances in New Zealand where significant damage to nontarget plants was recorded. That represents 0.5% of the nearly 400 insect, mite, or fungal species used in classical weed biocontrol. A considerable success has been achieved in the control of aquatic weeds. For example introduction of three South

American insect species could successfully control the alligator weed in Sothern United States [23]. Aquatic weed control through pathogenic fungi has also been achieved [24].

3. Importance of Botanicals in Biocontrol

The potential of botanicals in biological control has been demonstrated conclusively in various aspects. For example this can be applied to control populations of insect pests in the field and in storage or to repel the troublesome insects from the field or from our households. The potential uses of botanicals against mollusks and nematodes have also been studied [25], and a few selected examples are mentioned below. Plant extracts and products can be used to increase the vigour of plants and thereby protect them from bacterial and fungal diseases in addition to their direct effect on pathogens. Wood chips and cover crops are prevalently used to control weeds.

Many plants producing useful chemicals effective in biocontrol are very familiar to man as either medicinal plants or as spices. Therefore people readily accept products of plant origin as safe. As for all natural products, the residual effect, if any on the environment is minimal. The research today on botanicals is drifting towards purification of active compounds in botanical extracts with a view of commercially producing these in pure form. Yet the crude preparations are effective and contain a mixture compounds effective in varying degrees and acting in different mechanisms thus enormously reducing the rate of resistance development in target organisms. Therefore it may be pertinent to mention here that, while chemical characterization of active ingredients should progress, the use of holistic botanicals should be the intention of research. At present raw materials for botanical formulations are directly obtained from plants and this may not be viable if application of botanicals for pest and disease problems becomes the choice of public. At this point biotechnology can come in handy to develop methodologies to mass production of botanicals using bioprocessing approaches.

3.1. Botanicals in Insect Control

3.1.1. Attracting Predators and Parasites

Insects are generally regarded as a nuisance, but they also are an essential component in the ecosystem having an array of functions that are beneficial for mankind. Therefore the choice of strategies for the control of insect pests should find the correct balance between causing the maximum harm to pest while protecting as far as possible the useful insects. Advancing further, friendly insects may be attracted to control pests. Predators include animals such as frogs, lizards, birds, dragonflies and spiders.

If one considers the living plants as botanicals, there are reports where growing of selected plants has proven useful in control of pests, especially in integrated pest management setup [26]. Agroecosystems are subject to rigorous manipulations often favouring monoculture or a few selected crops at any given time causing a drastic ecological imbalance. Hence a more meaningful habitat management should be carefully thought about and implemented in order to approach a good balance between various forms of organisms [27].

This may be achieved as temporary measure and as such is suitable for fields with annual crops. Larval stages as well as adult of Ladybug or Ladybird beetle (*Hippodamia convergens*) feed on insect pests and is considered as a friendly insect in agroecosystems. Lind (1998) has described the biology of Ladybird beetles and provides a list of plant species which can be grown to attract them [28]. Lacewings, hoverflies, insidious flower bugs, spiders, parasitic wasps, tachinid flies, chalcids are other categories of beneficial insects. Some of them prey on harmful insects and some others are parasitoids. The idea is to provide food resources and habitat required by the beneficials. Plants bearing flowers for example provide food for parasitic wasps which migrate to crop in search of pest hosts for laying eggs. Jones and Gillett (2005) reported that establishing sunflower rows within the crop field increases populations of beneficial insects [29]. Long *et al.* (1998) studied the effect of 14 perennial plants and 12 annuals (flowering plants and grasses) which act as insectary plants and observed migration of seven dominant predators at least 100 feet away from where they feed [30].

Although growing plants other than the intended crop for control of pests is an attractive concept in biological control, this may not be feasible in large agricultural fields where monoculture is necessary for easy management and use of machinery for harvesting etc., because this requires leaving some rows or borders or hedges for plants intended for attracting natural enemies of pests.

It should be noted that not only beneficial insects, but also adults of some of the pests also feed on nectar. Hence use of such plants in biocontrol may not be as straight forward as it seems at first glance. Nevertheless different insects show selectivity or preference to different nectar sources and this allows room for choice of plants after careful studies of selectivity/ preference.

It is an interesting phenomenon that plants attracts carnivores and parasitoids of pests by releasing chemical signals consequent to herbivory. This is mutually beneficial for plant and the predators and parasitoids of herbivores as plants guide the predators and parasitoids towards their prey/ host, relieving the plants from the stress of infestation. Terpenoids are such chemicals [31]. Loivamakia *et al.* (2008) pointed out that isoprenes, also emitted by certain plants, may perceived by chemical receptors of different parasitic wasps [32]. The response shown by *Diadegma semiclausum* was be repelled, whereas *Cotesia rubecula* did not perceive it and did not respond. This finding however is important because one has to consider the background level of isoprenes, if attracting beneficial insects through this mechanism is desired. For example poplars and willows produce high amounts of isoprene so this method may not be suitable for farmyards close to such plantations.

3.1.2. Insecticides, Repellents and Feeding Deterrents

Isman (2006) identified pyrethrum, rotenone, neem and essential oils as the major botanicals used in insect control [25]. The source of pyrethrum, an oleoresin, is pyrethrum daisy, *Tanacetum cinerariaefolium* (Asteraceae). This is a neurotoxin and causes rapid death of insects and hence is very useful in controlling flying insects. Two main components of pyrethrum are pyrethrins I and II. The major problem in this compound is its very short half life particularly in the presence of light and therefore synthetic derivatives known as pyrethroids are developed. Pyrethrin preparations are used in indoors to control insect vectors diseases and other nuisance insects. Its incorporation in IPM is also limited due to its high toxicity on insects and hence both pests and their enemies can be killed. Pyrethrins killed both glasshouse whitefly and its parasitoid [33]. Duso *et al.* (2008), reported more toxicity towards the predator (*Phytoseiulus persimilis*) than the pest, two-spotted spider mite *Tetranychus urticae* [34]. The selectivity of pyrethrins is limited because it acts as a contact insecticide.

In contrast, aqueous leaf extract of syringa tree, *Melia azedarach* and a commercial formulation from the neem tree, *Azadirachta indica* (containing azadirachtin), showed high toxicity on the pest diamondback moth (*Plutella xylostella*), with no direct effect on the longevity of two parasitoids *Cotesia plutellae* and *Diadromus collaris* in laboratory studies [35]. Survival, fecundity, development, oviposition and feeding of *P. xylostella* had been severely affected. Furthering their investigation, they have demonstrated that application of the same preparations reduced the damage by the pest even at similar infestation level as controls and the parasitoids emerging from the hosts in treated plots were higher [36]. In an attempt to find nontarget effect of two neem formulations used in control of insects in forests, only one macroinvertebrate species out of eight tested had been affected by one preparation at 10 times higher concentration than expected environmental concentration [37]. Long term exposure of three detritivorous species to the expected environmental concentration, did not cause significant mortality increase or show any antifeedant effect.

Derraik and Slaney (2005) investigated a cost free approach for mosquito control using used coffee grounds (UCG) which has been promoted by public health authorities in Brazil [38]. The effect of caffeine on mosquitoes had been reported previously [39] using *Aedes aegypti* model. Both of these studies have raised the important issue of the concentration of caffeine or UCG. They noted that as caffeine and UGC at low concentrations can support, especially under low nutrient conditions, the growth of bacteria and other microorganisms resulting in an increase of the survival of mosquito larvae. The larvae will feed on those microorganisms. Therefore repeated application of UCG at high concentration is required for the control of mosquitoes and failing in this approach will have a negative impact. Further studies by Laranja *et al.* (2006) by following several generations of mosquitoes demonstrated the efficacy of UCG in prevention of resistance development to the treatment [40].

Neem seed oil (NSO) and azadirachtin have also been found effective as biocontrol agents. Schmutterer and Singh (1995) provided an extensive list of 413 insect pest species which are sensitive to neem products [41]. The only exception was oriental yellow scale, *Aonidiella orientalis*, which is a pest on neem. Main insecticidal active component in neem is azadirachtin which is a component of neem oil. Salannin and meliantriol are two other chemicals present mainly in formulations for biocontrol based on neem seed. These chemicals kill or suppress activities of pests by having different behavioral and physiological impacts on their targets [42]. This is important in slowing down the rate of resistance development by pests, which is one of the major problems associated with application of chemical pesticides. Azadirachtin inhibits moulting and thus inhibits growth and delays or prevents metamorphosis resulting in death at the larval stage. Another impact is azadirachtin acting as an antifeedant. However this effect is relatively inconsistent among different groups of insects in comparison to uniformity in physiological effect related to moulting controlled by ecdysteroids (moulting hormone). If the insects escape these effects, azadirachtin has the advantage of being able to reduce the reproductive ability of both males and females [43]. Chemical insecticides, other than systemic ones, mainly act via contact and therefore do not distinguish between pests and beneficial organisms. Azadirachtin must be ingested and therefore specifically controls the bugs causing damage to plants. As such neem based products containing azadirachtin can be safely used in IPM systems [42, 44].

In addition to NSO, neem seed cake prepared from residue after extracting oil is also very useful in pest control. Seed oils from other plants have also been studied with positive results. NSO has been a successful botanical insecticide in the field against aphids of pepper and strawberry, but failed to do so against ahipds of lettuce [45].

Unlike water extracts of neem (seed, leaves, bark etc.) containing azadirachtin and other active ingredients in pest control, NSO poses some problems for beneficial organisms. Being an oily preparation direct contact causes suffocation of the insects [25]. Raguraman and Singh (1999) found out that NSO have multiple negative effects on egg parasitoid *Trichogramma chilonis* by acting as a feeding deterrent, feeding and contact toxin and oviposition deterrent and therefore is not suitable for IPM practices [46].

The essential oils extracted from aromatic plants contain mixtures of complex chemicals and are known to repel insects. In addition they can also be used as contact and fumigant pesticides [47, 48]. Methyl bromide and phosphine are the mostly used chemical fumigants and several others such as Sulfuryl fluoride, Propylene oxide, Ethyl formate, Carbonyl sulfide, Hygrogen cyanide, Cyanogen, Methyl iodide and Ozone are being considered.

The most attractive feature of plant essential oils in pest control is the low mammalian toxicity. Majority of these compounds have been used traditionally as fragrances and flavours in perfume and food. Nevertheless commercial pesticides originating from plant essential oils are not common, except for some mosquito repellents. Citronella (*Cymbopogon nardus* and *Cymbopogon winterianus*), lavender (*Lavendula* sp.), cinnamon (*Cinnamomum* spp.), clove (*Syzygium aromaticum*), lemon (*Citrus limon*), rosemary (*Rosmarinus officinalis*), *Eucalyptus* spp., peppermint ((*Mentha* × *piperita*, also known as *M. balsamea*), thyme (*Thymus vulgaris*) and *Geranium* spp. essential oils are used in traditional as well as commercial preparations of mosquito repellents. The ingredients found in these essential oils are chemically identified as citronellal, d-limonene, menthol, 1,8-cineole, eugenol, p-menthane-3,8-diol and thymol [25]. The major form of preparations is sprays. Lighting candles containing those essential oils is also used to repel and reduce biting. Oils in liquid form can also be applied on the skin. Plant essential oils are also popular among village people as repellents of other biting flies, ants and ticks and fleas on their farm and pet animals. The extraction of oil is not necessary in those conditions as crushed plant material can easily be used instead.

3.2. Botanicals in Nematode Control

The lead for search of useful plants for extracting botanicals against nematodes also came from the observation that many plants show natural resistance to nematode attacks. Elbadri *et al.*, (2008) tested five essential oils and 15 extracts originated from such plants against *Bursaphelenchus xylophillus*, pine wood nematode responsible for wilting of pine trees [49]. The essential oils from clove plant (*Syzygium aromaticum*), mustard (*Brassica integrefolia*), thyme (*Thymus vulgaris*) and *Pelargonium inquinans* gave excellent control of the nematodes. Many plant extracts in methanol, hexane and hot water also showed mortality of the nematode, but methanol extract of *Magnolia officinalis* emerged as having significantly high impact. According to Martín and Magunacelaya (2005), commercial preparations of extracts of *Quillaja saponaria* containing both saponin and nonsaponin were much more effective in control of eight plant parasitic nematodes than saponin fraction or nonsaponin fraction alone [50]. This synergism is often observed when using botanicals.

Certain plants that are resistant to nematode secrete allelochemicals into the soil thus reducing nematode population by affecting their reproduction. Glucosinolates are such compounds produced by plants of mustard family, giving the characteristic odour. Therefore use of cover crops with such activities is an option where extraction of products is not required. Dufour (2003) reported use of suitable cover crops, oil radish as a trap crop and oil radish, rapeseed and sudangrass as green manure to control nematode attacks [51]. The allelochemicals released from live plants and during the decomposition of residues are suggested as responsible for nematode control.

Prasad *et al.* (2002) found several plant extracts were effective in killing J2 of *Meloidogyne incognita* and pre-adults of *Rotylenchulus reniformis in vitro* and in reducing nematode galls in roots of mung crop. *Calotropis procera* was most effective in reducing galls, followed by *Datura stramonium* and *Solanum surattense* leaf extracts [52]. In search of more botanicals against nematodes, root extracts of spotted spurge (*Chamaesyce maculata*); poinsettia (*Euphorbia pulcherima*); lantana (*Lantana camara* var. hybrida) and seed meal of a *Brassica* sp. were found to be effective in controlling sting nematode (*Belonolaimus longicaudatus*) [53].

Nematodes infestations in mushroom cultivation, more specifically button mushroom, cause economic losses. Use of chemical pesticides is prohibitive as delicate mushrooms will wither and distort. Farmers empirically plant Asteraceae plants around the sheds where mushrooms cultivated to keep away infestations. They also mix dried leaves of plants that they believe would keep the mushrooms healthy with the compost. Nagesh and Reddy (2000) reported the efficacy of leaves of neem, pongamia and castor as effective for nematodes in mushroom cultivation [54].

3.3. Botanicals in Mollusk Control

Mollusks are herbivores and therefore one could expect some kind of defense would evolve in plants against them. Mollusk species also are hosts of pathogens of animals and humans. These pathogens require snails in order to complete their life cycle. Schistosomiasis caused by trematodes is one of health concerns related to snails. Therefore much attention is given to control snails with a view of breaking the life cycle of such pathogens for the control of diseases caused by them.

Many plant species have been identified as having of protection against snails and slugs by village dwellers for centuries, but was not a popular subject among scientists until recently. Marston *et al.* (1996) screened 153 crude extracts of 28 Panamanian plant species against *Biomphalaria glabrata*, one of the intermediate hosts of schistosomiasis [55]. They identified *Cyperus luzulae* (Cyperaceae) and *Hymenaea courbaril* (Leguminosae) as plants having highest molluscicidal potent. Among different crude preparations and pure products obtained from *Carica papaya* and *Areca catechu*, lyophilized *C. papaya* latex powder found to be more effective against a vector snail, *Lymnaea acuminate* [56].

Commiphora molmol oil extract was proven to cause death of adult and young of six freshwater snails of medical and veterinary importance (*Lymnaea natalensis*, *Bulinus truncatus*, *Biomphalaria alexandrina*, *Physa acuta*, *Melania tuberculata* and *Cleopatra bulimoides*). Eggs of the first four species were also killed [10]. Extracts of unripe fruits of *Solanum nigrum villosum* had higher mollucicidal ability on *Galba truncatula* than its extracts of ripe fruits and leaves [57]. Additional piece of information provided by this study was the demonstration of the potential of using the same plant extracts to kill the parasite *Fasciola hepatica* living in the snail. The mollucicidal effect of *Solanum* spp., *S. nigrum* in particular has been confirmed using another test snail, *Biomphalaria alexandrina* [58].

As yucca is a plant known to repel slugs and snails, Hollingsworth and Armstrong (2003) used extracts of yucca as one of the test material against *Zonitoides arboreus* (Say), a nuisance mollusk which feeds on roots of orchids in pots and found effective in its control [59]. Neem extracts, *viz.* azadirachtin and neem oil, were phagostimulants. Neem, in many instances, was a promising pesticide for a variety of organisms. Later in 2002 Iglesias *et al.* demonstrated azadirachtin killed all the eggs of a different mollusk, *Deroceras reticulatum* at a concentration of less than 0.05 mg cm^{-2} after 24 h exposure in paper-contact toxicity test [60]. *Oncomelania hupensis* another host of schistosomiasis has been sensitive to leaf, stem and root extracts of *Eupatorium adenophorum* [61].

Martin *et al.* (2008) experimented with husks of quinoa (*Chenopodium quinoa*) grain for its activity against Golden apple snail (*Pomacea canaliculata*) [62]. These husks have no commercial value and taste bitter due to high saponin content, particulary bidesmosidic saponins. The husks were not effective, but after an alkaline treatment, which converts bidesmosidic saponins to bidesmosides, the product killed 100% snails at low concentration as 0.33 ppm in 24 h period. It was also reported that the product was not toxic to fish at 54 ppm level. As a general rule chemical molluscicides are toxic to fish at lower concentrations than those effective against mollusks.

3.4. Botanicals in Disease Control

Contamination by aflatoxins produced by *Aspergillus flavus* and *A. parasiticus*, are a major health issue associated with storage of botanical material such as spices, cereals, oil seeds and nuts. Aflatoxins are notorious carcinogens in addition to being toxic causing acute hepatic necrosis. As such, although fungi producing aflatoxins are not pathogens *per se*, their control is a burning issue in human health. Aflatoxin contamination in stored medicinal plant products raises concerns about the safety of such herbal raw material. The same is true for spices which are potent biocontrol agents of microorganisms. *A. flavus* and thus aflatoxin production was completely inhibited in stored material of six medicinal plants by essential oils of *Cymbopogon flexuosus* and its components [63]. The most effective component was eugenol. This problem in rice grains was addressed by Reddy *et al.* (2009), and established that complete inhibition of the fungal growth and toxin production by extracts of *Syzigium aromaticum* followed by *Curcuma longa*, *Allium sativum* and *Ocimum sanctum* at a rate of 5 g kg^{-1} [64]. According to Kumar *et al.* (2007), essential oils of *Chenopodium ambrosioides* has a broad spectrum of fungicidal effects, in addition to having inhibition of aflatoxin production also exhibiting antioxidant properties [65]. This plant is growing as a weed and it is used in herbal medicine. Therefore it is available in plenty in tropics and has been proven not only to have no toxic effect on human but even as beneficial. The essential oil of *Thymus vulgaris* L. was highly effective in inhibition of both *A. flavus* and aflatoxin production. In addition thyme oil was also suppressive against eight food contaminating fungi showing superiority as a fungitoxicant over commonly used synthetic fungicides [66].

Phytophthora spp. can cause not only root, stem and crown rots, but also fruit and foliar blights affecting severe crop losses under conditions favourable to the pathogen. Drenching soil with fungicides is done to reduce the propagules in soil, however is not very effective. Therefore once again scientists have experimented with botanicals in finding a solution to *Phytophthora* infections. In one such study [67], different formulations of plant extracts have been tested against *P. nicotianae* on Periwincle (*Catheranthus roseus*) in Petry dish and in green house. Certain formulae were very effective in keeping the propagules in soil as undetectable for at least 21 days and keeping a significantly higher number of healthy plants in inoculated soil under conditions favourable for disease development. It is noteworthy that different formulations of the same botanicals showed different reactions in pathogen control and plant protection (eg. formulations and 1 and 2 of pepper extract – mustard oil) showing the impact of other substances than the most prevalent active ingredients.

Different type of "disease" is permanent discolouration of sapwood by sapwood fungi. The most troublesome species in pine sapwood is *Sphaeropsis sapinea*. As fungicides are inefficient in controlling sapwood staining, various biological control has been attempted and *Trichoderma harzianum* was identified as most effective. On the lookout for better biological control, plant oils were included in the studies. Oils containing Oxygenated alcohol or phenolic monoterpenes were found to be effective and commercial formulations based on pine oil containing such secondary metabolites have been suggested [68].

Herbal medicine had been the remedy for all ailments in mankind as well his domestic and farm animals prior to the development of modern "Western" medicine. Therefore it is not surprising to find many medical uses including control of pathogens and parasites in medicine and veterinary medicine. Two of the most documented trees are neem in Asia and Africa and tea tree in Australia. As some information on neem and neem products are given above, the importance of tea tree will be briefly discussed here. Tea tree oil (TTO) or Melaleuca oil is distilled from leaves of *Melaleuca alternifolia*. It is commercially available and has been used as hair oil. A large number of reports are available for its medicinal use as a topical decolonization agent against many pathogens. This botanical is probably the one which has attracted the highest attention in Western medicine. This is obvious as judged by the number of reviews appeared in recent times. For example, antimicrobial, antiinflammatory and other medicinal properties of TTO were the subject of the reviews by Halcón and Milkus (2004) and Carson *et al.* (2006) [69, 70].

TTO has been proven to be active as anti-Legionella disinfectant, for control of water system contamination, especially in spas, in small waterlines or in particular respiratory medical devices, after experimenting with *Legionella pneumophila* of different serogroups and sources of isolation [71]. TTO can be used for efficient control of *Staphylococcus aureus* in both suspensions and biofilms [72]. Research at Macquarie University in order to rank the efficiency of plant essential oils as antimicrobial agents, recognized a unique place for TTO, because it provides good antimicrobial activity without accompanying problems of instability or irritation [73].

3.5. Botanicals in Weed Control

Traditional farmers use residues of plant species selected through their experience for control of weed in their crop fields. They also realize that some plant residues used as mulch not only reduce weed problem, but enhance the general plant health. Scientists too have taken interest to investigate the potential of botanicals in weed control. In their work published in 1983, Putnam and DeFrank showed the potential of residues of Sorghum (*Sorghum bicolor*) in the control of *Portulaca oleracea* and *Digitaria ischaemum* [74]. These weeds were reduced by 70% and 98%, respectively. Similar effects have been observed with residues of other crops; barley (*Hordeum vulgare*), oats (*Avena sativa*), wheat (*Triticum aestivum*) and rye (*Secale cereale*).

Among ten plant species tested, *Bidens pilosa* and *Tephrosia candida* were most effective against control of weeds in paddy fields with concomitant 20% increase in the harvest in Viet Nam [75]. They proposed several other plant species as potential candidates for weed control in paddy fields.

A very troublesome weed, particularly in cereal crops cultivation, is *Striga* spp. as some of these species occur as root parasites, and hence cannot be controlled using broad spectrum herbicides. The control depends mainly on cultural practices and cultivation of resistance varieties. Marley *et al.* (2004) demonstrated seed powder of neem, and to a lesser extent parkia fruit powder and parkia fruit peel powder, were effective in controlling seedling emergence of *Striga hermonthica* [76].

Attempts to use of eucalypt oil for control of *Parthenium hysterophorus* is a good example of botanical extracts in weed control [77] in contrast to decaying plant residues. Eucalypt oil acts by inhibiting seed germination, reducing cellular respiration and chlorophyll content and increasing water loss causing the plants to wilt by 15 days after exposure. The same group later reported the effect of eucalypt oil on seed germination of different plants including weeds as well as crops [78]. Nevertheless this may still have applications in weed control in the absence of crop (pre-planting) and also in field with standing crop by carefully planning the timing of weed control program.

4. Biotechnology and Bioprocessing in Botanicals for Biocontrol

4.1. Genetic Engineering

With the rapid development of genetic engineering it has now become a rather simple operation to express a peptide in a totally different organism from the one naturally producing that peptide. As in the development of famous golden rice where carotene is expressed in the rice grain, more complex expressions have also been achieved. Botanicals that are useful in biocontrol are in fact substances produced by plants for their own defense. Most of these are secondary metabolites having complex structures and formed as a result of chains of biochemical reactions or by chemical modifications of such substances. The development of biotechnological approaches for enhanced production is hindered by this complexity involving unknown regulatory genes and enzymes [79]. Therefore engineering of such traits from one plant to another seems unlikely at first glance. More feasible genetic engineering approaches would be over expression and constitutive expression of such chemicals. Expression is usually controlled by chemical signals by either keeping the genes switched off to arrest the expression or switching them on when necessary. Relieving suppression or induction can start a cascade of reactions leading to the production of the substance in question. Therefore if such suppressors or inducers can be identified, the expression of already present defense can be forced to over express constitutively. A single molecule may be able to control the expression of several related substances. If this can be achieved harvesting botanicals can be made more profitable. It may also lead to the production of crop varieties with high resistance to pest attacks. As these are theoretically feasible, the scientific research interest focused on this at present is on the increase.

Hain *et al.* (1993) provides an example for transferring the expression of a stilbene synthase from grape vine (*Vitis vinifera*) to tobacco (*Nicotiana tabacum*) [80]. This resulted in the accumulation of stilbenes and thereby to an increased resistance to *Botrytis cinerea.* Accumulation of cyanogenic glucosides is another defense mechanism employed by certain plants against herbivore pests. Bak *et al.* (2000) succeeded in transferring two genes responsible for cytochrome P450 enzymes from *Sorghum bicolor* to tobacco and *Arabidopsis thaliana* and observed accumulation of cynaogenic glucosides [81]. It is interesting to note that engineering a gene for chorismate pyruvatelyase from *Escherichia coli* in *N. tabacum* has resulted in 1000 fold increase 4-hydroxybenzoate (4HB) as β glucosides [82]. This metabolite is produced by plants as defense against pathogens when triggered by fungal elicitors and made in plants by a long chain of reactions. In bacteria 4HB can be produced by a simpler manner by converting chorismate to 4HB by the above enzyme in one step. This exemplifies the possibility of engineering more simple mechanisms of transforming one compound to another found in other organisms.

Verpoorte and Memelink (2002) and Oksman-Caldentey and Inzé (2004) provide a good account on the progress made in genetic manipulation of secondary plant metabolite production [83, 79]. In these articles they discussed the possible alternative approaches in metabolic engineering to alter metabolic pathways in order to obtain directed production of designer metabolites. The possibility of protoplast fusion of related or unrelated plant species provides a good avenue for combination enzymes in one species with substrate in another species leading to novel compounds of plant origin. This technology is known as combinatorial biochemistry. Combining cells of two *Solanum* species produced solanidine and tomatidine which are characteristic to individual species, but the hybrid also produced totally new compounds as well [84].

4.2. Plant Cell and Tissue Culture

Micropropagation for generation of plantlets through somatic embryogenesis can produce a large number of genetically identical plantlets within a short period and in a small space, once the protocols have been optimized. In addition to mass production of plantlets, tissue culture technology can be used to generate variations through somaclonal variation. Subsequent selection can lead to identification of better performing genotypes for further multiplication. Oraganogenesis is yet another possibility which enables production of tissue specific substances [85]. Genetic engineering protocols also depend on plant cell culture at one stage of development of transgenic plants.

Another promising aspect of *in vitro* technologies for the production of botanicals is the ability of plant cells to grow in cell suspension culture. *Catharanthus roseus* showed success in production of alkaloids in cell suspension culture [86]. Podophyllotoxin is a useful botanical extracted from *Podophyllum hexandrum* for treating general warts. Chattopadhyay *et al.* (2002) used 3 *l* bioreactor for the cell suspension culture of this plant and successfully recovered podophyllotoxin [87].

Cyclotides are small cyclic peptides stabilized by disulfide bonds between six conserved cystine residues. Their biological activities include antimicrobial and insecticidal actions. Dörnenburg *et al.* (2008) developed techniques to produce cyclotides using callus, suspension culture and hydroponic cultures of *Oldenlandia affinis* and evaluated for production of Kalata B1 accumulation [88]. *In vitro* culture produced only up to 15% of Kalata B1 in comparison with plants grown in hydroponics. Further improvement is probable by manipulating culture conditions and the same group reported higher rate of cell multiplication in 25 l photobioreactor (Seydel *et al.*, 2009) [89]. They claim this approach for harvesting Kalata B1 is more profitable than other methods, such as field cultivation and chemical synthesis.

It has been noted that rate of production and accumulation is enhanced in tissues rather than in cells in suspension. Therefore organogenesis *in vitro* is an alternative for cell suspension or callus cultures. It is a rather easy task to transform plant cells with *Agrobacterium rhizogenes* to induce hairy root formation, the main organ for the production of botanicals. Hairy roots secrete the product into the culture medium which then can be removed and extracted [90]. Introducing a trapping system to trap the secreted metabolite enhances both production and recovery. For example, recovery of betalaines (red natural pigment) produced by hairy roots of *Beta vulgaris* can be increased up to 97.2% by using alumina:silica (1:1) into the culture medium [91].

Srivastava and Srivastava (2007), while highlighting the problems associated with maintaining the delicate hairy root cultures, described different technologies available [92]. The types of reactors used are basically as liquid-phase, gas-phase, or hybrid reactors. Information on nature, applications, perspectives and scale up of hairy root cultures for the production of valuable secondary metabolites are presented in their review. It is also possible to provide substrates for biotransformation by hairy root cultures together with the nutrient medium to increase the efficiency of production. This is useful also in bringing down the cost in downstream processing as the cost of concentration and purification will be reduced. Addition of menthol or geraniol at 25 mg l^{-1} in to hairy root cultures of *Anethum graveolens* did not have impact on the growth and resulted in transformation of these in to volatile products [93]. In a recent experiment, Hernandez-Vazquez *et al.* (2010) were able to increase the production of centellosides in cell cultures of *Centella asiatica* by feeding with α-amyrin with DMSO to assist penetration of α-amyrin in to the cells [94].

Cells cultivated in suspension cultures can be immobilized in various matrices and such immobilized cells have been tested for production of bioactive compounds. When comparing yield of Scopadulcic acid B, a diterpene having antiviral and anti-tumour activity, production by *Scoparia dulcis*, Mathew and Jayachandran (2009) noted 350.57 mg g^{-1} of cells by the 19th day by cells of immobilized in *Luffa* sponge in contrast to 50.85 mg g^{-1} of cells after 30 days of incubation in suspension culture [95]. This is a seven fold increase with a shorter time of incubation. Continuous removal of products and inhibitory substances is an added advantage of immobilized cells.

5. Conclusion

The findings are highly encouraging and more scientific efforts will certainly be devoted into the development of more efficient bioprocess technologies for profitable production of bioactive botanicals including those useful in biological control of pests and diseases of plants and animals including humans. At present the technologies are developed principally for pharmaceutically important botanicals due to their high market value. These developments nevertheless are applicable to botanicals in biocontrol too and hence one can be hopeful that such botanicals produced through new biotechnologies will enter the market in the near future.

References

[1] Perry, A. S., Yamamoto, I., Ishaaya, I. & Perry, R. Y. 1998. *Insecticides in Agriculture and Environment: Retrospects and Prospects*, 261 p. Berlin: Springer-Verlag.

[2] Forget, G., Goodman, T. & de Villiers, A. 1993. *Impact of Pesticide Use on Health in Developing Countries*, 335 p. Ottawa: International Development and Research Centre.

[3] Koch, R.L. 2003. The multicolored Asian lady beetle, *Harmonia axyridis*: A review of its biology, uses in biological control, and non-target impacts, *Journal of Insect Science*, 3:32, 16, URL http://www. insectscience.org/3.32. Accessed 08/11/2010.

[4] Zimmermann, G. 2007. Review on safety of the entomopathogenic fungus *Metarhizium anisopliae*, *Biocontrol Science and Technology*, 17(9-10), 879-920.

[5] Miller, L. K., Lingg, A. J. & Bulla, L. A. J. 1983. Bacterial viral and fungal insecticides. *Science*, 219, 715-721.

[6] Hajek, A. E., McManus, M. L. & Delalibera Júnior, I. 2007. A review of introductions of pathogens and nematodes for classical biological control of insects and mites, *Biological Control*, 41(1), 1-13.

[7] Gashaw, F., Erko, B., Teklehaymanot, T. & Habtesellasie, R. 2008. Assessment of the potential of competitor snails and African catfish (*Clarias gariepinus*) as biocontrol agents against snail hosts transmitting schistosomiasis, *Transactions of Royal Society of Tropical Medicine and Hygiene*, 102(8), 774-779.

[8] de Oliveira, E. J., Rabinovitch, L., Monnerat, R. G., Passos, J. K. J., & Zahner, V. 2004. Molecular characterization of *Brevibacillus laterosporus* and its potential use in biological control, *Applied and Environmental Microbiology*, 70(11), 6657-6664.

[9] Rae, R. G., Robertson, J. F. & Wilson, M. J. 2009. Chemoattraction and host preference of the gastropod parasitic nematode *Phasmarhabditis hermaphrodita*, *Journal of Parasitology*, 95(3), 517-26.

[10] Abd-Allah, K. F., Negm-Eldin, M. M., Saleh, M. H., El-Hamshary, A. M., El-Gozamy, B. M. & Aly, N. S. 2009a. A study on biological control of six fresh water snails of medical and veterinary importance, *Journal of Egyptian Society of Parasitology*, 39(1), 121-139.

[11] Abd-Allah, K. F., Saleh, M. H., El-Hamshary, A. M., Negm-Eldin, M. M., El-Fakahany, A. F., Abdel-Tawab, A. H., Abdel-Maboud, A. I. & Aly, N. S. 2009b.

Helobdella nilae and *Alboglossiphonia conjugata* leeches as biological agents for snails control, *Journal of Egyptian Society of Parasitology*, 39(1), 289-304.

[12] Way, M. J. & Heong, K. L. 2009. Significance of the tropical fire ant *Solenopsis geminata* (hymenoptera: formicidae) as part of the natural enemy complex responsible for successful biological control of many tropical irrigated rice pests, *Bulletin of Entomological Research*, 99(5), 503-12.

[13] Duddington, C. L & Wyborn, C. H. E. 2008. Recent research on the nematophagous Hyphomycetes, *The Botanical Review*, 38(4), 545-565.

[14] Kerry, B. 1997. Biological control of nematodes: prospects and opportunities, In: Plant nematode problems and their control in the Near East region (FAO Plant Production and Protection Paper - 144). (edited by M.A. Maqbool & B. Herry), URL http://www.fao.org/docrep/v9978e/v9978e0b.htm. Accessed 08/11/2010.

[15] Tian, B., Yang, J. & Zhang, K-Q. 2007. Bacteria used in the biological control of plant-parasitic nematodes: populations, mechanisms of action, and future prospects, *FEMS Microbiology and Ecology,* 61, 197–213.

[16] Khan, Z. & Kim, Y. H. 2007. A review on the role of predatory soil nematodes in the biological control of plant parasitic nematodes, Applied *Soil Ecology*. 35(2), 370-379.

[17] Janisiewicz, W. J. & Korsten, L. 2002. Biological control of postharvest diseases of fruits, *Annual Review of Phytopathology*. 40, 411-441.

[18] Paulitz, T. C. & Bélanger, R. R. Biological control in greenhouse systems, *Annual Review of Phytopathology*, 39, 103-133.

[19] Bardin, S. D., Huang, H. C. & Moyer, J. R. 2004. Control of *Pythium* damping-off of sugar beet by seed treatment with crop straw powders and a biocontrol agent, *Biological Control*, 29(3), 453-460.

[20] McFadyen, R. E. C. 1998. Biological control of weeds, *Annual Review of Entomology*. 43, 369-393.

[21] Martin, N. & Paynter, Q. 2010. Assessing the biosecurity risk from pathogens and herbivores to indigenous plants: lessons from weed biological control. *Biological Invasions*, 12(9), 3237-3248.

[22] Fowler, S. V., Syrett, P. & Hill, R. L. 2001. Success and safety in the biological control of environmental weeds in New Zealand, *Austral Ecology*, 25(5), 553-562.

[23] Buckingham, G. R. 1996. Biological control of alligatorweed, *Alternanthera philoxeroides*, the world's first aquatic weed success story, *Castanea*, 61(3), 32-43.

[24] Barreto, R., Charudattan, R., Pomella, A. & Hanada, R. 2000. Biological control of neotropical aquatic weeds with fungi, *Crop Protection*, 19(8-10), 697-703.

[25] Isman, M. B. 2006. Botanical insecticides, deterrents and repellents in modern agriculture and in increasingly regulated world, *Annual Review of Entomology*. 51, 45-66.

[26] Lavandero, I. B., Wratten, S. D., Didham, R. K. & Gurr, G. 2006. Increasing floral diversity for selective enhancement of biological control agents: A double-edged sword? *Basic and Applied Ecology*, 7(3), 236-243.

[27] Landis, D. A., Wratten, S. D. & Gurr, G. M. 2000. Habitat management to conserve natural enemies of Arthropod pests in agriculture, *Annual Review of Entomology*, 45, 175-201.

[28] Lind P. 1998. Encouraging ladybugs, *Journal of Pesticide Reforms*, 18(3), 21-23.

[29] Jones, G. A. & Gillett, J. L. 2005. Intercropping with sunflowers to attract beneficial insects in organic agriculture, *The Florida Entomologist*, 88(1), 91-96.

[30] Long, R. F., Corbett, A., Lamb, C., Reberg-Horton, C., Chandler, J. & Stimmann, M. 1998. Beneficial insects move from flowering plants to nearby crops, *California Agriculture*, 52(5), 23-26.

[31] Mumm, R., Posthumus, M. A. & Dicke, M. 2008. Significance of terpenoids in induced indirect plant defense against herbivorous arthropods, *Plant Cell and Environment*, 31, 575–585.

[32] Loivamakia, M., Mumm, R., Dicke, M. & Schnitzlera, J-P. 2008. Isoprene interferes with the attraction of bodyguards by herbaceous plants, *PNAS.* 105(45), 17430-17435.

[33] Simmonds, M. S. J., Manlove, J. D., Blaney, W. M. & Khambay, B. P. S. 2002. Effects of selected botanical insecticides on the behaviour and mortality of the glasshouse whitefly *Trialeurodes vaporariorum* and the parasitoid *Encarsia formos*, *Entomologia Experimentalis et Applicata*, 102(1), 39-47.

[34] Duso, C., Malagnini, V., Pozzebon, A., Castagnoli, M., Liguori, M. & Simoni, S. 2008. Comparative toxicity of botanical and reduced-risk insecticides to Mediterranean populations of *Tetranychus urticae* and *Phytoseiulus persimilis* (Acari Tetranychidae, Phytoseiidae), *Biological Control*, 47(1),16-21.

[35] Charleston, D. S., Kfir, R., Dicke, M. & Vet, L. E. M. 2005. Impact of botanical pesticides derived from *Melia azedarach* and *Azadirachta indica* on the biology of two parasitoid species of the diamondback moth, *Biological Control*, 33(2), 131-142.

[36] Charleston, D. S., Kfir R., Dicke M. & Vet L. E. M. 2006. Impact of botanical extracts derived from *Melia azedarach* and *Azadirachta indica* on populations of *Plutella xylostella* and its natural enemies: A field test of laboratory findings, *Biological Control*, 39(1), 105-114.

[37] Kreutzweiser, D. P. 1997. Nontarget effects of neem-based insecticides on aquatic invertebrates, *Ecotoxicology and Environmental Safety*, 36(2), 109-117.

[38] Derraik, J. G. B. & Slaney, D. 2005. The toxicity of used coffee grounds to the larvae of *Ochlerotatus* (*Finlaya*) *notoscriptus* (Skuse) (Diptera: Culicidae), *The Annals of Medical Entomology*, 14, 14-24.

[39] Laranja, A. T., Manzatto, A. J. & de Campos Bicudo, H. E. M. 2003. Effects of caffeine and used coffee grounds on biological features of *Aedes aegypti* (Diptera, Culicidae) and their possible use in alternative control. *Genetics and Molecular Biology*, 26(4), 419-429.

[40] Laranja, A. T., Manzatto, A. J. & de Campos Bicudo, H. E. M. 2006. Caffeine effect on mortality and oviposition in successive generations of *Aedes aegypti*, *Revista de Saúde Pública*, 40(6), 1112-1117.

[41] Schmutterer, H. & Singh, R. P. 1995. In*: List of Insects Susceptible to Neem Products*, (edited by H. Schmutterer), Pp.326-365. New York: VCH Publication.

[42] Saxena, R. C. 1989. Insecticides from neem. In*: Insecticides of plant origin*, (edited by J. T. Arnason, B. J. R. Philogène & P. Morand). Pp. 110-115. ACS Symposium Series No. 387. Washington DC: American Chemical Society.

[43] Mordue (Luntz), A. J. & Nisbet, A. J. 2000. Azadirachtin from the neem tree *Azadirachta indica*: its action against insects, *Anais da Sociedade Entomológica do Brasil*, 29(4), 615-632.

[44] Rao, G. V. R., Visalakshmi, V., Suganthy, M., Reddy, P. V., Reddy, Y. V. R. & Rao, V. R. 2007. Relative toxicity of neem to natural enemies associated with the chickpea ecosystem: a case study, *International Journal of Tropical Insect Science*, 27, 229-235.

[45] Lowery, D. T., Isman, M. B. & Brard, N. L. 1993. Laboratory and field evaluation of neem for the control of aphids (Homoptera: Aphididae), *Journal of Economic Entomology*, 86(3), 864-870.

[46] Raguraman, S. & Singh, R. P. 1999. Biological Effects of Neem (*Azadirachta indica*) seed oil on an egg parasitoid, *Trichogramma chilonis*, *Journal of Economic Entomology*, 92(6),1274-1280.

[47] Regnault-Rogger, C. 1997. The potential of botanical essential oils for insect pest control, *Integrated Pest Management Reviews,* 2(1), 25-34.

[48] Isman, M. B. 2000. Plant essential oils for pest and disease management, *Crop Protection*, 19(8-10), 603-608.

[49] Elbadri, G. A. A., Lee, D. W., Park, J. C., Yu, H. B., Choo, H. Y., Lee, S. M. & Lim, T. H. 2008. Nematocidal screening of essential oils and herbal extracts against *Bursaphelenchus xylophilus*, *Plant Pathology Journal*, 24(2), 178-182.

[50] Martín, R. S. & Magunacelaya, J. C. 2005. Control of plant-parasitic nematodes with extracts of *Quillaja saponaria*, *Nematology*. 7(4), 577-585.

[51] Dufour, R., Guerena, M. & Earles, R. 2003. *Nematodes: Alternative Controls*, Pp. 1-15 URL http://www.attra.ncat.org/attra-pub/nematode.html. Accessed 20/02/2010.

[52] Prasad, D., Ram, D. & Imtiyaz, A. 2002. Management of plant parasitic nematodes by the use of botanicals, *Annals of Plant Protection Sciences*, 10(2), 355-359.

[53] Cox, C. J., McCarty, L. B., Toler, J. E., Stephen, A., Lewis, S. A. & Martin, S. B. 2006. Suppressing sting nematodes with *Brassica* sp., poinsettia, and spotted spurge extracts, *Agronomy Journal*, 98, 962–967.

[54] Nagesh, M. & Reddy, P.P. 2000. Status of mushroom nematodes and their management in India, *Integrated Pest Management Reviews*, 5(3), 213-224.

[55] Marston, A., Dudan, G., Gupta, M. P., Solis, P. N., Correa, M. D. & Hostettmann, K. 1996. Screening of Panamanian plants for molluscicidal activity, *Pharmaceutical Biology*, 34(1), 15-18.

[56] Jaiswal, P. & Singh, D. K. 2008. Molluscicidal activity of *Carica papaya* and *Areca catechu* against the freshwater snail *Lymnaea acuminate*, *Veterinary Parasitology*, 152(3-4), 264-270.

[57] Hammami, H. & Ayadi, A. 2008. Molluscicidal and antiparasitic activity of *Solanum nigrum villosum* against *Galba truncatula* infected or uninfected with *Fasciola hepatica*, *Journal of Helminthology*, 82(3), 235-239.

[58] El-Sherbini, G. T., Zayed, R. A. & El-Sherbini, E. T. 2009. Molluscicidal activity of some *Solanum* species extracts against the snail *Biomphalaria alexandrina*, *Journal of Parasitology Research*, URL http://www.hindawi.com/journals/jpr/2009/474360.html. Accessed 08/11/2010.

[59] Hollingsworth, R. G. & Armstrong, J. W. 2003. Effectiveness of products containing metaldehyde, copper or extracts of yucca or neem for control of *Zonitoides arboreus* (Say), a snail pest of orchid roots in Hawaii, *International Journal of Pest Management*, 49(2), 115-122.

[60] Iglesias, J., Castillejo, J. & Ester, A. 2002. Laboratory evaluation of potential molluscicides for the control of eggs of the pest slug *Deroceras reticulatum*

(Muumlller) (Pulmonata: Limacidae), *International Journal of Pest Management.* 48(1), 19-23.

[61] Zou, F. C., Duan, G., Xie, Y. J., Zhou, Y., Dong, G. D., Lin, R. Q., Zhu, X. Q. 2009. Molluscicidal activity of the plant *Eupatorium adenophorum* against *Oncomelania hupensis*, the intermediate host snail of *Schistosoma japonicum*. *Annnnals of Tropical Medical Parasitology*, 103(6), 549-53.

[62] Martín, R. S., Ndjoko, K. & Hostettmann, K. 2008. Novel molluscicide against *Pomacea canaliculata* based on quinoa (*Chenopodium quinoa*) saponins, *Crop Protection*, 27(3-5), 310-319.

[63] Kumar, A., Shukla, R., Singh, P. & Dubey, N. K. 2009. Biodeterioration of some herbal raw materials by storage fungi and aflatoxin and assessment of *Cymbopogon flexuosus* essential oil and its components as antifungal, *International Biodeterioration & Biodegradation*, 63(6), 712-716.

[64] Reddy, K. R. N.,Reddy, C. S. & Muralidharan, K. 2009. Potential of botanicals and biocontrol agents on growth and aflatoxin production by Aspergillus flavus infecting rice grains, *Food Control*, 20(2), 173-178.

[65] Kumar, R., Mishra, A. J., Dubey, N. K. & Tripathi, Y. B. 2007. Evaluation of *Chenopodium ambrosioides* oil as a potential source of antifungal, antiaflatoxigenic and antioxidant activity, International Journal of Food Microbiology. 115, 159–164.

[66] Kumar, A., Shukla, R., Singh, P., Prasas, C.S. & Dubey, N.K. 2008. Assessment of *Thymus vulgaris* L. essential oil as a safe botanical next term preservative against post harvest fungal infestation of food commodities, *Innovative Food Science & Emerging Technologies*, 9(4), 575-580.

[67] Bower, J.H. & Locke, J.C. 2004. Effect of formulated plant extracts and oils on population density of *Phytophthora nicotianae* in soil and control of Phytophthora blight in the greenhouse, *Plant Disease*, 88(1), 11-16.

[68] Joël, L., Vanneste, J. L., Hill, R. A., Kay, S. J., Farrell, R. L. & Holland, P. T. 2002 Biological control of sapstain fungi with natural products and biological control agents: a review of the work carried out in New Zealand, *Mycological Research*, 106(2), 228-232.

[69] Halcón, L. & Milkus, K. 2004. *Staphylococcus aureus* and wounds: A review of tea tree oil as a promising antimicrobial, *American Journal of Infection Control*, 32(7), 402-408.

[70] Carson, C. F., Hammer, K. A., & Riley, T. V. 2006. *Melaleuca alternifolia* (Tea Tree) Oil: a Review of Antimicrobial and Other Medicinal Properties, *Clinical Microbiology Reviews*, 19(1), 50-62.

[71] Mondello, F., Girolamo, A., Scaturro, M. & Ricci, M. L. 2009. Determination of *Legionella pneumophila* susceptibility to *Melaleuca alternifolia* Cheel (tea tree) oil by an improved broth micro-dilution method under vapour controlled conditions, *Journal of Microbiological Methods*, 77(2), 243-248.

[72] Kwieciński, J., Eick, S. & Wójcik, K. 2009. Effects of tea tree (*Melaleuca alternifolia*) oil on *Staphylococcus aureus* in biofilms and stationary growth phase, *International Journal of Antimicrobial Agents*, 33(4), 343-347.

[73] Williams, L. 1996. Ranking antimicrobial activity. *International Journal of Aromatherapy*, 7(4), 32-35.

[74] Putnam, A. R. & DeFrank, J. 1983. Use of phytotoxic plant residues for selective weed control, Crop Protection, 2(2), 173-181.
[75] Hong, H. N., Xuan, T. D, Eiji, T. & Khanh, T.D. 2004. Paddy weed control by higher plants from Southeast Asia, *Crop Protection*, 23(3), 255-261.
[76] Marley, P. S., Shebayan, J., Aba, D. A. & Idem, N. 2004. Possibilities for control of *Striga hermonthica* in Sorghum (*Sorghum bicolor*) using neem (*Azadirachta indica*) and parkia (*Parkia biglobosa*)-based products, *International Journal of Pest Management*, 50(4), 291-296.
[77] Kohli, R. K., Batish, D. R. & Singh, H. P. 1998. Eucalypt oils for the control of Parthenium (*Parthenium hysterophorus* L.), *Crop Protection*, 17(2), 119-122
[78] Batish, D. R., Setia, N., Singh, H. P. & Kohli, R. K. 2004. Phytotoxicity of lemon-scented eucalypt oil and its potential use as a bioherbicide, *Crop Protection*, 23(12), 1209-1214.
[79] Oksman-Caldentey, K-M.& Inzé,D. 2004. Plant cell factories in the post-genomic era: new ways to produce designer secondary metabolites, *Trends in Plant Science*. 9(9), 433-440.
[80] Hain, R., Reif, H. J., Krause, E., Langebartels, R., Kindl, H., Vomam, B., Wiese, W., Schmelzer, E., Schreier, P., Stocker, R. & Stenzel, K. 1993. Disease resistance results from foreign phytoalexin expression in a novel plant, *Nature*, 361, 153-156.
[81] Bak, S., Olsen, C. E., Halkier, B. A. & Møller, B. L. 2000. Transgenic tobacco and *Arabidopsis* plants expressing the two multifunctional *Sorghum* Cytochrome P450 Enzymes, CYP79A1 and CYP71E1, are cyanogenic and accumulate metabolites derived from intermediates in dhurrin biosynthesis, *Plant Physiology*, 123, 1437-1448.
[82] Siebert, M., Sommer, S., Li, S., Wang, Z., Severin, K. & Heide, L.1996. Genetic Engineering of Plant Secondary Metabolism: Accumulation of 4-Hydroxybenzoate glucosides as a result of the expression of the bacterial ubiC gene in tobacco, *Plant Physiology*, 112, 811-819.
[83] Verpoorte, R. & Memelink, J. 2002. Engineering secondary metabolite production in plants, *Current Opinion in Biotechnology*, 13(2), 181-187.
[84] Laurila, J., Valkonen, J. P. T., Hiltunen, R., Pehu E. & Laakso I. 1996. Formation of parental-type and novel glycoalkaloids in somatic hydrids between *Solanum brevidens* and *S. Tuberosum*, *Plant Science*, 118, 145–155.
[85] Rout, G. R., Samantaray, S. & Das, P. 2000. *In vitro* manipulation and propagation of medicinal plants, *Biotechnology Advances*, 18(2), 91-120.
[86] Kurz, W. G. W., Chatson, K. B., Constabel, F., Kutney, J. P., Choi, L. S. L., Kolodziejczyk, P., Sleigh, S. K., Stuart, K. L. & Worth, B. R. 1980. Alkaloid production in *Catharanthus roseus* cell cultures: Initial studies on cell lines and their alkaloid content, *Phytochemistry*. 19(12), 2583-2587.
[87] Chattopadhyay, S., Srivastava, A. K., Bhojwani, S. S., Virendra, S. & Bisaria V. S. 2002. Production of podophyllotoxin by plant cell cultures of *Podophyllum hexandrum* in bioreactor, *Journal of Bioscience and Bioengineering*, 93(2), 215-220.
[88] Dörnenburg, H., Frickinger, P. & Seydel, P. 2008. Plant cell-based processes for cyclotides production, *Journal of Biotechnology*, 135 (1), 123-126.
[89] Seydel, P., Walter, C. & Dörnenburg, H. 2009. Scale-up of *Oldenlandia affinis* suspension cultures in photobioreactors for cyclotide production, *Engineering in Life Sciences*, 9(3), 219-226.

[90] Guillon, S., Tre′mouillaux-Guiller, J., Pati, P. K., Rideau, M. & Gantet, P. 2006. Hairy root research: recent scenario and exciting prospects. *Current Opinion in Plant Biology*, 9, 341–346.

[91] Rudrappa, T., Neelwarne, B., Aswathanarayana, R. G. 2004. *In situ* and *ex situ* adsorption and recovery of betalains from hairy root cultures of *Beta vulgaris*, *Biotechnology Progress*, 20, 777–785.

[92] Srivastava, S. & Srivastava, A. K. 2007. Hairy root culture for mass-production of high-value secondary metabolites, *Critical Reviews in Biotechnology*, 27(1), 29-43.

[93] Faria, J. M. S., Nunes, I. S., Figueiredo, A. C., Pedro, L. G., Trindade, H. & Barroso, J. G. 2009. Biotransformation of menthol and geraniol by hairy root cultures of Anethum graveolens: effect on growth and volatile components, *Biotechnology Letters*, 31(6), 897-903.

[94] Hernandez-Vazquez, L., Bonfill, M., Moyano, E., Cusido, R. M., Navarro-Ocaña, A. & Palazon, J. 2010. Conversion of α-amyrin into centellosides by plant cell cultures of *Centella asiatica*, *Biotechnology Letters*, 32(2), 315-319.

[95] Mathew, A. J. & Jayachandran, K. 2009. Production of scopadulcic acid B from *Scoparia dulcis* Linn. using a *Luffa* sponge bioreactor, *Plant Cell, Tissue and Organ Culture*, 98(2), 197-203.

In: Bioprocess Sciences and Technology
Editor: Min-Tze Liong
ISBN 978-1-61122-950-9

Chapter 16

INSECT CHITIN-REMODELING ENZYMES: PHYSIOLOGICAL ROLES, APPLICATIONS AND PREPARATION

Q. Yang*[*], *J. Yan, T. Liu, M. Qu, F. Liu
School of Bioscience and Biotechnology, Dalian University of Technology, Dalian, P. R. China

ABSTRACT

Insects are the largest living group on earth consisting of 75% of all animal species. They play versatile roles that increase crop production via pollinators, cause injury to crops leading to economic losses and spread diseases to human. These creatures are rich yet unexploited resources for enzymes, which may be the key for crop protection, epidemic disease control and perhaps biomass transformation. Chitin metabolism, for example, recruits many important enzymes. This chapter aims offering researchers or students a general view of insect chitin-remodeling enzymes.

Chitin is a linear homopolymer of N-acetylglucosamines conjugated by β-1,4-glycosidic bonds. Insect exoskeleton consists of chitin highly complexed with protein, lipids and other components such as phenols, calcium carbonate, manganese and zinc. Structural rigidness limits organisms' growth, thus an elaborate process of molting is evolved, which depends on the existence and balance of some certain hormone, and subsequently catalyzed by chitin-remodeling enzymes including chitin synthases and chitinolytic enzymes. Coordination of chitin synthesis and its degradation requires strict control of the participating enzymes during development, which are thus ideal targets for novel insecticides. The preparation of these enzymes is essential for biochemical studies. Moreover, since chitin is the second most abundant but underutilized biomass, chitin degradation by chitinolytic enzymes is of biotechnological importance.

[*] Corresponding author. Email: qingyang@dlut.edu.cn

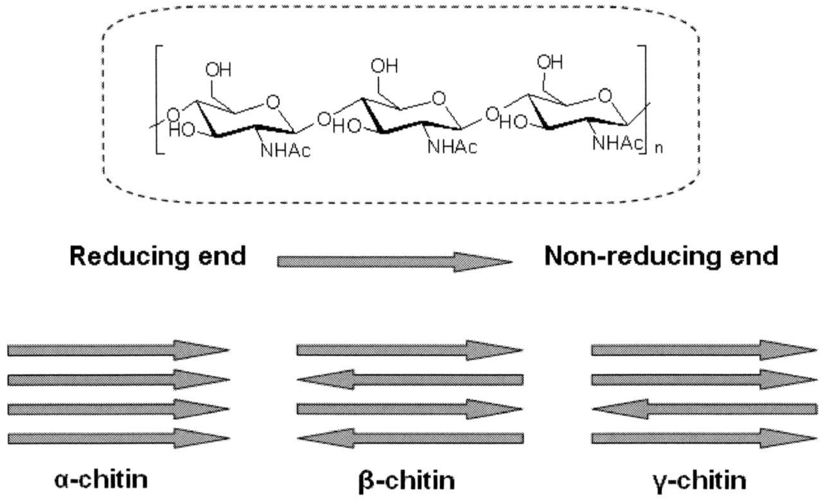

Figure 1. Schematic plot of chitin forms.

1. CHITIN

1.1. Structure

Chitin, an acetylated polysaccharide similar to cellulose, is a linear polymer of β-(1,4)-linked *N*-acetyl-D-glucosamine (GlcNAc) and a few residues of unacetylated glucosamine. Chitin with more than 40% acetyl off is termed chitosan. Adjacent chains of chitin are cross-linked by hydrogen bonds to form chitin microfibrils. Each chain contains both a reducing and non-reducing ends, because the sugar ring can be open to give an aldehyde group. The difference makes α-chitin the highest degree of strength and stability, while β- and γ-chitins are less tightly packed in contrast. The microfibrils are laid down in parallel within a single layer, but are rotated by a constant angle in successive layers to produce a plywood-like helicoidal arrangement to contribute its overall strength [1]. Additionally, the chitin microfibrils are covalently linked to the surrounding proteins.

1.2. Distribution

Chitin is the second most abundant biopolymer on earth (next to cellulose), and is estimated that ten-billion tons of chitin are synthesized and degraded in the biosphere annually [2]. Ocean is one remarkable environment with abundant biomass of chitin, and marine bacterium that may not metabolize chitin but are capable of degrading chitin and utilize it as nutrition. Chitin is widely spread in arthropod, including chelicerata, crustacea, myriapoda, and insecta, mainly as protective exoskeleton. The physiological function of chitin in arthropod is elaborated further with insects as an example. Cell wall of fungi constituted of chitin and other compounds is considered as exoskeleton somehow and involved in maintaining and protection, and in development and morphogenesis as well. In addition to its presence in the cuticle of arthropods and cell wall of fungi, chitin is involved in the formation of the skeletons of annelids and mollusks. Furthermore, chitin is found in the eggshells and pharynx of nematodes [3], and in the cyst wall of *Entamoeba histolytica* [4].

1.3. Physiological Function

An exoskeleton as the name implies, is an external skeleton, which is constructed of chitin highly complexed with proteins and other compounds such as phenols, calcium carbonate, manganese and zinc, and functions as protection and structural support. The presence of exoskeleton makes arthropod (esp. insects) an evolutionary success in animal kingdom and promise the perfect combination of rigidness and extensibility. For one thing exoskeleton covers the organisms completely like the skin of vertebrates, thus serves as an interface between the organism and the environment, an essential barrier for small animals (as insects) with a high surface-to-volume ratio. Additionally, exoskeleton provides an advantage mechanical support, considering the size-to-weight ratio of insects. Furthermore, the chitin content of the dry mass may be as high as 40% [5], much of which is re-adsorbed during molting and periods of starvation, the integument could also be viewed as a food reserve for insects.

Occurrence of chitin in insects is in particular as an instance. As high as 40% of the dry mass [5], two major roles of the chitin are related to the cuticle that overlays the epidermis and the peritrophic membrane that lines the midgut. In addition, chitin is also found in salivary glands and mouthparts, in the intima that lines the foregut and hindgut and in tracheae that forms spiral thickenings within the tracheal trunk (known as taenidia). However, the two major structures that occur in insects, epidermis and peritrophic membrane, are of remarkable mechanical properties and of vital function in digestion. Those physical functions are discussed in more details in the followings.

1.4. Chitin Metabolism

Arthropods, like insecta, chelicerata and myriapoda, are covered with rigid and resistant exoskeleton being constructed of chitin highly complexed with proteins and other compounds, while the cuticle of many crustaceans is further strengthened by calcification [6]. However, this structural rigidness limits organisms' growth, thus an elaborate process of molting (Figure 2) that depends on the existence and balance of certain hormone is required, and subsequently catalyzed by numbers of enzymes, chitin is degraded and 're-synthesized' to allow organisms to grow during a pale, soft-bodied after-molting period.

Generally, the biosynthesis of chitin and its degradation is an elaborate process which includes [7]: 1) the hormone-controlled synthesis, translocation and activation of chitin synthases; 2) the formation of UDP-GlcNAc as the substrate donor by biotransformation, phosphorylation, amination, acetylation; 3) polymerization of GlcNAc followed by 4) hydrogen bond formation, 5) microfibrils orientation and 6) covalent and noncovalent interaction with proteins or polysaccharides to form supramolecular structures; and 7) a physiologically necessary degradation by three enzymes, an endo-enzyme (chitinase) that randomly hydrolyze β-(1-4)-glycosidic bonds of chitin polymers to chito-oligomers, an exo-enzyme (β-N-acteyl-glucosaminidase) that subsequently hydrolyze the oligomers to monomers, and chitodextrinase which hydrolyzes small chitooligosaccharides. Insect chitin metabolism, as an example, is discussed in details in the following paragraph and shown in the figure below.

The molting begins when an insect reaches 'crucial' size or weight as well as other factors including nutritional status and adequate photoperiod. As a response to those factors, insect becomes inactive but is in preparation for ecdysis: the brain secretes prothoracicotropic hormone that causes peripheral prothoracic gland to synthesize and secrete the steroid hormone ecdysone. Then ecdysone is converted by peripheral tissues to the active 20-hydroxyecdysone (20E), the key hormone that regulates molting. Increase of 20E causes epidermal cell division to form new cuticle and separates new cuticle from the old cuticle by digestion, in the presence of molting fluid containing chitinolytic enzymes. Those enzymes include chitinase for yielding oligomers, β-N-acteyl-glucosaminidase for hydrolysis of dimmers or terminal polymers, and chitodextrinase for hydrolysis of small chitooligosaccharides. The proportion of exo-and endo-enzyme is 1:6. It is between the apolysis (the separation of old cuticle) and ecdysis (the casting off of the old cuticle) dose chitin synthesis occurs. It begins with the release of trehalose into the hemolymph, which acts by trehalase to produce glucose. Glycogen phosphorylase converts glucose to glucose-1-phosphate that is aminated and acetylated to form a monomer later. This monomer is activated by uridine diphosphate and added to the end of an existing chitin chain. Besides the exoskeleton, attachment on the cuticle like sensory bristles, alimentary tract and trachea are shed as well. Eventually, an increase in internal pressure aids rigidity of the new cuticle and leads to a crack that allows insects extricate themselves. The tanned impermeable new cuticle begins to function meanwhile a decrease of 20E titers occurs. Molting is finally completed. Metamorphosis, also directed by 20E, is much more complex than molting. The endocrine signal is transduced into tissue-specific pattern and bound to a heterodimer of ecdysone receptor (EcR) and ultraspiracle (USP). By turning on/off certain genes encoding related enzymes, metamorphosis is initiated and eventually it turns a larva to an adult, say caterpillar into a butterfly [90].

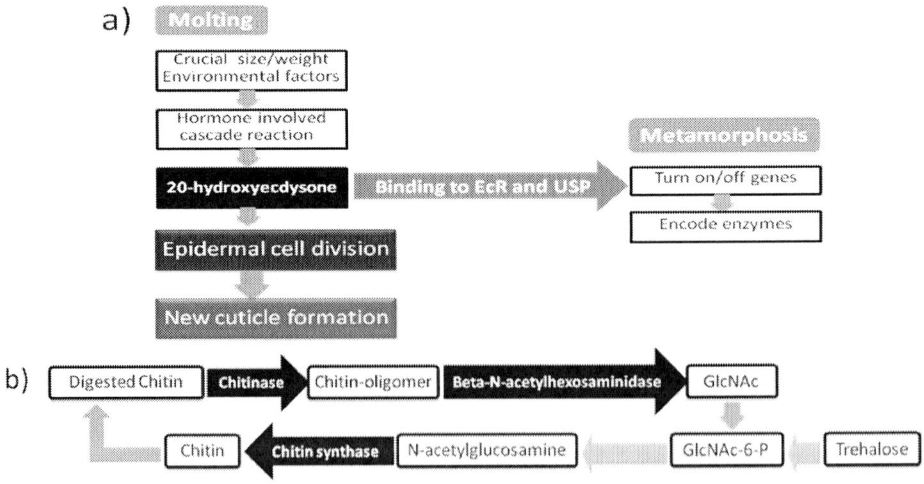

Figure 2. a) Regulation of the insect molting process and b) the biosynthesis and degradation of chitin.

Crustacean, supported by recent morphological and molecular comparison [8], is genetically linked with insects. The pattern of growth and hormones involved is analogous to that of insects, too. The hormones involved are molt inhibiting hormone [9], crustecdysone [10], hyperglycaemic hormone [8] and methyl farnesoate [11]. Besides hormone controlling, growth in crustacean is also effected by external factors like temperature and food supply. However unlike insect, the growth pattern of crustacean is of more diversity and of less understanding [12].

2. Chitin-Remodeling Enzymes

2.1. Key Enzymes in Insect Chitin Metabolism

Chitin is one of the most important structural components in insect. Upon the requirement of growth and development of insects, the chitin-containing tissues such as exoskeleton and peritrophic membrane are degraded and re-synthesized. For this purpose, insects have evolved to produce various enzymes with tissue specificity and function distinguishment. Both expression and activity of these enzymes are strictly controlled by different paths. It is currently considered chitin synthesis involves chitin synthase, while insect chitin degradation system contains two enzymes, chitinase (EC 3.2.1.14) and β-*N*-acetyl-D-hexosaminidase (EC 3.2.1.52). Chitinase initially degrades chitin into oligomers, and then *β*-*N*-acetyl-D-hexosaminidase cleaves these oligomers into monosaccharide GlcNAc. The individual role of the enzymes is discussed in the followings.

2.2. Insect Chitinases

Chitinases are widely distributed in almost all organisms. They can be classified into family 18 (GH18) and family 19 (GH19) glycosyl hydrolases on the basis of amino acid similarity. Insect chitinases belongs to GH18.

The first cDNA encoding insect chitinase was cloned from tobacco hornworm, *Manduca sexta* [13]. It is a 554 amino acid-residue protein composed of two domains, one N-terminal catalytic domain and one C-terminal chitin-binding domain. These two domains are connected by a serine/threonine-rich linker region. Since then, cDNAs encoding chitinases with the same domain composition have been obtained from many insect species of different orders including Lepidoptera (*Bombyx mori, Spodoptera exigua, Ostrinia furnacalis, Chilo suppressalis, Spodoptera litura, Bombyx mandarina, Spodoptera frugiperda, Agrotis ipsilon, Helicoverpa armigera, Choristoneura fumiferana, Mamestra brassicae, Mythimna separata, Hyphantria cunea, Lacanobia oleracea*), Coleoptera (*Tribolium castaneum*), Hymenoptera (*Nasonia vitripennis, Apis mellifera*), Diptera (*Drosophila melanogaster, Culex quinquefasciatus, Aedes aegypti, Anopheles gambiae*), Phthiraptera (*Pediculus humanus corporis*) and Hemiptera (*Acyrthosiphon pisum*) (Data are derived from GenBank database). TmChit5, the gene encoding chitinase from *Tenebrio molitor* is a 2839 amino acid-residue protein composed of more than one catalytic as well as chitin-binding domains [14]. Zhu *et al.* have comprehensively studied all putative genes encoding chitinases from genomes of three insects including *D. melanogaster*, *T. castaneum* and *A. gambiae* [15,16]. There are 17-24 genes encoding chitinases in these genomes. These chitinases can be divided into eight groups based on sequence similarity and domain architecture (Figure 3).

Why insect produces so many kinds of chitinases? One possibility is that these chitinases play different physiological roles. To be closer to the answer, the transcriptional level of all chitinases from *T. castaneum* at different developmental stages and different tissues were studied [16]. Through Group I to VIII chitinases were expressed at every developmental stages (embryonic, larval, pupal and adult stages), Group IV chitinases were expressed only in larval and adult stage. It was also observed that Group I to III chitinases are both expressed in epidermal tissues while Group IV and VIII enzymes are mainly in gut tissues. RNA interference has been then applied to clarify the physiological roles of each chitinase gene during larval-larval, larval-pupal and pupal-adult molts, respectively [16]. The Group II chitinase gene, *TcCHT10*, was found to be vital for every molting stage. Since it is only expressed in epidermal tissues and contains multiple catalytic domains and chitin-binding domains, it is proposed to function in disruption of the crystalline chitin fibrils of cuticle and then degradation of the resulting single chains in endo-pattern. The Group I chitinase gene, *TcCHT5*, was found important for pupal-adult molt though it was expressed at all developmental stages. Group III chitinase, *TcCHT7*, a membrane-associated protein, were found important for abdominal contraction and wing/elytra extension after pupation. RNA interference of genes encoding group IV chitinases was found to have no effect on the molts. Since they are predominantly expressed in gut tissues, they are presumed to degrade the midgut peritrophic membrane or chitin-containing materials.

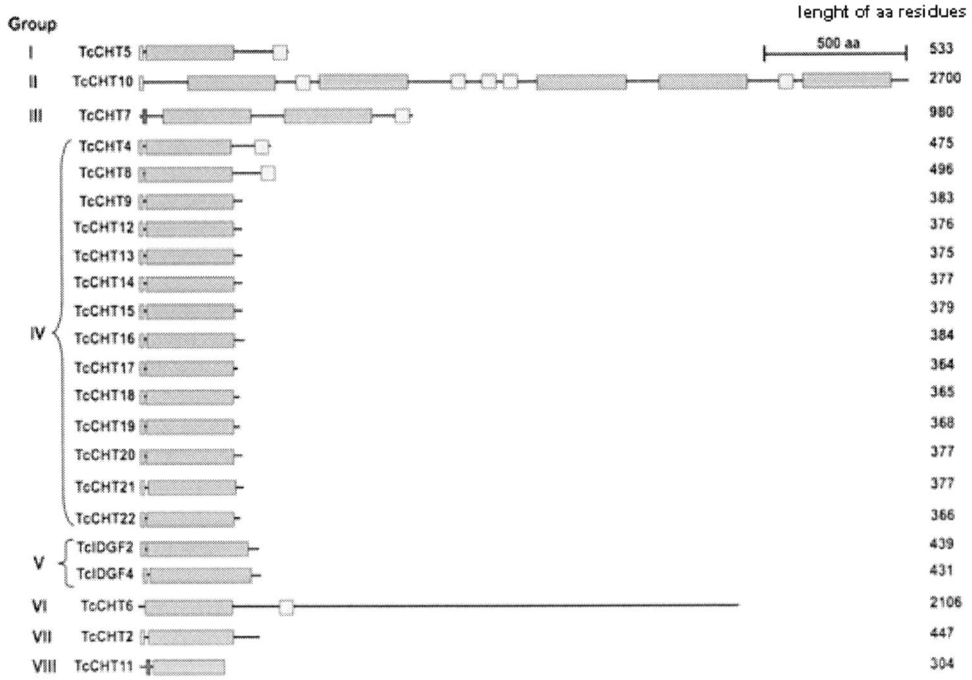

Figure 3. Classification by genes encoding insect chitinases.

Insect chitinases consist of two domains, catalytic domain and chitin-binding domain. The catalytic domain is a classical $(\beta/\alpha)_8$ TIM-barrel domain, eight-stranded β-barrel cored by eight α-helices outside with an orientation anti-barrel to the β-strands. There are four highly conserved motifs in the catalytic domain, which are K(F/V)M(V/L/I)AVGGW in β3, FDG(L/F)DLDWE(Y/F)P in β4, M(S/T)YDL (R/H)G in β6 and GAM(T/V)WA(I/L)D in β8. The amino acid triplet, DWE, in the second conserved motif β4 has been proved to be important for enzymatic activity by site-directed mutation [17, 18]. The glutamic acid residue (E) in DWE is likely to be the proton donor. A replacement of glutamic acid residue at the active center with an aspartic acid residue (D) will result in the total loss of activity. The chitin-binding domain is characterized by a conserved motif containing six cysteines ($CX_{11-24}CX_{5-6}CX_{9-19}CX_{10-17}CX_{4-14}C$), function to increase the affinity to polymeric instead of oligosaccharide substrates [19]. The linker is flexible and highly O-glycosylated [19], and proposed for structure stabilization.

Insect chitinases act on chitin or chitiooligosaccharides. Using oligosaccharide as substrate, the preferred cutting site for chitinase is between the second and the third *N*-acetyl-D-glucosamine moiety numbered from its non-reducing end. Some chitinases also hydrolyze chitosan (partially deacetylated chitin) and bacterial cell wall polysaccharides.

2.3. Insect β-N-acetyl-D-hexosaminidases

Like chitinases, β-*N*-acetyl-D-hexosaminidases are widely distributed in organisms ranging from bacteria, fungi, plants and animals. They can be grouped into glycosyl hydrolase family 3 (GH 3), GH 20 and GH 84 on the basis of amino acids sequence similarity. Most of the known insect β-*N*-acetyl-D-hexosaminidases belong to GH 20. An insect contains three or more β-*N*-acetyl-D-hexosaminidase genes. *T.castaneum*, for example, contains as many as seven β-*N*-acetyl-D-hexosaminidase genes [20].

Insect β-*N*-acetyl-D-hexosaminidases can be categorized into four groups (Figure 4) according to phylogenetic analysis [21]. Overall, the β-*N*-acetyl-D-hexosaminidases from the same insect species are not inclined to the same group. In other words, insect β-*N*-acetyl-D-hexosaminidases are not phylogenetically related but vary in enzymatic properties, tissue location and physiological functions.

Group I share the lowest sequence identities with the other three groups, but high identities with mammal β-*N*-acetyl-D-hexosaminidases. Biochemical studies of Group I chitinases from either *B. mori* or *S. frugiperda* showed their wide substrate spectrum to function as lysosomal enzymes [22,23].

Group II enzymes contain N-terminal transmembrane regions and are presumed to be membrane-bound proteins that are involved in *N*-glycan modification before the maturation of glycoproteins [24,25]. They could not cleave chitooligosaccharides but selectively cleave the terminal GlcNAc residues from one of the antennae of biantennary *N*-glycans [25].

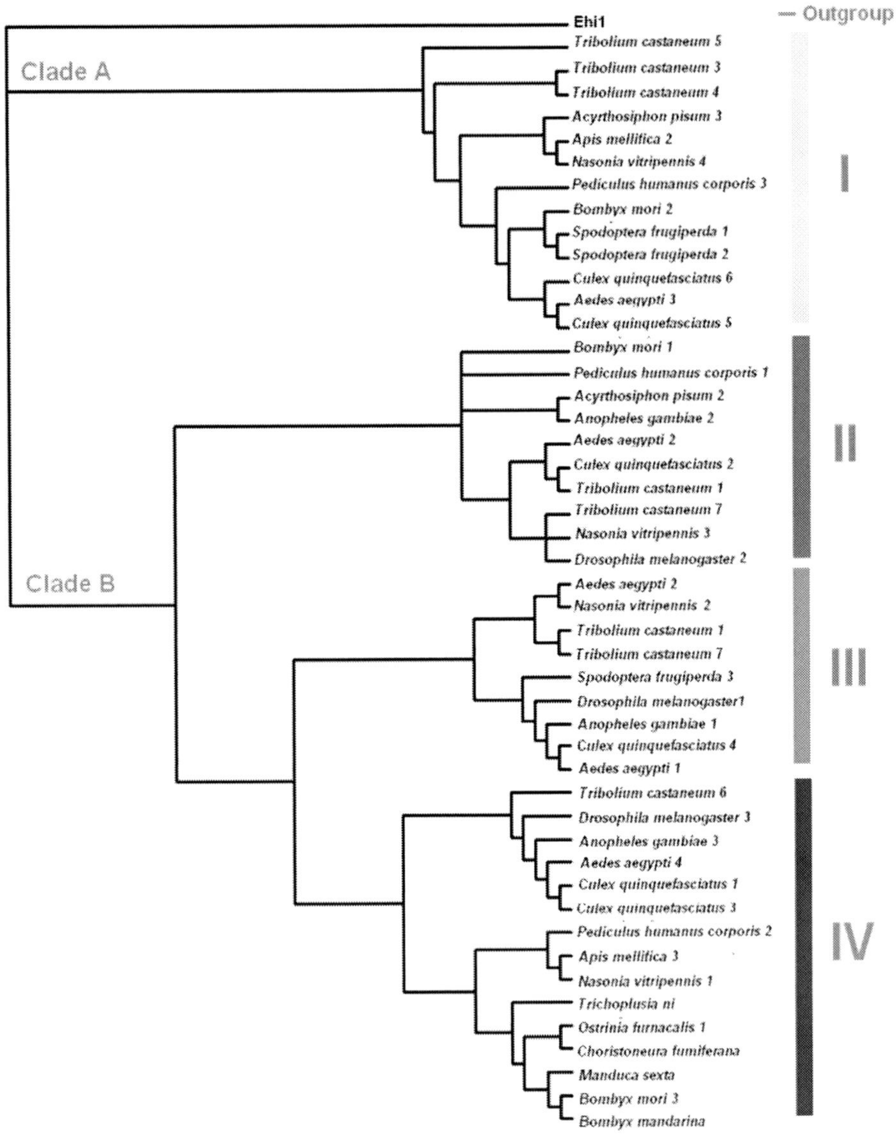

Figure 4. Phylogenetic tree of insect β-*N*-acetyl-D-hexosaminidases.

Group III enzymes are presumed to be involved in chitin degradation because of their localization in cuticle and high cleavage efficiency in chitooligosaccharides degradation [26]. However, enzymatic activity towards chitooligosaccharide substrates does not exclude other activities. We have purified and identified a group III member (OfHex1) from the Asian corn borer, *O. furnacalis* [21]. It efficiently degrades chitooligosaccharides but cannot cleave N-glycan substrates, and its transcriptional level is up-regulated at molting stage which indicates that Group III β-*N*-acetyl-D-hexosaminidases OfHex1 is not involved in glycoprotein degradation or N-glycan modification.

Insect β-*N*-acetyl-D-hexosaminidases contain two domains, one N-terminal zincin-like domain and one C-terminal catalytic TIM-barrel domain. Though the crystal structure of an insect hexosaminidase is unknown yet, the structures of their analogs from mammals and microorganisms have been revealed [27-34]. The N-terminal domain has an α+β fold comprising of five strands of parallel β-sheet, each of which is sandwiched by two approximately parallel α-helices. The C-terminal catalytic domain contains a classic $(\beta/\alpha)_8$ TIM-barrel fold but without α5 and α7, and the key catalytic residues Glu and Asp are highly conserved.

Both GH18 family chitinase and GH20 family β-*N*-acetyl-D-hexosaminidases follow the same catalytic mechanism, termed substrate-assisted retaining mechanism. In this mechanism, carbonyl oxygen of the 2-acetamido group of substrate acts as a nucleophilic group to attack C1 of the substrate itself, leading to the formation of an oxazolinium intermediate. Since the key catalytic amino acid residues are conserved in their respective family, human chitinase (HsCHT) and human β-*N*-acetyl-D-hexosaminidase B complexed with NAG-thiazoline (HsHexB) are often applied as structural templates respectively.

2.4. Chitin Synthases

Chitin synthase is the enzyme involved in the final step of chitin biosynthesis. It catalyzes the polymerization of N-acetylglucosamine through the formation of β-1,4 glycosidic linkage. According to the sequence homology analysis and the catalytic feature, it can be grouped into glycosyltransferase family 2, to which cellulose synthase and hyaluronan synthase belong .

Chitin synthase has been found in chitin containing organisms such as fungi, nematodes and insects. Every enzyme contains a highly conserved catalytic site and several transmembrane helices for binding to cell membrane. Chs3p, the chitin synthase from *Saccharomyces cerevisiae* is the best characterized one. Chs3p catalyzes the formation of more than 90% chitin in the yeast cell wall [35]. Many factors, such as chaperon [36], coat proteins [37,38] and activators [39], take parts in the intracellular transportation, activity control and regulation of Chs3p in *S. cerevisiae*. Insect chitin synthase attracts research interest in that it is an essential enzyme for the formation of cuticle as well as peritrophic matrix remodeling during metamorphosis. Compared to yeast, the knowledge about insect chitin synthase is very limited. Here we give a general description of its gene structure, deduced protein structure, classification, biofunction and regulation.

2.4.1. Gene and Deduced Protein Sequences of Insect Chitin Synthases

Insect chitin synthase is a large protein with a molecular weight of about 170 kDa and encoded by a gene that contains about 4,500 nucleic acids. The first insect chitin synthase gene was obtained from *Lucilia cuprina* (the sheep blowfly) in 2000 [40]. So far, more than 30 cDNAs encoding insect chitin synthases have been isolated and sequenced. The deduced protein of insect chitin synthase contains 16-17 transmembrane helices, and seven highly conserved motifs. These motifs are the K-(β-fold)-DG and R-(β-fold)-GK for UDP-binding; DXD and G(X)4(Y/F)R) region for donor saccharide binding; GEDRXX region for binding to saccharide acceptor; Q(Q/R)XRW for the product-binding and (S/T)WGT(K/R) for the transportation of newborn chitin chain across the plasma membrane [41].

2.4.2. Classification of Insect Chitin Synthases

All the species that can synthesize chitin contain more than one chitin synthases. The yeast *S. cerevisiae* encodes three chitin synthases that have different functions. Insects appear to possess two different kinds of chitin synthases according to the amino acid sequence analysis and Southern blot. According to the sequence homology analysis, these chitin synthases are closely related and can be grouped into two different phylogenetic classes [50], class A (CHSA) and class B (CHSB) (Figure 5). In terms of either gene or protein structure, *CHSA* is presumed to be responsible for the biosynthesis of cuticle chitin while *CHSB* is thought to be involved in the midgut chitin formation.

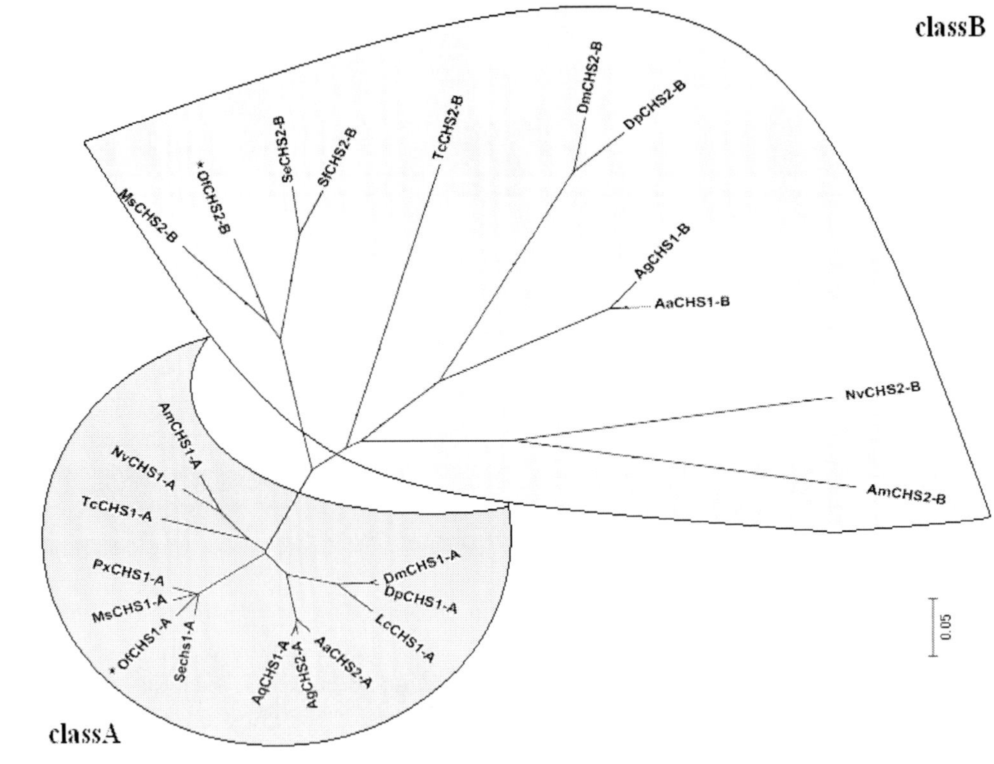

Figure 5. Phylogenic tree of insect chitin synthases. Insect chitin synthases can be divided into two classes, Class A and Class B.

2.4.3. Gene Structure of Insect Chitin Synthases

Both *CHSA* and *CHSB* contain multiple exons (Figure 6). Compared to *CHSB*, the genomic structure of *CHSA* is more complex. Genomic DNA sequence analysis reveals that the genome of *CHSA* contains an extra alternative exon, which is 177bp long and encodes the second to the last transmembrane helixes (Figure 7). The alternative exon results in two spliced variants, which vary in gene transcriptional pattern and post-translational glycosylation. Besides the 177bp splicing exons (19a and 19b for *OfCHSA*), other splicing site is also found. The genomic DNA of the insect chitin synthase from *Ostrinia furnacalis* as well as *Bombyx mori* carries an alternative transcriptional start site (Sa and Sb, Figure 6).

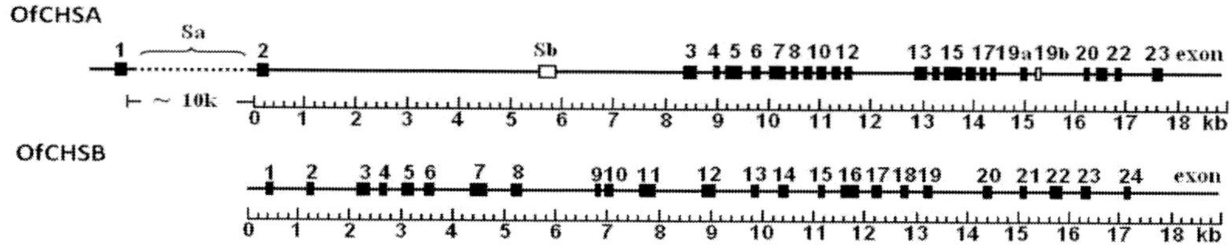

Figure 6. The genomic DNA structure of chitin synthases from the insect *Ostrinia furnacalis*. OfCHSA has 23 exons and two splicing sites: the transcriptional starting sites and the 19th exon. OfCHSB has 24 exons. The shaded boxes represent exons and lines represent introns. The length of each genome is scaled underneath.

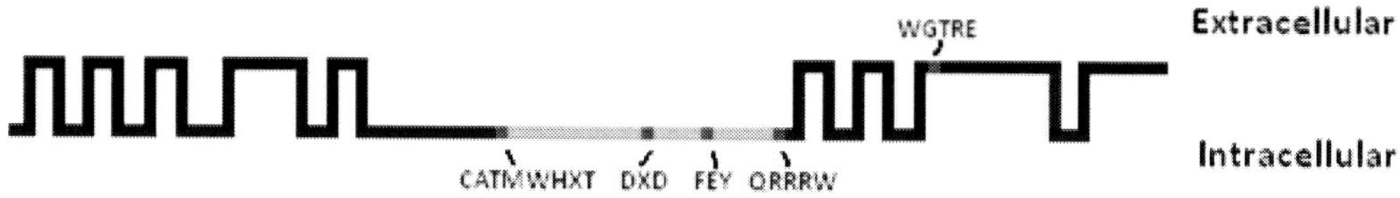

Figure 7. The predicted secondary structure of insect chitin synthases. The highly conserved amino acids are presented.

2.4.4. Structural View of Insect Chitin Synthases

Insect chitin synthase is a large transmembrane protein with 16-17 transmembrane helices. Based on the deduced amino acid sequence analysis, insect chitin synthases contain three domains (Figure 7): the N-terminal domain containing 9-10 transmembrane helices, the C-terminal domain containing seven transmembrane helices and a central catalytic domain. Six (K-(β-fold)-DG, R-(β-fold)-GK, DXD, G(X)4(Y/F)R), GEDRXX and Q(Q/R)XRW) of the seven highly conserved motifs mentioned before are located in the catalytic domain. The last conserved motif (S/T)WGT(K/R) is at the end of the five transmembrane helices. These five helices were believed to take part in either the translocation of nascent chitin chains to cross the cell membrane by forming a channel. Like cellulose synthase, the insect chitin synthase is thought to polymerize and was found to be trimerized when Merzendorfer lab attempted to purify insect chitin synthase from *M. Sexta* [42].

Although ChsA and ChsB possess structural motifs in common, slightly difference exists between them. ChsA has a coiled-coil domain at the end of the five transmembrane helices while ChsB chitin synthase does not. This domain is thought to be involved in protein-protein interaction.

2.4.5. Function of Insect Chitin Synthases

ChsA and ChsB functions differently. By using in situ hybridization, these two genes were found to be transcribed in different tissues. *CHSA* exists in the epidermis while *CHSB* is in the midgut [19,40]. In order to further confirm their function, RNAi is used. The dsRNA interference of *CHSA* inhibits the biosynthesis of insect cuticle during molting [43] and results in abnormal eggs [64], while RNAi of *CHSB* leads to death [43]. The expression pattern of *CHSA* and *CHSB* during insect development stages indicates that *CHSA* is closely related to the cuticle formation while *CHSB* is to midgut formation.

2.4.6. Transcriptional Regulation of Insect Chitin Synthases

Since insect molting and feeding are controlled by hormone, the expression level of insect chitin synthase changes during molting and feeding stage. In *Manduca*, *MsCHSA* is mainly expressed during larvae feeding and pupal stage in epidermis according to RT-PCT analysis [44]. However, *MsCHSB* is expressed at the whole feeding stage and the first day of wandering stage when peritrophic matrix formation is formed [45]. Similarly, *TcCHSA* is mainly expressed at the last instar larvae, prepupa and pupa stages, while *TcCHSB* is detected both at the late larvae stage and in adults. The phenomena is true for enzymes from insects *S. frugiperda* [46] and *S.exig*ua [47,48]. Since the molting of insect is controlled by ecdysteroid, the expression level of *CHSA* and *CHSB* are believed to be regulated by ecdysteroid.

2.4.7. Post Translational Regulation of Insect Chitin Synthases

It has been proved that the synthesis of insect chitin synthase is controlled by hormone at transcriptional level, but the discrepancies between mRNA amounts and enzymatic activities suggest that the regulation can be at post translational level too. Insect chitin synthase does not carry a signal peptide, it is expressed as a zymogen [49]. In *Manduca*, chitin synthase is activated by trypsin. However, the proteolytic activation could also take place when adding midgut extracts [49]. A chymotrypsin-like protein, which is presumed to have a protein-protein interaction with class B chitin synthase in *Manduca,* has recently been revealed by Yeast Two Hybrid [50]. This chymotrypsin-like protein carries a trypsin cleavage site, and may be the factor that participates in the activation of chitin synthase.

2.5. Application of Insect Chitinases and β-N-acetyl -D-hexosaminidases

2.5.1. Preparation of Chitooligosaccharides and GlcNAc from Chitin

Since chitin is the second most abundant biomass but underutilized in nature, the utilization of chitin is of significance in terms of environmental and economic benefits. Chitooligosaccharides and GlcNAc, the hydrolytic products of chitin, have a broad range of applications in biochemical, food, medicine and various chemical industries. Generally, chitooligosaccharides and GlcNAc are prepared by means of concentrated acid hydrolysis. But this procedure has many drawbacks such as low yield, environmentally toxic byproducts and acidic wastes. On the other hand, chitin biodegradation is existed in many organisms such as fungi, crustaceans and insects, which use chitin as main structural components. Chitinase (EC 3.2.1.14) and β-*N*-acetyl-D-hexosaminidase (EC 3.2.1.52) are two enzymes involved in chitin biodegradation, and the ratio of them can be controlled to regulate the chitin degradation efficiency and the degree of polymerization of hydrolytic products. So, hydrolysis of chitin by means of chitinolytic enzymes is a promising alternative.

2.5.2. Targets for Pesticide

Chitin is one of the most important structural components of the cuticle and peritrophic membrane of insects, the eggshell and sheath of nematodes and the cell wall of fungi. The chitin-containing 'exo/endo-skeleton' acts as physicochemical barriers against environmental hazards and predators. Pests die when chitin structures are destroyed or not properly formed.

Thus, the participating enzymes in chitin synthesis and degradation are ideal targets for developing pesticides.

Peptidyl nucleosides, isolated from bacteria *Streptomyces*, function as substrate analogs to inhibit chitin synthase. However, its low efficacy and toxicity to non-targeted animal systems are concerned. Acylureas, by an unknown mechanism, selectively inhibit chitin synthase and plays an important role in integrated pest management. Additionally, buprofezin, has been reported to affect on chitin synthesis by blocking incorporation of chitin precursors and act on sucking insects specifically.

As for chitinase inhibitors, allosamidin is a pseudotrisaccharide that mimics the cataylytic transition-state intermediate and exhibits a strong inhibitory activity in nano- to micromolar range. However, the widespread existence of chitinase in organisms ranging from microorganisms, plants, insects, mammals to human makes allosamidin have board activities toward non-targeted organisms, which limit the application to pest control.

β-N-acetyl-D-hexosaminidases, though widespread among species as well, have been evolved with different physiological functions. This functional variety makes it possible to develop specific inhibitors towards chitin-containing organisms. TMG-chitotriomycin, derived from bacteria, has been recently reported as such an inhibitor. Thus, an inhibitor designed to target functional specific *β-N*-acetyl-D-hexosaminidase could be a solution to develop selective and safe pest-control agents.

3. Preparation of Chitin-Remodeling Enzymes

Low expression level in vivo results in a challenge to obtain chitinase and β-*N*-acetyl-D-hexosaminidase from insects. However, due to their importance in both biochemistry studies and application, great efforts have been made aiming to solve this problem. A brief review of technologies herein to the preparation is given.

3.1. Purification of Insect Chitinases

Insect chitinases can be isolated from molting fluid, integument and pupal haemolymph. Smaller molecule-weight chitinases were initially believed to be derived from the larger one by restrict proteolysis, but now is known to be produced by multiple chitinase genes with different length. Insect chitinases have been obtained from *Stomoxys calcitrans*, *Pieris rapae*, *Tenebrio molitor*, and *Bombyx mori*. It is worthnoting that insect chitinases could also be purified from the digestive fluid of *Cupiennius salei* and the venom gland of *Chelonus sp.* Here the purification and characterization of *Bombyx mori* chitinases [51] is taken as an example. The supernatant of the homogenate of the fifth-instar larval integument was firstly purified by DEAE-Cellulofine A-500 and then by hydroxylapatite chromatography. Based on the difference in hydrophobicity, Butyl-Toyopearl 650M was used. The active fraction obtained was shown to contain two chitinases by SDS–PAGE. These two chitinase isozymes were then separated by Fractogel EMD DEAE 650(M) as the final step. Thus two chitinases were purified approximately 558 and 262 folds with activity recoveries of 3.7 and 31.1%, respectively.

3.2. Purification of Insect β-*N*-acetyl-D-hexosaminidases

Insect β-*N*-acetyl-D-hexosaminidases have been purified and obtained from both insect tissues and cultured insect cells. As for insect tissues, they have been purified from the hemolymph and integument of *B. mori*, the molting fluid, hemolymph and integument of *M. sexta*, the integument of *P. rapae*, and the integument of *O.furnacalis*. As for insect cells, they have been purified from the cells of insects *Drosophila melanogaster*, *C. quinquefasciatus* and *S. frugiperda*. Similar to chitinases, insects contain more than one β-*N*-acetyl-D-hexosaminidases. Here the purification and characterization of *O. furnacalis* β-*N*-acetyl-D-hexosaminidase (OfHex1) is taken as an example [21]. Natural OfHex1 was purified from the fifth instar larvae through a sequential combination of anion exchange chromatography on Source 30 Q column, hydroxyapatite chromatography on ceramic hydroxyapatite column, anion exchange chromatography on Mono Q column, and gel filtration chromatography on Superdex 200 column. Gel filtration gave the baseline separation of OfHex1 from the final impurities. A yield of 0.47 mg of OfHex1 with high purity was obtained from 300 insects. Finally, it was purified by 1,400 folds to homogeneity, with an overall activity yield of 20%, with a specific activity of 25 units· g^{-1}.

3.3. Preparation by Gene Recombination Technology

As mentioned in Section 2 of this chapter, both chitinases and β-N-acetylhexosaminidases are of importance in view of biochemical studies and applied biotechnologies as well. The availability of these enzymes by the production of insects limits the application. Many expression systems, such as *Escherichia coli and Pichia pastoris* are now available to produce recombinant chitinases and β-*N*-acetyl-hexosaminidases in active forms. Generally, these systems include prokaryotic expression systems, e.g. *Escherichia coli*, and eukaryotic systems, e.g. yeasts, baculovirus-insect cells and mammalian cells (Table 1).

3.3.1. Expression of Chitinases and β-N-acetyl-D-hexosaminidases by Prokaryotic Expression System

Escherichia coli (*E. coli*) is the most widely used prokaryotic expression system for the production of heterologous proteins. Recent progress in the fundamental understanding of transcription, translation, and protein folding in *E. coli* makes this bacterium more valuable than ever for the expression of complex eukaryotic proteins. The *E. coli* expression system allows rapid expression and subsequent large-scale, cost-effective manufacturing of recombinant proteins. For these reasons, *E. coli* system is ideal for antigen expression and functional expression of non-glycosylated proteins. However, *E. coli* expression system lacks eukaryotic molecular chaperones and post-translational modification (Table1). For this reason, some eukaryotic proteins cannot be expressed successfully in soluble or functional forms.

Table 1. General comparison of the features of each expression system

Characteristics	Prokaryotic	Eukaryotic		
	E. coli	Yeast	Insect cells	Mammalian cells
Doubling time	Rapid (30 min)	Rapid (90 min)	Slow (18-24 h)	Slow (24 h)
Cost of growth medium	Low	Low	High	High
Expression level	High	Low – high	Low - high	Low – moderate
Yield (mg/l)	50-500	10-200	10-200	0.1-100
Location of secreted protein	Periplasm	Medium	Medium	Medium
Protein refolding	Usually required	Maybe required	No	No
N-linked glycosylation	No	High mannose	Simple, no Sialic Acid	Complex
O-linked glycosylation	No	Yes	Yes	Yes
Phosphorylation	No	Yes	Yes	Yes
Acetylation	No	Yes	Yes	Yes
Acylation	No	Yes	Yes	Yes
gamma-carboxylation	No	No	No	Yes

3.3.1.1. Expression of Chitinases in E. coli

Different vectors have been used to express chitinases in *E. coli*, such as pET28a-c, pGEX4T-3, pQE30, pGEM and pRSET (Table 2). pET28a-c is a series of vectors with a strong T7 promoter, a N-terminal 6×His and T7 tag, and an optimal C-terminal His tag. The T7 tag can be used for Western blot to detect the expressed protein and subsequent purification using Ni^{2+} affinity chromatography. pET28a vector has been used to express chitinases from *Manduca sexta* and *Spodoptera litura* [52,53] in the form of inclusion bodies. pET28b vector has been used to express a chitinase from *Helicaoverpa armigera* [54]. Similarly to pET28b, pQE30 contains an N-terminal His tag and a T5 promoter. pQE30 has been used in the expression of a chitinase from *Choristoneura fumiferana* [55]. pGEX4T-3 contains a *tac* promoter and an N terminal GST tag to facilitate affinity purification by Glutathione Sepharose column. pGEX4T-3 has been used to express *Haemaphysalis longicornis* chitinase in the form of inclusion body [56].

Some fungi chitinases, such as chi33 and chis42 have also been reported to express in *E. coli* in the form of inclusion bodies [57]. Additionally, bacteria chitinases from *Serratia marcescens* [58] and Vibrio *harveyi* [59] have expressed functionally in *E. coli* by using the expression vector, pGEM-7zf, which contains a T5 promoter [60]. Since no insect chitinase has been expressed in *E.coli* in active forms but as inclusion bodies, *E. coli* as insect chitinase expression system is mostly used for the preparation of antigens or antibodies.

Table 2. Chitinases expressed in *E.coli*

Sources			Host cells	Vectors			Features	References
Species	pI	MW (kDa)		Names	Promoters	Tags		
Manduca sexta	5.32	62.2	BL21(DE3)	pET28a	T7	His	Inclusion body	[52]
Choristoneura fumiferana	5.39	62.2	G10003	pQE30	T5	His	No activity	[55]
Haemaphysalis longicornis	7.49	104.3	BL21(DE3)	pGEX4T-3	*tac*	GST	Inclusion body	[56]
Trichoderma harzianum	5.07	34.3	Origami™	pQE32	T5	His	Inclusion body	[57]
Trichoderma atroviride	3.93	32.5	BL21(DE3)	pRSET	T7	His	Inclusion body	[91]
Metarhizium anisopilae	5.60	45.9	BL21(DE3)	PT7-7	T7	-	Inclusion body	[92]
Vibrio harveyi	4.21	79.0	BL21(DE3)	pET-12a	T7	-	Soluble with activity	[59]
Serratia marcescens	5.93	55.5	DHl	pGEM-7zf	T7	-	Soluble with activity	[58]
Coccidioides immitis	5.87	47.6	TG-1	pGEX4T-1	*tac*	GST	Soluble with activity	[93]

3.3.1.2. Expression of β-N-acetyl-D-hexosaminidase in E. coli

E. coli is suitable for the expression of bacteria β-N-acetyl-D-hexosaminidase rather than insect β-N-acetyl-D-hexosaminidase. Many bacterium β-N-acetyl-D-hexosaminidase genes have been over-expressed functionally in *E.coli* (Table 3). *E.coli* BL21 (DE3) and pET are the mostly applied host strain and the mostly applied vectors series. The pET vectors contain a T7 promoter that drives high level expression of β-N-acetyl-D-hexosaminidases and a His tag in the C terminal to access affinity purification. pET 20b (+) has been used to express a pot ential β-N-acetyl-D-hexosaminidase from marine bacterium *Alteromonas* sp. strain O-7 in a soluble form [61]. pET 25b (+) has been successfully used for the expression of a β-N-acetyl-D-hexosaminidase from *Symbiobacterium thermophilum* [62]. pET 23a has been used to obtain two novel β-N-acetyl-D-hexosaminidases from *Paenibacillus sp.* Strain TS12 [34]. pET 29b has been used in the expression of a β-N-acetyl-D-hexosaminidase named Dispersin B from the Gram-negative periodontopathogen *Actinobacillus actinomycetemcomitans* CU1000 [31].

Other expression vectors, such as pLLP-OmpA, pThioHisBare and pUC19, are also used in β-N-acetyl-D-hexosaminidase expression. pLLP-OmpA contains a *lpp* promoter and a C-terminal His tag. It also provides *E.coli* OmpA signal peptide for the secretion of recombinant products. pLLP-OmpA has been used to express a β-N-acetyl-D-hexosaminidase from *Aeromonas caviae* CB101 [63]. pThioHisB carries a mutated thioredoxin sequence (HP-trx) to faciliate soluble fusion expression and affinity purification. The expression is induced by the *trc* promoter. pThioHisB has been used in the expression of a β-N-acetyl-D-hexosaminidase from *Streptomyces thermoviolaceus* [64].

Table 3. β-*N*-acetyl-D-hexosaminidases expressed in *E.coli*

Sources				Host cell	Vectors			Features	References
Species	Gene	pI	MW (kDa)		Names	Promo-ters	Tags		
Alteromonas sp.	*hex86*	*8.15*	86	BL21 (DE3)	pET20b(+)	T7	His	Soluble with activity	[61]
Symbiobacterium thermophilum	*nahA*	5.68	55.8	BL21 (DE3)	pET 25b (+)	T7	His	Soluble with activity	[62]
Paenibacillus sp.	*hex1* *hex2*	*5.13* *5.04*	105.2 107.7	BL21 (DE3)	pET23a	T7	His	Soluble with activity	[34]
Actinobacillus actinomycete-mcomitans	*dspB*	*6.11*	39.4	BL21 (DE3)	pET29b	T7	His S	Soluble with activity	[31]
Aeromonas caviae	*nagA1*	*5.57*	97.8	BL21 (DE3)	pLLP-OmpA	*lpp*	His OmpA	Soluble with activity	[63]
Streptomyces thermoviolaceus	*nagC*	*6.24*	63	TOP10	pThioHisB	*trc*	HP-trx	Soluble with activity	[64]
Streptococcus gordonii	*gcnA*	*5.08*	72.1	XL1-Blue MRF	pUC19 pREP4	*lac*	-	Soluble with activity	[94] [33]

3.3.2. Expression of Chitinases and β-N-acetylhexosaminidases in Yeast

Saccharomyces cerevisiae and the methylotrophic yeast *Pichia pastoris* are two mostly applied yeast strains. Since expression of exogenous genes is achieved by integration of foreign DNA into chromosomal DNA of the yeast genome, the integrated gene is stable for many generations. High expression level combined with high cell-density growth of yeast contributes higher yields of recombinant proteins. Furthermore, compared with insect or mammalian cells, yeast provides many eukaryotic molecular chaperones and post-translational modifications and is easier and less expensive to work with (Table 1). However, yeast lacks appropriate chaperones for proper folding of a number of proteins.

3.3.2.1. Expression of Chitinases in Yeast

Many insect chitinases have been expressed in the yeast Pichia pastroris. Vectors including pGAPZαA, pPICZαA and pPIC9K, are used (Table 4). pGAPZαA contains a strong *GAP* promoter, a C-terminal *myc* epitope for protein detection and a His tag for purification. It also provides an N-terminal *S. cerevisiae* α-factor secretion signal that makes expressed protein secret into the medium. pGAPZαA is used in expression of a chitinase from *Lacanobia oleracea* in *P. pastoris* strain SMD1168 [65]. pPICZαA contains *AOX1* promoter for high-level, methanol-induced protein expression in *P. pastoris*. pPICZαA has been used to express the chitinase from *S. cerevisiae* in *P. pastoris* strain X-33 [66]. pPIC9K has an additional HIS4 gene for spositive colon selection and has been successfully used to obtain recombinant chitinases from prokaryotic *Beauveria bassiana* [67] and *Thermomyces lanuginosus* [68].

Saccharomyces cerevisidae can also be used as host for chitinase expression. The expressing vector pYES2 contains a *URA3* gene as a selection marker and *Gal* promoter. Chitinases from *Trichoderma aureoviride* [69] and *Aspergillus nidulans* [70] have been expressed functionally in *S. cerevisidae*.

Table 4. Chitinases expressed in yeasts

Sources			Hosts	Vectors			Features	References
Species	pI	MW (kDa)		Names	Promo-ters	Tags		
Lacanobia oleracea	5.18	61.8	*P.pastoris* SMD1168H	pGAPZaA	*Gap*	His	High activity	[65]
Beauveria bassiana	6.20	36.7	*P.pastoris* GS115	pPIC9K	*AOX1*	-	With activity	[67]
Trichoderma aureoviride	5.06	45.9	*S.es cerevisiae* H158	pYES2	*Gal*	-	With activity	[69]
Thermomyces lanuginosus	6.46	49.1	*P.pastoris* GS115	pPIC9K	*AOX1*	-	With activity	[68]
Saccharomyces cerevisiae	4.22	59.0	*P.pastoris* X-33	pPICZaA	*AOX1*	His	High activity	[66]

3.3.2.2. Expression of β-N-acetyl-D-hexosaminidases in Yeast

Several β-N-acetyl-D-hexosaminidases have been produced functionally by the yeast *Pichia pastoris*. Many vectors are available, such as pPICZα series and pPIC9 (Table 5). pPICZαA has been used in expression of three β-N-acetyl-D-hexosaminidases (HEXO1, HEXO2 and fdl) from the insect *D. melanogaster* [25] in *P. pastoris* GS115. pPICZαB and pPICZαC, quite similar to pPICZαA vector, have been used to express two putative β-N-acetyl-D-hexosaminidases from Silkworm *B. mori* [23]. pPIC9 contains an *AOX1* promoter for high-level, methanol-induced expression and an N-terminal *S. cerevisiae* α-factor secretion signal peptide for secretory expression. pPIC9 vector has been used by our group to express OfHex1, from the Asian corn borer *O. furnacalis* in *P. pastoris* GS115 [71]. Through three steps, the recombinant OfHex1 is obtained with high purity and the same enzymatic activities as its native form. This is the first report of successful expression and purification of insect β-N-acetyl-D-hexosaminidase in yeast system.

Besides *P. pastoris*, the yeast *S. cerevisiae* is an alternative strain. The β subunit of human β-N-acetyl-D-hexosaminidase HsHexB was expressed in *S. cerevisiae* and the active dimeric HsHexB was localized to the vacuole, the yeast counterpart to the lysosome [72]. Two exochitinases from *T. harzianum* has also been over-expressed functionally in the yeast *S. cerevisiae* [73].

3.2.3. Expression of Chitinases and β-N-acetyl-D-hexosaminidases in Insect Cells

The insect baculoviruses represents an efficient path for producing complex eukaryotic proteins and has been put into use since 1990s. are a family of large rod-shaped viruses, such as the *Autographa californica* nuclear polyhedrosis virus (AcMNPV), and used to infect insect cells to express target proteins. The most widely used insect host cells are Sf9 and Sf21, both of which derived from the lepidopteran insect, *S. frugiperda*. Compared with *E. coli* and yeast, insect baculovirus expression system better resembles insect and mammalian cells in terms of post-translational processing and protein folding (Table 1).

Table 5. β-*N*-acetyl-D-hexosaminidases expressed in the yeast *Pichia pastoris*

Sources				Host	Vectors			Features	References
Species	Gene	pI	MW (kDa)		Name	Promoter	Tag		
Drosophila melanogaster	*HEXO1* *HEXO2* *fdl*	5.63 7.86 6.30	70.6 70.7 75.5	*P.pastoris* GS115	pPICZα A	*AOX1*	His c-*myc*	High activity	[25]
Bombyx mori	*BmGlcNAcase1* *BmGlcNAcase2*	5.77 5.47	69.7 61.7	*P. pastoris* GS115	pPICZα B pPICZα C	*AOX1*	His c-*myc*	High activity	[23]
strinia furnacalis	*OfHex1*	5.32	68.0	*P.pastoris* GS115	pPIC9	*AOX1*	-	High activity	[71]

3.2.3.1. Expression of Chitinases in Insect Cells

The baculovirus has successfully been used to express chitinases from *Manduca sexta*, [74,75] *Helicaoverpa armigera* [55], *Haemaphysalis longicornis*, *Spodoptera litura*, *D. melanogaster*[16], *Tribolium castaneum* and *Spruce budworm* (Table 6). pVL1393 is the general transfer vector with no additional tags. It contains a strong polyhedron promoter. pBlueBac transfer vector contains a strong polyhedron promoter for high lever expression, and a C-terminal V5 epitope as well as one His tag. Besides baculovirus, some other expression vectors are used. pMT-V5/Bip/His vector has been used to express a fat-body-specific chitinase from tsetse fly in S2 cell [76]. pMT-V5/Bip/His vector carries an induced promoter for *Drosphila* S2 cell. It contains an N-terminal *Drosphila* Bip secretion signal peptide, a C-terminal V5 and a His tag. Kraker's lab compared the expression levels of chitinase MsChi by using different insect cell lines and concluded the insect cell line High Five was better than Sf21 and Sf9 [77].

3.2.3.2. Expression of β-N-acetyl-D-hexosaminidases in Insect Cells

Several β-N-acetyl-D-hexosaminidases genes from insects and mammals have been expressed in insect cells with high activity (Table 7). The pVL1393 transfer vector has been used to express two putative Sf9 β-N-acetyl-D-hexosaminidase genes, *SfGlcNAcase-1* and *SfGlcNAcase-3*, in Sf9 cells [22]. The pBlueBac4.5 transfer vector has been used in expression of two Sf9 β-N-acetyl-D-hexosaminidase genes, *Sfhex1* and *Sfhex2* [78]. The BacPAK6 transfer vector has been used to express a membrane-associated β-N-acetyl-D-hexosaminidase from *S. frugiperda* (Sf-fdl) in Sf9 [34]. Human β-N-acetyl-D-hexosaminidase was firstly expressed functionally in Sf9 cells in 1990 [79]. Due to that successful expression, Schuette's group completed the analysis of glycosylation and disulfide bond pattern of HsHexB [80] and obtained the crystal structure [30]. However, Boose et al. revealed that the oligosaccharide locations and structures presented on the recombinant HsHexB differed from their native forms [79], because of lacking the mannose 6-phosphate recognition marker that target most degradative hydrolases to lysosomes.

Table 6. Chitinases expressed in insect cells

Sources			Host	Vectors			Features	References
Species	pI	MW (kDa)		Name	Promoter	Tag		
Manduca sexta	5.32	62.2	Sf9, Sf21, High Five	pVL1393	Polyhedron	-	High activity	[74,75]
Spodoptera litura	5.26	62.4	Sf9	pHcMU2	polyhedrin	-	High activity	[53]
Helicaoverpa armigera	6.02	65.7	Sf9	pBlueBac	polyhedrin	His, V5	High activity	[54]
Drosophila melanogaster	4.45	49.9	Sf9	pVL1393	polyhedrin	His	High activity	[16]
Glossina morsitans	7.45	51.1	S2	pMT-V5	metallothionein	His, V5	With activity	[76]

Table 7. β-*N*-acetyl-D-hexosaminidases expressed in insect cells

Source				Host	Vector			Features	References
Species	Gene	pI	MW (kDa)		Name	Promoter	Tag		
Spodoptera frugiperda	*SfGlcNAcase-1*	8.37	62.9	Sf9	pVL1393	polyhedrin	-	High activity	[22]
	SfGlcNAcase-3	5.53	63.5						
	Sfhex1	8.37	62.9	Sf9	pBlueBac 4.5	polyhedrin	His, V5	High activity	[78]
	Sfhex2	5.53	63.5						
	Fdl	6.75	70.5	Sf9	BacPAK6	polyhedrin	His	High activity	[24]

3.2.4. Expression of Chitinases and β-N-acetyl-D-hexosaminidases in Mammalian Cells

The mammalian expression system provides molecular chaperones and post-translational modifications (Table 1) with the help of which the expressed proteins are mostly similar to their native forms. Thus, the mammalian expression system is widely used in numerous areas as biomedicine, biochemical, and protein engineering.

3.2.4.1. Expression of Chitinases in Mammalian Cells

The mammalian expression system is used for mammal chitinases (Table 8). The pNUT vector contains a mouse metallothionein promoter and can be transformed into baby hamster kidney cells for protein expression. It has been used to express the human chitinase [81]. The pCMV-β-Gal vector contains a CMV promoter, and has been used for the expression of the chitinase from mouse in Cos1 cells [82]. The pcDNA3.1D vector carries CMV promoter and a C-terminal V5 and a His tag. The pcDNA3.1D vector has been used to express chitinase from rat in Cos7 cell [83].

Table 8. Chitinases expressed in mammalian cells

Source			Host	Vector			Features	References
Species	pI	MW (kDa)		Name	Promoter	Tag		
Homo sapiens	5.15	40.1	Hamster kidney cells	pNUT	Metallo-thionein	-	High activity	[81]
Mus musculus	5.60	43.4	Cos1	pCMV-β-Gal	CMV	-	High activity	[82]
Rattus norvegicus	4.87	52.0	Cos7	pcDNA3.1D	CMV	His, V5	High activity	[83]

3.2.4.2. Expression of β-N-acetyl-D-hexosaminidases in Mammalian Cells

Considering that defects or mutations in human β-*N*-acetyl-D-hexosaminidases cause serious diseases, mammalian cells expression system becomes a choice to produce adequate β-*N*-acetyl-D-hexosaminidases and many vectors have been used successfully (Table 9).

Table 9. β-*N*-acetyl-D-hexosaminidases expressed in mammalian cells

Source				Host	Vector			Features	References
Species	Gene	pI	MW (kDa)		Name	Promoter	Tag		
Homo sapiens	*HEXA*	5.04	60.7	TSD-NG	pREP4	RSV	-	High activity	[84]
	NEXB	6.29	63.1	CHO cell	pIRES2-EGFP	CMV	EGFP	High activity	[85]
					pcDNA 3.1D//V5-His-TOPO		His, V5		
					pEFneo	EF-1α	-		

The pRET4 vector is designed for high-level, constitutive expression under RSV promoter in mammalian cells. It has been used for expression of α subunits of human HsHexA in Tay-Sachs nueral glial cells (TSD-NG) [84]. Three different eukaryotic expression vectors containing two different promoters are used to express the β subunit of human HsHexB in CHO cells, while with quite different expression levels [85]. The pIRES2-EGFP vector contains the enhanced green fluorescent protein (EGFP) coding region to obtain protein of interest with EGFP fused at its C-terminal. The pcDNA3.1D/V5-His-TOPO vector provides a choice of C-terminal V5 and 6X His as detectable tags. Both pIRES2-EGFP and pcDNA3.1D/V5-His-TOPO vectors carry the same strong HCMV IE promoter. The pEFneo vector contains the promoter of elongation factor 1α (EF-1α) gene. Studies suggested that the human EF-1α promoter significantly increase the yield of HsHexB in CHO cells. Mammalian cell lines, such as human HeLa cells [86], monkey kidney COS cells [87-90] are alternatives for expression of mammalian β-*N*-acetyl-D-hexosaminidases.

REFERENCES

[1] Neville, B. W., Mishkin, D. J. & Traynham, R. T. 1984. The laterally positioned odontogenic keratocyst. A case report. *Journal of Periodontology*, 55, 98-102.

[2] Muzzarelli, R. A., Mattioli-Belmonte, M., Pugnaloni, A. & Biagini, G. 1999. Biochemistry, histology and clinical uses of chitins and chitosans in wound healing. *EXS*, 87, 251-264.

[3] Veronico, P., Gray, L. J., Jones, J. T., Bazzicalupo, P., Arbucci, S., Cortese, M. R., Di Vito, M. & De Giorgi, C. 2001. Nematode chitin synthases: gene structure, expression and function in *Caenorhabditis elegans* and the plant parasitic nematode *Meloidogyne artiellia*. *Molecular Genetics and Genomics*, 266, 28-34.

[4] Van Dellen, K. L., Bulik, D. A., Specht, C. A., Robbins, P. W. & Samuelson, J. C. 2006. Heterologous expression of an Entamoeba histolytica chitin synthase in *Saccharomyces cerevisiae*. *Eukaryotic Cell*. 5, 203-206.

[5] Kramer, K. J., Hopkins, T. L. & Schaefer, J. 1995. Applications of solids NMR to the analysis of insect sclerotized structures. *Insect Biochemistry and Molecular Biology*, 25, 1067-1080.

[6] Cohen, E. 1993. Chitin Synthesis and Degradation as Targets for Pesticide Action. *Archives of Insect Biochemistry and Physiology*, 22, 245-261.

[7] Hof, C. H. J. & Briggs, D. E. G. 1997. Decay and mineralization of mantis shrimps (*Stomatopoda: Crustacea*) - A key to their fossil record. *Palaios*, 12, 420-438.

[8] Boore, J. L., Lavrov, D. V. & Brown, W. M. 1998. Gene translocation links insects and crustaceans. *Nature*, 392, 667-668.

[9] Soyez, D. & Kleinholz, L. H. 1977. Molt-Inhibiting factor from crustacean eyestalk. *General and Comparative Endocrinology*, 31, 233-242.

[10] Dall, W. & Barclay, M. C. 1977. Induction of viable ecdysis in western rock lobster by 20-hydroxyecdysone. *General and Comparative Endocrinology*, 31, 323-334.

[11] Tamone, S. L. & Chang, E. S. 1993. Methyl Farnesoate Stimulates Ecdysteroid Secretion from Crab Y-Organs Invitro. *General and Comparative Endocrinology,* 89, 425-432.

[12] Hartnoll, R. G. 2001. Growth in Crustacea - twenty years on. *Hydrobiologia*, 449, 111-122.

[13] Kramer, K. J., Corpuz, L., Choi, H. K. & Muthukrishnan, S. 1993. Sequence of a cDNA and expression of the gene encoding epidermal and gut chitinases of *Manduca sexta*. *Insect Biochemistry and Molecular Biology*, 23, 691-701.

[14] Royer, V., Fraichard, S. & Bouhin, H. 2002. A novel putative insect chitinase with multiple catalytic domains: hormonal regulation during metamorphosis. *Biochemical Journal,* 366, 921-928.

[15] Zhu, Q. S., Deng, Y. P., Vanka, P., Brown, S. J., Muthukrishnan, S. & Kramer, K. J. 2004. Computational identification of novel chitinase-like proteins in the *Drosophila melanogaster* genome. *Bioinformatics*, 20, 161-169.

[16] Zhu, Q., Arakane, Y., Beeman, R. W., Kramer, K. J. & Muthukrishnan, S. 2008. Characterization of recombinant chitinase-like proteins of *Drosophila melanogaster* and *Tribolium castaneum*. *Insect Biochemistry and Molecular Biology*, 38, 467-477.

[17] Lu, Y. M., Zen, K. C., Muthukrishnan, S. & Kramer, K. J. 2002. Site-directed mutagenesis and functional analysis of active site acidic amino acid residues D142, D144 and E146 in *Manduca sexta* (tobacco hornworm) chitinase. *Insect Biochemistry and Molecular Biology*, 32, 1369-1382.

[18] Arakane, Y. & Muthukrishnan, S. 2010. Insect chitinase and chitinase-like proteins. *Cellular and Molecular Life Sciences*, 67, 201-216.

[19] Merzendorfer, H. & Zimoch, L.2003. Chitin metabolism in insects: structure, function and regulation of chitin synthases and chitinases. *The Journal of Experimental Biology*, 206, 4393-4412.

[20] Hogenkamp, D. G., Arakane, Y., Kramer, K. J., Muthukrishnan, S. & Beeman, R. W.2008. Characterization and expression of the [beta]-N-acetylhexosaminidase gene family of *Tribolium castaneum. Insect Biochemistry and Molecular Biology*, 38, 478-489.

[21] Yang, Q., Liu, T., Liu, F. Y., Qu, M. B. & Qian, X. H. 2008. A novel beta-N-acetyl-D-hexosaminidase from the insect *Ostrinia furnacalis* (Guenee). *FEBS Journal*, 275, 5690-5702.

[22] Aumiller, J. J., Hollister, J. R. & Jarvis, D. L. 2006. Molecular cloning and functional characterization of beta-N-acetylglucosaminidase genes from Sf9 cells. *Protein Expression and Purification*, 47, 571-590.

[23] Okada, T., Ishiyama, S., Sezutsu, H., Usami, A., Tamura, T., Mita, K., Fujiyama, K. & Seki, T. 2007. Molecular cloning and expression of two novel beta-N-acetylglucosaminidases from silkworm *Bombyx mori. Bioscience Biotechnology and Biochemistry*, 71, 1626-1635.

[24] Geisler, C., Aumiller, J. J. & Jarvis, D. L. 2008. A fused lobes gene encodes the processing beta-N-acetylglucosaminidase in Sf9 cells. *Journal of Biological Chemistry*, 283, 11330-11339.

[25] Leonard, R., Rendic, D., Rabouille, C., Wilson, I. B. H., Preat, T. & Altmann, F. 2006. The Drosophila fused lobes gene encodes an N-acetylglucosaminidase involved in N-glycan processing. *Journal of Biological Chemistry*, 281, 4867-4875.

[26] Nagamatsu, Y., Yanagisawa, I., Kimoto, M., Okamoto, E. & Koga, D. 1995. Purification of a chitooligosaccharidolytic beta-N-acetylglucosaminidase from *Bombyx-Mori Larvae* during metamorphosis and the nucleotide-sequence of its Cdna. *Bioscience Biotechnology and Biochemistry*, 59, 219-225.

[27] Tews, I., Perrakis, A., Oppenheim, A., Dauter, Z., Wilson, K. S. & Vorgias, C. E. 1996. Bacterial chitobiase structure provides insight into catalytic mechanism and the basis of Tay-Sachs disease. *Nature Structural Biology*, 3, 638-648.

[28] Mark, B. L., Vocadlo, D. J., Knapp, S., Triggs-Raine, B. L., Withers, S. G. & James, M. N. G. 2001. Crystallographic evidence for substrate-assisted catalysis in a bacterial beta-hexosaminidase. *Journal of Biological Chemistry*, 276, 10330-10337.

[29] Mark, B. L., Mahuran, D. J., Cherney, M. M., Zhao, D. L., Knapp, S. & James, M. N. G. 2003. Crystal structure of human beta-hexosaminidase B: Understanding the molecular basis of Sandhoff and Tay-Sachs disease. *Journal of Molecular Biology*, 327, 1093-1109.

[30] Maier, T., Strater, N., Schuette, C. G., Klingenstein, R., Sandhoff, K. & Saenger, W. 2003. The X-ray crystal structure of human beta-hexosaminidase B provides new insights into Sandhoff disease. *Journal of Molecular Biology*, 328, 669-681.

[31] Ramasubbu, N., Thomas, L. M., Ragunath, C. & Kaplan, J. B. 2005. Structural analysis of dispersin B, a biofilm-releasing glycoside hydrolase from the periodontopathogen *Actinobacillus actinomycetemcomitans*. *Journal of Molecular Biology*, 349, 475-486.

[32] Lemieux, M. J., Mark, B. L., Cherney, M. M., Withers, S. G., Mahuran, D. J. & James, M. N. G. 2006. Crystallographic structure of human beta-hexosaminidase A: Interpretation of Tay-Sachs mutations and loss of G(M2) ganglioside hydrolysis. *Journal of Molecular Biology*, 359, 913-929.

[33] Langley, D. B., Harty, D. W. S., Jacques, N. A., Hunter, N., Guss, J. M. & Collyer, C. A. 2008. Structure of N-acetyl-beta-D-glucosaminidase (GcnA) from the endocarditis pathogen *Streptococcus gordonii* and its complex with the mechanism-based inhibitor NAG-thiazoline. *Journal of Molecular Biology*, 377, 104-116.

[34] Sumida, T., Ishii, R., Yanagisawa, T., Yokoyama, S. & Ito, M. 2009. molecular cloning and crystal structural analysis of a novel beta-N-Acetylhexosaminidase from *Paenibacillus* sp TS12 capable of degrading glycosphingolipids. *Journal of Molecular Biology*, 392, 87-99.

[35] Valdivieso, M. H., Mol, P. C., Shaw, J. A., Cabib, E. & Duran, A. 1991. CAL1, a gene required for activity of chitin synthase 3 in *Saccharomyces cerevisiae*. *Journal of Cell Bio*logy, 114, 101-109.

[36] Trilla, J. A., Duran, A. & Roncero, C. 1999. Chs7p, a new protein involved in the control of protein export from the endoplasmic reticulum that is specifically engaged in the regulation of chitin synthesis in *Saccharomyces cerevisiae*. *Journal of Cell Biology*, 145, 1153-1163.

[37] Sanchatjate, S. & Schekman, R. 2006. Chs5/6 complex: A multiprotein complex that interacts with and conveys chitin synthase III from the trans-Golgi network to the cell surface. *Molecular Biology of the Cell*, 17, 4157-4166.

[38] Trautwein, M., Schindler, C., Gauss, R., Dengjel, J., Hartmann, E. & Spang, A. 2006. Arf1p, Chs5p and the ChAPs are required for export of specialized cargo from the Golgi. *Embo Journal*, 25, 943-954.

[39] Trilla, J. A., Cos, T., Duran, A. & Roncero, C. 1997. Characterization of CHS4 (CAL2), a gene of *Saccharomyces cerevisiae* involved in chitin biosynthesis and allelic to SKT5 and CSD4. *Yeast*, 13, 795-807.

[40] Tellam, R. L., Vuocolo, T., Johnson, S. E., Jarmey, J. & Pearson, R. D. 2000. Insect chitin synthase cDNA sequence, gene organization and expression. *European Journal of Biochemistry*, 267, 6025-6043.

[41] Merzendorfer, H. 2006. Insect chitin synthases: a review. *Journal of Comparative Physiology B-Biochemical Systemic and Environmental Physiology*, 176, 1-15.

[42] Maue, L., Meissner, D. & Merzendorfer, H. 2009. Purification of an active, oligomeric chitin synthase complex from the midgut of the tobacco hornworm. *Insect Biochemistry and Molecular Biology*, 39, 654-659.

[43] Arakane, Y., Muthukrishnan, S., Kramer, K. J., Specht, C. A., Tomoyasu, Y., Lorenzen, M. D., Kanost, M. & Beeman, R. W. 2005. The *Tribolium* chitin synthase genes TcCHS1 and TcCHS2 are specialized for synthesis of epidermal cuticle and midgut peritrophic matrix. *Insect Molecular Bio*logy, 14, 453-463.

[44] Hogenkamp, D. G., Arakane, Y., Zimoch, L., Merzendorfer, H., Kramer, K. J., Beeman, R. W., Kanost, M. R., Specht, C. A. & Muthukrishnan, S. 2005. Chitin

synthase genes in *Manduca sexta*: characterization of a gut-specific transcript and differential tissue expression of alternately spliced mRNAs during development. *Insect Biochemistry and Molecular Biology*, 35, 529-540.

[45] Zhu, Y. C., Specht, C. A., Dittmer, N. T., Muthukrishnan, S., Kanost, M. R. & Kramer, K. J. 2002. Sequence of a cDNA and expression of the gene encoding a putative epidermal chitin synthase of *Manduca sexta. Insect Biochemistry and Molecular Biology*, 32, 1497-1506.

[46] Bolognesi, R., Arakane, Y., Muthukrishnan, S., Kramer, K. J., Terra, W. R. & Ferreira, C. 2005. Sequences of cDNAs and expression of genes encoding chitin synthase and chitinase in the midgut of *Spodoptera frugiperda. Insect Biochemistry and Molecular Biology*, 35, 1249-1259.

[47] Chen, X., Yang, X., Senthil Kumar, N., Tang, B., Sun, X., Qiu, X., Hu, J. & Zhang, W. 2007. The class A chitin synthase gene of *Spodoptera exigua*: molecular cloning and expression patterns. *Insect Biochemistry and Molecular Biology*, 37, 409-417.

[48] Kumar, N. S., Tang, B., Chen, X., Tian, H. & Zhang, W. 2008. Molecular cloning, expression pattern and comparative analysis of chitin synthase gene B in *Spodoptera exigua. Comparative Biochemistry and Physiology - Part B: Biochemistry & Molecular Biology,* 149, 447-453.

[49] Zimoch, L., Hogenkamp, D. G., Kramer, K. J., Muthukrishnan, S. & Merzendorfer, H. 2005. Regulation of chitin synthesis in the larval midgut of *Manduca sexta. Insect Biochemistry and Molecular Biology*, 35, 515-527.

[50] Broehan, G., Zimoch, L., Wessels, A., Ertas, B. & Merzendorfer, H. 2007. A chymotrypsin-like serine protease interacts with the chitin synthase from the midgut of the tobacco hornworm. *Journal of Experimental Biology*, 210, 3636-3643.

[51] 51. Daizo, K., Yuji, S., & Youichiro, U. 1996. Purification and Characterization of Bombyx mori Chitinases. *Insect Biochem. Molec. Bio.*, 27,757–767.

[52] Hao, C. J., Chai, B. F., Wang, W., Sun, Y. & Liang, A. H. 2005. Polyclonal antibody against *Manduca sexta* chitinase and detection of chitinase expressed in transgenic cotton. *Biotechnology Letters*, 27, 97-102.

[53] Shinoda, T., Kobayashi, J., Matsui, M. & Chinzei, Y. 2001. Cloning and functional expression of a chitinase cDNA from the common cutworm, *Spodoptera litura*, using a recombinant baculovirus lacking the virus-encoded chitinase gene. *Insect Biochemistry and Molecular Biology*, 31, 521-532.

[54] Ahmad, T., Rajagopal, R. & Bhatnagar, R. K. 2003. Molecular characterization of chitinase from polyphagous pest *Helicoverpa armigera. Biochemical and Biophysical Research Communications*, 310, 188-195.

[55] Zheng, Y., Zheng, S., Cheng, X., Ladd, T., Lingohr, E. J., Krell, P. J., Arif, B. M., Retnakaran, A. & Feng, Q. 2002. A molt-associated chitinase cDNA from the spruce budworm, *Choristoneura fumiferana. Insect Biochemistry and Molecular Biology*, 32, 1813-1823.

[56] You, M., Xuan, X., Tsuji, N., Kamio, T., Taylor, D., Suzuki, N. & Fujisaki, K. 2003. Identification and molecular characterization of a chitinase from the hard tick *Haemaphysalis longicornis. Journal of Biological Chemistry*, 278, 8556-8563.

[57] Boer, H., Simolin, H., Cottaz, S., Soderlund, H. & Koivula, A. 2007. Heterologous expression and site-directed mutagenesis studies of two *Trichoderma harzianum*

chitinases, Chit33 and Chit42, in *Escherichia coli*. *Protein Expression and Purification*, 51, 216-226.

[58] Brurberg, M. B., Nes, I. F. & Eijsink, V. G. H. 1996. Comparative studies of chitinases A and B from *Serratia marcescens*. *Microbiology-Uk*, 142, 1581-1589.

[59] Bendt, A., Huller, H., Kammel, U., Helmke, E. & Schweder, T. 2001. Cloning, expression, and characterization of a chitinase gene from the Antarctic psychrotolerant bacterium *Vibrio* sp strain Fi : 7. *Extremophiles*, 5, 119-126.

[60] Felse, P. A. & Panda, T. 1999. Regulation and cloning of microbial chitinase genes. *Applied Microbiology and Biotechnology*, 51, 141-151.

[61] Tsujibo, H., Miyamoto, K. & Yoshimura, M. 2002. Molecular cloning of the gene encoding a novel β-N-acetylhexosaminidase from a marine bacterium, *Alteromonas* sp. strain O-7, and characterization of the cloned enzyme. *Bioscience, Biotechnology, and Biochemistry*, 66, 471-475.

[62] Ogawa, M., Kitagawa, M. & Tanaka, H. 2006. A β-N-acetylhexosaminidase from *Symbiobacterium thermophilum*; gene cloning, overexpression, purification and characterization. *Enzyme and Microbial Technology*, 38, 457–464.

[63] Lin J, Xiao, X. & Zeng, X. 2006. Expression, characterization and mutagenesis of the gene encoding β-N-acetylglucosaminidase from *Aeromonas caviae* CB101. *Enzyme and Microbial Technology*, 38, 765-771.

[64] KuBOTA, T., Miyamoto, K. & Yasuda, M. 2004. Molecular characterization of an intracellular β-N-acetylglucosaminidase involved in the chitin degradation system of *Streptomyces thermoviolaceus* OPC-520. *Bioscience, Biotechnolology Biochemistry*, 68, 1306-1314.

[65] Fitches, E., Wilkinson, H., Bell, H., Bown, D. P., Gatehouse, J. A. & Edwards, J. P. 2004. Cloning, expression and functional characterisation of chitinase from larvae of tomato moth (*Lacanobia oleracea*): a demonstration of the insecticidal activity of insect chitinase. *Insect Biochemistry and Molecular Biology*, 34, 1037-1050.

[66] Hurtado-Guerrero, R. & van Aalten, D. M. F. 2007. Structure of *Saccharomyces cerevisiae* chitinase 1 and screening-based discovery of potent inhibitors. *Chemistry & Biology*, 14, 589-599.

[67] Fan, Y. H., Zhang, Y. J., Yang, X. Y., Pei, X. Q., Guo, S. J. & Pei, Y. 2007. Expression of a *Beauveria bassiana* chitinase (Bbchit1) in *Escherichia coli* and *Pichia pastoris*. *Protein Expression and Purification*. 56, 93-99.

[68] Guo, R. & Li, D. 2006. Molecular cloning and expression of a chitinase gene from the thermophilic fungus *Thermomyces lanuginosus*. *Acta Microbiologica Sinica*, 46, 99-103.

[69] Song, J. Z., Yang, Q., Liu, B. D. & Chen, D. F. 2005. Expression of the chitinase gene from *Trichoderma aureoviride* in *Saccharomyces cerevisiae*. *Applied Microbiology and Biotechnology*, 69, 39-43.

[70] Kim, S., Matsuo, I., Ajisaka, K., Nakajima, H. & Kitamoto, K. 2002. Cloning and characterization of the nagA gene that encodes beta-N-acetylglucosaminidase from *Aspergillus nidulans* and its expression in *Aspergillus oryzae*. *Bioscience Biotechnology and Biochemistry*, 66, 2168-2175.

[71] Liu, T., Liu, F. Y. & Yang, Q. 2009. Expression, purification and characterization of the chitinolyticβ-N-acetyl-D-hexosaminidase from the insect *Ostrinia furnacalis*. *Protein Expression and Purification*, 68, 99–103.

[72] Prezant, T. R. 1990. Expression of human lysosomal β-hexosaminidase in yeast vacuoles. *Biochemical and Biophysical Research Communications*, 170, 383-390.

[73] Draborg, H., Kauppinen, S. & Dalboge, H. 1995. Molecular cloning and expression in *S. cerevisiae* of two exochitinases from *Trichoderma harzianum*. *Biochemistry and Molecular Biology International*, 36, 781-791

[74] Zhu, X. H., Zhang, H., Fukamizo, T., Muthukrishnan, S. & Kramer, K. J. 2001. Properties of *Manduca sexta* chitinase and its C-terminal deletions. *Insect Biochemistry and Molecular Biology*, 31, 1221-1230.

[75] Arakane, Y., Zhu, Q. S., Matsumiya, M., Muthukrishnan, S. & Kramer, K. J. 2003. Properties of catalytic, linker and chitin-binding domains of insect chitinase. *Insect Biochemistry and Molecular Biology*. 33, 631-648.

[76] Yan, J., Cheng, Q., Narashimhan, S., Li, C. B. & Aksoy, S. 2002. Cloning and functional expression of a fat body-specific chitinase cDNA from the tsetse fly, *Glossina morsitans morsitans*. *Insect Biochemistry and Molecular Biology*, 32, 979-989.

[77] Gopalakrishnan, B., Muthukrishnan, S. & Kramer, K. J. 1995. Baculovirus-mediated expression of a *Manduca-Sexta* chitinase gene - Properties of the recombinant protein. *Insect Biochemistry and Molecular Biology*, 25, 255-265.

[78] Tomiya, N., Narang, S. & Park, J. 2006. Purification, characterization, and cloning of a *Spodoptera frugiperda* Sf9 β-N-acetylhexosaminidase that hydrolyzes terminal N-acetylglucosamine on the N-Glycan core. *The Journal of Biological Chemistry*, 281 19545–19560.

[79] Boose, J. A., Tifft, C. J. & Proia, R. L. 1990. Synthesis of a human lysosomal enzyme, β-hexosaminidase B, using the baculovirus expression system. *Protein Expression and Purification*, 1, 111-120.

[80] Schuette, C.G., Weisgerber, J. & Sandhoff, K. 2001. Complete analysis of the glycosylation and disulfide bond pattern of human β-hexosaminidase B by MALDI-MS. *Glycobiology*, 11, 549-556.

[81] Fusetti, F., von Moeller, H., Houston, D., Rozeboom, H. J., Dijkstra, B. W., Boot, R. G., Aerts, J. M. F. G. & van Aalten, D. M. F. 2002. Structure of human chitotriosidase - Implications for specific inhibitor design and function of mammalian chitinase-like lectins. *Journal of Biological Chemistry*, 277, 25537-25544.

[82] Zheng, T., Rabach, M., Chen, N. Y., Rabach, L., Hu, X. X., Elias, J. A. & Zhu, Z. 2005. Molecular cloning and functional characterization of mouse chitotriosidase. *Gene*, 357, 37-46.

[83] Chen, X. H., Xie, Z. H., Sun, S. J. & Cai, G. P. 2006. Cloning of a rat lung fibrogenic factor. *Gene*, 384, 9-17.

[84] Zarghooni, M., Bukovac S. & Tropak, M. l. 2004. An α-subunit loop structure is required for GM2 activator protein binding by β-hexosaminidase A. *Biochemical and Biophysical Research Communications*, 324, 1048-1052.

[85] Sinici, I., Zarghooni, M. & Tropak, M. B. 2006. Comparison of HCMV IE and EF-1 α promoters for the stable expression of β-subunit of hexosaminidase in CHO cell lines. *Biochemical Genetics*, 44, 173-180.

[86] Yamanaka, S., Johnson, O. N. & Norflus, F. 1994. Structure and expression of the mouse β-hexosaminidase genes, Hexa and Hexb. *GENOMICS*, 21, 588-596.

[87] Nishimoto, J., Tanaka, A. & Nanba, E. 1991. Expression of the β-acetylglucosaminidase-alpha subunit gene with the 4-base insertion of infantile Jewish Tay-Sachs desease. *The Journal of Biological Chemistry*, 266, 14306-14309.

[88] Sagherian, C., Thorner, P. & Mahuran, D. 1994. The Pro-Peptide of the Proβ-Polypeptide Chain of Human β-hexosaminidase Is Necessary for Proper Protein Folding and Exit from the Endoplasmic Reticulum. *Biochemical and Biophysical Research Communications*, 204, 135-141.

[89] Wakamatsu, N., Benoit, G. & Lamhonwah, A. M. 1994. Structural organization, sequence, and expression of the mouse HEXA gene encoding the alpha-subunit of hexoaminidase-A. *GENOMICS*, 24, 110-119.

[90] John Ewer. 2005. How the Ecdysozoan Changed Its Coat. *PLoS Biology*, 3(10): e349.

In: Bioprocess Sciences and Technology
Editor: Min-Tze Liong
ISBN 978-1-61122-950-9

Chapter 17

THE DEVELOPMENT OF MAMMALIAN CELL CULTURE PLATFORMS FOR PRODUCTION OF RECOMBINANT PROTEIN THERAPEUTICS

J. C. Yee [1,*]**,** ***K. P. Jayapal*** [2,*] ***and R. Motani***
[1]Bioengineering Department, Technology Development Genzyme Therapeutics, 45 New York Ave., Framingham, MA 01701, USA
[2]Cell Culture Development, Global Biologics Development, Bayer HealthCare Pharmaceuticals, 800 Dwight Way, Berkeley, CA 94710, USA

ABSTRACT

The past two decades have witnessed a phenomenal rise in the prominence of mammalian cells as hosts for production of recombinant protein therapeutics such as monoclonal antibodies, cytokines, enzymes and growth hormones. The quality and efficacy of these proteins are far superior when produced in mammalian cells as opposed to microbial or yeast cells, mainly due to the inherent biological characteristics of mammalian cells which lead to synthesis of appropriate glycoforms. Just a decade ago, recombinant protein yields of 1 g/L from mammalian cell culture processes were considered as the industry norm. Today, however, production processes yielding 3-5 g/L are quite common with several claims exceeding 10 g/L. This review explores the current trends and some recent technological advancements in cell culture bioprocess engineering that enabled achievement such high volumetric productivities, while concurrently improving product quality and process robustness. Specifically, this review will focus on: (i) strategies for development of high-yielding protein expression cell lines; (ii) current trends in bioprocess development, including media development and process monitoring, and the concept of platform technologies to enable robust high-titer high-density cultures; (iii) challenges and opportunities facing biomanufacturing in the era of globalization; and (iv) the promise of 'omics' and cell engineering strategies for future drug development.

* Corresponding authors: *J.C. Yee* Email: joonchong.yee@gmail.com; *K.P.Japayal* Email: karthikjayapal@gmail.com

While these technological innovations have undoubtedly resulted in substantially favorable cost-benefit ratios for consumers, some significant hurdles remain to be overcome as recombinant protein therapy reaches the mass market. Overcoming these obstacles require a comprehensive understanding of bioprocess operation from a multi-disciplinary perspective encompassing the elements of cell biology, molecular genetics, biochemical engineering and process instrumentation in a synergistic fashion.

INTRODUCTION

The approval of tissue plasminogen activator, the first mammalian cell-based recombinant human protein, for clinical use in 1987 heralded a new era of modern medicine. Today, over 150 recombinant proteins have been approved worldwide for therapeutic use and numerous others are in clinical trials. Recombinant protein therapeutics can be broadly classified into two categories – replacement factors and antibodies. Replacement factors such as enzymes, hormones and blood clotting factors supplement low-activity or deficient forms of native proteins in patients. Therapeutic antibodies bind to specific antigens expressed on target cells, triggering destruction mechanisms or blocking key signals of proliferative disease pathology. This makes them an attractive therapeutic choice for targeting cancers or inflammatory diseases such as rheumatoid arthritis. The dominance of drug development and disease treatment by recombinant proteins is evidenced by the number of FDA (U.S. regulatory agency) approvals in the last two decades (Tables 1 & 2).

Table 1. Examples of non-antibody recombinant proteins approved in the USA

Trade name	Protein name	Therapeutic Use	Manufacturer	U.S. Approval Year	Host Cell
Myozyme®	alglucosidase α	Pompe disease	Genzyme	2006	CHO
Elaprase®	Iduronate sulfatase	Hunter's syndrome	Shire	2006	HT-1080 (human)
Luveris®	Luteinizing hormone	Infertility	EMD Serono	2004	CHO
Aldurazyme®	Laronidase	Mucopolysaccharidosis I	Genzyme, Biomarin	2003	CHO
Rebif®	Cytokine interferon β-1A	Multiple sclerosis	Merck Serono	2002	CHO
Avonex®	Interferon β-1A	Multiple sclerosis	Biogen Idec	1996	CHO
Cerezyme®	glucocerebrosidase	Gauche disease	Genzyme	1994	CHO
Kogenate®	Factor VIII	Hemophilia	Bayer Healthcare	1993	BHK
Epogen®	Erythropoietin	Anaemia resulting from kidney disease, or chemotherapy treatments	Amgen	1989	CHO
Activase®	Tissue plasminogen activator	Acute myocardial infarction	Genentech	1987	CHO

Table 2. Examples of therapeutic antibodies approved in the USA

Trade name	Protein	Therapeutic Use	Manufacturer	U.S. Approval Year	Host Cell
Simponi®	Human mAb specific for tumor necrosis factor	Rheumatoid arthritis, psoriatic arthritis	Centocor Ortho Biotech	2009	Murine SP2/0
Arzerra®	Human mAb specific for CD20	Chronic lymphocytic leukemia	Genmab, GlaxoSmithKline	2009	NS0
Arcalyst®	Human mAB targeting IL-β	Cryopyrin-associated periodic syndrome	Regeneron	2008	CHO
Soliris®	mAb binds to complement protein C5	Paroxysmal nocturnal hemoglobin	Alexion	2007	CHO
Herceptin®	mAb specific to Her2 receptor expressed in breast cancer cells	Breast cancer	Genentech	2006	CHO
Rituxan®	mAb binds to CD 20 expressed in B-cells	Non-hodgkin's lymphoma	Genentech, Biogen Idec	2006	CHO
Orencia®	CTLA4-IgG fusion protein	Rheumatoid arthritis	Bristol-Meyers Squibb	2005	CHO
Erbitux®	mAb binds to epidermal growth factor	Colorectal cancer	Bristol-Meyer's Squibb, Imclone	2004	SP2/0
Tysabri®	mAb against adhesion molecule α4-integrin	Multiple sclerosis	Biogen Idec	2004	CHO
Avastin®	mAb binds to vascular endothelial growth factor	Colorectal cancer	Genentech	2004	CHO
Humira®	mAb against tumor necrosis factor -α	Rheumatoid arthritis	Abott	2002	CHO
Enbrel®	Recombinant tumor necrosis factor fused to IgG fragment	Rheumatoid arthritis	Amgen	1998	CHO

Commercial production of recombinant proteins relies on the large-scale cultivation of mammalian cells that have been genetically engineered to overproduce recombinant proteins of interest. To be effective as therapeutics, recombinant proteins have to be synthesized with the proper protein folding and post-translational modification. In most cases, protein glycosylation – the addition and alteration of sugar chains on specific amino acid residues determine the efficacy of the drug product [1]. The ability to synthesize proteins with glycoform structures close to native human proteins is the advantage that mammalian cells hold over other yeast or microbial expression systems. Accordingly, about 60 – 70% of all recombinant protein therapeutics are produced in mammalian cells such as Chinese hamster ovary (CHO), baby hamster kidney (BHK) and mouse myeloma NS0 cells.

Significant improvements in the development of mammalian cell culture platforms over the past two decades have resulted in higher cell densities in production bioreactors as well as higher specific productivities. Simultaneously, research and development timelines for delivering robust cell culture processes that generate candidate proteins for clinical studies or commercial processes have shortened. This was accomplished mainly through innovations in cell engineering, the use of high-throughput screening technologies, and the capabilities of scale-down bioreactor systems that can be used to establish process parameters and predict the performance for large-scale manufacturing processes.

2. Early Stage Cell line Development

2.1. Choice of Parental Cell Lines

Host cells, prior to genetic engineering for therapeutic protein production are regarded as parental cell lines. Among the dozens of therapeutic cell culture products that have emerged in the past decade, the majority are produced by cells derived from rodents, such as the Chinese hamster, the Syrian hamster and mouse. Except for vaccine productions where primary cells such as human fibroblast cells are used, all cells used for protein production are continuous cell lines. Unlike primary cells, continuous cell lines have abnormal growth and can replicate infinitely without entering senescent state.

Mammalian cells used for recombinant protein production typically fall into two categories depending on their tissue of origin. The first group is hamster or human cell lines derived from tissues that are not specialized in secreting proteins such as ovary, (e.g. Chinese hamster ovary) and kidney (e.g. baby hamster kidney, or human embryonic kidney cell lines). The second group are derived from tissues that are natural secretors, for example myeloma cells NS0 and SP2/0, both of which originate from plasmacytoma cells that have undergone repeated cloning and selection steps to yield immortalized non-IgG secreting B cells [2-4].

The differences in tissue origin between the various cell lines also resulted in unique metabolic phenotypes and specific nutrient requirements. For instance, the cultivation of NS0 myeloma cells requires the supplement of exogenous cholesterol in media, which is usually delivered via the use of serum [5,6] or cyclodextrins to enhance solubility of cholesterol in cultivation medium [7]. In contrast, Chinese hamster ovary cells do not require cholesterol supplements. However, commercial production CHO cell lines that were developed in the 1980s, such as DUKX-B11 and DG-44 have mutated or deleted copies of the dihydrofolate reductase (*dhfr*) gene [8,9], which resulted in the requirement of nucleotide precursors in cultivation of these cell lines.

The choice of host cells used for recombinant protein therapeutics also dictates the glycosylation profiles of the recombinant protein, which in turn affect its *in vivo* biological function and efficacy when used for human therapy. The glycosylation machinery of CHO cells is almost similar to its human counterparts with some minor differences. However, normal CHO cells do not express certain isoforms of fucosyltransferases, galactosyltransferases, sialyltransferases or N-acetylglucosaminyltransferases [10]. Due to its inability to express α2,6-sialyltransferase, it has been reported that recombinant IgG expressed in CHO contain only α2,3-linked sialic acid residues whereas human IgG contain both α2,6- and α2,3-linked sialic acid residues [11].

The extent to which the parental cell affects the glycan structure is illustrated by comparing the same recombinant fusion protein (CTLA4-IgG) expressed in CHO cells and mouse NS0 myelomas [12]. The study showed that the discrepancy of sialic acid forms found on the outer arms of glycan structures has an affect on the plasma clearance of glycoproteins *in vivo*. When compared to CHO cells, oligosaccharides from myeloma expressed fusion IgG had terminal N-glycolylneuraminic acid but no detectable N-acetylneuraminic acid, which resulted in higher clearance rates when used for human therapy.

Whether alterations in glycosylation profile can affect product efficacy is dependent on the *in vivo* mechanism triggered by the 'foreign' oligosaccharide form, and the severity of heterogeneity issues in the expressed recombinant protein. Several excellent reviews are available in the literature [1,13-15] for an in-depth analysis of glycosylation variations in mammalian expression systems and their impact on therapeutic activity.

2.2. Cloning of Recombinant Gene into Parental Cell Line

The development of recombinant protein producing cell lines relies on the archetype of molecular biology and genetic manipulation. Figure 1 represents an overview of the typical steps involved in the development of a producing cell-line, and its associated timeline. In the first step, a plasmid DNA containing the desired recombinant protein and a selectable genetic marker are constructed and transfected into mammalian cells. The plasmid DNA typically contains the following elements: [1] a mammalian promoter or enhancer sequence to drive high transcription of the gene of interest (GOI), [2] a DNA sequence coding for the recombinant protein of interest that is optimized for efficient codon usage as well as elimination of secondary structure which potentially leads to mRNA instability (the sequence also typically lacks non-coding introns to eliminate chances of splice variants), and [3] a second selection marker gene that co-expresses with the GOI.

The second gene or selection marker is usually transferred to render recipient mammalian cells a selective advantage by either compensating a host cell's metabolic deficiency or conferring resistance to toxic agents added into the culture medium (Table 3). In the latter case, antibiotics such as hygromycin, neomycin or gentamycin is introduced into the cell culture medium a few days after the initial gene transfection, and only cells that express the antibiotic resistance marker survives.

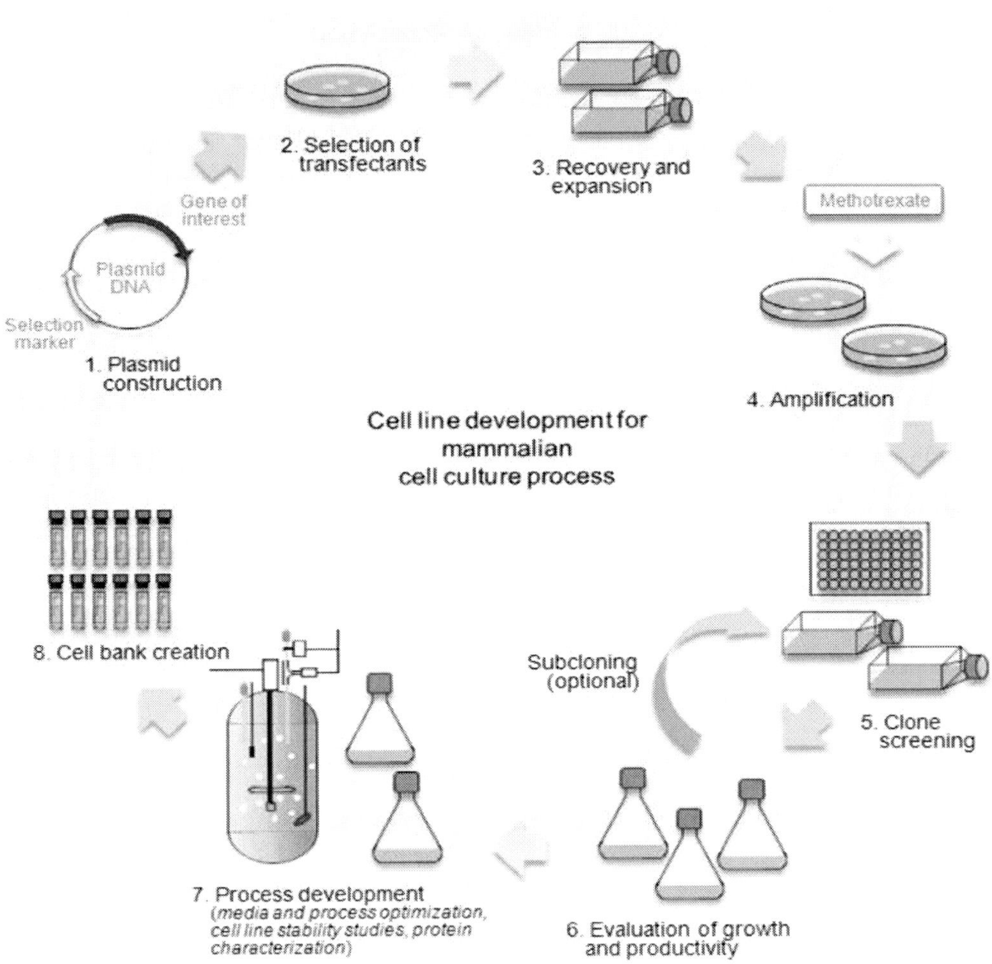

Figure 1. An overview of steps involved in the early stage of recombinant protein producing cell line development. Timely completion of this process is critical for meeting deadlines to generate clinical material and file for new drug application.

Table 3. Commonly used antibiotic selection and amplification markers used for selection of production cell lines

	Selectable gene marker	Selective conditions	Mechanism of action	References
Antibiotic selection marker	Neomycin phosphotransferase (Neo)	100 – 1000 μg/ml of G418 or kantamycin	Amino glycoside blocks polypeptide synthesis by inhibiting translation elongation steps. Neo gene inhibits G418 activity by phosphorylation.	[16,17]
	Hygromycin-B-phosphotransferase	50 – 200 μg/ml of	Amino glycoside antibiotic that	[18]

	(Hph)	hygromycin	interferes with protein translocation and causing mistranslation. Hph is a kinase that phosphorylates and inactivates hygromycin.	
	Zeocin resistance gene (Sh Ble)	0.1 – 50 µg/ml of zeocyin	Copper-chelating glycopeptide that causes cell death by binding to DNA and cleaving it. Zeocin resistance is conferred by the Sh Ble gene product which inactivates zeocin.	[19]
	Puromycin-N-acetyl transferase (Pac)	1 – 10 µg/ml puromycin	Amino-nucleoside antibiotic which inhibits peptidyl transfer on eukaryotic ribosomes. Pac gene expression detoxifies puromycin through acetylation.	[20]
Complementation of metabolic deficienncies	Dihydrofolate reductase (Dhfr)	5nm – 10 µM of methotrexate in glycine, hypoxantine, thymidine free media.	Dhfr is an enzyme that catalyzes the conversion of folic acid (vitamin B12) to tetrahydrofolate, THF, which is a co-factor involved in the synthesis of DNA precursors such as thymidine and purine. Methotrexate inhibits the activity of DHFR.	[21,22]
	Glutamine Synthetase (GS)	10 – 100 µM Methionine sulfoxamine	GS is an enzyme which catalyzes the conversion of glutamate to glutamine. Methionine sulfoxamine inhibits activity of GS.	[23,24]

One of the most commonly used selection and amplification systems in Chinese hamster ovary (CHO) cells is based on dihydrofolate reductase (DHFR), an enzyme which is required for the synthesis of DNA precursors. DHFR-deficient CHO cell lines like DUKX-B11 and

DG44 lack the endogenous expression of this enzyme. After being transfected with plasmid DNA coding for the recombinant gene of interest along with a copy of *dhfr* gene, the cells are then cultivated in a medium lacking glycine, hypoxanthine and thymidine so that only clones which are transfected with the gene survives. Clones which survived the first selection are further cultivated in media with increasing concentration of methotrexate (MTX), a drug that inhibits the activity of DHFR enzyme. After 2 – 3 weeks of cultivation in MTX supplemented media, the majority of cells die, and what is left is a surviving cell population which can contain several hundred copies of the integrated foreign plasmid. This process of 'amplification' typically produces clones that produce more recombinant protein than unamplified cells, thus improving specific productivity from 10- to 20- fold.

Another similar selection strategy is the glutamine synthetase (GS) system utilized in myeloma NS0 and GS-CHO cells which lack endogenous expression of glutamine synthetase [23,24]. GS catalyzes the conversion of glutamate and ammonia to glutamine, an essential amino acid required for biomass and energy production. Upon co-transfection of the recombinant gene and GS into host cells, the cells are cultivated in glutamine free media to select for producing clones. To amplify the copies of recombinant gene, methionine sulphoximine (MSX), an inhibitor of glutamine sythesis activity is typically added to the media at a concentration of 10-100 μM. In both cases of DHFR and GS amplification systems, the inhibitors MTX/MSX can be withdrawn in the later stages of process development when copies of the recombinant gene is stably integrated into host cell's genome.

While multiple steps of amplification can lead to hundreds of copies of inserts in the host cell's genome, this also leads to the risk of chromosome rearrangements and karyotypic instabilities in the host cell lines [22,25,26]. Moreover, studies have shown that high gene copy numbers does not necessarily correlate with high productivities of the recombinant protein [27]. Rather, the integration site of the recombinant gene in the host cell's genome, or the progressive loss in the protein expression over time has a major effect on the level of transgene expression. These incidences are attributed to epigenetic factors resulting in the condensation and transcriptional inactivation of the chromatin at the site of integration. Transgene expression could be enhanced and stabilized via genetic manipulations such as modification of histone acetylation [28], inclusion of specific DNA elements capable of blocking the formation of condensed chromatin (heterochromatin) [29-31], and targeting sites of high gene-expression [32].

2.3. High-throughput Screening Methods for Clone Selection

Transfection efficiency is characterized by the percentage of cell population which contains a copy of the recombinant gene stably integrated and expressed in the chromosome. This number typically ranges from 10% to 40% depending on the transfection protocols. Common methods such as polymer- or lipid- based carriers [33,34], electroporation [35] and calcium phosphate precipitation [36] are effective in delivering foreign DNA into cell's nucleus. The choice of one transfection method over the other depends on cellular phenotype. Typically, for the transfection of adherent cells, reagents containing calcium phosphate, lipid-based or polymer-based reagents are used. The ratio of DNA to lipid-based reagents is often optimized for efficient transfection.

Following transfection, the selection of a "clonally-derived" cell population from a single cell is performed. One of the most commonly used method owing to its relative simplicity and low cost is limiting dilution cloning. In this method, a low density cell suspension is dispensed into each well of 96 or 384 well micro-titer plates at an average of less than 3 cells per well. Wells are observed microscopically for any viable cells that could proliferate under the selection conditions. Supernatants from individual wells can be assayed for productivity by methods such as enzyme-linked immunosorbent assay (ELISA). Cell clones that yield high product titer are transferred into larger micro-well plates (12 or 24 well plates) and subsequently to T-flasks to be monitored for growth and productivity profiles. Clones which deliver the highest protein titer are typically selected for a second round of cloning (sub-cloning) to reduce population heterogeneity.

Being labor-intensive and time-consuming, limiting dilution cloning is not amenable to screening of more than a few hundred cell lines. A majority of the surviving cell population shows only an average level of protein production, whereas a minority shows productivity significantly above average. Biopharmaceutical industries employ other automated high-throughput technologies with increased screening capacities to evaluate thousands of clones [37]. Flow cytometry-based cell sorting techniques is facilitated by antibody targeted to the recombinant protein and ligand-conjugated fluorochromes, thus enabling isolation of cells based on cell-surface protein expression. Fluorescence-activated cell sorting (FACS) facilitate the screening of millions of cell clones, and usually predicts accurately the productivity levels of clones because it is not affected by differences in growth rates or medium volumes in wells, as is in the case with limited dilution cloning.

In a FACS-based clone selection report, a cell surface protein CD20, is co-expressed, as a reporter protein with the therapeutic protein and detected using fluorescently labeled antibody [38]. The therapeutic protein and reporter protein are linked by an internal ribosome entry site (IRES), thus both are transcribed on the same mRNA, but translated separately. Cells are screened by flow cytometry for high-level expression of CD20, and cells identified with high-level expression are sorted and subsequently cloned. This method provides an effective and rapid process for selecting high productivity clones at an early stage. Cell clones can also be subjected to re-screening to monitor instability in protein expression over several passages [38].

2.4. Cell-Banking

The selection of one producing clone amongst the many candidate cell clones is based on several factors. Robust cell growth in reactors and the achievement of high protein titers are the obvious criterion. Other factors such as the stability of the protein expression for each clone, typically evaluated over 50 – 80 population doublings, and the characterization of protein glycosylation profiles throughout different generations of the cell population have to be evaluated and considered in the early stage of cell line development.

Candidate cell clones are selected for cryopreservation or banking process. Cells destined for cryopreservation are usually cultivated in small vessels, harvested in mid-exponential growth phase, concentrated and mixed with media containing a cryo-protectant, before storing in liquid nitrogen tanks at cryogenic temperatures of -180 °C. In general, three categories of cell banks are made – the Research cell bank (RCB), the Master Cell bank

(MCB) and the Working Cell bank (WCB). Research cell banks (RCB) are created primarily for use in research and development, when early candidate clones are screened and being evaluated. In the manufacturing stage, a two-tiered cell banking system consisting of MCB and WCB are created to guarantee a constant supply of cell lines used for future manufacturing cycles. Under Good manufacturing practice (GMP) regulation, a thorough documentation of the history of the production cell line which includes details on the origin of the clone, its genotype as well as its passage history must be established. Furthermore, the MCB is characterized and tested extensively to be free of contaminants such as bacteria, virus and mycoplasma. Strict regulatory guidelines are enforced to validate the biosafety in the use of both WCB and MCB for manufacturing recombinant proteins [39].

MCB is used as the primary stock to generate the working stock of cryopreserved cells, termed the WCB. The WCB stock serves as the only cell source for the production process, and can only be used for a defined number of passages, or population doublings in the production process. This production timeline has to be defined to guarantee that the cell population remains genetically stable, i.e. cells do not lose expression of the recombinant protein.

3. Process Development

Processes developed at laboratory scale for cell line development can seldom be directly implemented in commercial manufacturing. Industries typically spend a significant amount of time and resources to develop robust and economically viable manufacturing processes which can yield consistent product quality and quantity for years. Given the competitive nature of the biopharmaceutical arena, there is an ever-increasing need for businesses to reach target markets as soon as possible after a promising candidate drug is discovered. This enforces shortened timelines for process development. Industries have evolved by adopting platform processes that work for their cell type and business needs. It is now common for industries to base the first round of process development largely on their platform processes to produce preclinical and clinical trial material.

Traditionally, the primary goals of mammalian cell process development are two-fold – (1) media development, which involves incremental improvements from the platform process media, and (2) determining the optimal process parameter range which can lead to higher cell density, productivity and product quality. Traditionally cultures are evaluated either in small scale flasks/roller bottles or scaled-down bioreactor models (1 L to ~20 L working volume). In bioreactors, the culture is carefully monitored and automated controls for pH, temperature and dissolved oxygen are used. Once optimal culture medium and parameter ranges are established, the process is verified at pilot-scale (50 L to 1,000 L reactors) and then transferred to manufacturing settings for cultivation at commercial scale (200 L to 20,000 L scale).

3.1. Cell Culture Media Development

The optimization of cell culture medium is one of the key steps for the development of a mammalian cell culture platform. Compared to bacterial fermentation, media used for the cultivation of mammalian cells are highly complex and enriched with essential nutrients because they do not synthesize certain key nutrients *in vitro*. Since 1950s, researchers have developed cell culture media for specific cell lines. One of the earliest formulation, Eagle's Minimal Essential Medium (MEM) contained minimum ingredients of 13 amino acids, 6 vitamins, 6 ionic species, glucose and certain undefined serum components [40,41]. Other more complex forms of basal media formulations such as Dulbecco's Modification of Eagle's medium (DMEM), Rosewell Park Memorial Institute (RPMI), and cell line specific media such as Ham's F10, F12 media were developed later. Today, there can be as many as 50-70 total components in a commercial cell culture media formulation.

3.1.1. Cell Culture Media Components

The key components of mammalian cell culture medium are summarized in the few categories described below:

Serum, Hydrolysates and Related Components

The composition of serum is undefined, comprising mainly sugars, amino acids, nucleosides, fatty acids, vitamins, growth factors, hormones, trace elements, and lipids. Depending on the cell line, the supplementation of 5 - 20% serum (vol/vol) may be required for cell growth and proliferation. Despite its benefits, serum usage in recombinant protein production is generally avoided because of its undefined animal origin source, lot-to-lot variability, high cost and the challenges in down-stream protein purification from the interference of serum proteins in media.

The process of formulating a serum-free media for a cell line requires extensive media development efforts and in some cases, adaptation of cell lines to serum free condition [42-44]. In the process of adaptation, cells could demonstrate poor growth, apoptotic cell death and lower protein productivities. These issues can be circumvented by supplementation of some of the serum components in a defined manner. Supplementation of either one, or a combination of these factors: insulin, transferrin, ethanolamine and selenium in serum-free media has been shown to be beneficial to cells in serum-free media [45-47]

Serum-free cultivation media typically requires the addition of complex plant or yeast hydrolysates in their formulation to replenish nutrients and growth promoting factors present in serum [48-50]. While hydrolysates can replace animal-derived components, the composition of hydrolysates is still relatively undefined and its composition can vary significantly between different lots.

Buffers

Sodium bicarbonate coupled with a controlled 5 – 10% carbon dioxide environment is the most commonly used pH buffer system in cell culture media. One of the drawbacks of the presence of bicarbonate in the media is that it increases dissolved CO2 concentration in media, simultaneously decreasing intracellular pH, which can potentially inhibit cell growth, metabolism and productivity [51]. Also, the pK value of a bicarbonate buffer is typically lower than the culture pH, yielding a suboptimal buffering system [52]. As an alternative to bicarbonate buffer, organic buffers which do not rely on dissolved CO2 for pH control can be used in the media formulation of many cell lines. Some of the examples for commonly used organic pH buffers are 2-[4-(2-hydroxyethyl)-lpiperazinyl] ethanesulfonic acid (HEPES) and 3-(N-morpholino)propanesulfonic acid (MOPS) [53].

Salts

Cell culture medium contains low concentrations of ionic salts to compensate for osmotic imbalances created by the metabolic by-products secreted by cultured cell lines. The salts that are mainly used in media are sodium (Na+) and pottasium (K+) ions, in the form of chloride salts. In addition to maintaining osmolality, sodium and potassium ions play a role in membrane transport and balancing proper ionic strength for the functionality of cellular enzymes.

Carbohydrates

Glucose is the primary 6-carbon carbohydrate source provided in the cell culture media, and is present in relatively high concentration (5- 30 mM). Glucose in cell culture media is actively transported across the cytoplasmic membrane and eventually converted into pyruvate in the cytosol through a series of enzymatic reactions in the glycolysis pathway. The break-down of glucose and the supply of carbon source to the TCA cycle is an important part of oxidative metabolism which generates cellular ATP energy. Glucose can also be metabolized via pentose phosphate pathway to generate reducing equivalents (NADPH) and DNA precurors. Continuous cell lines show an overactive glycolytic flux, resulting in a majority of pyruvate to be diverted to lactate instead of entering TCA cycle. Alternative sugar sources such as galactose and fructose are consumed by mammalian cells at a much slower rate, which reduces lactate formation and more efficient oxidative metabolism [54].

Amino Acids

Mammalian cell culture medium requires 12 – 14 essential amino acids, which provides carbon and nitrogen source for cellular needs. These primarily include arginine, histidine, isoleucine, leucine, lysine, methionine, phenylalanine, threonine, tryptophan, valine, glutamine, cysteine, tyrosine [41,55]. Depletion of essential amino acids in the cell culture medium may lead to malfunction of cell regulation causing cell death by apoptosis [56]. Some non-essential amino acids are required by specific cell lines, for instance proline is generally required by mutant CHO cells, and serine is frequently required at low cell densities during cell cloning [57]. Elevated consumption of non-essential amino acids leads to ammonia buildup and undesirable effects on product quality. Hence, during media development efforts, an optimum range of amino acids concentration in media has to be formulated based on spent media analysis of cell culture supernatants.

Glutamine is typically provided in the concentration range of 4 – 8 mM. Glutamine is actively transported by more than one transport systems into the cell and similar to glucose, is consumed at a higher rate in continuous cell lines compared to primary cells. Glutamine undergoes incomplete oxidation termed glutaminolysis to generate ATP, or it can be completely oxidized to CO_2. Over-consumption of glutamine results in the generation of by-products such as alanine, lactate and ammonium. Glutamine also generates metabolic precursors such as purines and pyrimidines in addition to other amino acids, such as, glutamate, asparigine and aspartate that either are incorporated into proteins or participate in energy metabolism via TCA cycle.

Vitamins

Vitamins are present at low concentrations in media where they are used as co-factors for enzymatic reactions. The amounts and specific vitamin requirements are cell line-dependent. Eagle's basal media contained seven vitamins and cofactors including choline, folic acid, nicotinamide, pantothenate, pyridoxal, riboflavin and thiamine which are all essential for cell growth [40]. Biotin, which is a co-factor for fatty acid metabolism is added in small quantities [58], and vitamin B12 might be required at low cell densities [59]. Depending on cell line of interest, vitamins can act synergistically with other lipid and nucleotide compounds, to sustain cell-growth. For example, the vitamin myo-inositol supplemented with the lipid constituent can stimulate the growth of keratinocytes [60].

Growth Factors, Hormones and Carrier Proteins

Growth factors are proteins that promote cell growth and proliferation. It should be noted that the role of proteins in serum-free media is usually speculative and highly cell line dependent. Transformed cells do not require as many growth factors as primary cell culture [61]. One such growth factor, insulin, plays a role in stimulating glucose uptake and synthesis of RNA, protein and lipids in mammalian cells [62]. A wide array of growth factors such as fibroblast growth factor (FGF), epithelial growth factor (EGF), transforming growth factor (TGF), platelet-derived growth factor (PDGF)and cytokines have shown positive growth effects in different cell lines [42]. Carrier proteins are needed to facilitate transport of nutrients or trace elements. In protein free medium formulation, carrier proteins are replaced by alternative non-animal derived compounds. Transferrin is a carrier protein that promotes iron uptake in cell culture [63,64]. Iron chelators could also partially replace the requirement of transferrin in cell culture media [65].

Lipids

Lipids are required for the cultivation of cell lines as they are incorporated into structural elements such as plasma membrane, mitochondrial and endoplasmic reticulum membranes. Mammals are unable to form unsaturated carbon bonds along fatty acyl chains beyond the C-9 position. Therefore, the presence of essential fatty acids (EFAs) such as linoleic acid and linolenic acids in culture medium is important [66,67]. Most cell culture media are provided with EFAs or some form of fatty acid complex solution, however some cells can grow without lipids supplementation even in serum-free media. In CHO cells, the addition of linoleic acid improved culture performance in spinner flasks due to a more robust membrane structure that is resilient towards environmental shear stress [66].

Nucleosides, cyclic nucelotides and polyamines

CHO cells that are DHFR mutants require the presence of hypoxanthine, thymidine, and glycine in the media, for the survival and growth due to its deficiency in synthesizing folic acid [8]. Organic cations or polyamines such as putrescine, spermidine and spermine are needed for replication and cell division in CHO cells [68-70]. Additionally, cyclic adenosine 3':5'-monophosphate (cAMP) and guanosine 3':5'-monophosphate (cGMP) have been known to improve antibody productivity in hybridoma cultures [71].

Trace Elements

Trace elements such as iron, manganese, zinc, molybdenum, selenium, vanadium and copper are provided in serum-free media in nano-molar concentrations. Trace elements are a source of co-factors for metabolic enzymes in the cells. Serum-free media developed for CHO cell line showed beneficial effects when supplemented with iron, copper, zinc [72]. Selenium has been shown to alleviate oxidative stress in mammalian cells [73-75], and functions as an effective iron-carrier in CHO cells [76]. Manganese (Mn^{2+}) is a co-factor for the glycosylation enzyme, β1,4-galactosyltransferase, and the addition of Mn^{2+} can modify the oligosaccharide structures on recombinant erythropotein produced in CHO cells [77].

3.1.2. Media Development Approaches

Specific types of media are developed to support cell growth at each process stage. During the seed expansion stage, the primary goal of the medium is to support active growth and generate high cell population in the shortest timeframe from the point of vial thaw to inoculation into production bioreactors. As such, nutrients are provided in excess and in some cases, growth promoting factors are included. In the later stage, where cells are growing in production bioreactors, the requirement of the media is shifted from the focus of stimulating fast growth, to sustaining growth and increasing specific protein productivities. In the case of high-density perfusion or fed-batch culture (discussed later in detail), a concentrated feed media is supplied to provide adequate nutrients for high-cell density cultures, and provide pH buffers. Over-feeding of nutrients in feed media can result in the formation of metabolic by-products such as ammonia or lactate that could inhibit growth and affect product quality. Amino acids and carbohydrate source is typically provided in stoichiometric proportions in the feed media to support oxidative metabolism and minimize by-product formation [78,79].

In the early stage of cell line development, clonal media is developed for screening of transfectants. It has been reported that media developed for low cell density cell cloning of hybridoma cells, must contain higher amounts of asparagine, glutamine, serine, inositol and pyurvate [57]. At low population densities, these nutrients were lost from the cell to the media environment in amounts which exceeded the biosynthetic capacity of the cell.

Today's challenge in the industry, is to develop chemically defined, protein free media formulations to improve process performance. Choosing an effective media development strategy and designing the right cell culture experiments is important to achieve results in a short time frame. In line with this discussion, rational media design encompassing several different approaches such as component titration, media blending studies, spent media analysis and automated high throughput screening has been adopted by the industry [80]. Statistical methods such as response surface design and factorial design are typically employed to understand and quantify the variable (i.e. media component) that affects the outcome of cell growth or protein titer [81-84].

3.2. Culture Systems for Mammalian Cells

Typical culture systems employed in mammalian cell culture development are discussed in the following section. Table 4 shows a list of these systems and highlights key characteristics of each option along with its pros and cons.

Table 4. Pros and cons of some commonly use cell culture platforms

Cultivation Unit	Pros	Cons
Small scale units		
- Static units (T-flask, petri dish)	Easy to use, disposable, high throughput operation, little or no optimization needed	Very small volumes, no process monitoring, severe gas transfer/mixing limitations
- Dynamic units (shake flask, spinner flask, roller bottle)	*Same as static units*; additionally - slightly larger volumes, better gas transfer/mixing from agitation	Small volumes, no process monitoring
Stirred tank bioreactors		
- Batch	Operationally very simple, large volumes, partial process control, homogenous, excellent gas transfer	Short campaign lengths, low peak cell density, nutrient limited
- Fed-batch	*Same as batch*; additionally - longer campaign lengths, high peak cell density, very large volumes, nutrient replenished through concentrated feed	Not suitable for labile proteins, requires feeding strategy optimization
- Perfusion	*Same as batch;* additionally - very long campaign lengths, very high peak cell density, best process control, true steady state, continuous nutrient replenishment and product withdrawal, suitable for labile products	Requires extensive process optimization, needs very stable cell clones, continuous feed requirements restrict operationally feasible volume
Disposable bioreactors	*Amenable to batch, fed-batch or perfusion modes*; additionally - short turnaround times, vendor guarantees sterility, no risk of cross contamination	Not yet available for very large volumes, requires training, expensive, robust disposable probe technologies not mature
Mini/micro bioreactors	Very high throughput, ideal for screening, partial process control	Very small volumes, operated mainly in batch mode, conditions may not represent scaled-up bioreactors

3.2.1. Small Scale Cultivation Units

Small scale cultivation units include a wide variety of systems for both static and dynamic (agitated) cultures (Figure 2). Static systems are characterized by unenforced external power inputs. Petri-dishes, multi-well plates and T-flasks are some examples of static cultivation systems. The working volumes of these systems are quite small, ranging from sub-milliliter scales to 50-100 mL. These systems, although commonly used during cell line development, are rarely employed during the process development stage.

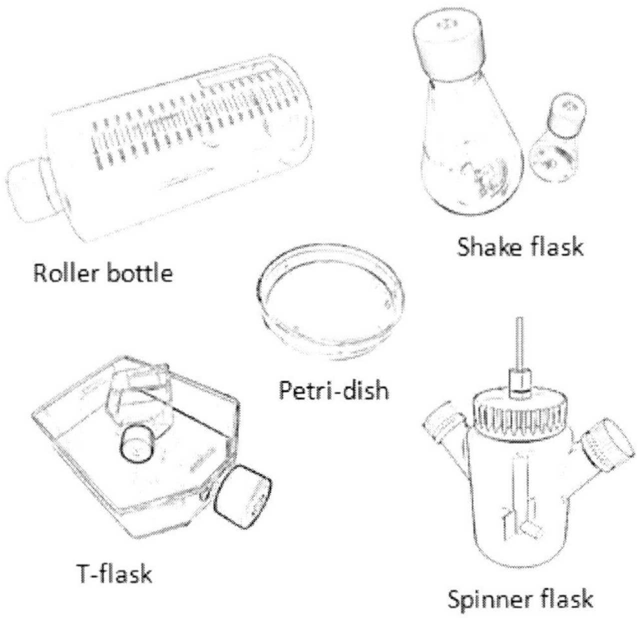

Figure 2. Commonly used small scale cultivation units for mammalian cell culture.

Small scale dynamic cultures are generally carried out with external mechanically driven units like shaking platforms, rolling or spinning units. These units drive glass or disposable flasks (screw capped with gas-permeable membranes) designed to fit the driving unit (Erlenmeyer flasks, roller bottles, spinner flasks). The working scale volumes for such systems range from a few milliliters to 1-2 L.

Temperature control for small scale systems is generally achieved by using humidified warm incubators. In addition, carbon dioxide may also be maintained at physiologically relevant levels in these cultures by adjusting incubator CO2 levels. Cell growth in these small scale systems is typically limited to $2\text{-}3\times10^6$ cells/mL, mainly due to O2 transfer rate and nutrient limitations. Oxygen transfer in such systems is through surface aeration and volumetric mass transfer coefficients (k_La) are generally quite low, although some moderately higher values up to ~30 h^{-1} have been reported for TubeSpin [85] and shake flask [86] systems at higher speeds. Thus, culture densities and productivities in small scale cultures may not accurately predict their behavior in larger systems.

Nevertheless, these systems continue to be used in the industry due to their ease of implementation, particularly in screening and media development studies. In addition, scale-up of cells from vials to seed trains and reactors require intermediate small scale cultivations which need to be validated for manufacturing processes. There are also a few examples of certain high-value products such as erythropoietin (Amgen, licensed in 1989, [87]) and vaccines (example: human influenza virus, [88]) which are produced in small scale units particularly roller bottles.

3.2.2. Stirred Tank Bioreactors

Despite the availability of a multitude of options, the stirred tank bioreactor remains as the industry's preferred workhorse for mammalian cell cultivation. The popularity of the stirred tank bioreactor stems from its simplicity, robustness and ease of implementation and scale-up. These bioreactors are generally equipped with aeration devices, impellers, baffles and several ports for process monitoring and culture sampling (Figure 3).

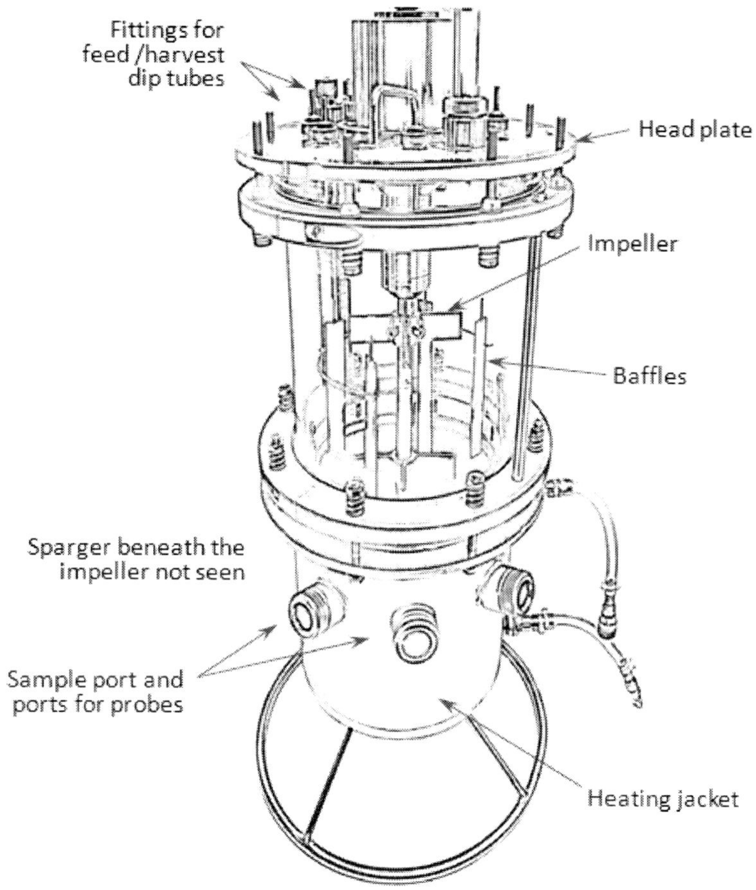

Figure 3. A typical stirred tank bioreactor with accessories for cultivation of mammalian cells.

Impellers and baffles serve to control adequate mixing of oxygen and nutrients, and to avoid cell damage. Aeration is typically achieved through sparging a mixture of gases (oxygen, nitrogen or air, and carbon dioxide, when needed). Additionally, gas sparging serves to strip off excess CO_2 from the reactor. Cell shear and damage due to bursting of gas bubbles can occur, and modeling of the fluid dynamics of gas aeration has to be considered when design stirred tank bioreactors [89,90].

Although stirred tank bioreactors are typically used for suspension cells, they can also quite easily be adapted for anchorage dependent cells on microcarriers [91,92]. Macroporous microcarriers provide a scaffold for anchorage dependent cells to grow on and provide some degree of protection from shear damage. Stirred tank bioreactors are generally operated in one of three principal modes (Figure 4) – batch, fed-batch or perfusion.

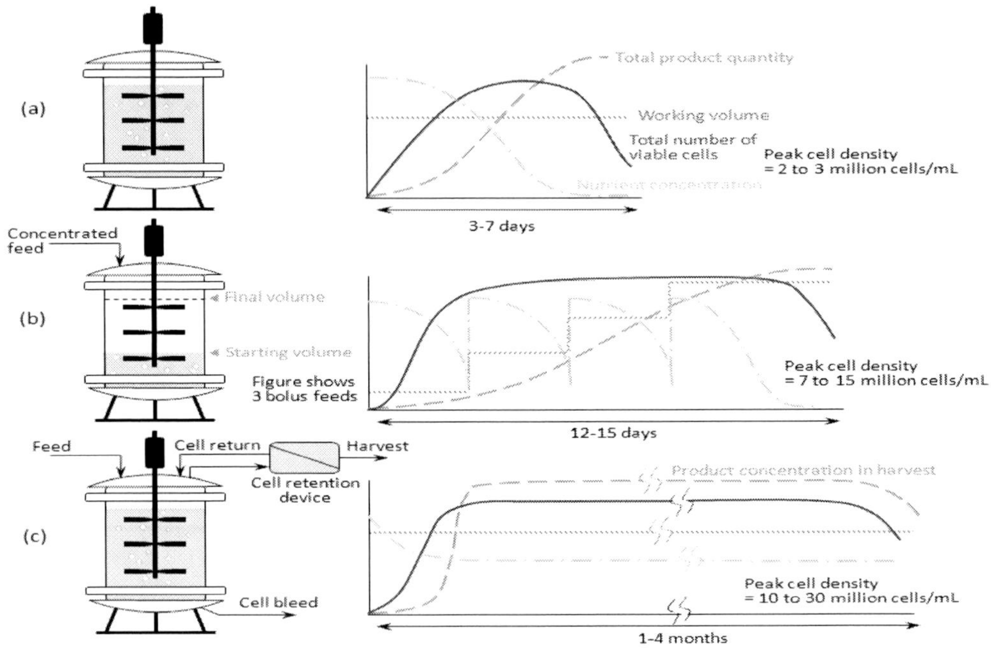

Figure 4. Bioreactor operation modes commonly employed in mammalian cell culture: (a) Batch, (b) Fed-batch and (c) Perfusion.

The *batch* reactor is the simplest where cells are inoculated into a reactor containing all the nutrients necessary for the entire course of cultivation. Operation typically occurs over 3-5 days, at the end of which cell death occurs due to accumulation of toxic metabolites or depletion of nutrients [93]. The culture broth is then isolated and the secreted protein of interest is purified. Batch cultures are typically characterized by increasing viable cell densities, followed by a peak and then a downward trend (Figure 4a). Although this mode of operation is quite easy to implement, peak cell densities rarely exceed 3-7x10^6 cells/mL and hence is not favored in the industry.

The *fed-batch* mode of operation (Figure 4b) can be viewed as an extension of the batch culture through periodic or continuous additions of certain nutrients to delay cell death. A typical fed-batch process will last about two weeks. In its simplest form, only basic nutrients such as glucose and glutamine are added to the culture. However, it is more common to feed a complex mix of nutrients [94]. Options include simply feeding a concentrated solution of the basal medium or adopting a more complex, but less heuristic approach incorporating spent medium analysis from various stages of culture [79]. The feeding strategy should also include considerations to mitigate build-up of toxic metabolites like lactate and ammonia. Previous studies have shown that, through a rational design of feed media composition, it is possible to elicit a shift in the cell's metabolism toward a more favorable lactate or ammonia producing state [79,95].

Today, the fed-batch mode is by far the most dominant culture method for animal cells in industry. Feeding strategies and logistics of operation are simple to understand and implement for operators. Besides, the duration of each campaign is short enough to mitigate risks associated with any adverse excursions such as equipment failure or contamination.

The limit of a fed-batch operation is generally dictated by the reactor volume. One way to overcome this limitation is to perform a continuous culture whereby a harvest stream is continuously collected at a rate equal to feed to maintain constant volume (chemostat cultures). This setup will theoretically allow for indefinite culture operation. However, high cell densities cannot be achieved because perfectly viable cells will be discarded in the harvest. This drawback can be overcome by using *perfusion* cultures (Figure 4c) [96-98]. In perfusion cultures, cells from the continuous harvest stream are separated and returned back to the reactor. This allows for processes to achieve much higher densities compared to fed-batch cultures (>$25x10^6$ cells/mL) which, in turn, implies that smaller scale reactors are needed to produce the same amount of protein. Since perfusion cultures can generally last for months at a time, labor costs associated with reactor turnaround are minimized. Perfusion cultures also reduce the mean residence time of fluid elements in the reactor since it is not necessary to wait until the end of campaign to harvest the product. This is sometimes a crucial and over-riding consideration for certain labile proteins which need to be stabilized as soon as possible [99].

Cell retention devices for perfusion cultures are generally based on principles of centrifugation, sedimentation, filtration or acoustic separation [97,98]. Besides retention efficiency, other important considerations for the design of perfusion cell retention devices include relatively small volume (compared to reactor volume), absence of cellular stress causing environments and scalability to manufacturing scales.

A drawback of the perfusion process is that cell lines with long-term stability characteristics (in terms of productivity) are necessary to implement successful operation. Also process development steps are prolonged because of the need to validate extended process durations.

3.2.3. Emerging Trends in Cell Culture Operation

Disposable bioreactors have been developed from polymers such as polystyrene, polyethylene, and polytetrafluorethylene [100,101]. The wave-mixed bag bioreactor performs mixing via means of a rocking or shaking platform for continuous surface renewal and aeration while minimizing bubble formation. These reactors come with disposable probes, minimizing efforts for construction and, thus, turnaround times between campaigns. Studies have claimed oxygen transfer rates rivaling those of stirred tank systems in these bioreactors (Technical Reference, GE Life Sciences). Successful applications of wave bioreactors for CHO, NS0 and hybridoma cells have been demonstrated for recombinant protein therapeutics [102-104]. Other disposable systems such as the stirred bag bioreactor and static cell bags have also been implemented for mammalian cell culture [100]. Together, these systems provide attractive alternatives to traditional systems for small to medium scale biotherapeutics manufacturing, and inoculum seed preparation for commercial manufacturing. The disposable units are generally available pre-sterilized, thus saving valuable time and resources for the end-users with minimal risk of cross-contamination.

Another emerging area in cell culture technology is the use of multi-well mini- or micro-bioreactor in scale-down or process development work. Micro-bioreactors could mimic conditions encountered in manufacturing systems while at the same time retaining the versatility of small scale units (in terms of amenability to high-throughput cultivation). Initially popularized by high-throughput screening studies for microbial and yeast process development, they are increasingly finding applications in mammalian cell cultivation [105-107]. Working volumes in these systems can range from 50 μL to a few mL.

3.3. Process Monitoring and Control

Commercial manufacturing of biological products necessitates the design of robust, well-defined and reproducible processes. Process monitoring and control form a quintessential part of such design. Typically, temperature, pH, dissolved oxygen (DO) and biomass concentration are some of the key variables monitored and controlled during bioreactor operation.

Temperature control is usually achieved by means of a heated water jacket surrounding a part of the reactor vessel. A thermocouple senses the temperature of a well-mixed zone in the bioreactor and the data is passed onto a PID-type controller which feeds back information to a system controlling the flow rate and temperature of jacket water. Optimal temperatures for animal cell cultures are generally around those that are physiologically relevant to living animals (35 °C – 37 °C).

pH control is another key aspect of bioprocess operation. Mammalian cells typically secrete lactic acid and ammonia as the by-products of glucose and glutamine metabolism respectively. Build up of these compounds can adversely affect the culture pH and cause deviations from physiologically relevant levels (6.8 – 7.0). Steam sterilizable glass electrodes are, perhaps, most commonly used to measure bioreactor pH. However, their low mechanical stability especially during heat sterilization cycles has led process engineers to explore other alternatives like optical pH sensors [108-111]. Optical sensors operate on the principle of absorbance or fluorescence from pH- sensitive dyes. They usually have a smaller dynamic range of operation compared to traditional electrochemical probes, but are more sensitive around the pKa of the dye. pH control is commonly achieved by means of a feedback control loop supplying CO2 or carbonate to the reactor.

Dissolved oxygen (DO) levels are also typically monitored and controlled in bioreactors because they can, sometimes, be a limiting factor preventing high cell densities. The steam strerilizable Clark-type sensor remains as the workhorse for industrial bioreactor oxygen monitoring [112]. However, fluorescence-based DO measurements are also gaining popularity [109-111]. These utilize a fluorescent dye which is quenched in the presence of oxygen in a concentration dependent manner. As with temperature and pH control, DO control is also achieved by means of a feedback control loop which adjusts the rate of oxygen sparging into the system to maintain constant DO.

Biomass concentration is yet another key process parameter that needs to be monitored carefully. Periodic offline cell density measurements utilizing trypan blue staining for viable/non-viable cell counts are still the most commonly employed method to estimate biomass concentration. Trypan blue is a diazo dye that can traverse through the membrane of dead cells but cannot enter live cells. This leads to a distinctive blue shade for dead cells which can be easily observed under a microscope. Numerous systems for automated counting of trypan blue stained live and dead cells are available (examples: Vi-Cell™ and Cedex™ cell counters). Online cell density measurements are also adopted for cases requiring cell density control (for e.g. perfusion systems). Indirect online cell density measurements through electrical or optical means are possible [113]. Electric probes rely on capacitance or permittivity measurements which depend on the fraction of fluids encapsulated within polarizable membranes. Optical methods for cell counts rely on absorbance or light scattering. Measurements are sometimes made at near infrared wavelengths where media components generally have the least absorbance [114].

Other variables are also monitored during bioreactor operation, such as nutrient and metabolite concentrations (glucose, lactate, ammonia, glutamine, dissolved carbon dioxide). Oxygen uptake rate (OUR) is sometimes estimated because it is a good indicator of cellular metabolism. Common methods for OUR measurements include stationary liquid phase balance which involves O_2 balance with an *apriori* knowledge of k_La (volumetric gas transfer coefficient) or a dynamic method which involves a short termination of O_2 supply to bioreactor and subsequent measurement of the rate of decrease in dissolved oxygen [115].

3.4. Process Scale-up

Scale-up of bioprocesses from laboratory to manufacturing scales remains a significantly challenging task. Oxygen supply and carbon dioxide stripping are, perhaps, the most important considerations for process scale-up [116]. The oxygen transfer rate is a function of a number of process parameters including physical parameters (like fluid viscosity, density and surface tension), bioreactor scale and geometry (height and diameter), operational conditions (sparging rate, agitation speed, design of impeller blades and baffles) as well as cell metabolism [117]. Gas transfer from surface aeration decreases markedly as reactors are scaled up due to reduced surface to volume ratio [118]. On the other hand, the volumetric gas transfer coefficient, k_La, has been shown to increase with increased sparging rates, determined by units of VVM (vessel volumes per minute) and the proportionality constant increases with vessel depths [116]. Both these parameters can be adjusted to maintain similar oxygen transfer rates during scale-up. Although both higher sparging and agitation rates can significantly improve gas transfer efficiency, shear damage of mammalian cells at higher mixing regimes is a concern. This problem can be addressed to some extent by addition of surfactants such as Pluronic F-68 which creates a protective layer around cells making them less susceptible to shear damage. A recent study employing Pluronic F-68 in fed-batch CHO cultures concluded that higher sparging may be more efficient than increased agitation rates to maintain mass transfer during scale-up [119]. However, surfactants need to be used sparingly in cultures because of two principle limitations – (1) the surfactants themselves are known to significantly reduce k_La in liquids and (2) they are hard to remove and are known to cause problems during downstream purification.

While oxygen transfer and carbon dioxide stripping are traditionally considered as the most critical engineering considerations for scale-up, bulk mixing is another important parameter. Poor mixing may lead to significant pH or nutrient gradients [120,121]. Perturbations in pH have been known to decrease specific productivity of cells [122]. However, although constant mixing times are desired during scale-up, it is often impractical as the specific power input required for larger scales become too high. Besides, mixing times are also not readily measurable, introducing significant errors in scale-up calculations. From a practical standpoint, it is more common to use one or a combination of several scale-up parameters such as constant specific power input (P/V), specific impeller pumping rate (Qs), impeller tip speed ($v$tip) and impeller shear rate during scale-up [123].

4. BIOMANUFACTURING

Typically, manufacturing processes for biopharmaceuticals begin with vial thaw from a working cell bank (Figure 5). Early stages of cell culture are conducted in T-flasks, shake flasks or spinner flasks with progressively increasing volumes. The inoculum seed is then scaled up through a succession of wave-type and/or small scale stirred tank bioreactors before being inoculated into a production scale reactor. Depending on the scale of the manufacturing bioreactor, this process may take two to six weeks after vial thaw.

Raw materials used for biomanufacturing are scrutinized to ensure quality and safety. They are tested rigorously to verify its identity, purity and safety before being used in the manufacturing process. However, this may not be entirely possible for certain complex medium components such as hydrolysates, peptones and serum. In such cases, a thorough validation and documentation of the origin of these components need to be established.

Following cell culture, the harvest from bioreactors is first clarified of cells and cell debris by centrifugation and filtration steps. The recombinant protein of interest is then purified from the clarified broth through a series of affinity chromatography, ultrafiltration/diafiltration (UF/DF) and/or membrane adsorption steps. The purification process also generally includes a viral inactivation step.

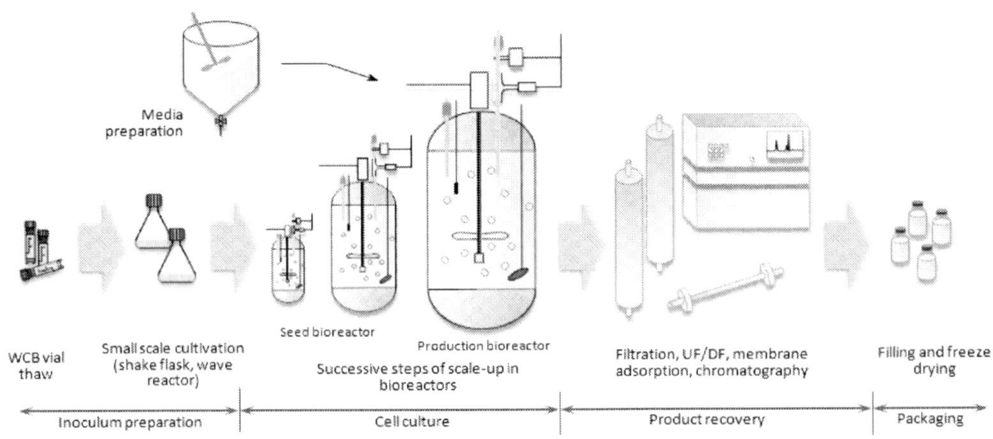

Figure 5. A schematic of various steps in manufacturing of recombinant proteins from mammalian cells.

4.1. Regulatory Oversight in Biomanufacturing

The biopharmaceutical industry is one of the most regulated industries in the world. The last few decades has seen an increase in regulations surrounding safety standards for biologic manufacturing. The history of pharmaceutical regulation traces back to the occurrence of tragic incidents in the 1930s when a mistake in the formulation of children's syrup caused deaths in the US, and to the side-effects of thalidomide synthetic drug develop in the 1960s. Both incidents triggered a review of drug approval practices in the US and Europe.

Since then, the growing awareness for the need of safe manufacturing processes has culminated in the establishment of certain Good Manufacturing Practices (GMP). GMP is part of a quality assurance system that covers the aspects of drug manufacturing and testing to ensure product safety and quality. Today, manufacturing of pharmaceutical compounds (including biodrugs) can be conducted only under GMP conditions. Table V shows a list of agencies responsible for GMP enforcement in different countries. One problem that arose from the increasing scrutiny of biopharmaceutical manufacturing was that, while drug manufacturers attempted to seek global markets, drug approval remained a national responsibility (Table 5). This often led to expensive, time-consuming and duplicative efforts to obtain drug approval in international markets. To overcome this problem and improve drug approval efficiency, regulatory agencies from the EU (EMA), US (FDA) and Japan initialized the International Conference on Harmonization of Technical Requirements for the Registration of Pharmaceuticals for Human Use (ICH) in 1990 [124]. Since then, the ICH has made significant contributions to the globalization of drug industry.

Table 5. List of major pharmaceutical regulatory agencies in some countries

Country	Regulatory Agency §
Australia	Therapeutic Goods Administration (TGA)
Brazil	National Health Surveillance Agency (ANVISA)
Canada	Therapeutic Products Directorate (TPD)
China	State Food and Drug Administration
France	Agence du Medicament *and* European Medicines Agency (EMA)
Germany	Federal Institute for Drugs and Medical Devices (BfArM) *and* European Medicines Agency (EMA)
India	Central Drugs Standard Control Organization (CDSCO)
Japan	Ministry of Health, Labor and Welfare
Malaysia	Department of Public Health
Russia	Public Health Institute, Ministry of Health
Singapore	Health Sciences Authority (HSA)
South Korea	Korea Food and Drug Administration (KFDA)
South Africa	Medicines Control Council (MCC)
Switzerland	Swiss Agency for Therapeutic Products (swissmedic)
UK	Medicines and Healthcare products Regulatory Agency *and* European Medicines Agency (EMA)
USA	Food and Drug Administration (FDA)

§ - Major regulatory agencies. Other auxiliary agencies may function in each country.

In the US, the FDA plays a key role in drug regulation. For biopharmaceuticals, preclinical testing of the candidate drug is followed by an Investigational New drug (IND) application before instituting clinical trials under FDA guidance [125]. If the results meet the clinical trial end-points, the manufacturer will submit a New Drug Application (NDA) and Biologics License Application (BLA) for commercialization. Following product launch, the safety and efficacy of the drug continues to be monitored through post market surveillance efforts.

4.2. Future Trends in Biomanufacturing

The past decade has witnessed rapid strides in the development of biologics which have, unsurprisingly, led to dramatic increases in the demand for manufacturing capacities. Yet, construction of a manufacturing facility remains a huge undertaking even for big pharmaceutical companies. Quite often, the need to capture a market share before other competitors means construction and commissioning of a manufacturing plant even before the completion of clinical trials. This presents a big risk, especially for smaller start-up companies which may simply become bankrupt if, after significant upfront investments, a promising drug candidate fails in clinical trials. Capacity sharing among biopharmaceutical industries and Contract Manufacturing Organizations (CMO) can, to some extent, mitigate such risks [126]. The philosophy of capacity sharing and contract manufacturing is built on a mutually beneficial partnership between (sometimes rival) companies. Recently, a number of new players have entered the field of contract manufacturing both in the traditional biopharmaceutical arena like the US and EU as well as in emerging Asian hubs like India, China, Singapore, Taiwan and South Korea [126,127]. Given the expected rise in demand for biotherapeutics, it appears that CMOs and capacity sharing strategies will play an inevitable role in biomanufacturing in the foreseeable future.

Another emerging trend in the field of biotherapeutics is the concept of follow-on biologics (sometimes referred to as "biosimilars" or "biogenerics"). A number of early biopharmaceuticals have lost patent protection and are now in production by non-innovator companies [128]. The costs associated with product innovation and development are generally absent or significantly reduced in biogenerics manufacturing. However, unlike its small molecule counterparts, biogenerics do face significant regulatory hurdles before approval because of the complexities associated with biological manufacturing. The key issue surrounding approval of biogenerics is the proof of similarity to the original product and how to demonstrate thereof. Nevertheless, we can be confident that these issues will be addressed as the demand for biotherapeutics increases especially in emerging Asian markets like India and China.

5. Future Prospects for Mammalian Cell Culture Processes

5.1. Genomic and Proteomic Applications in Cell Culture

Desirable cell line characteristics such as high specific productivities, robust and sustained growth in bioreactors, correct protein folding and glycosylation are mostly obtained by the lengthy screening of cell clones, using high throughput, small scale cell culture formats rather than by directed genetic or metabolic engineering. Tools such as DNA microarrays provide quantitative analysis of mRNA levels inside cells, while proteomic applications such as 2D gel separation, LC-MS/MS (liquid chromatography separation followed by mass spectrometry) reduce the complexity of protein pools, and can both identify and quantify protein expression with consistent results.

The use of genomic and proteomic tools have shed light on genes and gene sets that are relevant to bioprocessing traits such as the heterogeneity of productivity levels in industrial cell clones [129,130], the induced high protein productivity states under low temperature environment [131,132] and the addition of chemical inducer, sodium butyrate [133]. When applying genomic or proteomic tools to characterize complex traits such as high productivity, a large analytical dataset involving multiple biological replicates can increase the statistical significance of results. In most cases, the list of differentially expressed genes underlying two phenotypes (for e.g. low versus high productivity states) encompass several ontological classes. To extract biological information from these datasets, identifying altered pathways, as opposed to individual genes, has been found to aid in developing a hypothesis. The utilization of transcriptomic and proteomic analysis can also enhance physiological understanding of metabolic changes in cultured mammalian cells.

Although public genome resources and commercial microarrays are available for human- and mouse-derived production cell lines (such as HEK 293 and mouse NS0), the lack of genome sequence information for CHO has deterred the wide usage of these technologies in the field of recombinant protein production. Preliminary transcriptome profiling of CHO cells using cross-species mouse DNA arrays have provided limited success due to the sequence divergence between the two rodent species [134,135]. To address this issue, the Consortium on CHO cell genomics research was established in 2006 by members of industry and academia, with the goal of developing novel genomic resources for the Chinese hamster. Employing both methods of traditional Sanger sequencing and next-generation deep sequencing protocols, the consortium has to date sequenced more than 80,000 expressed sequence tags (ESTs) from Chinese hamster tissue and cultivated CHO cell lines [136,137]. The resulting bioinformatic resources have been used to develop custom CHO arrays for transcriptome profiling and provide a sequence database for CHO protein identification.

5.2. Cell Engineering

Many efforts have been invested in genetically modifying mammalian cells for the purpose of improving recombinant protein production. Some of the desirable cellular traits can be achieved through adaptation, or continuous passaging of cells in a defined culture environment. An example of successful adaptation is the cultivation of CHO cells from serum containing to serum –free media and simultaneously switching CHO cells from an attachment-dependent growth phenotype to full suspension growth [44]. In the adaptation process, cell population goes through a phase of low viability as only cells that possess the proper form of genetic traits would survive the new culture environment, while cells which lack the appropriate genetic traits would not survive.

A more direct and specific route in cell engineering involves the over-expression or suppression of one or more genetic elements to alter cellular behavior in the producer cell lines. Recent host cell engineering strategies have mainly focused on achieving these five desired traits (i) enhanced culture viability by delaying apoptosis (ii) control of cell proliferation (iii) higher protein secretion capacity (iv) alteration of cellular metabolism to reduce growth inhibitory metabolic by-products, such as lactate, and (v) modulation of protein glycosylation to enhance protein efficacy.

A strategy for enhancing final protein titer has been to seize the progression of cell cycle when cells achieve a high cell density to extend the production phase in bioreactors. The over-expression of negative cell cycle regulators such as the cyclin-dependent kinase inhibitor (CDK) p21 in CHO cells resulted in G1-phase cell cycle arrest and a reported 10-fold increase in specific productivity of the reported secreted alkaline phosphatase [138].

To prolong the production phase of cell culture reactors, both approaches of over-expressing anti-apoptotic genes, and suppressing the transcript levels of apoptotic genes have been employed. The anti-apoptotic B–cell lymphoma (Bcl) family proteins can inhibit the release of pro-apoptotic factors from mitochondria through the maintenance of mitochondrial membrane integrity [139]. Two of the Bcl family members, Bcl-2 and Bcl-xl have been transfected into NS0 [140], CHO [141,142], BHK [143] and hybridoma [144] cells. In adherent CHO-K1 and BHK cells, overexpression of Bcl-2 and Bcl-xl prolonged their viability in glucose free, serum-free or 50 mM ammonium supplemented medium [143].

Engineering efficient metabolic traits in cultured mammalian cells have become one of the paradigms for process improvements. Cultured mammalian cells typically consume large amounts of glucose and glutamine, in turn churning out growth inhibitory metabolic by-products such as lactate and ammonia. Attempts to reduce lactate production were performed by decreasing the expression of lactate dehydrogenase, which catalyze the formation of lactate from pyruvate [145] or by over-expressing pyruvate carboxylase, an enzyme that directs pyruvate to TCA cycle [146]. In both cases, the outcomes have been reduced specific glucose consumption and concurrent decrease in lactate production.

Large-scale gene expression and proteomic studies on producing cell lines have generally identified that high-level production of recombinant proteins is often associated with up-regulation of ER (endoplasmic reticulum) proteins involved in protein secretion [130]. Therefore, the engineering of protein folding machinery in the ER provides a plausible approach to enhance cellular capacity for protein secretion. However, the overexpression of chaperones: binding immunoglobulin protein (BiP) or protein disulfide isomerase (PDI) in recombinant cell lines has only been marginally successful in some experiments [147], and failed in other cases [148,149]. Recently the X-box binding protein-1 (Xbp-1) a transcription factor which regulates the expression of multiple genes involved in the unfolded protein response, has been overexpressed in erythropoietin-producing CHO cells [150]. Interestingly, the expression of XBP-1 was only effective in boosting production in CHO cells with high copies of the exogenous erythropoietin, but did not have any impact on low producers. This demonstrates that upregulation of XBP-1 can further improve protein production only when the accumulated amount of protein product exceeds the secretory capacity of host cells.

Improving and ensuring consistent quality of the recombinant protein produced is of utmost importance since it translates into better therapeutic efficiency and product safety. Recombinant glycoproteins produced in non-human host cells need to resemble its native form in humans. As mentioned earlier, the sialic acid on human glycans has α(2,3) and α(2,6) linkages, whereas CHO cells only produced α(2,3) sialic acid linkages. To circumvent this protein quality defect in CHO, a cDNA coding for the rat isoform of α(2,6) sialyltransferase was transfected into CHO cells to alter the terminal glycosylation structures on recombinant tissue plasminogen activator [151]. Similar strategies has been employed to alter glycan structures of recombinant human erythropoietin produced in CHO [152] and recombinant human antithrombin in baby hamster kidney cells [153].

6. Conclusion

The increasing interest for biologic drugs has amplified the demand for therapeutic proteins manufactured via cell culture technology. At the same time, concerns for the safety and quality of such products have also been raised. This has resulted in the development of more efficient and robust cultivation and production technology. Validation procedures and the adoption of current good manufacturing practice (cGMP) guidelines are enforced on all aspects of the manufacturing process; from the upstream cell culture process, to protein harvest, purification and product fill finishing. Also critical to the success of a biopharmaceutical company, is the presence of a formidable process development team that takes exciting target proteins from the drug discovery group to trials in the clinic. The path of discovery, development and commercialization of a therapeutic protein is lengthy and technically challenging. Currently, the high cost of therapeutic proteins reflects the economic burdens and associated risks encountered during development and production. Having started back in the late 1980s, the biopharmaceutical industry is reaching an important milestone where the maturity of technologies and the accumulation of process knowledge (as described in this review) bring hopes of reducing the cost of production and making protein therapeutics available to patients.

References

[1] Walsh, G. & Jefferis, R. 2006. Post-translational modifications in the context of therapeutic proteins. *Nature Biotechnology*, 24, 1241-1252.

[2] Kearney, J. F., Radbruch, A., Liesegang, B. & Rajewsky, K. 1979. A new mouse myeloma cell line that has lost immunoglobulin expression but permits the construction of antibody-secreting hybrid cell lines. *The Journal of Immunology*, 123, 1548-1550.

[3] Kohler, G. & Milstein, C. 1976. Derivation of specific antibody-producing tissue culture and tumor lines by cell fusion. *European Journal of Immunology*, 6, 511-519.

[4] Potter, M. 1972. Immunoglobulin-producing tumors and myeloma proteins of mice. *Physiological Reviews*, 52, 631-719.

[5] Sato, J. D., Cao, H. T., Kayada, Y., Cabot, M. C., Sato, G. H., Okamoto, T. & Welsh, C. J. 1988. Effects of proximate cholesterol precursors and steroid hormones on mouse myeloma growth in serum-free medium. *In Vitro Cellular & Developmental Biology*, 24, 1223-1228.

[6] Sato, J. D., Kawamoto, T. & Okamoto, T. 1987. Cholesterol requirement of P3-X63-Ag8 and X63-Ag8.653 mouse myeloma cells for growth in vitro. *Journal of Experimental Medicine*, 165, 1761-1766.

[7] Walowitz, J. L., Fike, R. M. & Jayme, D. W. 2003. Efficient lipid delivery to hybridoma culture by use of cyclodextrin in a novel granulated dry-form medium technology. *Biotechnology Progress*, 19, 64-68.

[8] Urlaub, G. & Chasin, L. A. 1980. Isolation of Chinese hamster cell mutants deficient in dihydrofolate reductase activity. *Proceedings of the National Academy of Sciences of the United States of America*, 77, 4216-4220.

[9] Urlaub, G., Kas, E., Carothers, A. M. & Chasin, L. A. 1983. Deletion of the diploid dihydrofolate reductase locus from cultured mammalian cells. *Cell*, 33, 405-412.

[10] Stanley, P., Raju, T. S. & Bhaumik, M. 1996. CHO cells provide access to novel N-glycans and developmentally regulated glycosyltransferases. *Glycobiology*, 6, 695-699.

[11] Shah, P., Reece-Ford, M., Dong, S., Goodall, M., Pidaparthi, S., Jefferis, R. & Jenkins, N. 1998. Physiological influences on recombinant IgG glycosylation. *Biochemical Society Transactions*, 26, S114.

[12] Flesher, A. R., Marzowski, J., Wang, W. C. & Raff, H. V. 1995. Fluorophore-labeled carbohydrate analysis of immunoglobulin fusion proteins: Correlation of oligosaccharide content with in vivo clearance profile. *Biotechnology and Bioengineering*, 47, 405.

[13] Hossler, P., Khattak, S. F. & Li, Z. J. 2009. Optimal and consistent protein glycosylation in mammalian cell culture. *Glycobiology*, 19, 936-949.

[14] Jenkins, N. 1996. Role of physiology in the determination of protein heterogeneity. *Current Opinion in Biotechnology*, 7, 205-209.

[15] Raju, T. S. 2003. Glycosylation variations with expression systems and their impact on biological activity of therapeutic immunoglobulins. *Bioprocess International*, 1, 44 - 53

[16] Southern, P. J. & Berg, P. 1982. Transformation of mammalian cells to antibiotic resistance with a bacterial gene under control of the SV40 early region promoter. *Journal Molecular and Applied Genetics*, 1, 327-341.

[17] Berry, J. M., Barnabe, N., Coombs, K. M. & Butler, M. 1999. Production of reovirus type-1 and type-3 from Vero cells grown on solid and macroporous microcarriers. *Biotechnology and Bioengineering,* 62, 12-19.

[18] Gritz, L. & Davies, J. 1983. Plasmid-encoded hygromycin B resistance: the sequence of hygromycin B phosphotransferase gene and its expression in *Escherichia coli* and *Saccharomyces cerevisiae. Gene*, 25, 179-188.

[19] Mulsant, P., Gatignol, A., Dalens, M. & Tiraby, G. 1988. Phleomycin resistance as a dominant selectable marker in CHO cells. *Somatic Cell and Molecular Genetics*, 14, 243-252.

[20] de la Luna, S., Soria, I., Pulido, D., Ortin, J. & Jimenez, A. 1988. Efficient transformation of mammalian cells with constructs containing a puromycin-resistance marker. *Gene*, 62, 121-126.

[21] Gandor, C., Leist, C., Fiechter, A. & Asselbergs, F. A. 1995. Amplification and expression of recombinant genes in serum-independent Chinese hamster ovary cells. *FEBS Lett*ers, 377, 290-294.

[22] Pallavicini, M. G., DeTeresa, P. S., Rosette, C., Gray, J. W. & Wurm, F. M. 1990. Effects of methotrexate on transfected DNA stability in mammalian cells. *Molecular and Cellular Biology*, 10, 401-404.

[23] Bebbington, C., Renner, G., Thomson, S., King, D., Abrams, D. & Yarranton, G. T. 1992. High-level expression of a recombinant antibody from myeloma cells using a glutamine synthase gene as an amplifiable selectable marker. *BioTechnology*, 10, 169-175.

[24] Bell, S. L., Bebbington, C., Scott, M. F., Wardell, J. N., Spier, R. E., Bushell, M. E. & Sanders, P. G. 1995. Genetic engineering of hybridoma glutamine metabolism. *Enzyme and Microbial Technology*, 17, 98-106.

[25] Bacsi, S. G. & Wejksnora, P. J. 1986. Effect of increase in ploidy on the activation of nucleolar organizer regions in Chinese hamster ovary (CHO) cells. *Experimental Cell Research*, 165, 283-289.

[26] Flintoff, W. F., Livingston, E., Duff, C. & Worton, R. G. 1984. Moderate-level gene amplification in methotrexate-resistant Chinese hamster ovary cells is accompanied by chromosomal translocations at or near the site of the amplified DHFR gene. *Molecular and Cellular Biology*, 4, 69-76.

[27] Lattenmayer, C., Trummer, E., Schriebl, K., Vorauer-Uhl, K., Mueller, D., Katinger, H. & Kunert, R. 2007. Characterisation of recombinant CHO cell lines by investigation of protein productivities and genetic parameters. *Journal of Biotechnology*, 128, 716-725.

[28] Kwaks, T. H., Sewalt, R. G., van Blokland, R., Siersma, T. J., Kasiem, M., Kelder, A. & Otte, A. P. 2005. Targeting of a histone acetyltransferase domain to a promoter enhances protein expression levels in mammalian cells. *Journal of Biotechnology*, 115, 35-46.

[29] Antoniou, M., Harland, L., Mustoe, T., Williams, S., Holdstock, J., Yague, E., Mulcahy, T., Griffiths, M., Edwards, S., Ioannou, P. A., Mountain, A. & Crombie, R. 2003. Transgenes encompassing dual-promoter CpG islands from the human TBP and HNRPA2B1 loci are resistant to heterochromatin-mediated silencing. *Genomics*, 82, 269-279.

[30] Benton, T., Chen, T., McEntee, M., Fox, B., King, D., Crombie, R., Thomas, T. C. & Bebbington, C. 2002. The use of UCOE vectors in combination with a preadapted serum free, suspension cell line allows for rapid production of large quantities of protein. *Cytotechnology*, 38, 43-46.

[31] Kim, J. -M., Kim, J. -S., Park, D. -H., Kang, H. S., Yoon, J., Baek, K. & Yoon, Y. 2004. Improved recombinant gene expression in CHO cells using matrix attachment regions. *Journal of Biotechnology*, 107, 95-105.

[32] Wirth, D., Gama-Norton, L., Riemer, P., Sandhu, U., Schucht, R. & Hauser, H. 2007. Road to precision: recombinase-based targeting technologies for genome engineering. *Current Opinion in Biotechnology*, 18, 411-419.

[33] Gulick, T. 2003. Transfection using DEAE-dextran. *Current Protocols in Cell Biology,* Chapter 20, Unit 20 24.

[34] Hawley-Nelson, P., Ciccarone, V. & Moore, M. L. 2008. Transfection of cultured eukaryotic cells using cationic lipid reagents. *Current Protocols in Cell Biology,* Chapter 9, Unit 9 4.

[35] Potter, H. 2003. Transfection by electroporation. *Current Protocols in Cell Biology,* Chapter 9, Unit 9 3.

[36] Kingston, R. E., Chen, C. A. & Okayama, H. 2003. Calcium phosphate transfection. *Current Protocols in Cell Biology,* Chapter 20, Unit 20 3.

[37] Browne, S. M. & Al-Rubeai, M. 2007. Selection methods for high-producing mammalian cell lines. *Trends in Biotechnology*, 25, 425-432.

[38] DeMaria, C. T., Cairns, V., Schwarz, C., Zhang, J., Guerin, M., Zuena, E., Estes, S. & Karey, K. P. 2007. Accelerated clone selection for recombinant CHO CELLS using a FACS-based high-throughput screen. *Biotechnology Progress,* 23, 465-472.

[39] Hesse, F. & Wagner, R. 2000. Developments and improvements in the manufacturing of human therapeutics with mammalian cell cultures. *Trends in Biotechnology*, 18, 173-180.

[40] Eagle, H. 1955. Nutrition needs of mammalian cells in tissue culture. *Science*, 122, 501-514.

[41] Eagle, H. 1959. Amino acid metabolism in mammalian cell cultures. *Science*, 130, 432-437.

[42] Barnes, D. 1987. Serum-free animal cell culture. *BioTechniques*, 5, 534-541.

[43] Hewlett, G. 1991. Strategies for optimising serum-free media. *Cytotechnology*, 5, 3-14.

[44] Sinacore, M. S., Drapeau, D. & Adamson, S. R. 2000. Adaptation of mammalian cells to growth in serum-free media. *Molecular Biotechnology*, 15, 249-257.

[45] Katsuta, H. & Takoaka, T. 1976. Improved synthetic media suitable for tissue culture of mammalian cells. In: *Methods in cell biology* (edited by D. Presscott) Pp. 145-158. New York, USA: Academia Press,

[46] Murakami, H., Masui, H., Sato, G. H., Sueoka, N., Chow, T. P. & Kano-Sueoka, T. 1982. Growth of hybridoma cells in serum-free medium: ethanolamine is an essential component. *Proceedings of the National Academy of Sciences of the United States of America*, 79, 1158-1162.

[47] Sanfeliu, A., Chung, J. D. & Stephanopoulos, G. 2000. Effect of insulin stimulation on the proliferation and death of Chinese hamster ovary cells. *Biotechnology and Bioengineering*, 70, 421-427.

[48] Gu, X., Xie, L., Harmon, B. J. & Wang, D. I. 1997. Influence of Primatone RL supplementation on sialylation of recombinant human interferon-gamma produced by Chinese hamster ovary cell culture using serum-free media. *Biotechnology and Bioengineering*, 56, 353-360.

[49] Rasmussen, B., Davis, R., Thomas, J. & Reddy, P. 1998. Isolation, characterization and recombinant protein expression in Veggie-CHO: A serum-free CHO host cell line. *Cytotechnology*, 28, 31-42.

[50] Schlaeger, E. J. 1996. The protein hydrolysate, Primatone RL, is a cost-effective multiple growth promoter of mammalian cell culture in serum-containing and serum-free media and displays anti-apoptosis properties. *Journal of Immunological Methods*, 194, 191-199.

[51] Gray, D. R., Chen, S., Howarth, W., Inlow, D. & Maiorella, B. L. 1996. CO2 in large-scale and high-density CHO cell perfusion culture. *Cytotechnology*, 22, 65-78.

[52] Leibovitz, A. 1963. The Growth and Maintenance of Tissue-Cell Cultures in Free Gas Exchange with the Atmosphere. *American Journal of Hygiene*, 78, 173-180.

[53] Nagira, K., Hayashida, M., Shiga, M., Sasamoto, K., Kina, K. Y., Osada, K., Sugahara, T. & Murakami, H. 1995. Effects of organic pH buffers on a cell growth and an antibody production of human-human hybridoma HB4C5 cells in a serum-free culture. *Cytotechnology*, 17, 117-125.

[54] Altamirano, C., Paredes, C., Illanes, A., Cairo, J. J. & Godia, F. 2004. Strategies for fed-batch cultivation of t-PA producing CHO cells: substitution of glucose and glutamine and rational design of culture medium. *Journal of Biotechnology*, 110, 171-179.

[55] Naylor, S. L., Busby, L. L. & Klebe, R. J. 1976. Biochemical selection systems for mammalian cells: the essential amino acids. *Somatic Cell and Molecular Genetics*, 2, 93-111.

[56] Franěk, F. & Chládková-Šrámková, K. 1995. Apoptosis and nutrition: Involvement of amino acid transport system in repression of hybridoma cell death. *Cytotechnology*, 18, 113-117.

[57] Eagle, H. & Piez, K. 1962. The population-dependent requirement by cultured mammalian cells for metabolites which they can synthesize. *Journal of Experimental Medicine*, 116, 29-43.

[58] Bhullar, R. P. & Dakshinamurti, K. 1985. The effect of biotin on cellular functions in HeLa cells. *Journal of Cellular Physiology*, 123, 425-430.

[59] Matsuya, Y. & Yamane, I. 1986. Population-dependent requirements of vitamin B12 and metabolically related substances of several mouse cell types in serum-free, albumin-fortified medium. *Cell Structure and Function*, 11, 9-19.

[60] Gordon, P. R., Gelman, L. K. & Gilchrest, B. A. 1988. Demonstration of a choline requirement for optimal keratinocyte growth in a defined culture medium. *The Journal of Nutrition*, 118, 1487-1494.

[61] Buhl, S. N. & Regan, J. D. 1972. Growth of a human leukemia cell line on protein-free medium. 1. *Proceedings of the Society for Experimental Biology and Medicine*, 140, 1224-1227.

[62] Mather, J. P. & Sato, G. H. 1979. The growth of mouse melanoma cells in hormone-supplemented, serum-free medium. *Experimental Cell Research*, 120, 191-200.

[63] Pollack, S. 1992. Receptor-mediated iron uptake and intracellular iron transport. *American Journal of Hematology*, 39, 113-118.

[64] Sunstrom, N. A., Gay, R. D., Wong, D. C., Kitchen, N. A., DeBoer, L. & Gray, P. P. 2000. Insulin-like growth factor-I and transferrin mediate growth and survival of Chinese hamster ovary cells. *Biotechnology Progress*, 16, 698-702.

[65] Yabe, N., Kato, M., Matsuya, Y., Yamane, I., Iizuka, M., Takayoshi, H. & Suzuki, K. 1987. Role of iron chelators in growth-promoting effect on mouse hybridoma cells in a chemically defined medium. *In Vitro Cellular & Developmental Biology*, 23, 815-820.

[66] Butler, M., Huzel, N., Barnabe, N., Gray, T. & Bajno, L. 1999. Linoleic acid improves the robustness of cells in agitated cultures. *Cytotechnology*, 30, 27-36.

[67] Grammatikos, S. I., Subbaiah, P. V., Victor, T. A. & Miller, W. M. 1994. Diverse effects of essential (n-6 and n-3) fatty acids on cultured cells. *Cytotechnology*, 15, 31-50.

[68] Holtta, E. & Pohjanpelto, P. 1982. Polyamine dependence of Chinese hamster ovary cells in serum-free culture is due to deficient arginase activity. *Biochimica et Biophysica Acta*, 721, 321-327.

[69] Olle, H. 1981. Role of Polyamines in the Control of Cell Proliferation and Differentiation. *Differentiation*, 19, 1-20.

[70] Pegg, A. E. & McCann, P. P. 1982. Polyamine metabolism and function. *American Journal of Physiology*, 243, C212-221.

[71] Dalili, M. & Ollis, D. F. 1988. The influence of cyclic nucleotides on hybridoma growth and monoclonal antibody production. *Biotechnology Letters*, 10, 781-786.

[72] Lee, G. M., Kim, E. J., Kim, N. S., Yoon, S. K., Ahn, Y. H. & Song, J. Y. 1999. Development of a serum-free medium for the production of erythropoietin by

suspension culture of recombinant Chinese hamster ovary cells using a statistical design. *Journal of Biotechnology*, 69, 85-93.

[73] Baker R, Baker S. & Rao, R. 1998. Selenium deficiency in tissue culture. Implications for oxidative metabolism. *Journal of Pediatric Gastroenterology and Nutrition,* 27, 387-392.

[74] Helmy, M., Ismail, S., Fayed, H. & El-Bassiouni, E. 2000. Effect of selenium supplementation on the activities of glutathione metabolizing enzymes in human hepatoma Hep G2 cell line. *Toxicology,* 144, 57-61.

[75] McKeehan, W. L., Hamilton, W. G. & Ham, R. G. 1976. Selenium is an essential trace nutrient for growth of WI-38 diploid human fibroblasts. *Proceedings of the National Academy of Sciences of the United States of America*, 73, 2023-2027.

[76] Zhang, J., Robinson, D. & Salmon, P. 2006. A novel function for selenium in biological system: Selenite as a highly effective iron carrier for Chinese hamster ovary cell growth and monoclonal antibody production. *Biotechnology and Bioengineering*, 95, 1188-1197.

[77] Crowell, C. K., Grampp, G., Rogers G., Miller, J., Scheinman, R. I. 2007. Amino acid and manganese supplementation modulates the glycosylation state of erythropoietin in a CHO culture system. *Biotechnology and Bioengineering*, 96, 538-549.

[78] Fike, R. 2009. Nutrient supplementation strategies for biopharmaceutical production, part 2. *Bioprocess International*, 7, 46-52.

[79] Wlaschin, K. F. & Hu, W. S. 2006. Fedbatch culture and dynamic nutrient feeding. *Advances in Biochemical Engineering/Biotechnology*, 101, 43-74.

[80] Fletcher, T. 2005. Designing culture media for recombinant protein production: a rational approach. *Bioprocess International*, 3, 30-36.

[81] Sandadi, S., Ensari, S. & Kearns, B. 2005. Heuristic optimization of antibody production by Chinese hamster ovary cells. *Biotechnology Progress*, 21, 1537-1542.

[82] Sandadi, S., Ensari, S. & Kearns, B. 2006. Application of fractional factorial designs to screen active factors for antibody production by chinese hamster ovary cells. *Biotechnology Progress*, 22, 595-600.

[83] Parampalli, A., Eskridge, K., Smith, L., Meagher, M., Mowry, M. & Subramanian, A. 2007. Development of serum-free media in CHO-DG44 cells using a central composite statistical design. *Cytotechnology*, 54, 57-68.

[84] Liu, C., Chu, I. & Hwang, S. 2001. Factorial designs combined with the steepest ascent method to optimize serum-free media for CHO cells. *Enzyme and Microbial Technology*, 28, 314-321.

[85] Zhang, X., Stettler, M., De Sanctis, D., Perrone, M., Parolini, N., Discacciati, M., De Jesus, M., Hacker, D., Quarteroni, A. & Wurm, F. 2010. Use of orbital shaken disposable bioreactors for Mammalian cell cultures from the milliliter-scale to the 1,000-liter scale. *Advances in Biochemical Engineering/Biotechnology*, 115, 33-53.

[86] Gupta, A. & Rao, G. 2003. A study of oxygen transfer in shake flasks using a non-invasive oxygen sensor. *Biotechnology and Bioengineering*, 84, 351-358.

[87] Kunitake, R., Suzuki, A., Ichihashi, H., Matsuda, S., Hirai, O. & Morimoto, K. 1997. Fully-automated roller bottle handling system for large scale culture of mammalian cells. *Journal of Biotechnology*, 52, 289-294.

[88] Kalbfuss, B., Genzel, Y., Wolff, M., Zimmermann, A., Morenweiser, R. & Reichl, U. 2007. Harvesting and concentration of human influenza A virus produced in serum-

free mammalian cell culture for the production of vaccines. *Biotechnology and Bioengineering*, 97, 73-85.

[89] Chisti, Y. 2000. Animal-cell damage in sparged bioreactors. *Trends in Biotechnology,* 18, 420-432.

[90] Chisti, Y. 2001. Hydrodynamic damage to animal cells. *Critical Reviews Biotechnology,* 21, 67-110.

[91] Nam, J. H., Ermonval, M. & Sharfstein, S. T. 2009. The effects of microcarrier culture on recombinant CHO cells under biphasic hypothermic culture conditions. *Cytotechnology*. 59, 81-91.

[92] Koller, M. R. & Papoutsakis, E. T. 1995. Cell adhesion in animal cell culture: physiological and fluid-mechanical implications. *Bioprocess Technology*, 20, 61-110.

[93] Butler, M. & Jenkins, H. 1989. Nutritional aspects of the growth of animal cells in culture. *Journal of Biotechnology*, 12, 97-110.

[94] Bibila, T. A. & Robinson, D. K. 1995. In pursuit of the optimal fed-batch process for monoclonal antibody production. *Biotechnology Progress*, 11, 1-13.

[95] Petch, D. & Butler, M. 1996. The effect of alternative carbohydrates on the growth and antibody production of a murine hybridoma. *Applied Biochemistry and Biotechnology*, 59, 93-104.

[96] Voisard, D., Meuwly, F., Ruffieux, P. A., Baer, G. & Kadouri, A. 2003. Potential of cell retention techniques for large-scale high-density perfusion culture of suspended mammalian cells. *Biotechnology and Bioengineering*, 82, 751-765.

[97] Castilho, L. R. & Medronho, R. A. 2002. Cell retention devices for suspended-cell perfusion cultures. *Advances in Biochemical Engineering/Biotechnology*, 74, 129-169.

[98] Woodside, S. M., Bowen, B. D. & Piret, J. M. 1998. Mammalian cell retention devices for stirred perfusion bioreactors. *Cytotechnology*, 28, 163-175.

[99] Jiang, R., Monroe, T., McRogers, R. & Larson, P. J. 2002. Manufacturing challenges in the commercial production of recombinant coagulation factor VIII. *Haemophilia*, 8 Suppl 2, 1-5.

[100] Eibl, R. & Eibl, D. 2009. Application of disposable bag bioreactors in tissue engineering and for the production of therapeutic agents. *Advances in Biochemical Engineering/Biotechnology*, 112, 183-207.

[101] Eibl, R., Kaiser, S., Lombriser, R. & Eibl, D. 2010. Disposable bioreactors: the current state-of-the-art and recommended applications in biotechnology. *Applied Microbiology and Biotechnology*, 86, 41-49.

[102] Okonkowski, J., Balasubramanian, U., Seamans, C., Fries, S., Zhang, J., Salmon, P., Robinson, D. & Chartrain, M. 2007. Cholesterol delivery to NS0 cells: challenges and solutions in disposable linear low-density polyethylene-based bioreactors. *Journal of Bioscience and Bioengineering*, 103, 50-59.

[103] Singh, V. 1999. Disposable bioreactor for cell culture using wave-induced agitation. *Cytotechnology*, 30, 149-158.

[104] Ye, J., Kober, V., Tellers, M., Naji, Z., Salmon, P. & Markusen, J. F. 2009. High-level protein expression in scalable CHO transient transfection. *Biotechnology and Bioengineering*, 103, 542-551.

[105] Legmann, R., Schreyer, H. B., Combs, R. G., McCormick, E. L., Russo, A. P. & Rodgers, S. T. 2009. A predictive high-throughput scale-down model of monoclonal

antibody production in CHO cells. *Biotechnology and Bioengineering,* 104, 1107-1120.

[106] Chen, A., Chitta, R., Chang, D. & Amanullah, A. 2009. Twenty-four well plate miniature bioreactor system as a scale-down model for cell culture process development. *Biotechnology and Bioengineering*, 102, 148-160.

[107] Amanullah, A., Otero, J. M., Mikola, M., Hsu, A., Zhang, J., Aunins, J., Schreyer, H. B., Hope, J. A. & Russo, A. P. 2010. Novel micro-bioreactor high throughput technology for cell culture process development: Reproducibility and scalability assessment of fed-batch CHO cultures. *Biotechnology and Bioengineering*, 106, 57-67.

[108] Ge, X., Hanson, M., Shen, H., Kostov, Y., Brorson, K. A., Frey, D. D., Moreira, A. R. & Rao, G. 2006. Validation of an optical sensor-based high-throughput bioreactor system for mammalian cell culture. *Journal of Biotechnology*, 122, 293-306.

[109] Hanson, M. A., Ge, X., Kostov, Y., Brorson, K. A., Moreira, A. R. & Rao, G. 2007. Comparisons of optical pH and dissolved oxygen sensors with traditional electrochemical probes during mammalian cell culture. *Biotechnology and Bioengineering*, 97, 833-841.

[110] Naciri, M., Kuystermans, D. & Al-Rubeai, M. 2008. Monitoring pH and dissolved oxygen in mammalian cell culture using optical sensors. *Cytotechnology*, 57, 245-250.

[111] Lam, H. & Kostov, Y. 2010. Optical instrumentation for bioprocess monitoring. *Advances in Biochemical Engineering/Biotechnology*, 116, 1-28.

[112] Johnson, M. J., Borkowski, J. & Engblom, C. 2000. Steam sterilizable probes for dissolved oxygen measurement. Reprinted from Biotechnology and Bioengineering, Vol. VI, Issue 4, Pages 457-468 (1964). *Biotechnology and Bioengineering*, 67, 645-656.

[113] Kiviharju, K., Salonen, K., Moilanen, U. & Eerikainen, T. 2008. Biomass measurement online: the performance of in situ measurements and software sensors. *Journal of Industrial Microbiology and Biotechnology*, 35, 657-665.

[114] Olsson, L. & Nielsen, J. 1997. On-line and in situ monitoring of biomass in submerged cultivations. *Trends in Biotechnology*, 15, 517-522.

[115] Ruffieux, P. A., von Stockar, U. & Marison, I. W. 1998. Measurement of volumetric (OUR) and determination of specific (qO2) oxygen uptake rates in animal cell cultures. *Journal of Biotechnology*, 63, 85-95.

[116] Matsunaga, N., Kano, K., Maki, Y. & Dobashi, T. 2009. Culture scale-up studies as seen from the viewpoint of oxygen supply and dissolved carbon dioxide stripping. *Journal of Bioscience and Bioengineering*, 107, 412-418.

[117] Garcia-Ochoa, F. & Gomez, E. 2009. Bioreactor scale-up and oxygen transfer rate in microbial processes: an overview. *Biotechnology Advances*, 27, 153-176.

[118] Marks, D. M. 2003. Equipment design considerations for large scale cell culture. *Cytotechnology*, 42, 21-33.

[119] Xing, Z., Kenty, B. M., Li, Z. J. & Lee, S. S. 2009. Scale-up analysis for a CHO cell culture process in large-scale bioreactors. *Biotechnology and Bioengineering*, 103, 733-746.

[120] Bylund, F., Collet, E., Enfors, S. O. & Larson, G. 1998. Substrate gradient formation in the large-scale bioreactor lowers cell yield and increases by-product formation. *Bioprocess and Biosystems Engineering*, 18, 171-180.

[121] Langheinrich, C. & Nienow, A. W. 1999. Control of pH in large-scale, free suspension animal cell bioreactors: alkali addition and pH excursions. *Biotechnology and Bioengineering*, 66, 171-179.
[122] Osman, J. J., Birch, J. & Varley, J. 2002. The response of GS-NS0 myeloma cells to single and multiple pH perturbations. *Biotechnology and Bioengineering*, 79, 398-407.
[123] Varley, J. & Birch, J. 1999. Reactor design for large scale suspension animal cell culture. *Cytotechnology*, 29, 177-205.
[124] Sahoo, N., Choudhury, K. & Manchikanti, P. 2009. Manufacturing of biodrugs: need for harmonization in regulatory standards. *BioDrugs*, 23, 217-229.
[125] Walsh, G. 2004. Drug approval in the European Union and the United States. In: *Pharmaceutical biotechnology, drug discovery and clinical applications* (edited by O. Kayser & R. H. Muller). Pp. 201-210. Weinheim, Germany: Wiley-VCH Verlag GmbH & Co.
[126] Thiel, K. A. 2004. Biomanufacturing, from bust to boom...to bubble? *Nature Biotechnology*, 22, 1365-1372.
[127] Kamarck, M. E. 2006. Building biomanufacturing capacity--the chapter and verse. *Nature Biotechnology*, 24, 503-505.
[128] Walsh, G. 2006. Biopharmaceutical benchmarks 2006. *Nature Biotechnology*, 24, 769-776.
[129] Charaniya, S., Karypis, G. & Hu, W. S. 2009. Mining transcriptome data for function-trait relationship of hyper productivity of recombinant antibody. *Biotechnology and Bioengineering*, 102, 1654-1669.
[130] Dinnis, D. M., Stansfield, S. H., Schlatter, S., Smales, C. M., Alete, D., Birch, J. R., Racher, A. J., Marshall, C. T., Nielsen, L. K. & James, D. C. 2006. Functional proteomic analysis of GS-NS0 murine myeloma cell lines with varying recombinant monoclonal antibody production rate. *Biotechnology and Bioengineering*, 94, 830-841.
[131] Baik, J. Y., Lee, M. S., An, S. R., Yoon, S. K., Joo, E. J., Kim, Y. H., Park, H. W. & Lee, G. M. 2006. Initial transcriptome and proteome analyses of low culture temperature-induced expression in CHO cells producing erythropoietin. *Biotechnology and Bioengineering*, 93, 361-371.
[132] Yee, J. C., Gerdtzen, Z. P. & Hu, W. S. 2009. Comparative transcriptome analysis to unveil genes affecting recombinant protein productivity in mammalian cells. *Biotechnology and Bioengineering*, 102, 246-263.
[133] Kantardjieff, A., Jacob, N. M., Yee, J. C., Epstein, E., Kok, Y. J., Philp, R., Betenbaugh, M. & Hu, W. S. 2009. Transcriptome and proteome analysis of Chinese hamster ovary cells under low temperature and butyrate treatment. *Journal of Biotechnology*, 145, 143-159.
[134] Ernst, W., Trummer, E., Mead, J., Bessant, C., Strelec, H., Katinger, H. & Hesse, F. 2006. Evaluation of a genomics platform for cross-species transcriptome analysis of recombinant CHO cells. *Biotechnology Journal*, 1, 639-650.
[135] Yee, J. C., Wlaschin, K. F., Chuah, S. H., Nissom, P. M. & Hu, W. S. 2008. Quality assessment of cross-species hybridization of CHO transcriptome on a mouse DNA oligo microarray. *Biotechnology and Bioengineering*, 101, 1359-1365.
[136] Jayapal, K., Wlaschin, K. F., Hu, W. S. & Yap, G. S. 2007. Recombinant protein therapeutics from CHO cells - 20 years and counting. *Chemical Engineering Progress*, 103, 40-46.

[137] Kantardjieff, A., Nissom, P. M., Chuah, S. H., Yusufi, F., Jacob, N. M., Mulukutla, B. C., Yap, M. & Hu, W. S. 2009. Developing genomic platforms for Chinese hamster ovary cells. *Biotechnology Advances*, 27, 1028-1035.

[138] Fussenegger, M., Schlatter, S., Datwyler, D., Mazur, X. & Bailey, J. E. 1998. Controlled proliferation by multigene metabolic engineering enhances the productivity of Chinese hamster ovary cells. *Nature Biotechnology*, 16, 468-472.

[139] Harris, M. H. & Thompson, C. B. 2000. The role of the Bcl-2 family in the regulation of outer mitochondrial membrane permeability. *Cell Death Differ*, 7, 1182-1191.

[140] Tey, B. T., Singh, R. P., Piredda, L., Piacentini, M. & Al-Rubeai, M. 2000. Bcl-2 mediated suppression of apoptosis in myeloma NS0 cultures. *Journal of Biotechnology*, 79, 147-159.

[141] Meents, H., Enenkel, B., Eppenberger, H. M., Werner, R. G. & Fussenegger, M. 2002. Impact of coexpression and coamplification of sICAM and antiapoptosis determinants bcl-2/bcl-x(L) on productivity, cell survival, and mitochondria number in CHO-DG44 grown in suspension and serum-free media. *Biotechnology and Bioengineering*, 80, 706-716.

[142] Tey, B. T., Singh, R. P., Piredda, L., Piacentini, M. & Al-Rubeai, M. 2000. Influence of bcl-2 on cell death during the cultivation of a Chinese hamster ovary cell line expressing a chimeric antibody. *Biotechnology and Bioengineering*, 68, 31-43.

[143] Mastrangelo, A. J., Hardwick, J. M., Zou, S. & Betenbaugh, M. J. 2000. Part II. Overexpression of bcl-2 family members enhances survival of mammalian cells in response to various culture insults. *Biotechnology and Bioengineering*, 67, 555-564.

[144] Simpson, N. H., Singh, R. P., Perani, A., Goldenzon, C. & Al-Rubeai, M. 1998. In hybridoma cultures, deprivation of any single amino acid leads to apoptotic death, which is suppressed by the expression of the bcl-2 gene. *Biotechnology and Bioengineering*, 59, 90-98.

[145] Jeong, D., Kim, T. S., Lee, J. W., Kim, K. T., Kim, H. J., Kim, I. H. & Kim, I. Y. 2001. Blocking of acidosis-mediated apoptosis by a reduction of lactate dehydrogenase activity through antisense mRNA expression. *Biochemical and Biophysical Research Communications*, 289, 1141-1149.

[146] Irani, N., Wirth, M., van Den Heuvel, J. & Wagner, R. 1999. Improvement of the primary metabolism of cell cultures by introducing a new cytoplasmic pyruvate carboxylase reaction. *Biotechnology and Bioengineering*, 66, 238-246.

[147] Borth, N., Mattanovich, D., Kunert, R. & Katinger, H. 2005. Effect of increased expression of protein disulfide isomerase and heavy chain binding protein on antibody secretion in a recombinant CHO cell line. *Biotechnology Progress,* 21, 106-111.

[148] Dorner, A. J., Wasley, L. C. & Kaufman, R. J. 1992. Overexpression of GRP78 mitigates stress induction of glucose regulated proteins and blocks secretion of selective proteins in Chinese hamster ovary cells. *The EMBO Journal*, 11, 1563-1571.

[149] Kitchin, K. & Flickinger, M. C. 1995. Alteration of hybridoma viability and antibody secretion in transfectomas with inducible overexpression of protein disulfide isomerase. *Biotechnology Progress*, 11, 565-574.

[150] Ku, S. C., Ng, D. T., Yap, M. G. & Chao, S. H. 2008. Effects of overexpression of X-box binding protein 1 on recombinant protein production in Chinese hamster ovary and NS0 myeloma cells. *Biotechnology and Bioengineering*, 99, 155-164.

[151] Minch, S. L., Kallio, P. T. & Bailey, J. E. 1995. Tissue plasminogen activator coexpressed in Chinese hamster ovary cells with alpha(2,6)-sialyltransferase contains NeuAc alpha(2,6)Gal beta(1,4)Glc-N-AcR linkages. *Biotechnology Progress*, 11, 348-351.

[152] Zhang, X., Lok, S. H. & Kon, O. L. 1998. Stable expression of human alpha-2,6-sialyltransferase in Chinese hamster ovary cells: functional consequences for human erythropoietin expression and bioactivity. *Biochimica et Biophysica Acta*, 1425, 441-452.

[153] Grabenhorst, E., Hoffmann, A., Nimtz, M., Zettlmeissl, G. & Conradt, H. S. 1995. Construction of stable BHK-21 cells coexpressing human secretory glycoproteins and human Gal(beta 1-4)GlcNAc-R alpha 2,6-sialyltransferase alpha 2,6-linked NeuAc is preferentially attached to the Gal(beta 1-4)GlcNAc(beta 1-2)Man(alpha 1-3)-branch of diantennary oligosaccharides from secreted recombinant beta-trace protein. *European Journal of Biochemistry*, 232, 718-725.

In: Bioprocess Sciences and Technology
Editor: Min-Tze Liong
ISBN 978-1-61122-950-9

Chapter 18

MICROORGANISMS IN REMOVAL OF HEAVY METALS: MECHANISMS, EVALUATION AND CASE STUDIES

***W. A. Ahmad*[1,*]*, Z. A. Zakaria*[2] *and J. Jaapar*[3]**
[1]Department of Chemistry, Faculty of Science, Universiti Teknologi Malaysia, 81310 Skudai, Johor, Malaysia
[2]c/o Chemical Engineering Pilot Plant, Universiti Teknologi Malaysia, 81310 Skudai, Johor, Malaysia
[3]Faculty of Chemical Engineering, Universiti Teknologi MARA, 40450 Shah Alam, Selangor, Malaysia

ABSTRACT

The continuing presence of heavy metals both in the solid and liquid industrial wastes is of great concern. Currently practiced methods to remove heavy metals include landfill, cementation and incineration (for solid wastes) while carbon adsorption, ion exchange, filtration, precipitation, neutralization and solvent extraction are normally used for liquid waste. Even though these technologies offers cheap and efficient mode of waste management, the generation of large volumes of toxic sludge, the liberation of obnoxious gases and the possibility of chemical spillage makes it imperative to have an alternative treatment technique. The exploitation of microbial ability to survive in harsh environments such as wastewaters polluted with heavy metals is an alternative solution for heavy metals removal. These microbes would normally nullify the toxicity of heavy metals via the combination of one or more of the following processes; oxidation, reduction, accumulation or adsorption. However, before any attempt of using this technology at the industrial scale, various process arrangements need to be investigated at the laboratory, pilot and semi – commercial scales. Testing should also be carried out using simulated heavy metal solutions followed by the real industrial wastes. Besides this, the overall heavy metal biotransformation process can be elucidated using instrumental analysis such as using electron microscopy (metal deposition/ translocation/

* Corresponding author. Email: azlina@kimia.fs.utm.my; drwanazlina@gmail.com

accumulation, formation of microbial biofilm), EDAX (intracellular/ extracellular metal deposition), EPR (paramagnetic condition, useful to justify biooxidation/ reduction process), UV - vis, XAFS, voltammetry and FTIR (involvement of functional groups from the microbes in metal binding).

1. INTRODUCTION

Metal ions are involved in all aspects of microbial life. This include the stabilization of biological structures such as cell walls and protein conformation, as a catalyst in different biochemical processes and to initiate, moderate and inhibit reactions. However, when a metal with no known biological function competes with or replaces a functional metal, toxicity results [1]. Although toxicity is generally viewed as a property of soft heavy metals, detrimental effects can also result from the surplus or deficiency of essential metals. Indeed, a number of metals, at first investigated because of concern over their toxic nature, were subsequently found to have biological roles. Fe and Co are examples of the reverse situation; their biological roles were appreciated prior to the discovery of toxic effects [2, 3]. Several factors have been identified to be responsible for the toxicity of heavy metals towards microorganisms. The more important factors include pH, oxidation-reduction potential (ORP), inorganic anions, inorganic cations, water hardness, clay minerals, organic matter and temperature. This underlying effect of toxicity has prompted the study of a new field of research where most of the focus was directed towards the mechanism of microbial resistance to toxic metals. The interaction between metals and microorganisms has resulted in the accumulation of a variety of pollutants, both organic and inorganic, by means of modifying their chemical and/ or physical characteristics. Amongst the suggested mechanisms used by microorganisms for metal removal include bioaccumulation, precipitation, biosorption, oxidation / reduction and volatilization.

2. MECHANISMS

2.1. Bioaccumulation

Bioaccumulation is a mode of metal accumulation by living cells. The process is dependent on the metabolic activity of the cell, which in turn can be affected significantly by the presence of metallic ions [4]. It also refers to the adsorptive and metabolism - dependent mechanism in bacteria. This type of metal uptake can occur actively or in passive mode. In active uptake, the metals are transported into the cell across the cell membrane through the cell metabolic cycle. Passive uptake refers to uptake that is independent of the biological metabolic cycle [5]. The bioaccumulation of metals by bacteria may occur intracellularly or extracellularly. The three major classes of biopolymer namely proteins, nucleic acids and polysaccharides, all provide sites at which metal ions will bind. The ligand groups available include negatively charged groups such as carboxylate, thiolate or phosphate, and groups such as the amine function which coordinate to the metal centre through lone pairs of electrons. Metal ions are also bound by specialized macrocyclic ligands such as the porphyrins, while transition metals, notably Fe, are often present in enzymes in dimmers or clusters with bridging sulphide or oxide ligands [1].

2.1.1. Intracellular Accumulation

Microbial cells have the ability to accumulate intracellularly both the metabolically essential metals (Na, Mg, K, Fe, Ca) and the non - essential metals (Ag, Au, Hg, Cr). Intracellular accumulation is an energy - dependent process and requires active respiration by the microbial cells. Active uptake usually requires a specific transport system. Microorganisms have well - developed transport systems capable of accumulating metals against a concentration gradient. When a metal is taken into the cell, ions of an equivalent charge are released by the cell. For example, the essential Ca^{2+} ions will be excreted from the cell when the non - essential Cd^{2+} is transported into the cell. This will cause the normal intracellular concentration of Ca^{2+} to decrease hence, prompting a series of reactions involving energy consumption where the Cd^{2+} would ultimately either be precipitated inside the cells as Cd microparticles or bind to metal - binding proteins such as the metallothioneins. pH and the presence of other ions can substantially affect intracellular metal uptake because uptake of specific metals usually occurs at an optimal pH value. For example, Ni uptake is optimum at pH 4 whereas Zn uptake occurs optimally at pH 6.5 [1].

2.1.2. Extracellular Accumulation

Extracellular metal accumulation basically proceeds via the initial complexation followed by the subsequent accumulation. These processes occur either on the microorganisms cell wall or cell membrane. Microbial cell walls are anionic owing to the presence of carboxyl, hydroxyl, phosphoryl, and other negatively charged site. Cationic metals would rapidly bind to these sites by metabolically - independent reaction. Bacteria and fungi are known to produce chemicals that have a high metal-binding capacity. One example of this 'chemical' is the siderophore, which is a low molecular weight molecule that have a high specificity to bind or chelate Fe compounds. It is secreted under low Fe stress condition by fungi and also by aerobic and facultative anaerobic bacteria. Some well-characterized siderophores are as follows; siderophore (*bacteria/ fungi*) – Enterobactin (*Escherichia coli*), Agrobactin (*Agrobacterium tumefaciens*), Arthrobactin (*Arthrobacter* sp.), Ferrioxamines (*Actinomyces* sp.), Ferrichromes (*Aspergilli, Penicillia*), Fusarinines (*Fusaria*) and Rhodotorulic acid (*Rhodotorula* sp.) [3].

2.1.2.1. Bacteria

The Gram - positive bacteria are particularly suited for metal ion removal. Different charged functional groups present in the cell wall region mediate the interaction between the cell walls and metal ions. Anionic sites include carboxylate (from peptidoglycan) and phosphate (from teichoic acid) while the positive charged sites are exclusively made up of ammonium, D - alanine (teichoic acid), amino sugar (glycan) and diaminopimelic acid (the peptide section of peptidoglycan). The carboxyl and phosporyl groups are the major sites for metal accumulation while the amine groups usually particpitates in the binding of anionic metal compounds such as chromate, arsenate and chloroaurate ions. For the Gram - negative bacteria, metal depositions normally occur at the polar head regions of the constituent membranes or along the peptidoglycan layer. The phosphoryl substituents in the lipopolysaccharide layer are responsible for the binding of divalent cations (eg. Mg, Ca, Cd, Co, Ni) compared to the role of carboxyl groups [6].

2.1.2. Fungi

Fungal cell wall is rich in chitin which is responsible for extracellular metal accumulation. One example on the exploitation of fungal cell wall in metal removal is the accumulation of uranium by *Aspergillus fumigatus*. This three - stage process is started when a complex between the uranyl ions in solution and the nitrogen from the chitin is formed. Then, additional U is adsorbed by the three - dimensional network of the chitin around the uranyl - chitin complex formed earlier. This uranyl - chitin complex acts as the nucleation site for uranium deposition. Finally, the uranyl - chitin complex hydrolyses resulting in the precipitation of uranyl hydroxide within the chitin network [7].

2.1.2.3. Algae

In the metal ion - algae interaction, the electrostatic binding is widely accepted as being the major mechanism taking place i.e. the cationic metal species would be attracted to anionic sites in or on the algal biomass while the anionic metal species will be attracted to the cationic sites. The following functional groups has been suggested to be responsible for metal ion binding by algal biomass; carboxyl, amide, hydroxyl, phosphate, amino, imidazole, thiol and thioether moieties which are present in proteins, lipids and carbohydrates structures. Amongst the main algal genus showing high metal removal capacity include *Chlorella*, *Chlamydomonas* and *Ulothrix*.

2.2. Precipitation

Metal precipitation results from the reaction between excreted substances from living bacteria with metallic species present in the solution. Living bacteria produce and excrete substances that chemically react with metals present in solution to produce an insoluble metal compound. An example of metal precipitation is the production of hydrogen sulphide (H_2S) by sulphate - reducing bacteria. The H_2S reacts with metals to form insoluble metal sulphide compounds. Another example is the deposition of Au particle inside the cell wall of *Sargassum natans* [8].

2.3. Biosorption

Biosorption is a way of metal sequestration by cell, or rather by its external components [4]. It utilises the ability of biological materials to accumulate heavy metals from wastewater. Biosorption is also an independent process, in terms of metabolic activity of the cell and should be strictly addressed as the chemical processes that take place on the cell wall [4, 5]. This passive physico - chemical uptake of heavy metals occur in different ways and is dependent much on the affinity of a metallic species towards the active site provided by the structural features of the cell wall. This relates to the different types of binding mechanisms for each metallic species with anionic binding sites [4].

2.4. Oxidation and Reduction

In general, oxidation and reduction is a function of living microorganisms. This also acts as the mechanism for the microorganisms to acquire resistance or tolerance to the toxic heavy metals. Highly specific mechanisms for resistance to toxic metallic ions such as Hg^{2+}, Cr^{6+}, Ag^{+}, Cd^{2+}, Pb^{2+} and the arsenical compounds, are normally determined by genes on plasmids or transposons. Example for the role of microorganisms, in particular bacteria, in changing the form of metals via the redox reactions are as shown in Equations 1 – 4.

$$CH_3Hg^+ \xrightarrow{\text{(organomercurial lyase)}} CH_4 + Hg^{2+} \xrightarrow{\text{(Hg(II) reductase)}} Hg^o$$ (Equation 1)

$$As^{3+} - OH \xrightarrow{\text{(arsenite oxidase)}} As^{5+}(OH)_3O$$ (Equation 2)

$$C_6H_{12}O_6 + 8\,CrO_4^{2-} + 34\,H^+ \xrightarrow{\text{(chromate reductase)}} 8\,Cr^{3+} + 6\,HCO_3^- + 20\,H_2O$$ (Equation 3)

$$4\,FeS_2 + 15\,O_2 + 14\,H_2O \rightarrow 4\,Fe(OH)_3 + 8\,SO_4^{2-} + 16\,H^+$$ (Equation 4)

The reduction of mercury as shown in Equation 1 has been reported for *Pseudomonas putida* [9], *Escherichia coli* and *Staphylococcus aureus*. The enzyme cofactor is generally NADPH, but some Hg(II) reductases can use either NADH or NADPH. Arsenite, As(III) can be oxidized to the less toxic form of arsenate, As(V) from the function of arsenite oxidase (Equation 2). This condition has been reported for several bacteria namely *Achromobacter*, *Xanthomonas* and *Pseudomonas spp*. One strain namely *Acidithiobacillus ferrooxidans* can oxidize the arsenic minerals such as arsenopyrite ($FeS_2.FeAs_2$), energite ($3Cu_2S.As_2S_5$) and orpiment (As_2S_3) [1]. Reduction of arsenate to arsenite has also been reported in *Chlorella* and *Micrococcus lysodeikticus*. The role of chromate reductase in *Acinetobacter haemolyticus* to reduce Cr(VI) into the Cr(III) form has been reported by Pei et al. (2008) [10]. Complete breakdown of 1 mole glucose would yield sufficient energy to reduce 8 moles Cr(VI) (Equation 3). For example, complete breakdown of 5 g L^{-1} glucose (28 mM) is expected to yield sufficient energy to reduce 500 mg L^{-1} Cr(VI) (192 mM). Based on the reaction stoichiometry between glucose and Cr(VI) of 1 : 8, breakdown of 28 mM glucose would yield enough energy to reduce approximately 204 mM Cr(VI) [11]. The oxidation of Fe from its minerals can be carried out by *Acidithiobacillus ferrooxidans* as shown in Equation 4.

2.5. Volatilization

Volatilization occurs when living microorganisms methylate metals which primarily as a form of detoxification. Volatilization is important in metal transformations in the environment, particularly in solids and sediments. One example for microbial metal volatilization is the methylation of mercury, whereby the mercuric ion, Hg(II), is converted to the methylmercury compounds such as dimethylmercury. Other metals such as Se, Te, As and Sn are also subject to volatilization by bacteria and fungi.

3. Evaluation Tools for Microbial Heavy Metal Removal

3.1. Fourier – Transform Infrared (FTIR) Analysis

Infra - red analysis is useful especially in conducting analytical work involving the determination of functional groups present in a sample. This is because the IR spectrum for any chemical compound is one of its very specific characters. For example, the near infra - red region (4,000 - 15,000 cm^{-1}) is very useful for quantitative analysis in the food industry, the far infra-red region (30 – 200 cm^{-1}) for studies on spin and vibrational mode for bones and crystal lattice while the most common and useful region used is the middle infra-red or simply known as IR (200 – 4,000cm^{-1}) as almost all of the basic vibration occurs in this region [12]. The infra - red radiation interacts only with molecules in its vibrational mode. According to Hooke's Law, the frequency or wavenumber for this stretching vibration is influenced by the mass and binding strength of the vibrating atoms. Hence, a stronger binding strength would result in a higher position of wavenumber recorded but for a higher mass of bonded atoms, the absorbance would occur at a relatively lower wavenumber [12]. One example for the application of FTIR in elucidating the metal - microbes interaction is the analysis on the Cr(VI) - loaded cells of *Bacillus cereus* [13]. The involvement of weakly basic amine group of amino acids is depicted in Figures 1 and 2.

There was a notable shift for the amine group upon contact with Cr(VI) i.e. 3433 cm^{-1} to 3131 cm^{-1} as opposed to the C=O based groups (1657 cm-1 to 1654 cm-1). This can be explained as follows; Cr(VI) exists as negatively charged oxoanions in aqueous envFement. Since the experiment was conducted at pH 2.0, complete protonation of the amine groups (NH_2 to NH_3^+) would occur allowing an electrostatic interaction with the negatively charged Cr(VI) ions. The binding of Cr with a considerably high atomic number would literally reduce the vibration of the RNH-Cr complex, thus shifting the stretching vibration to a lower position [8, 14]. At the same time, the carboxyl groups would be fully dissociated to COO^-, hence eliminating its role in binding of the negatively charged Cr(VI) ions.

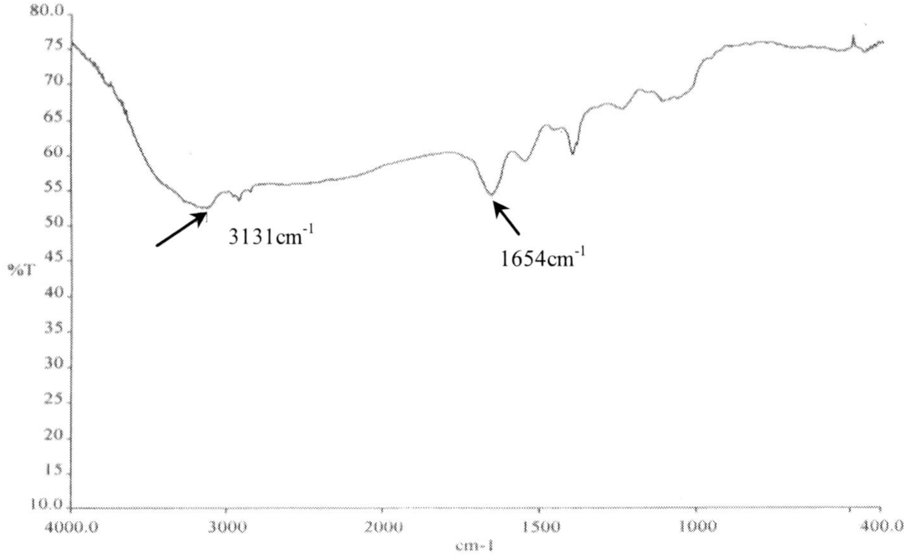

Figure 1. FTIR spectrum of heat-treated cells of *B. cereus* in 200mg L^{-1} Cr(VI).

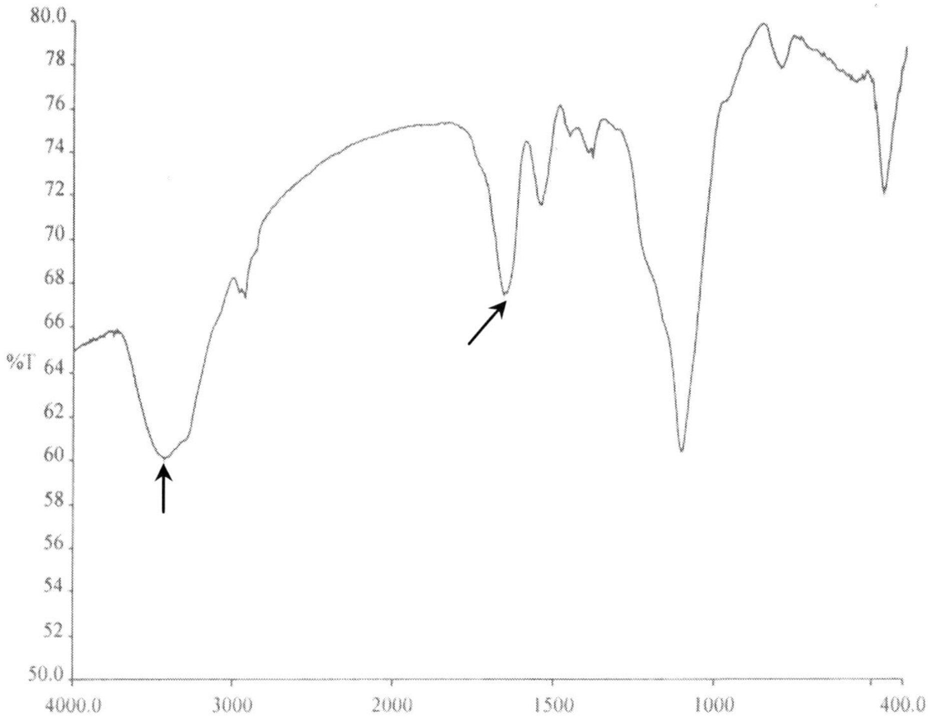

Figure 2. FTIR spectrum of heat-treated cells of *B. cereus* without Cr(VI).

3.2. Cyclic Voltammetry

Cyclic voltammetry is useful in determining the ability of a microorganism to change the oxidation state of a metal via the oxido-reduction process. One example is the determination of Au reducing potential for the gram negative cells of *Acithiobacillus ferrooxidans* (*A. ferrooxidans*) [15]. The profile was obtained using a voltammeter (Mini and Microelectrode System UME) via the following parameters: scan rate - 4 mV s^{-1}; 0 – 1500 mV; current sensitivity - 0 – 100 A; background correction - 0.1 M HCl; purging, N_2 gas; no. of cycle, 1; working electrode - glassy carbon; reference electrode – Ag / AgCl and platinum as the auxiliary electrode. Result from the analysis is shown in Figure 3.

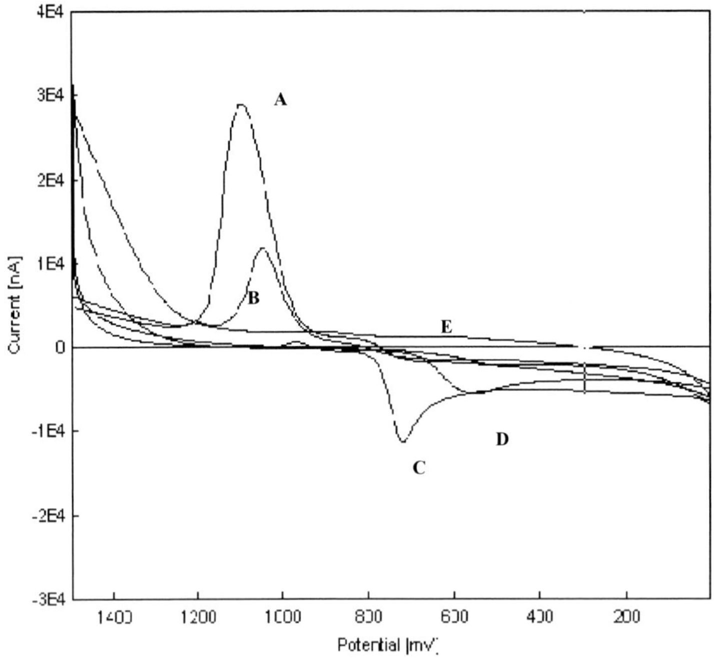

Figure 3. Profiles of Au solution in the absence (peaks A and C) and presence (B and D) of *A. ferrooxidans* with 0.5 M HCl as background correction (peak E).

Figure 3 showed the voltammogram obtained for Au solution only and Au solution in the presence of *A. ferrooxidans* cell, with 0.5 M HCl acting as the background electrolyte. A sharp peak (A) was observed at E_{app} = 1.098 V. This would give an actual E value of 0.876 V after taking into considerations the standard potential of the Ag / AgCl electrode used, 0.222 V. This value correlates well with the value suggested for the reduction of Au (III) to Au (I) which was around 0.9 V [8]. No other peaks were observed in both profiles ruling out the presence of any inorganic impurities in the solution which could probably result from the N_2 purging of the solution. Upon addition of the *A. ferrooxidans* cell, a similar profile (peak B) was obtained for the Au solution with $E_{app.}$ = 1.053 V which would later gives an actual value of 0.831 V. In the presence of *A. ferrooxidans* cell, a decrease in peak's height was observed which could probably indicate the dominancy of Au (I) species in solution. It was also observed that for solution containing Au only, the peak indicating Au (III) reduction to Au (I) was obtained at 0.876 V and 26.67 mA (peak A). However, upon addition of *A. ferrooxidans* cell the peak was slightly shifted to 0.831 V but the height was significantly reduced to around 12 mA (peak B). This clearly indicated the role of *A. ferrooxidans* as a reducing agent for Au which was probably through the involvement of amino groups from the cell wall. The role of amino groups as a potent chelator for Au complexes has been reported by other researchers [16, 17]. However the effect of spontaneous Au reduction by UV - light from the surrounding environment should not be discounted either. No other peak was observed in one flow of electrical current which could indicate that further reduction of Au from Au (I) to Au (0) had not taken place. Hence, this would suggest that during the experiment, Au was present as its ionic form and not in its colloidal form. HCl as the background electrolyte, did not show any interfering effect on the experiment. In this study, Ag / AgCl were used as the reference electrode while glassy carbon was used as the working electrode due to its inert characteristic. In the presence of an electrical current, the following reactions take place:

$$AuCl_4^- \longrightarrow AuCl_2^- + 2e \qquad \text{(Anode; Equation 5)}$$

$$H^+ + 2e \longrightarrow H_2\,(g) \qquad \text{(Cathode; Equation 6)}$$

3.3. X-ray Diffraction (XRD)

X - ray diffraction is a method of determining the arrangement of atoms within a crystal, in which a beam of X - rays strikes a crystal and diffracts into many specific directions. From the angles and intensities of these diffracted beams, a crystallographer can produce a three-dimensional picture of the density of electrons within the crystal. From this electron density, the mean positions of the atoms in the crystal can be determined, as well as their chemical bonds, their disorder and various other information. In metal-microbe interactions, this technique is extremely useful to elucidate on the possible deposition of heavy metals onto/ into the bacterial cells region after the initial adsorption/ accumulation processes. For example, XRD was used to determine the structural form of Au deposited onto the cells surface of *A. ferrooxidans* [15]. Results obtained (diffractograms not shown) did not reveal any peaks showing the presence of organized structured Au particle. This could be indicative on the absence of crystalline Au structure formed on the bacterial surface. Amongst possible explanation for the situation is the low concentrations of Au used i.e. 0.05 - 0.36 mM. This concentration range was much lower than the Au concentration used by other researchers such as Kuyucak and Volesky (1989) who reported distinctive peaks for Au deposited on *Sargassum natans* at 38° 20', 44° 56' and 64° 6' when exposed to Au at 1.81 mM [8]. Sanchez-Loredo et al. (2001) has reported the formation of Au particle in the aqueous phase at Au concentrations of more than 0.5mM [18].

3.4. Electron Microscopy

Two of the most widely used instruments in electron microscopy to study the metal - microbe interactions are the Scanning Electron Microscope (SEM) and Transmission Electron Microscope (TEM). The SEM allows the visualization of surface at the cellular and subcellular level. Prior to viewing, the biological specimens has to be fixed, dehydrated, frozen and covered with a conductive layer (Au or Pt) to ensure that the specimens remain stable after exposure to the electron beam. One way of reducing the impact of chemical and thermal stress on the specimen during viewing is by using the Environmental scanning electron microscope such as the FESEM (Field Emission Scanning Electron Microscope). The low – dose electron beams used in FESEM enables more than structural information of the specimen to be gained. TEM provides resolutions as high as 0.3 nm which offers the operator to view microstructures such as cellular appendages, cell wall and small inclusion bodies. In contrast to TEM, a specimen must be electron translucent i.e. of a limited mass thickness. Specimens therefore either have to be small enough for imaging as a whole unit (protein molecule, phage particle, cellular appendage) or a larger specimen has to be cleaved or sanctioned. This requires the specimen to be fixed, dehydrated, embedded in resin, frozen and finally cut into ultra-thin sections [19].

One example on the use of FESEM to study environmental sample is to illustrate the formation of a microbial biofilm inside a bioreactor used in biotechnological applications such as in a Cr(VI) bioremediation system, Figure 4 [20].

After 5 days of column operation, biofilm - containing microorganisms in various shapes and sizes had grown massively and filled the substratum cavity (Figure 4A – 4D). At this stage, it can be said that the biofilm has developed into a stable microbial unit. It is advantageous to have a uniform colonization of the column with the bacteria to minimize clogging from the effect of dead - zones and to have a uniform flow over the entire cross-sectional area of the column. On example on the use of is to monitor the effect of Cr(VI) on the ultrastructure of the *A. haemolyticus* cells. It allows the evaluation on the possible Cr deposition inside the bacterial cell wall or cytoplasmic regions, Figures 5A – 5C [21].

When *A. haemolyticus* were grown without Cr (VI) (Figure 5A), the cells appeared as rod and cocci with inner and outer sections of bacteria clearly discernible. However, upon exposure to either 10 (Figure 5B) or 50 mg Cr (VI) L^{-1} (Figure 5C), the cells loses its shape and increases in size. The cells also remain intact and did not rupture from partial decompression. One interesting observation is cell wall for control cells i.e. cell grown without Cr (VI), appears thicker than cells exposed to Cr (VI). For example, exposing the cells to 10 mg Cr(VI) L^{-1} decreased cell wall thickness to 50 – 100 nm compared to 100 – 300 nm for the control cells. Fine electron - dense particles (indicative of Cr precipitation) was observed for *A. haemolyticus* grown in 50 mg Cr(VI) L^{-1} only.

3.5. X - Ray Absorption Fine Structure (XAFS) Spectroscopy

XAFS is an element specific and non - destructive technique which is applicable to natural and hydrated samples [22]. This makes it particularly useful for studying specific element in bacterial samples. Basically, the XAFS spectra are split into two distinct regions: a near-edge region that includes the fine structure associated with the edge itself, and an extended fine - structure region that consists of the weak oscillatory structure that may persist to as much as 1,000 eV above the edge. These two regions are referred to as the x - ray absorption near - edge structure (XANES) spectra and the extended X - ray absorption fine structure (EXAFS) spectra, respectively. XAFS has been reported to address the oxidation states and chemical environments of Cu in *Anabaena flosaquae* and the biosorption of Cr by *Ecklonia* sp. [23]. Our previous work with the XAFS spectroscopy revealed direct evidence on the form of Cr deposited inside the bacterial cells inner region [10]. XANES provides information on the final oxidation state of Cr in *A. haemolyticus* from the reduction of Cr(VI) and EXAFS provides information on the coordination envFement such as the nearest neighbouring atoms. The XANES spectra of Cr in *A. haemolyticus* cells, together with the XANES spectra of the reference compounds of Cr metal foil, Cr - acetate, $Cr(NO_3)_3$, $Cr(NO_3)_3(aq)$ and $K_2Cr_2O_7$ are shown in Figure 6.

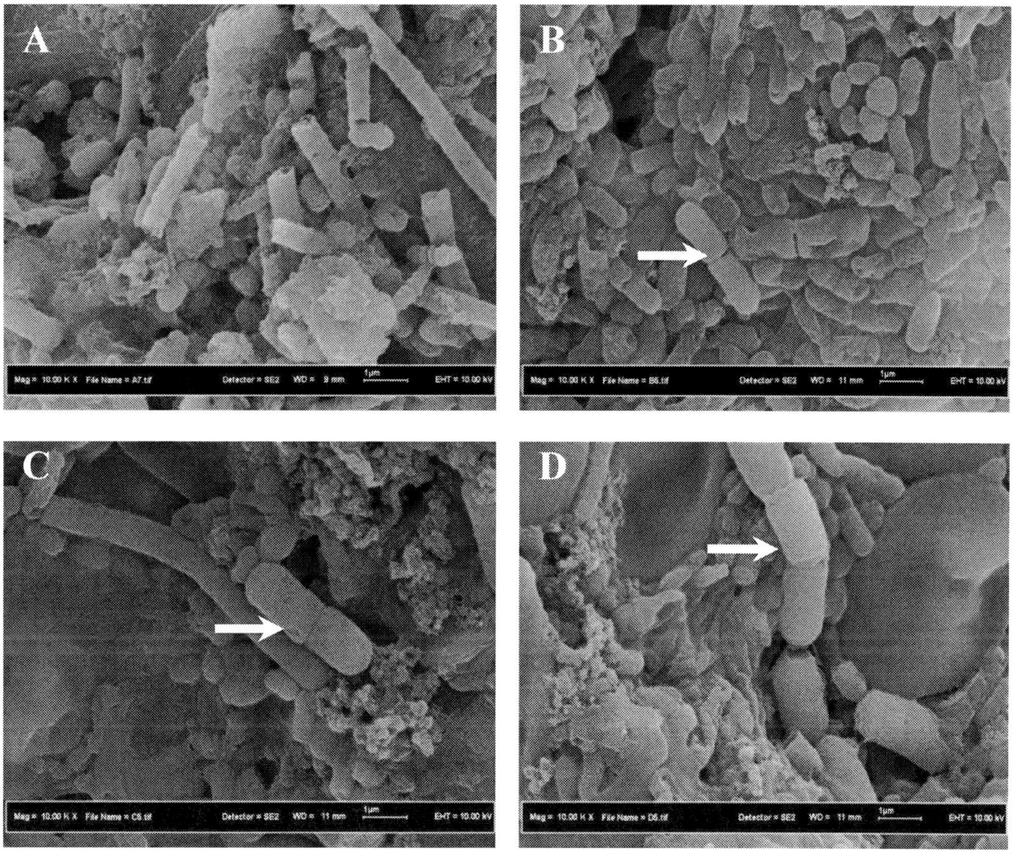

Figure 4. Scanning electron micrographs showing the formation of microbial biofilm on the surface of a rubber wood sawdust packed inside a bioreactor used for the bioremediation of Cr(VI), (A) clusters of bacterial population formed after 5 days of operation at *point a i.e.* nearest to the inlet point (B) *point b* (C) *point c* (D) *point d* – bioreactor section furthest from the inlet point; arrow indicates the possible presence of the primary bacterium *A. haemolyticus* which is known to form a chain-like structure or occurs in pair (Holt et al., 1994); bar represents 1 μm.

A similar finding was obtained by Park et al., (2008), where the XANES data for Cr(VI) reference, $K_2Cr_2O_7$ shows a well - defined Cr(VI) pre - edge peak at 5.984 keV (Figure 6) [23]. The pre-edge peak is attributed to the electron transition from Cr 1s orbital to 3d [24]. The lack of the pre - edge feature for any of the other reference compounds indicates that the Cr was in its reduced valence state, Cr(III). As seen in the XANES spectra, various Cr reference compounds with different oxidation states are distinguishable from each other. The small pre - edge feature for Cr(III) compounds is indicative of Cr(III) bound to oxygen ligands in an octahedral arrangement of atoms. The XANES spectra of the bacterial biomass were identical to that of the Cr(III) reference compounds and the absence of a well - defined pre - edge peak at 5.984 keV and the chemical shift demonstrate that only octahedral - coordinated Cr(III) existed in the *A. haemolyticus* cells.

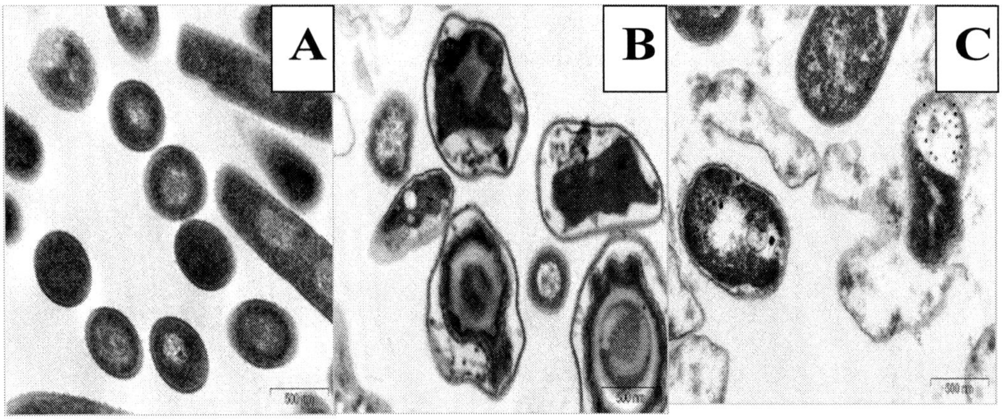

Figure 5. *A. haemolyticus* grown for 24 h in nutrient broth ; A – without Cr (VI), B - in 10 mg Cr(VI) L^{-1}, C - in 50 mg Cr(VI) L^{-1}; bar represents 500 nm.

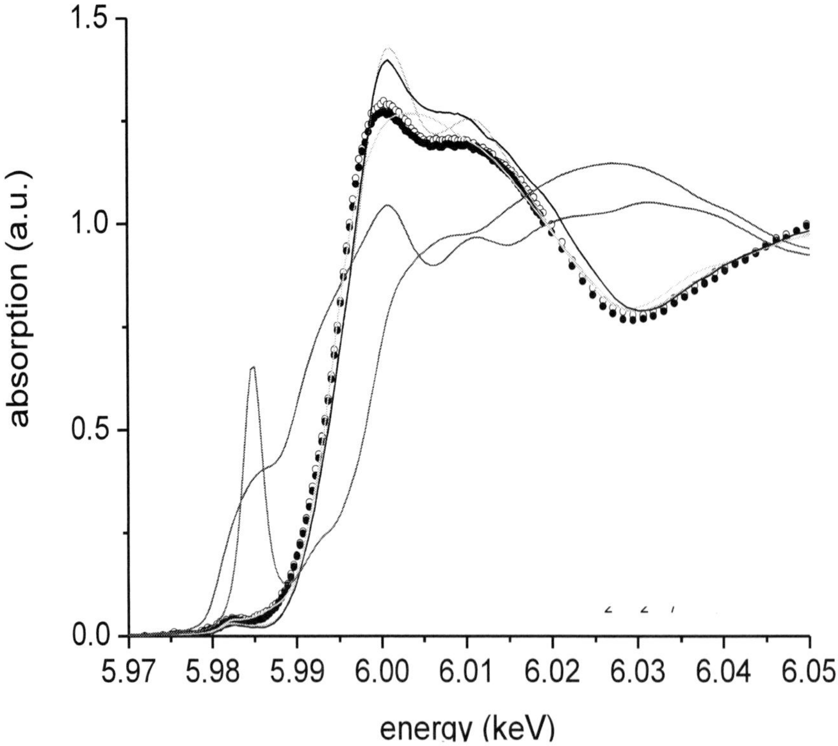

Figure 6. XANES spectra at Cr K-edge in unwashed (○) and washed (●) cells of *A. haemolyticus*, Cr foil (—), $Cr(NO_3)_3$ (—),$Cr(NO_3)_3(aq)$ (—), Cr-acetate (—), and $K_2Cr_2O_7$ (—) standards.

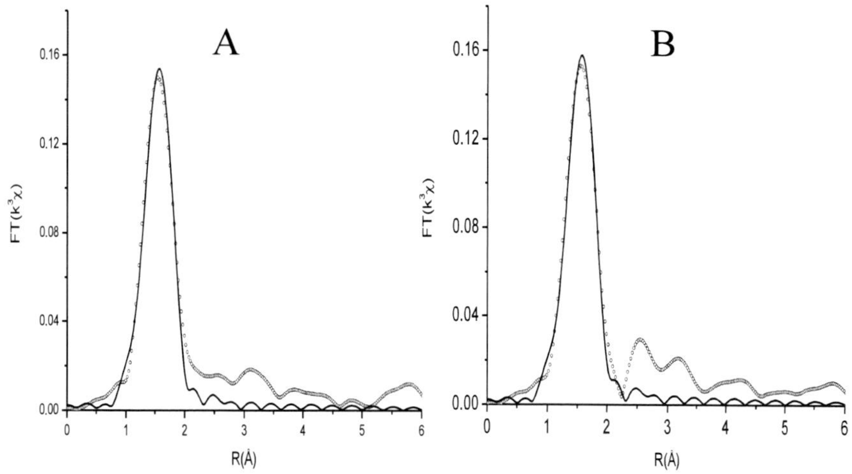

Figure 7. One shell fittings of the Fourier transforms of EXAFS of the Cr in (a) unwashed, and (b) washed *A. haemolyticus* cells (○) with Cr-O (—).

Interestingly, the washed and unwashed *A. haemolyticus* biomass appeared to be similar and thus, reflecting the strong adsorption / absorption of Cr onto the cell surface or intracellularly despite washes with PBS buffer which served to strip any adventitiously bound Cr ions. Figure 7 demonstrates that the first - shell peak of each sample is satisfactorily fitted. The Cr-O distance were approximately the same (Table 1) in the biomass and Cr(III) reference compounds; 1.98 Å for both Cr(III) reference compounds ($Cr(NO_3)_3.4H_2O$ and Cr - acetate) whereas, 1.97 - 1.98 Å for the washed and unwashed biomass. The longer Cr-O interatomic distance in the biomass than in $K_2Cr_2O_7$ reveals that the Cr in these samples is in the form of Cr(III). The fact that the oxidation state associated with Cr(VI) is greater than that with Cr(III) results in the shorter first - shell interatomic distance because of the stronger attractive force between Cr(VI) and its neighbouring oxygen [25]. The interatomic distance of Cr(III) and oxygen reported are between 1.96 - 1.99 Å [26] whereas for Cr(VI) and oxygen are between 1.64 - 1.67 [27].

The longer interatomic distance of Cr-O in Table 1 confirmed the results of the XANES analysis where only Cr(III) is present in the biomass. In addition to that, the results of the EXAFS fittings for the Cr(III) in the biomass also confirmed the results of the XANES analysis. This analysis proved that the Cr(VI) was not merely sorbed as Cr(VI) in *A. haemolyticus* but rather was reduced to Cr(III). The fitting results show only Cr(III) was present and that Cr(III) was neighbouring with oxygen ligands. The similarity of the biomass spectra to Cr-acetate spectra indicates that the Cr(III) was bound to more oxygen from bacteria than H_2O as in the case of Cr-acetate. Since the XAFS analysis suggests that the neighbouring atoms of Cr(III) is O and that Cr(III) coordinates less with H_2O, it is suggested that Cr(III) formed complexes more with the carboxyl (COO^-) group from *A. haemolyticus*.

Table 1. Fitting results in *A. haemolyticus* grown in LB broth with 60 mg L^{-1} of Cr(VI) and reference compounds

Sample	Bond	CN[a]	R(Å)	σ^2(Å^2)
Washed biomass	Cr-O	5.6	1.97	0.0018
Unwashed biomass	Cr-O	6.0	1.98	0.0027
$Cr(NO_3)_3$ (aq)	Cr-O	5.7	1.98	0.0024
Cr-acetate	Cr-O	5.4	1.98	0.0030

[a] CN represents the coordination number (number of neighbouring atoms), R is the interatomic distances given in angstroms, and σ^2 is the Debye-Waller factor given in angstroms squared.

4. Case Studies

4.1. Bioremediation of Cr(VI) From Electroplating Wastewater Using A 0.2 M^3 Pilot-Scale Bioreactor

4.1.1. Design of the Pilot Plant

The schematic diagram for the 0.2 m^3 pilot plant to remove Cr from industrial electroplating wastewater is shown in Figure 8.

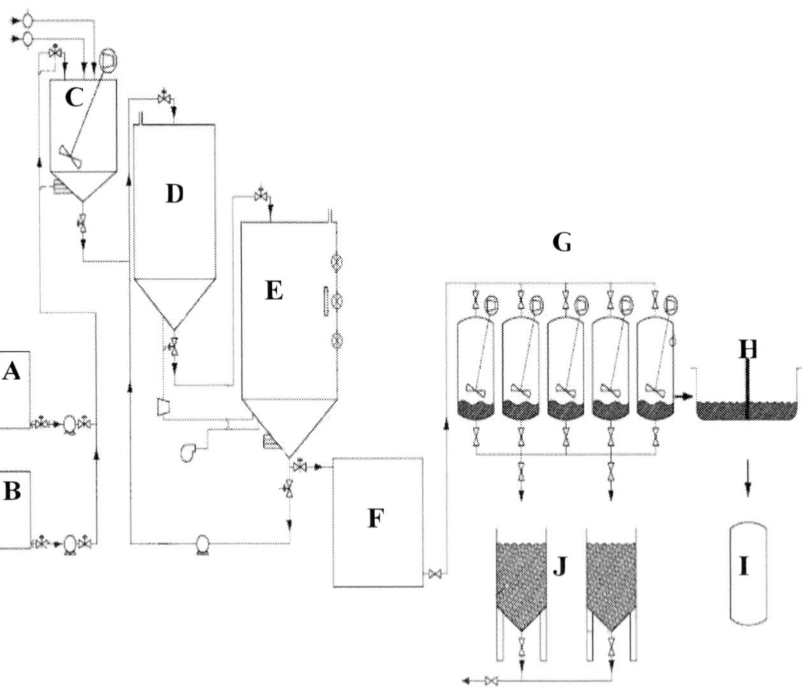

Figure 8. Schematic representation of the experimental setup for the Cr(VI) reduction system using a 0.2 m^3 bioreactor; *A* – nutrient tank, *B* – raw Cr(VI) wastewater tank, *C* – mixing tank, *D* – holding tank, *E* – bioreactor, *F* – receiving tank, *G* – flocculation and coagulation section, *H* – sludge drying bed, *I* – sludge collecting tank, *J* – powdered activated carbon column.

In the mixing tank, raw Cr(VI) wastewater was mixed with liquid pineapple waste (LPW). The pH of the mixture was adjusted to 7.0 0.5 using controller - regulated dosing of NaOH prior to transferring into the holding tank. Upon reaching the capacity of the holding tank, the neutralized and LPW amended Cr(VI) wastewater entered the 0.2 m^3 bioreactor gravitationally in a down flow mode. Effluent collected from the bioreactor was then passed through a flocculation and coagulation setup to remove Cr, other heavy metals, color, odor and organic contents from the wastewater. Sludge formed was dried at the sludge drying bed prior to safe disposal to a local waste management company, Kualiti Alam Sdn. Bhd, Bunit Nenas, Negeri Sembilan. The treated effluent was collected and filtered using powdered activated carbon prior to discharge.

4.1.2. Operation of the Pilot - Plant

The pilot-plant was operated for a period of three months with successful treatment of around 11 m^3 of the Cr(VI) – containing electroplating wastewater. The Low fluctuation in Cr(VI) concentrations (17 – 81 mg L^{-1}) assisted in the exceptional Cr(VI) reduction by the bioreactor (Figure 9).

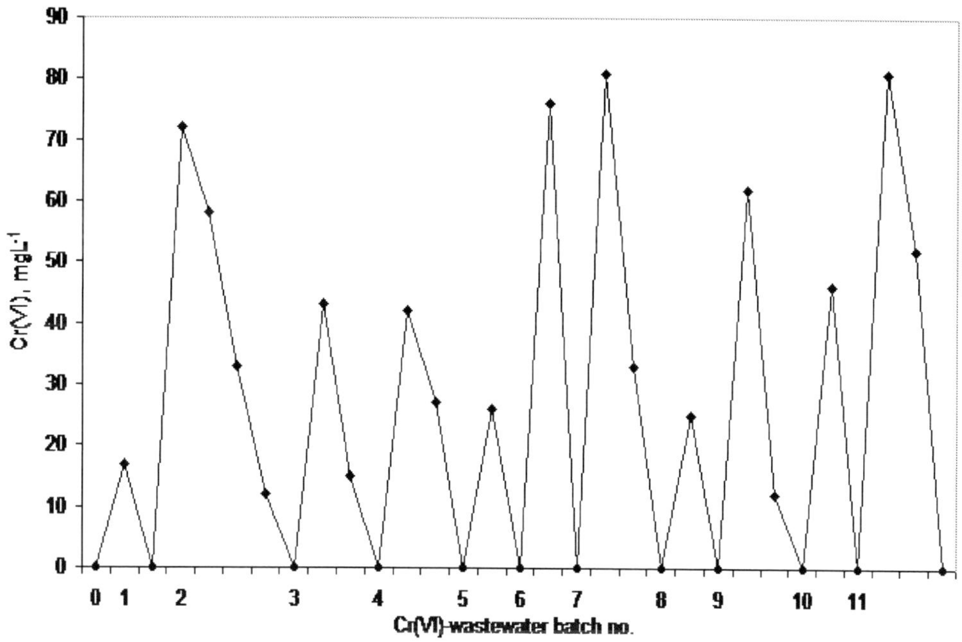

Figure 9. Profile for Cr(VI) reduction by the bioreactor; peaks indicate initial Cr(VI) concentration; bullets in – between peaks indicate no. of cycles needed to complete the Cr(VI) reduction process for each batch of wastewaters.

Complete Cr(VI) reduction was achieved at the first Cr(VI) inflow of 17 mg L^{-1} at 0.33 m^3 h^{-1} without the need to recirculate the Cr(VI) wastewater into the bioreactor. This shows fast start – up for the bioreactor i.e. complete Cr(VI) reduction was achieved immediately after inoculation without the need to adapt the bioreactor's content to the Cr(VI) wastewater. However, three cycles was needed to completely reduce 72 mg Cr(VI) L^{-1} in the second batch of wastewater even though the flowrate was decreased to 0.16 m^3 h^{-1}. Decreasing the incoming flowrate is normally applied to enhance the Cr(VI) reduction process via increasing the contact time between cells in the bioreactor and the wastewater contents (Cr(VI), DO, nutrients). With time, the number of cycles needed for Cr(VI) reduction was reduced that eventually leads to a single cycle of treatment only (batch 5). Wastewater batches of 6 and 7 demonstrates the full capacity of the bioreactor where although more than 70 mg Cr(VI) L^{-1} were present in the incoming wastewater, no recirculation was needed to completely convert Cr(VI) to the undetectable level i.e. complete conversion to the Cr(III) forms. One thing worth mentioning is the ability of the bioreactor content to withstand a long standby period without losing its Cr(VI) reduction capacity. Throughout the bioreactor run, no Cr(VI) was detected in the final discharged water, complete removal of total Cr after the coagulation / flocculation steps, more than 95 % removal of organic matters (in the forms of COD and BOD) and satisfactory removal of color and odor after passing through the PAC units. Table 2 shows example for pollutants removal profile using the coagulation / flocculation techniques:

Table 2. Profile for the bioreactor effluent treated using the combination of coagulation / flocculation and PAC filtration at initial Cr(VI) in wastewater of 62.38 mg L^{-1}

	Bioreactor effluent	Post-coagulation/ flocculation	After PAC	Final discharge	Percent removal
Cr(VI), mg L^{-1}	0	0	0	0	100
[a]Total Cr,mg L^{-1}	25.53	0.76	0.48	0.21	99.21
Turbidity (FAU)	36	18	43	27	25
Color (ADMI)	178	66	144	67	62.36
pH	-	-	-	7.22	-

[a] Total Cr is the sum of Cr(VI) and Cr(III) present in the solution; if no Cr(VI) was detected in a particular wastewater fraction, total Cr determined should constitute of Cr(III) species only.

This study demonstrates the technical feasibility of one locally developed Cr(VI) bioremediation system, tested at the 0.2 m^3 pilot – plant scale. The high percentage conversion of Cr(VI) to Cr(III) suggests the feasibility of using a bacterial system as an alternative treatment for Cr(VI) contamination in the aqueous system. The ability of *Acinetobacter haemolyticus*, as a primary Cr(VI) – reducing microorganisms in the bioreactor, to utilize liquid pineapple wastewater as a nutrient is an excellent example of the substitution of a cheap and readily available industrial waste in place of expensive growth medium and could be a significant factor in the commercial use of a process such this.

4.2. Recovery of Heavy Metals from Spent Dry Cell Batteries Using the Combination of Chemical Leaching and Bioleaching

The sequential leaching processes to recover heavy metals (Zn, Mn, Fe) from spent dry cell batteries was carried out in two 0.2 m^3 bioreactors (Figure 10). The process can be divided into four steps namely chemical leaching, bioleaching, combination of acid and bioleaching and the combination of ferric and bioleaching. The bioreactor contains a mixture of 50 kg of mixed batteries and 45 kg of D - size Zn - C type batteries. The bioreactor content was subjected to continuous sequential leachings for a period of one year. Six different bacterial cultures were used during the bioleaching process namely the *SL5B* culture (isolated from the Sungai Lembing Au mine in Pahang, Malaysia), *Acidithiobacillus ferrooxidans*, *Leptospirillum ferrooxidans, Thiobacillus thiooxidans*, *Acidianus brierleyi* and *Sulfobacillus thermosulfidiooxidans* (all obtained from the German Culture Collection, DSMZ, GmbH (Deutsche Sammlung von. Mikroorganismen und Zellkulturen). Using the multi-stage leaching processes, percent recovery for Fe was the highest i.e. (75 %) followed by Zn (43 %) and Mn (38 %) as shown in Figure 11.

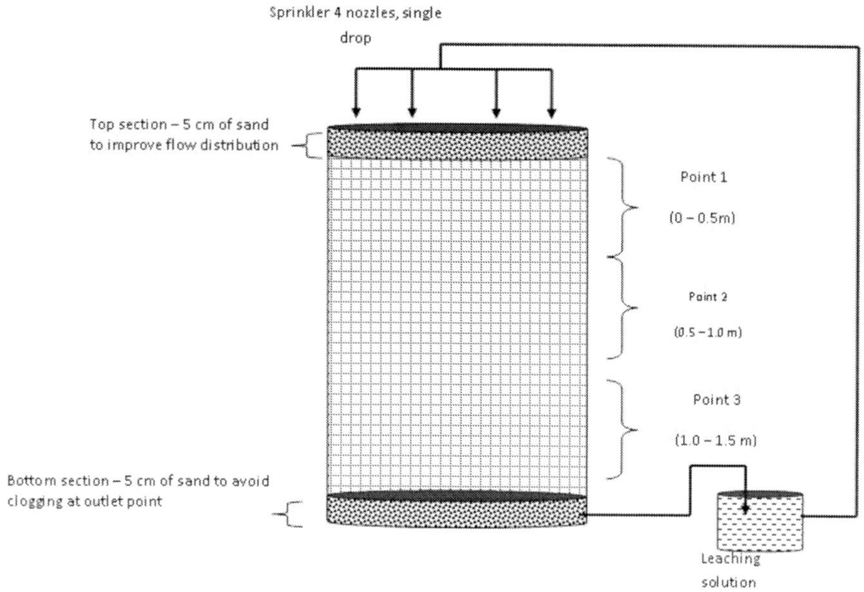

Figure 10. Operation of the sequential chemical and bioleaching processes to recover Zn, Mn and Fe from spect dry cell batteries.

Choice of Zn, Mn and Fe for recovery was based on its highest content in the batteries as shown in Table 3. The recovered metals were then purified using commercially available solvent extraction and jarosite precipitation techniques. Ni is mostly recovered via solvent extraction and electrowinning. Whereby, Zn and Cd can be recovered via precipitation followed by the smelting process. The two - step precipitation was conducted at pH 6 – 7 (for Zn) and pH 9 (for Cd and other heavy metals). In this bioleaching process, elements such as Ni, Cd other heavy metals were oxidized after 14 days of leaching, where the multi - stage bioleaching process helps accelerates the bacterial oxidation and therefore reducing the operation time.

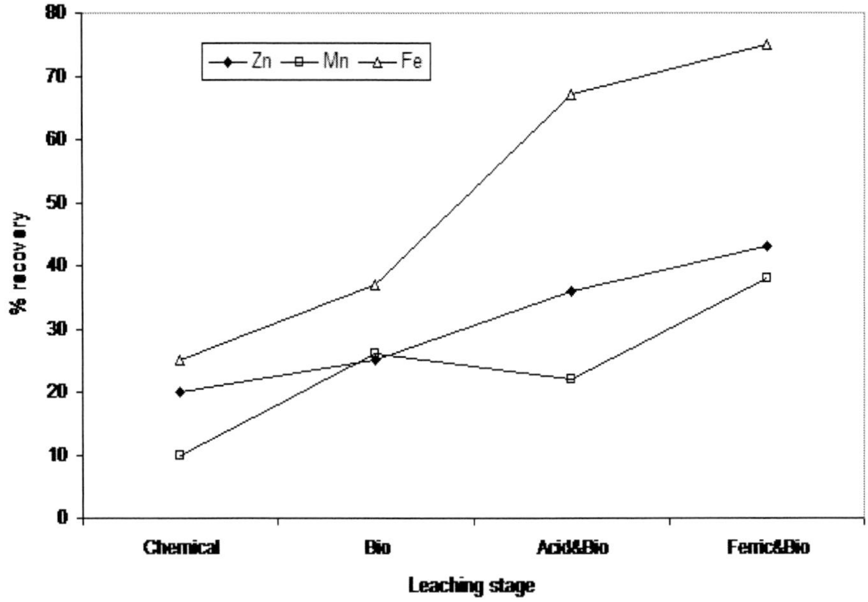

Figure 11. The percentage recovery of Zn - ◆, Mn - □ and Fe - △ from the spent dry cell batteries (0.2 m^3) after sequential treatments using chemical leaching and bioleaching.

Table 3. Metal content of batteries from digestion in aqua regia[a]

Element	Content
As (mg L^{-1})	5.02
Cd (%)	0.38
Co (%)	2.14
Cr (mg L^{-1})	26.14
Cu (mg L^{-1})	242.47
Hg (mg L^{-1})	7.02
Mn (%)	22.84
Ni (%)	5.15
Pb (mg L^{-1})	31.53
Fe (%)	5.31
Zn (%)	6.04
Al (%)	0.28
Li (mg L^{-1})	0.15
La (mg L^{-1})	0.30
V (mg L^{-1})	0.47

[a]18.36 kg batteries / 50 L aqua regia.

4.3. Biooxidation of Au from refractory Au ore and Mines Tailings

The biooxidation of Au from mines tailings obtained from the Lubuk Mandi gold mine was carried out using the 2 L bioreactor followed by 80 tonnage semi – commercial pilot - scale heap leaching mode. The stirred reactor was first operated batch – wise. A batch leaching run was initiated by inoculating a 10 % (v/v) mixed cultures consisting of *Thiobacillus thiooxidans* (TT) : *Thiobacillus ferrooxidans* (TF) : *Leptospirillum ferrooxidans* (LF) : *Caldibacillus ferrooxidans* (CF) into the stirred tank reactor containing the bacterial medium supplemented with 10 % (w/v) of the mines tailings. The mixture was allowed to stabilize for three days before switching to continuous operation (CSTR). For the Au recovery study, 100 kg of mines tailings was first oxidized in CSTR where profiles for the metal contents in the mines tailings before and after biooxidation are as depicted in Table 4.

From the results obtained, it was observed that besides Fe, more than 50 % of Cd, Cu and Zn were removed from the mines tailings after the biooxidation process by the mixed cultures. This indicates the capability of the mixed cultures to carry out the bioremediation of mines tailings. The 'biooxidized' mines tailings were then subjected to the cyanidation process for Au recovery. Table 5 shows the percentage of gold recovered from the 'biooxidized' mines tailings using cyanidation carried out in CSTR at different operating conditions.

Table 4. Metal analysis for the mines tailings before and after biooxidation

Element	Concentration, ppm		% solubilised
	Before biooxidation	After biooxidation	
Fe	2.18*	0.37*	83.02
Au	1.24	1.24	-
Cd	1.13	0.53	53.09
Cu	66.8	23.38	65.00
Pb	43.76	30.56	30.16
Zn	199.35	90.25	54.72

Table 5. Percentage of Au recovery for the 'biooxidized' mines tailings using cyanidation in CSTR

Parameter	Condition	% Au recovered
Pulp density (w/w)	30 %	98.8
	50 %	79.6
Cyanide (ppm)	600	72.2
	1000	96.8
	5000	98.8

The highest percentage of Au recovery (98.8 %) was achieved using 30 % (w/w) pulp density and 5000 ppm of cyanide. It is worthy to note that the Au recovered from the mines tailings was of the refractory type i.e. cannot be extracted using the conventional cyanidation method. The amount of Au recovered increased with the amount of Fe solubilised from the biooxidation process which clearly demonstrates the significant role of Fe – oxidizers such as TF and LF, in the mixed cultures.

Following the biooxidation test in the CSTR, the semi – commercial pilot scale heap leaching study was carried out on - site one gold mine in Lubuk Mandi, Kuala Terengganu, Malaysia. This study was carried out with the collaboration of Permint Minerals Sdn. Bhd. and Sumber Lubuk Sdn. Bhd. The pilot scale bio - heap leaching study was carried out as follows; a marked area was dug to a 30 cm depth using an excavator. The pad area was flattened and sharp objects were removed. The pad was then covered with two layers of geotextiles with one geomembrane (HDPE, 0.5 mm thick) placed in between the geotextiles. The geotextile provides protection for the geomembrane from sharp objects. Slope of the liner was then determined by running water from behind the heap. A perforated PVC pipe was placed over the liner and connected to the air pump and airflow was checked to ensure good distribution of air for aeration. Figure 12 shows the schematic diagram of the heap lining and pad.

A total of 80 tonnes mixture of refractory gold ores and mines tailings was dumped into the heap. Freshly prepared mixed bacterial cultures (8000 L) were then sprayed onto the surface of the refractory gold ores – mines tailings mixture. An excavator was then used to reconstruct the ore into a proper heap - shape. Perforated HDPE pipes were placed on top of the heap and used to spray the remaining bacterial cultures (previously used for agglomeration). Continuous aeration was supplied until a constant DO level of 6.00 mg O_2 L^{-1} was achieved. Basal salts solution (pH 2.0; 3.0 g $(NH_4)_2SO_4$, 1.0 g KH_2PO_4, 1.0 g K_2HPO_4, 0.1 g $MgSO_4$ per L of solution) were then irrigated over the heap to initiate bacterial growth via the oxidation of Fe^{2+} ions present on the refractory gold ores and mines tailings into the Fe^{3+} form. The Fe^{3+} - rich solution was collected in a collecting pond containing small volumes of basal salts before being recycled back into the heap. The process was carried out for two months with daily replenishing of the basal salts to maintain the pond's volume. The overall spraying system was also periodically monitored for clogging. During the biooxidation process, plastic sheet was used to cover the heap from possible leaching by the rainfall. Figure 13 shows some of the works involved during the construction of the heap.

After the biooxidation process, the Fe^{3+} - metal rich – basal salts solution pooled in the collecting pond was transferred into the holding pond. The heap was then 'rinsed' using tap water until a clear effluent was observed. Lime was then sprayed over the heap to increase the pH of the heap to greater than 9.0 for thiosulphate extraction. Sequential extraction was carried out using the following lixiviant solutions; 0.2 M $Na_2S_2O_3.5H_2O$, 0.05 M $NaSO_3$, 0.003 M $CuSO_4.5H_2O$ and 0.1 M NH_4OH. The pregnant solution (containing Au extracted using the lixiviant) was then precipitated for Au using Cu strips followed by the sedimentation of Au in the sedimentation drum. The Au – less pregnant solution was then transferred back into the pond and recirculated continuously onto the heap for three months. Tap water was added daily into the pond to maintain the water volume while the solution pH was adjusted using NH_4OH. Similarly, the concentration of the thiosulphate solution was also maintained. At the end of the three months period of thiosulphate leaching, 31.8 % of Au was extracted, where the amount of Au in the refractory ores and mines tailings mixture was reduced to 0.15 ppm from the initial value of 0.27 ppm. Experimental results showed biooxidation condition strongly affected the solubilization of heavy metals during the bioleaching process. After the bio – heap leaching process, the heavy metals content of the mines tailings and the refractory Au ores was relatively low. Most of the metals were also present in its stable form, hence negligible impact to the environment.

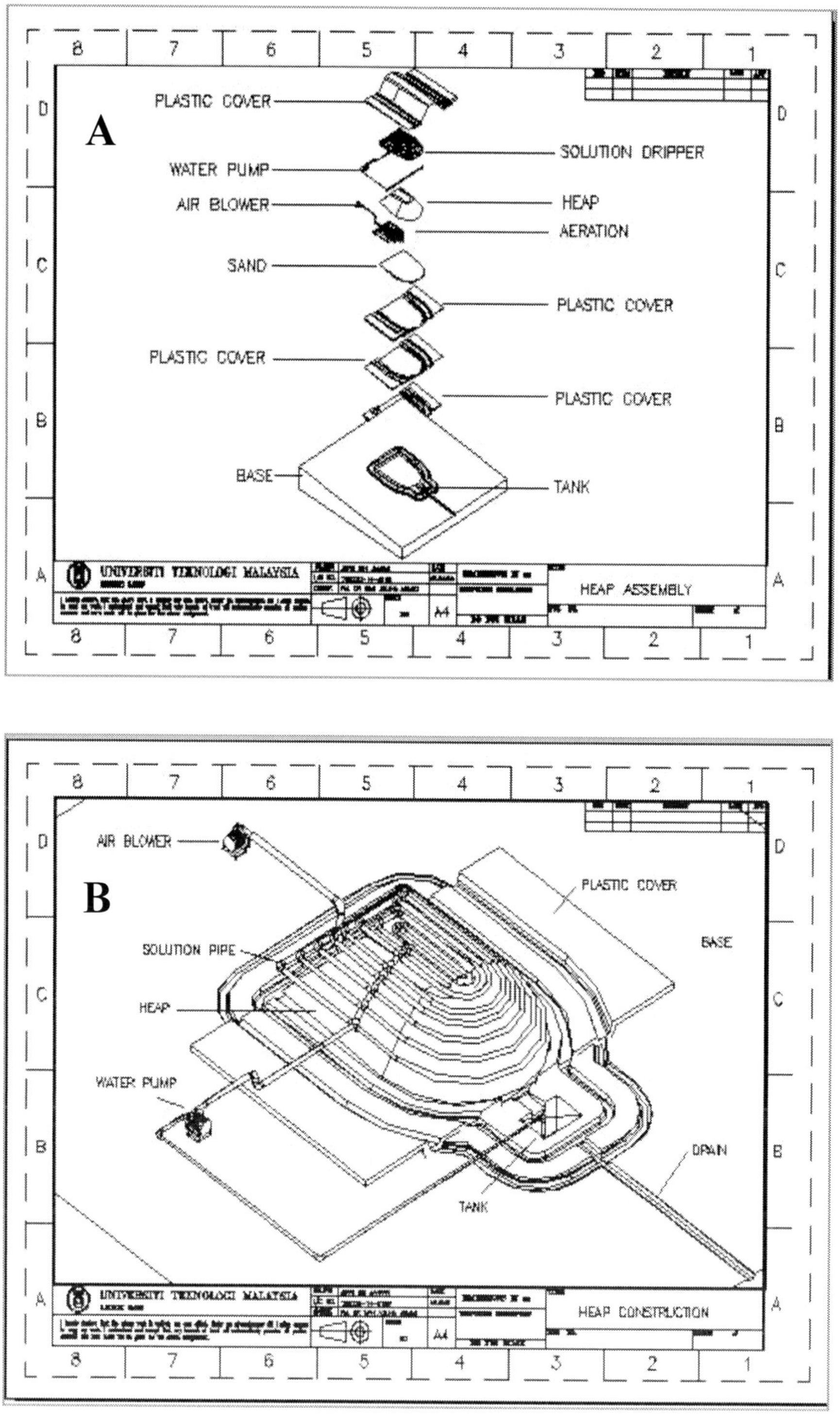

Figure 12. Schematic diagram of (A) heap lining and (B) pad.

Figure 13. The construction of the heap for the recovery of Au from refractory Au ores and mines tailings using the bio – heap leaching process.

REFERENCES

[1] Hughes, M. N. & Poole, R.K. 1989. *Metals and Microorganisms*. London, UK: Chapman and Hall.

[2] Silver S, Misra T. K. & Laddaga R. A. 1989. Bacterial resistance to toxic heavy metals. In: *Metal Ions and Bacteria* (edited by Beveridge, T. J. & Doyle R. J.) Pp. 122. New York, USA: John Wiley and Sons.

[3] Nies, D. H. & Silver, S. 1995. Ion efflux systems involved in bacterial metal resistances. *Journal of Industrial Microbiology*, 14, 186-199.

[4] Vivaraghavan, T., Kapoor, A. & Cullimore, D. R. 1999. Removal of heavy metals using the fungus *Aspergillus niger*. *Bioresource Technology*, 70, 95-104.

[5] Volesky, B. 2003. *Sorption and biosorption*. Montreal, Canada : BV- Sorbex, Inc.

[6] Collins, Y. E. & Stotzky, G. 1989. Factors affecting the toxicity of heavy metals to microbes. In: *Metal Ions and Bacteria* (edited by T. J. Beveridge & Doyle R.J.). Pp. 31. New York, USA: John Wiley and Sons.

[7] D'Souza, S. F. & Bhainsa, K. K. 1999. Biosorption of uranium (VI) by *Aspergillus fumigatus*. *Biotechnology Techniques*, 13, 695-699.

[8] Kuyucak, N. & Volesky, B. 1989. The mechanism of gold biosorption. *Biorecovery*, 1, 219-235.

[9] Von Canstein, H., Li, Y. & Wagner-Dobler, I. 2001. Long-term performance of bioreactors cleaning mercury-contaminated wastewater and their response to temperature and mercury stress and mechanical pertubation. *Biotechnology and Bioengineering*, 74, 212-219.

[10] Pei, Q. H., Shahir, S., Tao, L. & Ahmad, W.A. 2008. Determination of chromium (VI) reduction by *Acinetobacter haemolyticus* using X-ray absorption fine structure spectroscopy. *Journal of Fundamental Sciences*, 4, 415-422.

[11] Zakaria, Z. A., Zakaria, Z., Surif, S. & Ahmad, W. A. 2007c. Biological detoxification of Cr(VI) using wood-husk immobilized *Acinetobacter haemolyticus*. *Journal of Hazardous Materials*, 148, 164-171.

[12] Hitam, R. 1996. Teknik Spektroskopi Inframerah. Fakulti Sains. UTM

[13] Zakaria, Z. A., Aruleswaran, N. Kaur, S. & Ahmad, W. A. 2007a. Biosorption and bioreduction of Cr(VI) by locally isolated Cr – resistant bacteria. *Water, Science and Technology*, 56, 117–123.

[14] Sudha Bai, R. & Abraham, E. T. 2002. Studies on enhancement of Cr(VI) biosorption by chemically modified biomass of *Rhizopus nigricans*. *Water Research*, 36, 1224-1236.

[15] Zakaria, Z. A. 2002. A study on the removal of gold by *Thiobacillus ferrooxidans*. MSc. Thesis, Universiti Teknologi Malaysia.

[16] Niu, H. & Volesky, B. 2000. Gold-cyanide biosorption with L-cysteine. *Journal of Chemical Technology and Biotechnology*, 75, 436-442.

[17] Lack, B., Duncan, J. & Nyokong, T. 1999. Adsorptive cathodic stripping voltammetric determination of gold (III) in the presence of yeast mannan. *Analytica Chimica Acta*, 385, 393-399.

[18] Sanchez-Loredo, M. G., Cabrera-Robledo, A. & Grote, M. 2001. Preparation of gold powders by means of redox-active extractive systems. *Materials Chemistry and Physics*, 9257, 1-6.

[19] Hoppert, M. 2003. *Microscopic techniques in biotechnology*. Weinheim, Germany: Wiley-VCH Verlag GmbH & Co.

[20] Ahmad, W. A., Zakaria, Z. A., Razali, F. & Samin, J. 2009. Evaluation of the combined Cr(VI) removal capacity of sawdust and sawdust-immobilized *Acinetobacter haemolyticus* supplied with brown sugar. *Water, Air and Soil Pollution*, 204, 195-203.

[21] Zakaria, Z. A., Zakaria, Z., Surif, S. & Ahmad, W.A. 2007b. Hexavalent Chromium Reduction by *Acinetobacter haemolyticus* Isolated From Heavy-metal Contaminated Wastewater. Journal of Hazardous Materials, 146, 30-38.

[22] Huggins, F. E., Najih, M. & Huffman, G. P. 1999. Direct speciation of chromium in coal combustion by-products by X-ray absorption fine-structure spectroscopy. *Fuel*, 78, 233–242.

[23] Park, D., Yun, Y. & Park, J. M. 2008. XAS and XPS studies on chromium-binding groups of biomaterial during Cr(VI) biosorption. *Journal of Colloid and Interface Science*, 317, 54-61.

[24] Peterson, M., Brown, G. E., Parks, G. A. & Stein, C. L. 1997. Differential redox and sorption of Cr(III/VI) on natural silicate and oxide minerals: EXAFS and XANES results. *Geochimica et Cosmochimica Acta*, 61, 3399-3412.

[25] Wei, Y., Hsu, L., Wang, H. P. & Chen, K. 2007. XAS study of chromium recoverable from plating sludge. *Journal of Electron Spectroscopy and Related Phenomena*, 156–158, 204-207.

[26] Parsons, J. G., Hejazi, M., Tiemann, K. J., Henning, J. & Gardea-Torresdey, J. L. 2002. An XAS study of the binding of copper(II), zinc(II), chromium(III) and chromium(VI) to hops biomass. *Microchemical Journal*, 71, 211–219.

[27] Sawalha, M. F., Gardea-Torresdey, J. L., Parsons, J. G., Saupe, G. & Peralta-Videa, J. R. 2005. Determination of adsorption and speciation of chromium species by saltbush (*Atriplex canescens*) biomass using a combination of XAS and ICP–OES. *Microchemical Journal*, 81, 122–132.

In: Bioprocess Sciences and Technology
Editor: Min-Tze Liong
ISBN 978-1-61122-950-9

Chapter 19

BIOREMEDIATION OF HEAVY METALS IN THE ECOSYTEM

J. M. R. S. Bandara*
Department of Agricultural Biology, University of Peradeniya, Peradeniya, Sri Lanka

ABSTRACT

Heavy metals are those having a specific density of more than 5 g/cm^3. The chemical elements that cause health hazard to humans on chronic or acute exposure namely lead (Pb), cadmium (Cd), mercury (Hg) and arsenic (As) are the most concerned heavy metals. Although arsenic is a metalloid, is usually classified as a heavy metal. Heavy metals in the environment invariably enter the food chain and cause health hazard mainly through bioaccumulation. Some of the most studied cases of heavy metal poisoning are, Minamata disease caused by mercury toxicity from contaminated fish in Japan, Itai Itai disease due to elevated dietary cadmium in rice in the Jinzu River Basin in Japan as a result of pollution from the waste of a Zinc ore, most recent case of General Arsenic toxicity in Bangladesh due to contaminated ground water. Heavy metals find their way into the food chain through natural and anthropogenic activity. Although, the most desirable action is to avoid heavy metals entering into food chain, it is not practical as the agricultural practices and industrial activities release many undesirable heavy metals into the environment. Toxic levels are developed gradually with bioaccumulation. Therefore corrective remedial measures are mandatory. The physical or chemical remedial measures are expensive in large scale programs. Bioremediation where microorganisms are playing a major role is the most acceptable, cheaper and efficient system that is available at present. Bioremediation is the more environmental friendly and energy efficient alternative to chemical or physical filtration such as Reverse Osmosis in many heavy metal contaminations. Heavy metal binding peptides of microbial origin, chemisorption or biosorption, complexation, adsorption on microbial surfaces or chelator assisted extraction of heavy metals, use of genetically engineered microorganisms (GEM)

* Corresponding author. Email: bandara.sarath@gmail.com

to sequester cadmium using metallothionein on outside surface of bacteria and microbe – plant combined remediation techniques are the currently used technology. *Desulfovibrio* spp is known to intracellular and periplasmic accumulation of Cd. Surface complexation models (SCMs) for metal ion adsorption by thermophilic microorganisms has been established. The thiosulfate reductase gene (*phsABC*) from *Salmonella enterica* serovar *Typhimurium* was expressed in *Escherichia coli* to overproduce hydrogen sulfide from thiosulfate for heavy metal removal (or precipitation). containing polyhistidyl Recombinant Improvement of biosorption through nanotechnology is being studied at present for efficient bioremediation of heavy metals.

1. Heavy Metals

1.1. What Are Heavy Metals

Heavy metals are any metallic chemical elements with a specific gravity greater than 5 g/cm^3. They are not degradable and exist in nature in many organic and inorganic forms. Free ionic heavy metals are toxic at low concentrations to all forms of life. The most commonly known heavy metals for their toxicity to humans are mercury (Hg), cadmium (Cd), arsenic (As), and lead (Pb). Some of the heavy metals such as copper (Cu), zinc (Zn) and iron (Fe) are used by human and plant cells to control significant metabolic and signaling functions making them essential for life. The metal elements when taken in, particularly by human cells can develop toxicity are Pb, Cd, Hg and thallium (Tl).

Routine exposure to heavy metals has become an increasingly important health hazards in the modern world. The most known events of extensive poisoning due to regular dietary or occupational exposure are the well documented Minamata disease due to Hg toxicity in Minamata city in Kumamoto Prefecture in 1956 in Japan [1], Itai-Itai chronic renal failure and osteomalacia in the Jinzu River basin of Toyama Prefecture in Japan due to exposure to elevated dietary Cd [2], and arsenic poisoning in Bangladesh that started in 1970 with the introduction of 8-12 million tube wells as a solution to bacterial contaminated drinking water by the United Nations Children Emergency Fund (UNICEF) the biggest mass poisoning in the modern history, where 30 million people in Bangladesh is affected with As poisoning [3, 4].

1.2. Why Are they Important

Heavy metals are natural components of the Earth's crust, they are ubiquitous pollutants of the environment and their toxicity is often linked with the bioaccumulation associated with their long biological half life. Bioaccumulation refers to the increase in concentration of an element over time in a biological organism, in comparison to the concentration of the element concern in the surrounding environment. Accumulation of compounds in the biological system occurs as the amounts taken in are neither metabolized nor excreted totally. The toxic effect of heavy metals is often due to an interaction between the heavy metal ions and a specific target protein leading to a change in protein structure and function. Heavy metals in organic and inorganic forms are usually disseminated within the ecosystem through the water supply or as gases in the environment. Heavy metals enter water supply by industrial and consumer waste, through agricultural inputs or even from polluted environments that favors acidic rain which releases heavy metals in the earth crust into streams, rivers, natural lakes, water reservoirs, irrigation water and ground water table.

1.2.1. Toxicity Mechanism in Plants and Human

Heavy metals once in circulation in plants interfere in variety of processes in plant cells. Heavy metals often interrupt the activities of enzymes. The enzymatic activities of a cell are managed subject to homeostasis metabolism. Homeostasis is the ability of a cell to maintain a condition of stability within its internal environment by adjusting its physiological processes when dealing with external changes. Heavy metal stress disrupts the homeostasis metabolism leading to unstable internal environment in the cell. The heavy metal elements such as Cu, Cd, Cr, Ni, Pb and Zn are known to disturb chromatin structure, mutagenesis and cytokinensis. They are reported to increase the activity of DNase and cause a decline in DNA synthesis. It is also reported that heavy metals in the cellular environment cause chemical alteration or mutagenesis of nucleotide bases either by direct action of metals elements or by generation of reactive oxygen species. The mechanism of inhibition of enzyme activity in mitochondria and photosynthetic process occurs by the damage cause to the active center, the proteins assembled into active complexes, and dissociation of active complexes. The essential elements and the toxic heavy metals often behave within the plant in the same manner. However it is important to transport the metal elements in a protected form to prevent the proteins affected by the oxidizing properties of the metals. Therefore it is the natural phenomenon that metals are transported across cellular environment by metal transporter proteins e.g. divalent metal transporter (DMT) proteins. The stress generated by heavy metal toxicity is managed by stress adaptation and tolerance process by the activation of genes for γ-glutamylcysteine synthase, dipeptidyltranspeptidase also known as phytochelatin synthase, peroxidases and catalases, and Zn – transporter [5]. Zinc finger type proteins that are rich in cysteine in both plant and animal cells will trap and sequester heavy metals in plant cells they referred to as Phytochelatin and in animal cells they are metallothionein.

1.2.2. Well Studied Cases of Heavy Metal Poisoning

Minamata disease due to Hg poisoning, Itai Itai caused by elevated dietary Cd leading to chronic renal failure, Arsenic in ground water in Bangladesh, Cd from smoking tobacco and Pb poisoning from paints used in toys are the most known heavy metal toxicity cases reported.

1.2.2.1. Minamata Disease

Minamata disease was first reported in Minamata City, Kumamoto Prefecture of Japan in 1956 and later in 1965 in Niigata City in Niigata Prefecture. Mercury poisoning was attributed to methyl mercury that was generated from the mercury catalyst used in the conversion of acetylene to acetaldehyde in the chemical manufacturing industry. In the process of waste water management the resultant methyl mercury was transported to waterways. Methyl mercury found their way into fishes and shellfishes through the food chain, as the methyl mercury was also released into water from bottom sediments by bacteria.

Neurotoxic Methyl mercury then enters the humans through the food chain along with the consumption of fish and shellfish.

Mercury from different sources, either as elemental mercury from earth resources, or mercury vapor or organic methyl or ethyl mercury is converted to divalent inorganic mercury (Hg^{++}) by biomethylation [6]. Hg^{++} is the toxic form to human cells. The recommended safe daily intake of mercury is 7μg/day for a person with a body weight 69.85 kg which amounts to an intake of 0.1 μg/kg/day [6].

1.2.2.2. Lead Poisoning

Fadrowski et al. (2010) working on teen age US children reported that trace levels of lead (1.5 μg/dl) in blood could inflict damage to kidneys [8]. Lead is considered a neurotoxin that is reported be associated with central and peripheral nervous system disorders and visual deterioration in humans [9, 10]. Lead is also known to be nephrotoxic causing renal failure and also cause hypertension [11]. American Association of Psychological Science reported that Attention Deficit Hyperactivity Disorder (ADHD) in children is associated with exposure to lead even to trace levels often from, drinking water, children's costume jewelry and painted wooden toys, imported candies. Nearly all children in the US are reported to have measurable levels of lead in blood.[7]

1.2.2.3. Itai –Itai: Through Elevated Dietary Exposure to Cadmium

Itai-itai (in Japanese means ouch-ouch) disease was found in the cadmium (Cd) polluted Jinzu River basin in Toyama Prefecture, of Japan in 1950. Itai-itai disease was officially recognized in 1968 as the first disease induced by environmental pollution in Japan after legal proceedings initiated by the Ministry of Health and Welfare. Jinzu River which provided the irrigation water for rice cultivation was polluted from a Zinc mine waste in the region. Cadmium toxicity was more prevalent among women above 50 years of age. Prominent symptoms were the chronic renal failure with dysfunction of the proximal interstitial tubule and bone defect osteomalacia [12]. Iron deficiencies were modifying factors for Itai-Itai disease [13]. Since then many countries have reported occurrence of chronic renal failure especially among rice farmers. Mae Sot, Tak Province in Thailand [14, 15], North Central Province in Sri Lanka [16], Bangladesh [17], and Ni-Cd workers in Belgium and Sweden [18], Mollusks used for food in Australia and New Zealand [19].

1.3. Where do they Come from

Hg, As, Cd and Pb are the most dangerous elements, that could cause serious health hazards, by contaminating the food chain. It is now well established that these elements enter into the food chain through non point source pollution of soil and water ways by agricultural practices or by direct pollution due to unconcerned addition of industrial or domestic/urban waste into the aquatic environment. The first step in the entry to food chain is water. The degree of accumulation of heavy metals in crop plants depends on the plant transfer coefficients of heavy metals. Low land rice farming systems are more vulnerable to contamination and plant transfer through irrigation water. However, the soil organic matter, soil pH, redox potential of soil (anaerobic/aerobic status of soil), rain fall, amount of surface runoff and the impact of whole ecosystem determines the potential contamination of food crops with toxic heavy metals from soil or irrigation water [20].

A more convenient grouping of heavy metals based on their ability to transfer into crop plants was proposed by Chaney (1980) [21]. Elements with a negligible uptake and translocation in plants due to the lower solubility in soil were group as low health risk Group I, elements namely Ag, Cr, Ti, Y and Zr. The Group II, comprised of those metal elements (As, Hg, and Pb) with minimal health risk to human, which are strongly bound to soil colloids. They may be absorbed by plant roots under environmental conditions that favor their absorption. Chaney grouped soil contaminant fluorine also in group II. However, it was observed that F in water table can cause serious health problem such as dental fluorosis or act as a modifying factor in the Cd absorption [20]. Group III elements namely, B, Cu, Mn, Ni, and Zn are readily absorbed and translocated by plants but reach phytotoxic levels at lower concentration that can pose any human health risk if used as food. The group IV metal elements Cd, Co, Mo and Se can cause serious health risk in the food chain to both human and animals because their plant tissue concentrations that are not phytotoxic are capable of bioaccumulating in humans to cause health hazard.

1.4. How do they Contaminate Drinking Water

Arsenic a metalloid, was a commonly used weedicide, and was popular as an insecticide and a defoliant in crop production in the early days of application of chemicals for plant protection in farming. Arsenic was used in agriculture as inorganic salts or as organic compounds of arsenite (As3+) and arsenate (As5+). These organic compounds are gradually degraded by soil microorganisms, a stable enrichment culture of bacteria (*Geothrix fermentas)* to release elemental As into soil water/ water table [22]. They reported the release of As into water table by iron reducing anaerobic bacteria, because when ferric compounds are exhausted arsenite is used for respiration eventually to release soluble form of As into water table in Bangladesh. The poisoning by As in Bangladesh occurred due to release of elemental As from sedimentary soil into water table by microbial activity. As is poorly translocated to shoots because the toxicity levels are initially felt in the roots and resulting root damage interrupt in As absorption. Aerial parts of plants are low in As content as such link with the food chain is broken. McLaughlin et al., 1999 reported that elevated dietary intake of As is mostly due to intake of water contaminated with As than from ingestion of As containing food. Widely documented Blackfoot disease in Taiwan, a peripheral vascular disease with gangrene of the extremities and skin cancer is reported to be due to long term chronic exposure to low levels of As often through drinking water.

Cadmium carried by contaminated phosphate fertilizer enters water ways along with surface runoff water and eventually enters reservoirs. Though the soluble form of Cd++ remains in circulation in irrigation water and reached crop plants and enter the food chain as it is easily taken up by many varieties of crop plants especially rice and legumes [20], Cd in colloidal form settles down to anaerobic sediment layers in reservoirs and canals. Bandara et al. (2010) reported the release of Cd in colloidal form to soluble form by an anaerobic bacterium species tolerant to Cd *Sphingomonas macrogoltabida* [20]. They also reported elevated dietary cadmium in the region due to higher level of cadmium in all crop plants cultivated in the region as well as the cow's milk and fresh water fish Tilapia raised under elevated aquatic levels of Cd [16, 20].

2. How to Manage Heavy Metals in the Environment

Phytoremediation, the environment cleanup using plants and their associated microorganism, is considered cost effective and environmental friendly non invasive alternative to clean up of pollution caused by heavy metals. Plants can be used as constructed wetland, or strategically planted along the canals and rivers for rhizofiltration, phytostabilizatin, phytostimulation, phytovolatilization, phytoextraction or for phytodegradation. However for phytoremediation of heavy metals only rhizofiltration, phytostabilization and phytoextraction are useful. In our discussion we are concerned with the role of microorganisms in phytoremediation of heavy metals. In this regard phytostimulation of microbial degraders of organic compounds in the root zone need to be mention though they are of no use in the remediation of heavy metal problem because of the elemental nature of the pollutant. Grasses such as *Festuca* sp., rye grass and *Panicum* sp are used for this purpose because of the dense root zone of these plants. Mulberry tree is also used because of their ability to produce phenolic compounds that stimulate expression of microbial genes that produces compounds for degradation of polychlorinated biphenyls (PCB) and PAH. The rhizosphere bacteria are an asset in the bioremediation of heavy metals and many organic pollutants. Most rhizosphere bacteria could amend the bioavailability of metals. Some bacteria are capable of excreting biosurfactants such as rhamnolipids that make hydrophobic pollutants more water soluble. The most attractive example of role of rhizosphere associated microorganism in the bioremediation is the *Pseudomonas fluroescens* that produced metal chelators such as siderophores, organic acids. We have observed the ability of *Pseudomonas fluroescens* to chelate Fe^{3+} in the rhizosphere of *Eleusine coracana* plant [23].

2.1. Heavy Metals in Water Bodies

Heavy metals in water bodies are a more burning problem that needs immediate attention. This has lead to more research activities in the bioremediation of heavy metals in water bodies. Most heavy metals in the natural waters are distributed as dissolved metal ions, bound to suspended particulate matter (SPM) such as clay and silt mostly in colloidal forms, and in complexes in sediments. The cadmium concentration in water bodies rarely exceeds the WHO minimum recommended value of 3μg/l [24]. However under polluted environment where the metal contamination has caused health hazard such as chronic renal failure the Cd levels in water varied from 30-50 μg/l [16]. The type of phosphate fertilizer used, the soil pH and rainfall pattern determines the amount of Cd in water ways. The range of Cd in River Mahaweli in Sri Lanka was found to vary from 3 – 23.5 μg/l [20]. Water bodies are the main source of contamination of food chain with heavy metals. Metals in water are only removed from solution when it is immobilized. The methods used to immobilize are adsorption, chemical precipitation & flocculation, membrane filtration and ion exchange. Conventionally these are the methods used in the management of heavy metal pollutants in waste water.

However, the high cost involved and the high energy requirement are the main disadvantages in addition to low efficiency in removing the pollutants completely. In the modern era however, energy independent more efficient biological process of metal immobilization that is commonly referred to as biosorption is gaining popularity. The biosorption is a property of certain types of inactive, dead, microbial biomass to bind and concentrate heavy metals from very dilute aqueous solutions. This property in biomass is very much similar to that of chemical substances, as an ion exchange of biological origin. It is the cell wall structure of some bacteria, fungi and algae have this ability. The bioaccumulation we discussed earlier is an exactly opposite phenomenon of active accumulation of heavy metals by living cells. Biosorption is the removal of organic, inorganic, gaseous, soluble or insoluble forms of substances from solution by biological material. Biosorption is a physico-chemical process and includes such mechanisms as absorption, adsorption, ion exchange, surface complexation and precipitation [25]. In this discussion we are focusing on the use of microbial biomass as a sorbent for heavy metals. It is not only used for decontamination of water from heavy metals but also in the enrichment of precious metals. However since the information reported in the field of biosorption of heavy metals is extensive, this chapter will discuss the principle involved in the biosorption of selected more important heavy metals cadmium, arsenic and lead.

Large-scale applications are efficient in passive biosorptive process than in the active bioaccumulative processes. This is mostly due to the fact that active metal uptake that occurs in living systems require the addition of nutrients and hence increase biological oxygen demand (BOD) or chemical oxygen demand (COD) in the water body. It is difficult to maintain a reasonably higher population of active microorganisms with the *in situ* metal toxicity [26, 27].

Many aquatic bacteria, yeast and algae has the ability to uptake dissolved metals from their surroundings onto their cell walls [28]. The amounts of cadmium and lead adsorbed by the cell walls of algae, bacteria and fungi vary among the species. The range of quantity of metals bound varies from few milligrams to few hundred milligrams per gram of biomass (Table 1) These differences are strongly related to the molecular structure of the cell wall of microorganisms concerned. Algal species *Ascophyllum nodosum*, yeast *Saccharomyces cerevisiae*, fungus *Phomopsis* sp. Bacterium *Bacillus laterosporus* are known to be the leading sorbent microorganisms for cadmium. The cell surface of algae used in biosorption are composed of sulfated polysaccharides and alginate [29], whereas, in fungi it is mainly β-1-3 glucan and glycoproteins. The Gram negative bacteria with lipopolysaccharide walls and G positive bacterial walls are characteristically rich in peptidoglycan with lipo teichoic acid. These various functional groups decide the quantities sorbed by the walls of microorganisms.

The sorption process of metal ions or sorbate, dissolved in water occurs onto a solid phase on the walls of the microorganism usually referred to as sorbent. Depending on the degree of affinity that the sorbent has for the metal ions, they are bound to the sorbent solid structure and bound by specific mechanism. This process takes place until an equilibrium is reached between the amount of metal ions (sorbate), bound to the solid surface of microorganisms and its fraction of sorbate that is left in solution (at a final concentration $\mathbf{C_f}$).

Metal removal ability of microorganisms is generally assessed using Langmuir isotherm equation [29].

Table 1. Removal capacities of heavy metal cadmium and lead by microorganisms

	Cadmium	Lead	References
Sorbant Biomass	pH q_{max} mg/g	pH q_{max} mg/g	
Algae			
Ascophyllum nodosum	4.9 215	3.5 272	[30, 31]
Caulerpa lentillifera	5 4.6	5 28.7	[32]
Chlorella vulgaris	4 86.6		[33]
Ecklonia maxima	6 88.5	6 232.6	[34]
Fucus vesiculous	3.5 73	3.5 229	[30, 31]
Laminaria japonica	4.5 136	4.5 348.1	[35]
Sargassum fluitans	4.5 79.8	3.5 266	[31, 36]
Sargassum natans	3.5 132	3.5 253	[30, 31]
Sargassum vulgare	4.5 88.8	3.5 228	[31, 36]
Sargassum vulgaris	6 118	6 248.6	[37]
Bacteria			
Athrobacter	6 13.4		[38]
Bacillus laterosporous	7 159.5		[39]
Bacillus subtilis	40.5	55.95	[40]
Citrobacter MCMB181	6 43.5	4.5 58.8	[41]
*Enterobacter sp.*J1	5 46.2	6 50.9	[42]
Escherichia coli	11.2	43.5	[40]
Pseudomonas putida	6 8.0	6.5 56.2	[43]
P. aeruginosa	57.4	110	[44]
Bifidobacterium lactis	6 32.1	5 111.3	[45]
B.longum	6 32.0	5 175.7	[45]
Lactobacillus fermentum	6 26.7	5 136.1	[45]
Fungi			
Aspergillus flavus		5 13.5	[46]
Aspergillus niger	6 4.38	5 10.19	[47]
Mucor rouxii	6 20.31	6 53.75	[48]
Phomopsis sp.	6 29.2	6 140.9	[49]
Rhizopus arrhizus	5 30.4	3.6 103.6	[50]
Saccharomyces cerevisiae	15.4	85.6	[51]

$$q = \frac{q_{max}\, b\, C_f}{1 + bC_f}$$

Where C_f= free concentration of metal in equilibrium state
q = bound concentration of metal in equilibrium state
q_{max} = Maximum binding capacity at given conditions
b = Coefficient related to the initial slope of the curve and to the affinity of binding
The q_{max} can be estimated by plotting C_f vs q.

In industrial water purification process, irrespective of the origin of the biomass, it is immobilized on to suitable carrier material for better and efficient handling. This immobilization process is an important step in increasing the mechanical strength of the biomass. However free biomass is more efficient in metal removal than the immobilized biomass. The free biomass is entrapped within a polymer matrix of agar, polyacrylamide, silica gel, sodium alginate, polysulfone etc. [53]. In the past immobilization was done by allowing microorganisms to get adsorbed onto inert support. Removal of Ni from polluted water by *Pseudomonas fluorescens* biomass immobilized on agar beads was 37mg/g compared to free biomass, 145 mg/g [54]. This decrease in efficiency is due to the reduction of surface area of the bacterium. Portion of the surface area is used for binding on to the polymer matrix which is otherwise used for binding metals. Raize et al. (2004) and Chojnacka et al. (2005) observed that the main functional groups responsible for heavy metal binding in bacteria are carboxyl, phosphoryl, sulfonate and amino groups [37, 55]. The anionic groups on the cell surface of microorganisms are the metal binding agents. This was confirmed by the fact that cationic uptake by *Bacillus subtilis* is reduced by the extraction of teichoic acid moieties (phosphodiester groups) and reduction of the number of free carboxyl groups [56]. Decline in the amount of lead and cadmium binding was observed by Halttunen (2007) by neutralizing carboxyl and phosphoryl groups on the bacterial cell surface in *Lactobacillus fermentum* and *Bifedobacterium longum* 46 (Table 1) [57]. The removal of Arsenic (V), Cd and Pb is a rapid, pH dependent and a reversible process that is restricted to the cell surface of the bacteria. It is also known that once the metals are adsorbed on to the bacterial cell surface, they can eventually interfere in bacterial metabolism by inactivating enzymes in the cytosol by binding to thiol groups in enzymes. This is one of the major limiting factors in anaerobic microbial digesters (Figure 1).

The forgoing discussion was mainly on the passive and cheaper process of heavy metal sorption by microbial biomass onto the cell surface through ionic binding to functional groups. Recent studies on alternative mechanism has suggested use of metal chelating compounds on marine algae *Sargassum sp.* specifically in *Sargassum fluitans* namely alginate and fucoidan the algal chelating agents on the cell surface. The efficient chelating activity of algal cell surface is due to the abundance of homopolymeric α-L- guluronic acid G-blocks in alginates. Presence of these G-blocks results in enhanced selectivity for cadmium binding [29, 58].

3. Sequestering Cd by Genetically Modified Microorganisms

Although Genetically Modified Organisms are not encouraged in the food chain, bioremediation of heavy metals and other pollutants in the environment using genetically modified microorganisms has extensive potential. Valls et al. (2000) reported a new approach to sequestering soil Cd on bacterial surfaces preventing absorption of Cd by plant roots, thus effectively curtailing entry of Cd into the food chain [59]. Mammalian metal chelating protein, metallothionein gene extracted from mouse was genetically transferred to express on the cell surface of Cd tolerant *Ralstonia eutropha* CH34. Mouse gene encoding Metallothionein was fused to the auto transporter β-domain of the IgA protease of *Neisseria gonorrhoeae.* This allowed the hybrid protein to be expressed on bacterial outer membrane. Modified Mtb gene (*mtb*) was transferred to *R. eutropha* which immobilized soil Cd++. A recombinant *E.coli* bacteria produce a hybrid pili, CS3 containing the hexahistidine peptide chelated and lead to accumulation of high concentrations of Cd^{2+} and Ni^{2+} at 656.2 and 276.5 nmol per mg dry weight of bacterial cell, respectively [60].

Chen Yen-Lin et al. (2009) proposed the use GMO for nanoparticle synthesis using Cd in the substrate and could help in the sequestering of Cd from polluted water [61]. Glutathione (GSH) is an important baseline chemical in cadmium sulfide (CdS) nanoparticle formation in yeasts and its synthesis needs two enzymes: γ-glutamylcysteine synthetase (γ-GCS) and glutathione synthetase (GS). Genetically modified *E. coli* strains that over-express either γ-GCS or GS are used in nanoparticle production. Upon exposure of the recombinant cells to cadmium chloride and sodium sulfide, GS over-expression augmented GSH synthesis and ameliorated CdS nanoparticles formation.

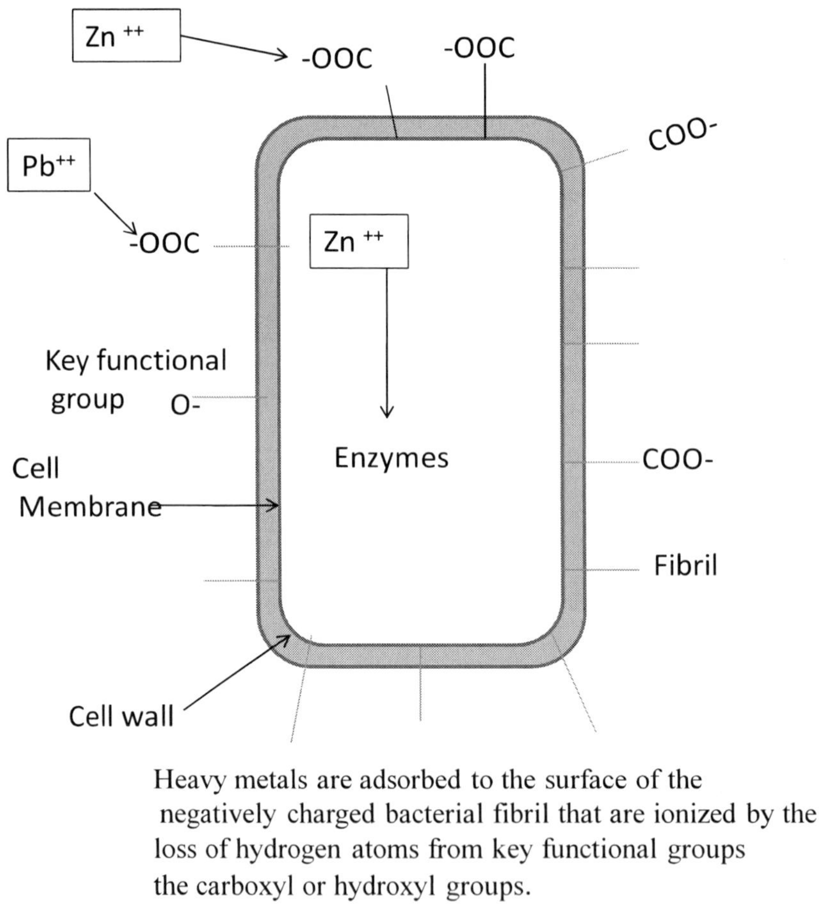

Figure 1. Heavy metal ion adsorption on to bacterial cell wall.

References

[1] Harada, M. 1995. Minamata disease: Methyl mercury poisoning in Japan caused by environmental pollution. *Critical Reviews in Toxicology*, 25, 1-24.

[2] Watanabe, Y., Kobayashi, E., Okubo, Y., Suwazono, Y., Kido, T. & Nogawa, K. 2002. Relationship between cadmium concentration in rice and renal dysfunction in individual subjects of the Jinzu River basin determined using a logistic regression analysis. *Toxicology*, 172, 93-101.

[3] Mehovic, J. & Blum, J. 2004. Arsenic Poisoning in Bangladesh. South Asia Research Institute for Policy and Development (SARID), September 08, 2004 http:// www.sarid.net/sarid-archives/04/040908-arsenic-poisoning.htm. Retrieved February 2010.

[4] Paul Bimal, K. 2004. Arsenic contamination awareness among the rural residents in Bangladesh. *Social Science and Medicine,* 59, 1741-1755.

[5] Prasad, M.N.V. and K.Strzalka (Eds) 2002. Physiology and biochemistry of metal toxicity and tolerance in plants. Kluwer Academic Publishers. Dordrecht. pp. 432.

[6] Clarkson, T. W., Magos, L. & Myers, G. J. 2003. The toxicology of mercury-current exposure and clinical manifestations. *New England Journal of Medicine*, 349, 1731-1737.

[7] Nigg, J.T, Nikolas, M., Mark Knottnerus, G., Cavanagh, K., Friderici, K. 2010. Confirmation and extension of association of blood lead with attention deficit/hyperactivity disorder (ADHD) and ADHD symptom domains at population typical exposure levels. *Journal of Child Psychology and Psychiatry.* 51, 58-65

[8] Fadrowski, J. J., Acien, A. N., Plaza, M. T., Guallar, E., Weaver, V. M. & Furth, S. L. 2010. Blood lead level and kidney function in us adolescents: the third national health and nutrition examination survey. *Archives of Internal Medicine*, 170, 75-82.

[9] Fox, D. A., Campbell, M. L & Blocker, Y. S. 1997. Functional alterations and apoptotic cell death in the retina following developmental or adult lead exposure. *Neurotoxicology*, 18, 645–664.

[10] Bressler, J., Kim, K. A., Chakraborti, C. & Goldstein, G. 1999. Mechanism of lead neurotoxicity. *Neurochemical Research*, 24, 595– 600.

[11] Khalil-Manesh, F., Gonick, H.C., Cohen, A.H., Alinovi,R., Bergamaschi,E., Mutti,A., Rosen, V.J.1992. Experimental model of lead nephropathy.I. Continous high- dose lead administration. *Kidney International.* 41, 1192-11203.

[12] Jarup, L. 2002. Cadmium overload and Toxicity. *Nephrology Dialysis transplantation*, 17, 35-39.

[13] Shigematsu, I., Minowa, M., Yoshida, T. & Miyamoto, K. 1979. Recent results of health examination on the general population in cadmium polluted and control areas in Japan. *Environmental health perspectives*, 28, 205-216.

[14] SwaddiwudhiponG, W., Limpatanachote, P., Mahasakpan, Krintratun, S. & Padungtod, C. 2007. Cadmium exposed population in mae sot district, tak province: 1. prevalence of high urinary cadmium levels in the adults. *Journal of Medical Association of Thailand*, 90, 143-8.

[15] Teeyakasem, W., Nishijo, M., Honda, R., Soisungwan, S., Swaddiwudhipong, W. & Ruangyuttikarn, W. 2007. Monitoring of cadmium toxicity in a Thai population with high-level environmental exposure. *Toxicology Letters*, 169, 185–195.

[16] Bandara, J. M. R. S., Senevirathna, D. M. A. N., Dasanayake, D. M. R. S. B., Herath, V., Bandara, J. M. R. P., Abeysekara, T. & Rajapaksha, K. H. 2008. Chronic renal failure among farm families in cascade irrigation systems in Sri Lanka associated with elevated dietary cadmium levels in rice and freshwater Fish (*Tilapia*). *Environmental Geochemistry and Health*, 30, 465–478.

[17] Kippler, M., Akesson, A. & Vahter, M. 2007. Cadmium exposure in swedish and rural bangladeshi women. presntation at the 2nd international conference on forensic science and medical science. Naresuan University, Phitsanulok, Thailand.

[18] Jarup, L., Hellstrom, L., Alfven, T., Carlsson, M. D., Grubb, A., Persson, B., Pettersson, C., Spang, G., Schutz, A. & Elinder, C. 2000. Low level exposure to cadmium and early kidney damage:OSCAR study. *Occupational and Environmental Medicine*, 57, 668-672.

[19] Australia New Zealand Food Authority. 1998. The Australian Market Basket Survey 1996. Melbourne, NSW: Information Australia.

[20] Bandara, J.M.R.S., Wijewardena, H.V.P., Liyanege, J., Upul, M.A. & Bandara, J.M.U.A.. 2010. Chronic Renal Failure in Sri Lanka Caused by Elevated Dietary Cadmium: Trojan Horse of the Green Revolution. *Toxicology Letters*. doi:10.1016/j.toxlet.2010.04.016

[21] Chaney, R. L. 1980. Health risks associated with toxic metals in municipal sludge. In: Bitton, G., Damro, B. L., Davidson, G.T. & Davidson, J. M. (Eds.), Sludge - Health Risks of Land Application. *Ann Arbor Science Publishers Inc, Ann Arbor*, MI, USA. pp. 59 -83. (not sure the full name of publisher-Ann Arbor Science Publishers)

[22] Islam, F. S., Gault, A. G., Boothman, C., Polya, D. A., Charnock, J. M., Chatterjee, D. & Lloyd, J. R. 2004. Role of metal-reducing bacteria in arsenic release from Bengal delta sediments. *Nature, 430, 68-71.*

[23] *Rajapaksa, R. G. A. S. 1990. Occurrence of some beneficial bacteria associated with roots of Finger Millet (Eleusine coracana* (L) Gaertn.) and their effects on nutrient uptake,growth and yield of Finger Millet. M.Phil. Dissertation. University of Peradeniya, Peradeniya, Sri Lanka.

[24] WHO. 2006. Guidelines for drinking water quality. Vol 1, Recommendations- 3rd Edition.URL: http://www.who.int/water_sanitation_health/dwg/gdwq3rev/en/index.html (Retrieved: Jan, 2007)

[25] Gadd, G. F. 2008. Biosorption: critical review of scientific rationale, environmental importance and significance for pollution treatment. *Journal of Chemical Technology & Biotechnology*, 84, 13 – 28.

[26] Brown, M. J. & Lester, J. N. 1982. Role of bacterial extracellular polymers in metal uptake in pure bacterial culture and activated sludge. *Water Research*, 16, 1539-1548.

[27] Dilek, F. B., Gokcay, C. F. & Yetis, U. 1998. Combined effects of Ni (II) and Cr (VI) on activated sludge. *Water Research*, 32, 303-312.

[28] Asku, Z., Kutsal, T., Gun, S., Haciosmanoglu, N. & Gholminejad, M. 1991. Investigation of biosorption of Cu (II), Ni (II), and Cr (VI) ions to activated sludge bacteria. *Environmental Technology*, 12, 915-921.

[29] Davis, T. A., Volesky, B., & Mucci, A. 2003. A review of the Biochemistry of heavy metal biosorption by brown algae. *Water Research*, 37, 4311-4330.

[30] Holan, Z. R., Volesky, B. & Prasetyo, I. 1993. Biosorption of cadmium by biomass of marine algae. *Biotechnology & Bioengineering*, 41, 819-825.

[31] Holan, Z. R. & Volesky, B. 1994. Biosorption of lead and nickel by biomass of marine algae. *Biotechnology & Bioengineering*, 43, 1001-1009.

[32] Pavasent, P., Apratikul, R., Sungkhum,V., Suthipariniyanont, P., Wattanachira, S. & Marhaba, T. F. 2006. Biosorption of Cu^{2+}, Cd^{2+}, Pb^{2+} and Zn^{2+} using dried marine greenalga *Caulerpa lentillifera. Bioresource Technology*, 97, 2321-2329.

[33] Aksu, Z. & Donmez,G. 2006. Binary biosorption of cadmium (II) and nickel (II) onto dried *Chlorella vulgaris*:Co-ion effect on mono component isotherm parameters. *Process Biochemistry*, 41, 860-868.

[34] Feng, D. & Aldrich, C. 2004. Adsorption of heavy metals by biomaterials derived from the marine alga *Ecklonia maxima*. *Hydrometallurgy*, 73, 1-10.

[35] Lee, H. S., Suh, J. H., Kim, I. B. & Yoon, T. 2004. Effect of aluminium in two metal bisorption by an algal biosorbent. *Minerals Engineering*, 17, 487-493.

[36] Davis, T. A., Volesky, B. & Vieira, R. H. S. F. 2000. *Sargassum* seaweed as biosorbent for heavy metals. *Water Research*, 17, 4270-4278.

[37] Raize, O., Argaman, Y. & Yannai, S. 2004. Mechanisms of biosorption of different heavy metals by brown marine macroalgae. *Biotechnology & Bioengineering*, 87, 451-458.

[38] Pagnanelli, F., Petrangelipapani, M., Toro, L., Trifoni, M. & Vegilio,F. 2000. Biosorption of metal ions on *Arthrobacter sp.*: Biomass characterization and biosorption modeling. *Environmental Science & Technology*, 34, 2773-2778.

[39] Zouboulis, A. I., Loukidou, M. X. & Matis, K. A. 2004. Biosorption of toxic metals from aqueous solutions by bacteria strains isolated from metal polluted soils. *Process Biochemistry*, 39, 909-916.

[40] Kulczycki, E., Ferris, F. G. & Fortin, D. 2002. Impact of cell wall structure on the behavior of bacterial cells as sorbents of cadmium and lead. *Geomicrobiology Journal*, 19, 553-565.

[41] Puranik, P. R. & Paknikar, K. M. 1999. Biosorption of lead, cadmium and zinc by *Citrobacter* strain MCM B-181:Characterization studies. *Biotechnology Progress*, 15, 228-237.

[42] Lu, W. B., Shi, J. J., Wang, C. H. & Chang, J. S. 2006. Biosorption of lead, copper and cadmium by an indigenous isolate *Enterobacter* sp. J1 possessing high heavy metal resistance. *Journal of Hazardous Materials*, 134, 80-86.

[43] Pardo, R., Herguedas, M., Barrado, E. & Vega, M. 2003. Biosorption of cadmium, copper, lead and zinc by inactive biomass of *Pseudomonas putida*. *Analytical & Bioanalytical Chemistry*, 376, 26-32.

[44] Chang, J., Law, R. & Chang, C. 1997. Biosorption of lead, copper and cadmium by biomass of *Pseudomonas aeruginosa* PU21. *Water Research*, 31, 1651-1658.

[45] Halttunen, T., Salminen, S. & Tahvonen, R. 2007. Rapid removal of lead and cadmium from water by specific lactic acid bacteria. *International Journal of Food Microbiology*, 114, 30-35.

[46] Akar, T. & Tunali, S. 2006. Biosorption characteristics of *Aspergillus flavus* biomass for removal of Pb(II) and Cu(II) ions from an aquous solution. *Bioresource Technology*, 97, 1780-1787.

[47] Kapoor, A., Viraraghavan, T. & Cullimore, D. R. 1999. Removal of heavy metal using the fungus *Aspergillus niger*. *Bioresource Technology*, 70, 95-104.

[48] Yan, G. Y. & Viraraghavan, T. 2003. Heavy metal removal from aqueous solution by fungus *Mucor rouxii*. *Water Research*, 37, 4486-4496.

[49] Saiano, F., Ciofalo, M., Cacciola, S. O. & Ramirez, S. 2005. Metal ion adsorption by *Phomopsis sp*. biomaterial in laboratory experiments and real waste water treatments. *Water Research*, 39, 2273-2280.

[50] Tobin, J. M., Cooper, D. G. & Neufeld, R. J. 1984. Uptake of metal ions by *Rhizopus arrhizus* biomass. *Applied and Environmental Microbiology*, 47, 821-824.

[51] Chen, C. & Wang, J. 2007. Influence of metal ionic characteristics on their biosorption capacity by *Saccharomyces cerevisiae*. *Applied Microbial Biotechnology*, 74, 911-917.

[52] Volesky, B. 2004. Sorption and Biosorption: Equilibrium Biosorption Performance 6.1. BV-Sobex Inc. Montreal,Canada http://biosorption.mcgill.ca/publication/book/6.1-4w(103-16).pdf (Retrieved Jan 2010)

[53] Mehta, S. K. & Gaur, J. P. 2005. Use of algae for removing heavy metal ions from waste water: Progress and prospects. *Critical Review in Biotechnology*, 25, 113-152.

[54] Lopez, A., Lazaro, N., Morales, S. & Marques, A. M. 2002. Nickel biosorption by free and immobilized cells of *Pseudomonas fluorescens* 4F39: a comparative study. *Water Air Soil Pollution*, 135, 157-172.

[55] Chojnacka, K., Chojnacki, A. & Gorecka, H. 2005. Biosorption of Cr^{+++}, Cd^{++} and Cu^{++} ions by blue green algae Spirulina sp.: Kinetics,equilibrium and the mechanism of the process. *Chemosphere*, 59, 75-84.

[56] Beveridge, T. J. & Murray, R. G. E. 1980. Site of metal deposition in the cell wall of *Bacillus subtilis*. *Journal of Bacteriology*, 141, 876-887.

[57] Halttunen, T. 2007. Removal of cadmium, lead and arsenic from water by lactic acid bacteria. 153 pp. University Turku, Finland.

[58] Davis, T. A., Llanes, F., Volesky, B. & Mucci, A. 2003. Metal selectivity of *Sargassum spp*. and their alginates in relation to their α-L-guluronic acid content and conformation. *Environmental Science & Technology*, 37, 2, 261-267.

[59] Valls, M., Atrian, S., de Lorenzo, V. & Fernández, L. A. 2000. Engineering a mouse metallothionein on the cell surface of *Ralstonia eutropha* CH34 for immobilization of heavy metals in soil. *Nature Biotechnology*, 18, 661 – 665.

[60] Saffar, B., Yakhchali, B. & Arbabi, M. 2007. Development of a bacterial surface display of hexahistidine peptide using cs3 pili for bioaccumulation of heavy metals. *Current Microbiology*, 55, 273-277.

[61] Chen, Y. L., Tuan, H. Y., Tien, C. W., Lo, W. H., Liang, H. C. & Hu, Y. C. 2009. Augmented biosynthesis of cadmium sulfide nanoparticles by genetically engineered *Escherichia coli*. *Biotechnology Progress*, 25, 1260-1266.

INDEX

B

C

E

F

G

H

I

M

N

O

P

Q

R

S

T

W

X

Y

Z